Pershore Coll

Value Addition of Horticultural Crops: Recent Trends and Future Directions

Amit Baran Sharangi · Suchand Datta
Editors

Value Addition of Horticultural Crops: Recent Trends and Future Directions

 Springer

Editors
Amit Baran Sharangi
Spices and Plantation Crops
Bidhan Chandra Krishi Viswavidyalaya
  (Agricultural University)
Mohanpur, West Bengal, India

Suchand Datta
Vegetables and Spices
Uttar Banga Krishi Viswavidyalaya
Pundibari, West Bengal, India

ISBN 978-81-322-2261-3   ISBN 978-81-322-2262-0 (eBook)
DOI 10.1007/978-81-322-2262-0

Library of Congress Control Number: 2015933683

Springer New Delhi Heidelberg New York Dordrecht London
© Springer India 2015

Printed on acid-free paper

Springer (India) Pvt. Ltd. is part of Springer Science+Business Media (www.springer.com)

To
Our
Beloved Parents

Foreword

No production system and production process can be viable without value addition. Value addition is a process that elevates a production into a product. So everyone has to go inclusive and critical on value addition process for creating new market demands or indulging renewed demand from the set of conventional customers. The problem of Indian horticulture is that it goes far for biological production, can be called a super green horticulture and at the same time presents a bleak story for golden horticulture that is value added horticulture to bring more returns for the growers and more satisfaction to the customers. A value added agriculture and horticulture can ignore the inflicts of seasonality in food availability and market vagaries. We are having a huge horticultural diversity that sprawls from sea to mountain ranges and more so we are having a huge pool of indigenous skill and acumen that need to be dovetailed to this massive process of value addition. The other essential ingredient for this renewed revolution shall be the creation and functioning of supply chains.

We are, through different missions viz., NHM and FSM, well triggered up to make agri-horticultural production into mega agri-horticultural entrepreneurships to go for global competitions as our external policy and create economic buoyancy as a fiscal policy. This book, I believe, has been a milestone in presenting different concepts, in very lucid and readable forms, to ultimately offer a kaleidoscopic vision of value addition in horticulture. The approach will generate the much needed reforms in our market policy, industrial outfits and a new treats for the livelihood of millions of farmers reeling under social and economic stresses. The targeted readers, especially students and teachers, will be much educated on the current discussion on secondary horticulture. I congratulate Dr Amit Baran Sharangi at BCKVV, Mohanpur, West Bengal, for the excellent compilation, editing and final check of the tables and figures. I also congratulate the publisher Springer for accepting the book for publication.

KAU, World Noni Research Foundation
Chennai, India

(K.V. Peter)

Preface

We feel delighted to bring out this volume entitled *Value Addition of Horticultural Crops: Recent Trends and Future Directions*. Value addition is both the science and art of transforming a production into a product. The connotation 'value addition' receives a new dimension, whenever prefixed or suffixed with 'horticulture'. Horticultural crops being different in nature as compared to agricultural ones pose a considerable challenge to the stakeholders in various stages, right from production to processing. Knowingly or unknowingly, the process of value addition is on since time immemorial, and thanks to the technological grooming that it is receiving currently, the horizon of our imagination is getting wider and wider.

The book with comprehensive coverage on the subject is a state-of-the-art compilation of background information, principles, research works, and scientific discussion coupled with adequate references on basic and applied aspects on the subject. The main features of the book is an in-depth narration of the scope of quality horticultural crops as a product in influencing present-day global export market arising out of renewed interest in these crops throughout the world. The book will also take care of the rational scientific approaches to post-harvest management of quality horticultural crops. It covers the traditional as well as present-day techniques, along with judicious blending of environment, society, technology and market compulsions in related areas viz., biodiversity, microbiology, ecology, biotechnology and the past as well future knowledge base regarding the subject of interest.

We would like to convey our deep appreciation to all the contributors and well wishers, especially Prof. K.V. Peter, who has been kind enough to write the foreword for the book. Dr. Mamta Kapila and her team from Springer deserve all praise for their best efforts in publishing the book.

Kalyani, India   Amit Baran Sharangi
25 December, 2014   Suchand Datta

Contents

About the Editors

Amit Baran Sharangi is an Associate Professor of eminence in the discipline of Horticultural Sciences and presently acting as the Head, Department of Spices and Plantation Crops, Faculty of Horticulture, Bidhan Chandra Krishi Viswavidyalaya (Agricultural University), India. He is in the profession of teaching for about 17 odd years. He is associated with the process of coconut improvement leading to the release of a variety Kalpamitra from CPCRI. One of his papers has ranked among the top 25 articles in ScienceDirect. He has published about 50 research papers in peer-reviewed journals, 40 conference papers, 12 reputed books as well as several book chapters published from CRC Press (USA), Nova Publishers (USA) and scores of popular scientific articles. Presently he is associated with 30 international and national journals as regional editor, technical editor, editorial board member and reviewer. Dr. Sharangi has visited abroad extensively on academic mission and obtained several international awards viz., ENDEAVOUR Post-doctoral Award-2010 (Australia), INSA-RSE Visiting Scientist Fellowship (UK, 2011), FULBRIGHT Visiting Lecturer Fellowship (USA, 2013), Achiever's Award (SADHNA) etc. He has delivered a couple of invited lectures in UK, USA, Australia, Thailand, Israel and Bangladesh on several aspects of herbs and spices. He is associated with a number of research projects as Principal and Co-Principal Investigators having academic and empirical implications. He is active member of several science academies and societies like NASI, NABS, ISNS, HSI, CWSS, SAH including the New York Academy of Science (NYAS), World Academy of Science, Engineering and Technology (WASET), to name a few.

Suchand Dutta did Ph. D. from Uttar Banga Krishi Viswavidyalaya. He joined his service during 2001 as the post of Lecturer in Uttar Banga Krishi Viswavidyalaya. Presently, he is serving as Associate Professor in Vegetable and Spice Crops in the Uttar Banga Krishi Viswavidyalaya, Pundibari, Cooch Behar, West Bengal, India. During his academic career he guided a number of M.Sc. students as chairman and published 45 research papers in national and international journals, 5 popular articles in English and a large number of popular articles in local languages. He has participated and presented more than 25 research papers in different national- and international-level seminar/symposia/conference/World Congress in India and abroad. He was associated as Co-Principal Investigator in Central Scheme for the Development of

Medicinal Plant Sponsored by National Medicinal Plants Board Dept. of ISM & H, Ministry of Health and Family Welfare, Govt. of India and associated as Scientist Integrated Programme for Development of Spices and now it is renamed Mission for Integrated Development of Horticulture from 2002 to till date. He organized one national-level workshop and recently he has taken the responsibility of In-Charge, All India Co-ordinate Research Project on Spices, Uttar Banga Krishi Viswavidyalaya, Pundibari. He wrote four books and three book chapters. He has actively participated in different training programmes for the benefit of the farmers.

Introduction

1

S.K. Acharya, K. Pradhan, P. Choudhuri,
and A.B. Sharangi

Abstract

There are three basic phases of agricultural growth and revolution and
these are: inductive, stimulative and simulative. The inductive phase of
agriculture is characterized with intensive crop production with a support
from several basic critical inputs viz., the magic seed, the fertilizer and
agrochemicals and irrigation water leading to the agrarian magnum opus
called Green Revolution. This inductive phase has a clear thrust on highest
possible production within shortest possible time and all done just to
tackle the threat of famines, the growth of industry or to keep supply line
well-loaded with food and fruits attuned to war field. In the second phase,
subsidies and allied incentives were integrated with the production process
so that the target farmers can be benefitted, empowered and ascribed with
a sense of social dignity. This phase has been characterized with huge
corporate social responsibility and incubation of agro-based small and big
ventures to invite the inevitable transformation of green agriculture into a
silvery agriculture. Here the value addition process took a quantum jump
to generate a belligerent market for agri-horti products. The simulative
phase of agricultural growth and development has been characterized with
future projections, digitized configuration and precision production factors. It started and went functionally geared up during 1980s, 1990s and
2000 onwards. This has been the value addition genera, keeping focuses
on and with branding, market segments, and value chain management, and
inclusive growth takes place through a denial to the geographic, temporal
and spatial barrier.

S.K. Acharya
Department of Agriculture Extension,
Bihan Chandra Krishi Viswavidyalaya,
Mohanpur, Nadia 741252, West Bengal, India

K. Pradhan
Assistant Professor, Department of Agricultural
Extension, Uttar Banga Krishi Viswavidyalaya,
Pundibari, Cooch Behar, West Bengal, India

P. Choudhuri
Department of Vegetable and Spice Crops,
Uttar Banga Krishi Viswavidyalaya,
Pundibari, Cooch Behar 736101, West Bengal, India

A.B. Sharangi (✉)
Department of Spices & Plantation Crops,
Bihan Chandra Krishi Viswavidyalaya,
Mohanpur, Nadia 741252, West Bengal, India
e-mail: dr_absharangi@yahoo.co.in

A.B. Sharangi and S. Datta (eds.), *Value Addition of Horticultural Crops: Recent Trends
and Future Directions*, DOI 10.1007/978-81-322-2262-0_1, © Springer India 2015

The history of agriculture transcends 10,000 years, and in the beginning, it has to support some few thousands of human beings across the globe. Now the same agriculture has to sustain the hunger and nutrition of 6.5 billion people across the world. The nomadic life starts not with agriculture but of course with horticulture, livestocks and fisheries. The Neanderthal's Diaspora kept on exploring roots, nuts, fruits and modified vegetative organs as their food; the source of animal proteins had been complemented by wild animals, ichthyofauna and even jungle birds. In experimenting or trying with the palatable fruits or capsules, some ancient people have to sacrifice their lives. This has been perhaps the oldest human experiments in screening out palatable food from poisonous food. After that and with the rapid improvisation in tools and techniques, a meagre transformation had been there, from Stone Age civilisation to Metal Age civilisation. The hunting economy has been transformed into a primitive production economy, and thus, the first agriculture started appearing for different clans and Diaspora some 10,000 years back.

With special reference to India and anywhere in the world, we can find three basic phases of agricultural growth and revolution, and these are (a) inductive, (b) stimulative and (c) simulative.

The inductive phase of agriculture is characterised with intensive crop production with a support from three basic critical inputs, (a) the magic seed, (b) the fertiliser and agrochemicals and (c) irrigation water, leading to agrarian *magnum opus* called *Green Revolution*. This has been done with a desperate attempt to rejuvenate and reconstruct the war-ravaged institutions and production system. The decades of 1940s, 1950s and 1960s across the world are the history of war, invention, industrialisation and agriculturisation. In India, a score of boost-up programmes were undertaken to ensure the intensive growth in agriculture and rural economy. These are community development programme (1952), intensive agricultural district programme (1960), intensive agricultural area programme (1964), high yielding variety programme (1965) and training and visit programme (1974):

So, this inductive phase has a clear thrust on highest possible production within shortest possible time and all done just to tackle the threat of famines, the growth of industry or to keep supply line well-loaded with food and fruits attuned to war field. Value addition, the concept and process, still has been a far lying proposition.

The second phase was the stimulative phase wherein subsidies and allied incentives were integrated with the production process or factor productions so that the target farmers or populace can be benefitted, empowered and ascribed with a sense of social dignity. The concept of welfare state and welfare economy were bound to operate so as to remove the stigma of famines and brunt of political economy is transforming social ecology which could be more resilient to absorb the conflict and resentment. This phase encompasses the decades of the 1970s and 1980s wherein the predominating programmes were IRDP, TRYSEM, DCWRA, ITDP, SFDA and so on. This phase has also been characterised with huge corporate social responsibility at one hand and on the other hand incubation of agro-based small and big ventures to invite the inevitable transformation of green agriculture into a silvery agriculture and that started creating the ground for agri-hort-preneurships:

Albeit, a basket of diverse agriculture started entering the global scenario in its gross production forms, the value addition process took a quantum jump to generate a belligerent market for agri-horti products and also a reinforced and renewed attempt to create and make functional product-based supply chain vis-a-vis value chains sprawling between site of production to site of consumption.

The simulative phase of agricultural growth and development has rightly been characterised with future projections, digitised configuration and precision production factors. It started and went functionally geared up during 1980s, 1990s and 2000 onwards. The beginning of globalisation era in the form of LPG (i.e. liberalisation, privatisation and globalisation) regime and the opening up of fiscal markets to the global one added a new favour to the production system and value addition. Now, every kilogramme of vegetables has to go globally saleable; the apple of a country with a value sticker can travel all islands and territories beyond the aqua geography. So, an ICT (information and communications technology)-driven and satellite-supported agriculture enters a virtual space of data sharing and product manoeuvring across the geography and sociology of the global market. Future tradings and online marketing have become the go and eye of modern agri-horti-entrepreneurships:

This has been right across the value addition genera, keeping focuses on and with branding, market segments, value chain management and inclusive growth takes place through a denial to the geographic, temporal and spatial barrier.

1.1 Issues of Food Security: The Role of Creating Value-Based Entrepreneurship

Every farm economy should be based on the basic pledge for ensuring food security, especially for the weaker section of the society. Food security includes physical volume (sufficient amount), its proper absorption (easy metabolic processes) and affordability (at cheaper price). It is very difficult to get all these three things operating at a common point. A lot of endeavours are there to integrate these three aspects by increasing the productivity through genetic modification and hybridisation, incorporating minerals and vitamins through golden rice project and mobilising community to access cheap but nutritionally rich natural foods.

Food security is a condition related to the ongoing availability of food. The concerns over food security are from time immemorial. The term 'food security' was established at the 1974 World Food Conference. A new definition was given at the 1996 World Food Summit with the emphasis on individuals enjoying food security, rather than the nation. According to the Food and Agriculture Organization (FAO), food security 'exists when all people, at all times, have physical and economic access to sufficient, safe and nutritious food to meet their dietary needs and food preferences for an active and healthy life':

Entrepreneurship cannot simply shy away with the responsibility of food and nutritional security. What is needed is to have as change from volume of green production to volume of golden production.

1.2 Global Food Requirement and Present Status

Crop yields worldwide are not increasing quickly enough to support estimated global needs in 2050. It has been estimated in the past that global agricultural production may need to increase 60–110 % to meet increasing demands and provide food security. In the current study, researchers assessed agricultural statistics from across the world and found that yields of four key crops (viz. maize, rice, wheat and soybean) are increasing 0.9–1.6 % every year. At these rates, the production of these crops would likely to increase 38–67 % by 2050, rather than the estimated requirement of 60–110 %. The top three countries that produce rice and wheat were found to have very low rates of increase in crop yields (CGIAR 2014).

The overall demand for agricultural products (including food, feed, fibre and biofuels) is expected to increase 1.1 % per year from 2005/2007 to 2050, from 2.2 % per year in the past four decades.

Increases in food demand are due to population growth and changes in diets.

As the population grows and more countries and population groups attain per capita food consumption with little scope for major increases, global food demand will grow at much lower rates. But for a long time to come, some countries might have difficulty increasing food consumption due to low income and significant poverty (Alexandratos and Bruinsma 2012).

A recently published WHO/FAO report recommends a minimum of 400 g of fruits and vegetables per day (excluding potatoes and other starchy tubers) for the prevention of chronic diseases such as heart disease, cancer, diabetes and obesity, as well as for the prevention and alleviation of several micronutrient deficiencies, especially in less developed countries (WHO 2014).

1.3 Inclusive Growth

Inclusive growth is a concept which advances equitable opportunities for economic participants during the process of economic growth with benefits incurred by every section of society.

The definition of inclusive growth is related to the macroeconomic and microeconomic determinants of the economy and economic growth. The microeconomic dimension captures the importance of structural transformation for economic diversification and competition, while the macroeconomic dimension refers to changes in economic aggregates such as the country's gross national product (GNP) or gross domestic product (GDP), total factor productivity and aggregate factor inputs (World Bank 2014).

Sustainable economic growth requires inclusive growth. The inclusive growth approach takes a longer-term perspective, as the focus is on productive employment as a means of increasing the income of poor and excluding groups and raising their standards of living (Ianchovichina and Lundstrom 2009).

The creation, refinement and addition of new skills into the depletive domain of classical skills, suffering from redundancy, will help eliminate the gender and community disparity and disparages. The value addition process centring around locally available resources and skills needs to be upgraded, and a market-linked enterprise cafeteria shall help the marginalised and weaker farmers to come up stronger and compete with the belligerent market and thus help them go stronger with inclusive growth.

1.4 Market for Field Crops Are Gradually Turning Inelastic

Food being the primary consumers goods and basic requirement, it is vulnerable to face the market inelasticity for a given market demand. The farmers, across the world, are stressed to divert from food crops to commercial crops in order to respond to a belligerent market for biofuels, ethanols, etc. This has set a new dimension for value addition to the agri-horticultural production system both at micro and mega levels and has been good enough to have redesignated agri-horticultural policy:

- The seasonal nature of agriculture leads to a lagged supply response. As demand has outpaced supply, prices have increased significantly, particularly for maize (corn), rice and wheat. Fertiliser prices have also increased dramatically over the last 2 years as increased supply to match rising demand has been held back by the limited production capacity.
- Biofuel policies. Surging demand for food crops has increased faster than supply due primarily to biofuel policies in industrialised countries and to a lesser extent changing diets in rapidly growing developing countries. Biofuel policies have diverted food crops from traditional export markets to the production of ethanol and biodiesel. The growing demand for livestock products, particularly in Asia, has increased the consumption of grain for feed. Erratic weather, trade policies and seasonal lags have slowed producers' response to the higher prices.

1.5 Brunt of Climate Change

The brunt of climate change has already been reflected through the decline of productivity, shifting of critical growth zones, mutagenesis of physiological expression of different genomics and intrusion of exotic genes to tantalise the classical functioning of sui generis characters, and these all are posing newer threats to food security. 1°C change in night temperature will invite a 20 % downfall of wheat productivity and 12 % for rice. That is again rendering a stage for juxtaposition between population growth and productivity fallout.

Horticultural crops are gone unmanaged in the brunt of climate change. Even a protected condition may provide higher deterrent climate change effects.

Climate change presents a major concern, often interacting with existing problems. It makes new demands for adaptation and coping strategies and presents new challenges for the management of the environment and agroecosystems. The Intergovernmental Panel on Climate Change (IPCC) report (Adger et al. 2007) ignores the role of diversity in production systems and the central role that agrobiodiversity will have to play in both adaptation and mitigation at the country, landscape, community and farmer levels:

- Nearly 20 % of all US food is imported, so climate extremes elsewhere will also have an effect. In 2011, 14.9 % of US households did not have secure food supplies, and 5.7 % had very low food security.
- Latest EU projections suggest that the most severe consequences of climate change will not be felt until 2050. But significant adverse impacts are expected earlier from more frequent and prolonged heatwaves, droughts and floods. Many crops now grown in southern Europe, such as olives, may not survive high temperature increases. Southern Europe will have to change the way it irrigates crops.
- In 2011, Russia banned wheat and grain exports after a heatwave. Warming will increase forest fires by 30–40 %. This will affect soil erosion and increase the probability of floods.
- In the Middle East and North Africa, declining yields of up to 30 % are expected for rice, about 47 % for maize and 20 % for wheat.
- Egypt expects to lose 15 % of its wheat crops if the temperatures rise by 2 °C and 36 % if the increase is 4 °C. Morocco expects crops to remain stable up to about 2030, but then to drop quickly later. Most North African countries traditionally import wheat and are therefore highly vulnerable to price shocks and droughts elsewhere.

1.6   Gender Empowerment

Gender empowerment has got two basic components, gender enabling and gender mainstreaming. While gender enabling is basically an economic process, gender mainstreaming is exclusively a social process. These two processes are subject to policy implications and operational role on the part of government and institutions. In economic enabling process, the postharvest processing and quality management initiatives can go in compliance with unique gender requirements and natural skills. Both horticulture and livestock, across the length and breadth of society, have got intrinsic and systemic property to go attuned with gender skills and behaviour:

- As identified by UNICEF, women are the potentially vulnerable group in this process. In some of the poorest areas of South Asia, cultural restrictions on women's ability to participate fully in food production activities have left them particularly vulnerable in times of economic crisis.
- Despite considerable cross-cultural variations in the gender division of labour, there does appear to be a general predominance of women in the household-based processing stage of crop production as well as in the transformation of raw food into edible form.
- There appears to be a fairly widespread gender differences in the involvement in livestock rearing. Larger stocks (cattle, buffalo, horse, etc.) have to be grazed over larger distances and often require male labour at least for some stages of livestock care. Smaller stock and poultry can be cared for with female labour alone. This often leads to greater male rights over cattle and female rights over smaller stock. In Bangladesh and India, 'share rearing' of goats and poultry is a common means by which poor women transform their only resource – labour power – into a productive

asset. Other common examples of primarily female activities are cultivation of homestead plots and 'backyard' gardening. If household food security is an objective of project intervention, then clearly targeting these spheres of activity will help to enhance women's control over household food resources.

1.7 Value Addition of Horticultural Crops World Scenario vis-a-vis Indian Scenario

After independence, India recorded faster growth in food-processing sector specifically during the early 1980s. After the Green Revolution, the country had increased agricultural production needed for its postharvest management. The importance of the sector was realised by the business community leading to diversification from grain trading to processing (Kachru 2006). In some areas like the solvent extraction industry, the growth in installed processing capacity has been far higher than the supply of the raw materials. However, in areas like horticultural crops especially fruit and vegetable processing, the growth has not been satisfactory due to poor demand for processed products by the consumers. In such cases, the industry has also not been able to develop the demand adequately. This is due to the food habits of the population. Indians mainly prefer fresh fruits and vegetables, spices, etc., over processed fruits and vegetables. India has experienced a considerable degree of crop diversification in terms of changes in the area under various crops since the Green Revolution, which mainly targeted the increased food grain production to resolve the country's food security problem. In the past one decade, the change in cropping pattern is more towards the horticulture sector and commercial crops (Mittal 2007). Horticultural crops comprising of fruits, vegetables, flowers, medicinal and aromatic plants, spices and plantation crops play a leading role in the food and livelihood security of India. Though these crops occupy only 8.5 % of arable land,

they contribute 29.5 % of the GDP in agriculture (Economic Survey 2007–2008). This calls for technology-led development. Cultivation of these crops is labour intensive, and as such, they generate a lot of employment opportunities for the rural population. India produced 261.98 million tonnes of horticultural crops from an area of 23.4 million hectare (NHB 2013). The performance in the production is lucrative, but in value addition part, India's share is not as expected. India only contributes 2 % in the world horticulture trade. The concept of agri-export zones and mega food parks has been promoted by the Indian government to promote food-processing industry and also its subsector like fruit and vegetable processing industry in India. Indian government has sanctioned US $ 22.97 million in establishing around 10 M food parks and offered the tax benefits to the concerned subsector of the food-processing industry. The share of food-processing industry in GDP has gone up to Rs.44, 93,743 crore in 2009–2010 with compound annual growth rate (CAGR) of 8.40 %. The current horticultural crop processing scenario in India, compared to the developed countries, is not satisfactory. The factors responsible for this are so many.

1.7.1 Extent of Postharvest Losses

India is the second largest producer of fruits and vegetables in the world. The country has emerged as the world's largest producer of spices, coconut and tea and the second largest producer and exporter of tea, coffee and cashew. Nanda et al. (2012) reported that in the post-Green Revolution era, even though food grains have been taken care of, horticultural crops, mainly fruits and vegetables, because of the need for simple processing, preservation and transport technologies, have suffered postharvest losses, estimated to be more than 25 %, amounting to a revenue loss of Rs. 500 billion (Tables 1.1 and 1.2). About 10–15 % fresh fruits and vegetables shrivel and decay, lowering their market value and consumer acceptability. Minimising these losses can increase their

Table 1.1 Postharvest losses in horticultural crops during different channels of handling/farm operations in India

Postharvest handling channel	Fruits, %	Vegetables, %	Plantation crops and spices
Harvesting	0.92–4.56	0.84–3.61	0.16–3.66
Collection	0.23–1.20	0.23–1.77	0.16–0.86
Sorting/grading	0.93–4.79	1.54–3.30	0.31–1.36
Packaging	0.08–0.94	0.10–1.64	0.06–0.24
Transportation	1.06–2.77	0.44–3.14	0.01–0.31
Total loss in farm operations	4.18–13.92	4.61–11.03	0.89–7.89
Total loss in storage	1.20–4.13	1.51–3.04	0.23–1.66
Overall total loss	5.77–18.05	6.88–12.98	1.12–8.64

Table 1.2 Postharvest losses in horticultural crops during different levels of storage in India

Level of storage	Fruits, %	Vegetables, %	Plantation crops and spices, %
Farm level storage	0.84–5.54	1.18–4.62	0.13–1.94
Godown/cold storage	0.00–3.34	0.30–2.18	0.21–0.61
Wholesaler level storage	0.99–5.91	1.32–3.87	0.22–1.24
Retailer level storage	1.10–3.79	1.70–2.62	0.14–2.09
Processing unit level storage	0.03–5.71	0.09–2.34	0.03–1.37
Total loss in storage	1.20–4.13	1.51–3.04	0.23–1.66
Overall total loss	5.77–18.05	6.88–12.98	1.12–8.64

supply and improve general nutrition of the common Indian without bringing additional land under cultivation. Improper and faulty handling, storage and marketing cause physical damage due to tissue breakdown. Mechanical losses include bruising, cracking, cuts and microbial spoilage by fungi and bacteria, whereas physiological losses include changes in respiration, transpiration, pigments, organic acids and flavour. On account of poor postharvest management, the losses in farm produce in India have been assessed to be of a very high order. It has been studied that the extent of losses could be curtailed to less than 50 % through the adoption of proper agro-processing technology. For reducing the rest of the losses, new initiatives need to be called for. Hence, long-term attention should be focussed on such as proper grain storage structures, cold stores and processing systems to avoid the losses. India is very ambitious to increase the processing level to 20 % by 2015 (MOFPI 2011).

1.8 Commodity-Wise Value Addition

1.8.1 Fruits and Vegetables

Joint efforts by the R&D institutions, farmers, government agencies and the trade has resulted in India being the second largest producer of fruits and vegetables after China. In the year 2012–2013, the country produced about 77.7 million tonnes of fruits and 159.5 million tonnes of vegetables contributing 25 % of the total world production (NHB 2013). However, the growth in postharvest sector has not kept pace with the production. Only about 2.2 % of the total fruits and vegetables produced are processed as compared to countries like the USA (65 %), Malaysia (83 %), the Philippines (78 %), France and Brazil (70 % each), etc. The fruit and vegetable processing in India is highly decentralised, small-scale industries accounting for 33 %, organised 25 %, unorganised 42 % and large number of units in cottage/household and small-scale sector having capacities of up to 250 tonnes/year. In organised sector of India, there are about 5,000 units, and several thousands in unorganised sector of fruit and vegetable processing are in the job. Significant developments in technology include better understanding of the process of ripening of fruits, optimum harvesting time, pre-cooling of freshly harvested produce, cold storing of the raw fruits and vegetables and sorting, cleaning, waxing and packaging technology for fruits. At CFTRI, DFRL and IIHR, Bangalore; IARI, New Delhi; GBPUA&T, Pantnagar; IIVR, Varanasi; and HPKV, Palampur, a number of technologies have been developed. Over the last few years, there has been a positive growth in ready-to-serve beverages, fruit juices and pulps, dehydrated and frozen fruit and vegetable products, tomato products, pickles, convenience veg-spice pastes, processed mushrooms and curried vegetables. The most significant work has been recorded in the technology for ripening of the fruits under controlled conditions. The production of juices and value-added products including jams, jellies,

pickles, canned products, etc., has become a commercial success. The industry using indigenous technology includes units engaged in juice extraction, concentration of juices, canning and production of several of the products like jams, jellies, canned fruits, dried vegetables, etc. Technology is still being imported for the establishment of large-scale exported oriented units for the production of items like banana paste, concentrates of various fruit juices and sorting, cleaning, washing, waxing and packaging of raw fruits and vegetables. Presently, India is processing so many fruit and vegetable processed products (Table 1.3).

1.8.1.1 Export Scenario of Indian Processed Fruits and Vegetables

Fresh fruits and vegetables comprise almost 35 % of the world trade in horticulture. The major product category in processed fruits and vegetables which are exported are mangoes (fresh and pulp), grapes, other fresh fruits, dried and preserved vegetables, pickles and chutneys and fruit beverages. But almost two-thirds is accounted for

Table 1.3 List of existing and new value-added products of fruits and vegetables in India

Fruit/vegetable	Existing products	New products
Apple	Juice, AJC, jam, jelly, cider, wine, pulp	Osmotically dried rings, canned apple, vinegar, carbonated juice, apple seed for nurseries, pectin, fibre from pomace
Apricot	Pulp, squash, RTS, jam, appetiser, dried apricot	Osmotically dried apricot, oil, apricot oil-based cream, etc.
Plum	Pulp, squash/appetiser, RTS, chutney, jam, wine/brandy	Plum sauce, seed oil
Peach	Canned peach, pulp, jam/chutney	Wine, kernel oil
Pear	Canned pear, pulp, jam	Apple pear blend, sand pear candy, vermouth
Mango	Pulp, RTS, squash, powder (amchur), slices in brine, pickle	Pulp/juice from in situ mangoes, pectin from just-ripe fruits
Apple	Juice, AJC, jam, jelly, cider, wine, pulp	Osmotically dried rings, canned apple, vinegar, carbonated juice, apple seed for nurseries, pectin, fibre from pomace
Apricot	Pulp, squash, RTS, jam, appetiser, dried apricot	Osmotically dried apricot, oil, apricot oil-based cream, etc.
Grapes	Raisin, juice	Carbonated juice/RTS
Litchi	Juice, squash, nectar/RTS	Carbonated drink
Cauliflower	Pickle, slices in brine	Frozen cauliflower heads, left over for drying powder
Carrot	Pickle, slices in brine, preserve, candy	Freezing and drying
Chillies	Green chilli puree, powder	Green chilli paste, oleoresin
Tomato	Juice, sauce, ketchup, paste	Drying, powder

by four items, namely, citrus, banana, apple and grape. India exported 724,178.09 million tonnes of processed fruits and vegetables valuing 383,616.21 lakh rupees (APEDA 2014). The major destinations for processed fruits and vegetables like dried and preserved vegetables are exported to Sri Lanka, the USA, the UAE, Germany, France and the Netherlands; mango pulp goes to the UAE, Saudi Arabia, Kuwait, the Netherlands and Hong Kong; pickle and chutney are taken up by the UK, the USA, the UAE, Germany, Canada, the Netherlands and Saudi Arabia; other processed fruits and vegetables (tomato paste, jams, juices, etc.) are imported by the USA, the Netherlands, the UK, the UAE, Indonesia, the Philippines and Russia. Russia is a major importer of processed fruits and vegetables from India, and the country imported 13,477 million tonnes of dehydrated vegetables and fruits worth Rs 5,963 lakh in 2010–2011. The second major importer is the USA with 11,164 million tonnes of dehydrated fruits and vegetables in 2010–2011. The consumption of processed fruits and vegetables is low in India compared to the primary foods because they are available fresh in the market to the consumer. The demand for processed foods is mostly lied in the urban market due to the lifestyle and purchasing power of the urban population. Thus, there is a large demand for processed food in the export market, and India can capture this market by restructuring and strengthening its infrastructure.

1.8.2 Value Addition in Spices

India is traditionally known as the spice bowl of the world. According to the Bureau of Indian Standards, about 63 spices are widely grown in our country. India is the largest producer, consumer and exporter of spices in the world with a 46 % share by volume and 23 % share by value, in the world market. India produced 57.44 lakh tonnes of spices from an area of 3.08 million hectare (NHB 2013). These spice sectors also play a significant role in the development of Indian economy. The Indian story of value-added spice products began in the early 1970s. Over the

years, with investments in quality and supply chain management, it has emerged to be the largest in the world. It is not only a local sourcing and processing point but has also changed itself as a value addition hub for the spice crops from the Asia-Pacific region and elsewhere across the globe. Different value-added products of spices available in India are spice oils and oleoresin, dehydrated pepper, freeze-dried green pepper, ginger candy, ginger beer/in brine/squash, ginger flakes, garlic pickle and paste, chilli powder, paste, puree, oleoresin, etc. (Table 1.4). The contribution of R&D to PHT of spices includes equipment and processes for cleaning, grading and packaging of whole spices and production of value-added products such as oleoresins and spice oils. Institutions like CFTRI, DFRL, Indian Institute of Spices Research and some of the SAUs including TNAU, Coimbatore, have contributed significantly to this development.

1.8.2.1 Global Market of Value-Added Spice Products

India is the largest exporter of spices in the world. The share of spices in total agricultural export is 6 %. Indian spices flavour foods in over 130 countries, and their intrinsic values make them distinctly superior in terms of taste, colour and fragrance. More than 150 spice-based value-added products from India are available across the globe. The USA, Canada, Germany, Japan, Saudi Arabia, Kuwait, Bahrain and Israel are the main markets for Indian spices and spice products. We have near monopoly in spice oils and oleoresins. India supplies more than 70 % of the total world supply of spice oils and oleoresins. Indian spices have obtained geographical indicators such as Malabar pepper, Alleppey green cardamom, Coorg green cardamom and Naga chilli. The demand for organic products is steadily increasing in the western markets at 20–25 % every year and that of organic spices is about 2 %. Spice exports have registered substantial growth during the last 5 years, registering a compound annual average growth rate of 23 % in value and 11 % in volume, and India commands a formidable position in the World Spice Trade during 2012–2013; a total of 8,62,542 tonnes of spices and spice products

Table 1.4 List of existing value-added products of spices in India

Spices	Existing products
Black pepper	Oleoresin, green pepper in brine, dehydrated green peppers, canned green pepper, frozen green pepper, cured green pepper, pepper oil, freeze-dried green pepper, white pepper powder, etc.
Ginger	Powder, wines, dry ginger, starch from spent ginger, preserves, gingiberin oil, oleoresin
Turmeric	Natural pigments, curcuminoids, oleoresins
Chillies	Powder, pickles, paste, oleoresin, oil, brined chilli, sauces
Paprika	Colour, paprika flavour
Coriander	Powder, oleoresins
Cumin	Powder, oleoresin
Fennel	Sugar-coated fennel, oleoresin, whole, etc.
Fenugreek	Powder, dried fenugreek leaves, etc.
Tree spices (cinnamon, cassia nutmeg, cloves)	Obesity regulators, stimulators, nutraceuticals
Paprika	Colour, paprika flavour
Cardamom	Encapsulated cardamom, cardamom tea, cardamom soft drink mix, cardamom oil and oleoresin
Garlic	Garlic powder, garlic paste, garlic oleoresin
Turmeric	Natural pigments, curcuminoids, oleoresins
Chillies	Powder, pickles, paste, oleoresin, oil, brined chilli, sauces
Paprika	Colour, paprika flavour
Coriander	Powder, oleoresins
Cumin	Powder, oleoresin
Fennel	Sugar-coated fennel, oleoresin, whole, etc.
Fenugreek	Powder, dried fenugreek leaves, etc.
Tree spices (cinnamon, cassia nutmeg, cloves)	Obesity regulators, stimulators, nutraceuticals

valued Rs.12,112.76 crore (US $ 2,212.13 million) has been exported from the country as against 5,75,270 tonnes valued Rs.9,783.42 crore (US $ 2,037.76 million) in 2011–2012, registering an increase of 26 % in volume and 24 % in rupee terms and 8.5 % in dollar terms of value (Table 1.4). During this period, the achievement in export earning is high, and it is mainly due to the rigorous focus and initiatives taken by the board for value addition and higher end processing of spices (Spices Board of India 2014). Value-added products like curry powder/paste, mint products and spice oils and oleoresins registered a growth of 46 % in volume and 37 % in value. During this period 8,665 t of spice oils and oleoresins valued at Rs 1,242 crore have been exported. Curry powder/paste followed suit with a 44 % increase in volume and 53 % increase in value. Turmeric marked an increase of 17 % and 45 % in volume and value of exports, respectively.

1.8.3 Processing of Plantation Crops

Plantation crops contribute substantially to the national economy with an export earning of Rs. 12.4 billion. Coconut alone contributed Rs. 1.72 billion by way of exports during 1996–1997. However, the coconut-based industry in India has been in the infancy stage. There is a considerable scope of product diversification, viz. production of coconut milk and milk powder, coconut cream, shell powder, shell charcoal, etc. Coconut wood utilisation needs more attention. In case of other crops, financially viable technologies for product diversification need to be developed. Such products are areca nut fat, tannin, arecoline, other chemicals from areca nut, honey-/chocolate-coated or salted kernels from cashew nuts and value-added products from by-products. The postharvest operations in these crops need to be mechanised. Though the technology has been developed for desiccated coconut, coconut cream and other products, it needs refinement. At CPCRI, Kasargod, a coconut dehusker has been developed for manually opening the nuts. Another motorised unit is under development. Copra drier using LDPE cover and batch-type hot air copra drier using agricultural waste as source of fuel have also been developed at CPCRI, Kasargod; KAU, Thrissur; and TNAU, Coimbatore. In case of the plantation crops like oil palm, necessary efforts are required for processing and value addition, especially with regard to the quality of products, energy inputs, packaging, etc., to meet the international quality standards and to reduce the cost of production. Processing of cocoa beans at small scale also needs attention.

1.9 Constraints and Opportunities in Value Addition of Horticultural Crops in India

1.9.1 Constraints

1. Processing units of horticultural crops are established far away from production catchments resulting high cost of transportation and quality loss perishable horticultural crops.
2. Lack of contract farming system to ensure a uniform supply of raw material.
3. Concept of storage and cold chains and cargo facilities at airports and ports are very much inadequate.
4. Lack of availability of appropriate machinery and equipments based on the quality and quantity of indigenous Indian raw material.
5. High capital investment for the establishment of processing units and high operational costs.
6. Taxes on processed products in India are among the highest in the world. No other country imposes excise duty on processed food.
7. Majority of the processing units are operated by the unskilled and semi-skilled workers.

8. The need for modernisation of existing units as outdated machinery used in small- and medium-scale enterprises leads to poor quality and low yield of the finished product.
9. Lack of quality control facilities for small- and medium-scale enterprises leading to non-uniform quality of finished products resulting in poor market acceptability.
10. Poor linkage between R&D institutions and food industries/entrepreneurs.
11. Cooperatives and other semi-government organisations are weak, and people's participation, either through Panchayat Raj institutions, NGOs, farmer organisations or industries' associations in food sector, remains extremely inadequate.

1.9.2 Opportunities

1. Contract farming system to ensure uniform supply of raw material.
2. Precision postharvest technologies using automation of postharvest handling, packaging, transportation and storage operations.
3. Introduction of low-cost bulk storage structures for the horticultural produce at production and despatch areas.
4. Development of complete cool chain on road and air for maintaining the optimum quality of the perishable produce from farm to fork.
5. Adoption of non-thermal/non-chemical processing and preservation of food products.
6. Production of safe food products with a maximum nutrient retention.
7. Faster detection of adulterants and chemical residues in processed horticultural products using bio-sensors/nano-bio-sensors.
8. Application of robotics, artificial neural networking, nutrigenomics, non-destructive and/or online testing techniques, supercritical fluid extraction for production of high-value products.
9. Recycling of horticultural wastes specially fruit, vegetable and plantation crops wastes into newer products.

10. Nanotechnological interventions in the development of bio-polymers for packaging and bio-composite for structures.

References

Adger WN, Agrawala S, Mirza MMQ, Conde C, O'Brien K, Pulhin J, Pulwarty R, Smit B, Takahashi K (2007) Assessment of adaptation practices, options, constraints and capacity. In: Parry ML, Canziani OF, Palutikof JP, van der Linden PJ, Hanson CE (eds) Climate change 2007: impacts, adaptation and vulnerability. Contribution of Working Group II to the fourth assessment report of the Intergovernmental Panel on Climate Change. Cambridge University Press, Cambridge, pp 717–743

Alexandratos N, Bruinsma J (2012) World agriculture towards 2030/2050, the 2012 revision. ESA working paper no. 12–03, June 2012. Food and Agriculture Organization of the United Nations (FAO), Rome. Available from http://www.fao.org/docrep/016/ap106e/ap106e.pdf

APEDA (2014) Promoting fruit and vegetable consumption around the world. http://www.apeda.gov.in/apeda-awebsite. Accessed 17 Oct 2014

CGIAR (2014) Current global food production trajectory won't meet 2050 needs. http://ccafs.cgiar.org/bigfacts/globalfood-demand/. Accessed 17 Oct 2014

Inclusive Growth Analysis. http://siteresources.worldbank.org/INTDEBTDEPT/Resources/4689801218567884549/What IsInclusiveGrowth20081230.pdf

Ianchovichina E, Lundstrom S (2009) Inclusive growth analytics: framework and application, Policy research working paper series 4851, The World Bank. http://wwwwds.worldbank.org/external/default/WDS ContentServer/IW3P/IB/2009/03/03/000158349_200 90303083943/Rendered/PDF/WPS4851.pdf

Kachru RP (2006) Agro-processing industries in India-growth. Status and prospects. Indian Council of Agricultural Research, New Delhi

Mittal S (2007) Can horticulture be a success story for India? Working paper no. 197. Indian Council for Research on International Economic Relations. ICRIER (Indian Council for Research on International Economic Relations), New Delhi

MOFPI (2011) List of project assisted case(s) under cold chain scheme. http://mofpi.nic.in/mofpiweb/list_sanction_cc.aspx?yr=2011

Nanda SK, Vishwakarma RK, Bathla HVL, Rai A, Chandra P (2012) Harvest and postharvest losses of major crops and livestock produce in India. AICRP on PHT, CIPHET (ICAR), Ludhiana

National Horticulture Board (2013) National Horticulture Board Data Base. http://nhb.gov.in/

Spices Board of India (2014) Spice products. http://www.indianspices.com/. Accessed 7 Sept 2014

WHO (2014) Horticulture: post harvest management. http://www.who.int/dietphysicalactivity/fruit/en/. Accessed 10 Sept 2014

World Bank (2014) Annual report 2014, World Bank. http://www.worldbank.org/en/about/annual-report. Accessed 17 Jan 2015

Preparation of Value-Added Products Through Preservation

2

M. Preema Devi, N. Bhowmick, M.R. Bhanusree, and S.K. Ghosh

Abstract

Fruits and vegetables provide an abundant and inexpensive source of energy, body-building nutrients, vitamins, and minerals. However, most fruits and vegetables are only edible for a very short time unless they are promptly and properly preserved. To make foods available throughout the year, humans have developed methods to prolong the storage life of products, that is, to preserve them. The rotting process can be postponed by adding preservatives, optimizing storage conditions, or applying modern techniques. Preservation in one form or another has been practised in all parts of the world since time immemorial, although scientific methods of preservation were developed only about a hundred years ago. Preservation also assures a stable market to farmers and horticulturists and enables them to expand their production without fear of a fall in demand. Fruits and vegetable preservation industry are still in its infancy in this country. Until about 50 to 60 years ago, other well-known methods of preservation such as jam, jellies, marmalades, etc. were confined to only a fewer larger industries. One of the main difficulties in the path of the growth of the fruit and vegetable industry has been the inadequacy of knowledge of the modern methods and techniques of preservation. So, to overcome these difficulties, an attempt is made in this chapter to highlight various aspects of the importance of various preservation methods and limitations to be considered during preservation of fruits and vegetables.

2.1 Introduction

Preservation is a process of keeping food materials in an altered condition for a long time without impairing their quality to the utmost extent, with the objectives to preserve fruits and vegetables at the stage of maximum palatability, taste, colour, flavour, quality, and nutritive value; to check

M.P. Devi (✉) • N. Bhowmick • M.R. Bhanusree • S.K. Ghosh
Uttar Banga Krishi Viswavidyalaya,
Pundibari, Cooch Behar, West Bengal, India
e-mail: preema.horti@gmail.com

A.B. Sharangi and S. Datta (eds.), *Value Addition of Horticultural Crops: Recent Trends and Future Directions*, DOI 10.1007/978-81-322-2262-0_2, © Springer India 2015

wastage of local or seasonal surplus; to make the product available for a longer period even in places where it is not produced; to preserve food materials during transit from producer to consumer; and to facilitate handling of food materials, which is done primarily through various methods of packaging (Lal et al. 1959).

2.2 Importance of Fruit and Vegetable Preservation

Horticulture is concerned with perishable crops such as fruits and vegetables. Unless the preservation industry develops with the development of horticulture, it will be uneconomical. Preservation takes care of surplus produce and thereby checks wastage from rotting: that is, the more we preserve, the more we are able to consume in the future. Preservation helps the farmers to obtain a better return by checking wastage during a market glut. It keeps the products in proper condition, which is not possible under ordinary or cold storage conditions. Hence, preservation is a suitable substitute for storage. It allows fruits and vegetables to be available during off seasons and in locations where these are not grown. Also, preservation has the additional benefit of foods being more palatable. Several ancillary industries such as the productions of cans, bottles, caps, cardboard, etc. may be established, which will generate employment opportunities.

2.3 Scope of Fruit and Vegetable Preservation in India

Fruits and vegetables are important supplements to the human diet, as they provide minerals and vitamins essential for maintaining health and protecting us from different diseases and disorders. According to human dieticians, an adult person working moderately requires 85 g of fruits and 300 g of vegetables every day, in addition to cereals, fish, meat, milk, etc. Fortunately, India, with its wide range of soil and climatic conditions, is ideal for growing varieties of fruits and vegetables and is the second largest producer of fruits and vegetables, producing about 81.2 million tonnes of fruits and 162.2 million tonnes of vegetables, of which about 25 % to 30 % of the total produce is wasted because of spoilage (Anonymous 2013). Most fruits and vegetables are seasonal and perishable in nature. During the peak harvesting time there may be a market glut, but because of insufficient transport facilities and poor availability of packaging materials, the surplus cannot be taken quickly to the markets in urban areas. Moreover, the surplus often cannot be stored for sale in the off season because of inadequate local cold storage facilities; thus, the cultivators do not get a good price for their produce because of the glut, and some of it is spoiled, resulting in complete loss. Preservation of fruits and vegetables can help to solve these problems. Small, poorly shaped, overripe, and infected fruits and vegetables that are unacceptable in the market and fetch a lower price can be utilized successfully in the preservation industry. With increased urbanization, rise in middle-class purchasing power, changes in food habits, and decrease in the practice of making preserved products in individual homes, there is increasing demand for industry-made products in the domestic market. Moreover, some of these preserved products such as canned mangoes, fruit juices, salted cashews, dehydrated foods, and frozen fruits are gaining popularity in the foreign market and are good foreign exchange earners. In spite of all these reasons, only 2.2 % of the total produce is processed in India as compared to 40 to 83 % in developed countries.

Thus, there is considerable scope for expansion of the fruit and vegetable preservation industry in India, which in turn will help in the development of horticulture and in earning more foreign exchange.

2.4 Limitations of the Fruit and Vegetable Preservation Industry in India

Lack of coordination between growers and processing units. It is necessary to supply fruits and vegetables continuously for the processing industries to run effectively. A contract between growers and the processing units would ensure continued availability of good-quality raw materials to the industry.

Lack of skilled manpower. Although India is an advantageous position in having a large reservoir of manpower, skilled manpower is in short supply. Workers should be properly trained.

Lack of awareness. Because most of the available knowledge regarding preservation is scattered in scientific papers, departmental reports, and other highly technical publications, people in general are not aware of the modern methods and techniques of preservation. To overcome this problem, this information has to be disseminated on a country-wide scale so that full advantage of it can be taken by all those interested in the industry.

Lack of marketing facilities. Although there is a demand for preserved products, there are not readily available in small towns because shopkeepers are unwilling to stock such items. The establishment of a growers' cooperative society would help in the marketing of such products.

Difficulty in the availability of containers. Bottles and cans are the two major types of containers required by the food processing industry. The initiative has already been taken up by a number of factories for manufacturing bottles of required specification, but there is great difficulty in the availability of cans because there are few factories for their manufacture. There is a need to set up more factories to meet the demand for cans.

with one another and spoil the taste and aroma; (4) air coming in contact with the product may react with glucosidal materials present in it and render the product bitter; and (5) traces of metal from the processing equipment may get into the product and spoil its taste and aroma.

All that inactivates the enzymes as well as microorganisms to control spoilage forms the basis of preservation techniques. In the preservation of foods by various methods, the following general principles are involved (Khurdiya and Roy 1986).

1. *Prevention or delay of microbial decomposition:*
    (a) By keeping out microorganisms (asepsis)
    (b) By removal of microorganisms, such as by filtration
    (c) By hindering the growth and activity of microorganisms, for example, by low temperature, drying, anaerobic conditions, chemicals, or antibiotics
    (d) By killing the microorganisms, for example, by heat or radiation
2. *Prevention or delay of self-decomposition of the food:*
    (a) By destruction or inactivation of enzymes, as by blanching
    (b) By prevention or delay of chemical reactions, for example, prevention of oxidation by means of an antioxidant
3. *Prevention of damage by insects, animals, mechanical causes, etc:*
    If all these principles are followed properly, the preserved products will remain in good condition by retaining the natural taste and aroma for a longer period.

2.5 Principles of Preservation

If prepared products of fruits and vegetables are kept for some time, the taste, aroma, and appearance of the products deteriorate rapidly (Amerine et al. 1965) for several reasons: (1) fermentation caused by microorganisms such as molds, yeasts, and bacteria; (2) enzymes present in the product may affect the colour and flavour adversely; (3) chemicals present in the pulp/juice may react

2.6 Different Processes or Forms or Methods of Fruit and Vegetable Preservation

2.6.1 Liquid Form

2.6.1.1 Beverage
All drinks, unfermented or fermented, sweetened or unsweetened, are designated as beverages. Among these, fruit juices have an eminent place

as they are rich in essential minerals and vitamins and other nutritive factors. At present, synthetic beverages are becoming available and are produced in large quantities by aerated water bottlers in this country.

2.6.1.1.1 Unfermented Beverages

Fruit juices that do not undergo alcoholic fermentation are termed unfermented beverages.

1. *Pure fruit juice:* The natural juice pressed out of a fruit, then strained, that remains particularly unaltered in its composition during its preparation and preservation. Edible acid may be added before use for improving taste (citric acid), and the juice may be diluted also: for example, mango, pineapple, citrus, grape, apple, pomegranate, mulberry, jamun, phalsa, passion fruit.

2. *Fruit juice beverage:* A natural, coarsely strained fruit juice, pressed from a fruit, with a moderate quantity of fruit pulp, that is considerably altered in composition by adding water and a small amount of sugar during processing and preservation. Acid and chemical preservatives are also added according to requirements. It may be further altered in composition and diluted before consumption, as pineapple and papaya.

3. *Squash:* Squash is a type of fruit beverage containing 25–33 % fruit juice or pulp, 40–50 % total soluble solids (TSS), 1.0 % acid, and 350 ppm sulfur dioxide. This beverage is diluted with chilled water before serving (Jood and Ketarpaul 2002); for example, orange squash, lemon squash, mango squash, pineapple squash.

4. *Cordial:* A sparkling clear fruit juice derived either from fruit juice or from squash, from which all the pulp and suspended materials are removed completely by the siphon method. It may be sweetened by adding sugar. It contains at least 25 % fruit juice, 30 % TSS, 1.5 % acid, and 350 ppm SO_2: examples are lime, orange, almond.

5. *Crush:* A fruit squash or fruit beverage that contains at least 25 % fruit juice or pulp and 55 % TSS. It also contains 1.0 % acid, and is diluted before use, such as pineapple crush.

6. *Fruit juice concentrate:* Fruit juice that has been concentrated by removal of water by either heat or freezing. Carbonated beverages and other products are made from the concentrate, which contains at least 32 % TSS.

7. *Ready-to-serve (RTS):* This fruit beverage contains at least 10 % fruit juice and 10 % TSS plus about 0.3 % acid. It is not diluted before use: ber, jamun, custard apple.

8. *Nectar:* Also a coarsely strained fruit beverage that contains at least 20 % fruit juice/pulp, 15 % TSS, and about 0.3 % acid. It is not diluted before use: jamun, bale, custard apple.

9. *Syrup:* A fruit beverage containing at least 25 % fruit juice or pulp and 65 % TSS. It also contains 1.3–1.5 % acid and is diluted before serving (Sharma et al. 1988): grape, pomegranate, jamun, pineapple, orange, strawberry, raspberry, mulberry, etc.

10. *Synthetic syrup or sarbat:* A heavy sugar syrup of 70–75 % strength when flavoured and coloured with artificial essence of fruits, herbs, and colours is known as synthetic syrup or sarbat.

11. *Barley water:* A fruit beverage that contains at least 25 % fruit juice, 30 % TSS, and 0.25 % barley starch.

     Barley water is prepared from citrus fruits such as lime, lemon, grapefruit, and orange: lime and lemon are mostly used.

12. *Carbonated beverage:* When fruit juice or syrup is preserved in CO_2 gas, it is then called a carbonated beverage. Orange juice preserved in this method is known as orangeade, and, similarly, lemon juice as lemonade.

2.6.1.1.2 Fermented

Fruit juices that have undergone alcoholic fermentation by yeasts include wine, champagne, port, sherry, tokay, muscat, perry, orange wine, berry wine, nira, and cider.

(a) *Alcohol:* When any fruit juice having 10 %–12 % fermentable sugar is allowed to ferment with yeast (*Saccharomyces ellipsoideus, S. malei, S. cerevisiae*) in anaerobic conditions and at 25–27 °C, after its sterilization ethyl alcohol is produced, which is then filtered and stored in an airtight container to check further infection and thereby fermentation with vinegar bacteria. It is also a sparkling clear liquid. The product contains varying quantities of alcohol. The alcohol content of wine varies from 7 to 20 %. Wines with 7–9 % alcohol are called light liquor or light wine, those with 9–16 % are termed medium wines, and those with 16–20 % are strong wines. Liquors contain 40–60 % alcohol, although in fruit brandy only 4–6 % alcohol is present: grape wine, apple cider, cashew apple feni, palm tree nira, aonla wine, etc.

(b) *Vinegar:* Vinegar is perhaps the oldest known fermentation product. It contains about 5 % acetic acid in water, a varying amount of fixed fruit acids, colouring matter, salts, and a few other fermentation products that impart a characteristic flavour and aroma to it.

Vinegar is a liquid derived from various substances containing sugar and starch by alcoholic, and subsequently acetic acid, fermentation. In the trade, vinegar is labelled according to the material used in its manufacture: vinegar made from malt is called malt vinegar, that from apple juice is called apple cider vinegar, and that from grape is called grape vinegar.

2.6.1.2 Other than Beverages

1. *Puree:* Puree is a concentrated fruit/vegetable pulp without seed and skin; 3–10 % sugar and 1 % salt are added with the chemical preservative, so that the concentration of total solid should be 12 %, of which 8.37 % is the salt-free fruit/vegetable solid.
2. *Sauce:* Sauce is the concentrated fruit/vegetable pulp without seed and skin. Sugar, salt, and various spices are added to the content so that one should not be dominant over another. The finished products should have not less than 18 % total solids. Vinegar may or may not be added along with a requisite amount of chemical preservatives: tomato, aonla.

3. *Ketchup:* Ketchup is made by concentrating fruit/vegetable juice or pulp without seeds and skins. Spices, salt, sugar, vinegar, onion, garlic, etc. are added to the extent that it contains not less that 12 % fruit/vegetable solid and 28 % total solid. Chemical preservatives may be added (30 mg/l of product)

2.6.2 Semisolid Form

(a) *Pulp:* Pulp of low pectin content fruit when concentrated with acid and a sufficient amount of sugar without addition of water is known as pulp. A chemical preservative is added.
(b) *Jam:* Jam is a mixture of fruit and sugar cooked to the consistency of a jelly, firm enough to hold the fruit tissues. It contains all the fruit pulp in its composition and is therefore not clear. A good jam must have a bright colour and true fruit flavor; it should be neither syrupy nor stiff, but of a proper jelly consistency, with evenly distributed fruit particles. It should be free from crystallization of sugar and must keep well. Examples: mango, pineapple, aonla, apple, pear, peach, plum.
(c) *Jelly:* Jelly is prepared by cooking essentially a clear fruit extract, strained, free from insoluble matter and sugar, as in case of a jam. A perfect jelly should be sparkling, transparent, and attractive in colour and should have a strong flavour of the original fruit. It should not be gummy, sticky, or syrupy or have sugar crystallized on it. It may be thick or soft set, but should be firm enough to retain a sharp edge when cut with a knife. Examples: guava, karonda, sour apple, jamun, wood apple, plum, loquat, papaya.
(d) *Marmalade:* A jelly in which pieces of fruits are suspended. The term marmalade in this country is usually associated with a product made from citrus fruit (orange, lemon, grapefruit, etc.); in this case, the suspension in the jelly is the shredded peels of the fruit. A good marmalade must have the shreds (10–15 %) evenly distributed in the whole mass, in addition to all the characteristics of a good jelly.

(e) *Chutney:* When fully matured green fruits or matured but tender vegetables are peeled, boiled, crushed, and mixed with acid, sugar, salt, coarsely powdered spices, and herbs and cooked to a reasonably thick consistency, this is known as chutney. Vinegar may or may not be added. Here the high percentage of sugar, acid, and spices with vinegar collectively act as preservatives. Hence, addition of chemical preservatives is not necessary. Finished product should contain 40 % fruit juice and 50 %TSS: mango, ber, aonla, jack fruit.

2.6.3 Solid Form

(a) *Canning:* Whole fruits or pieces of fruits are placed in a 33–50 % sugar solution, which is known as syrup; vegetables are placed in a 3–5 % salt solution after blanching, known as brine solution; or the vegetables may be suspended in juice of that vegetable containing 0.5 % salt and 1 % acid. When packed in cans, this is known as canning: mango, orange, papaya, pineapple, apple, pear, peach, etc.

(b) *Drying:* When 85–88 % moisture is evaporated artificially from fruits or vegetable slices, either by keeping them in the sun or under controlled temperature and humidity conditions inside an oven, this is known as drying. Drying done by sun exposure is sun drying; when done under controlled temperature and humidity in a closed chamber (dehydrator), this is known as dehydration: dehydrated ber, banana, plum, apricot.

1. *Leather:* Drying of strained fruits or vegetable pulp after adding a small amount of sugar then spreading on an aluminum plate until dried produces leathers. Fruit leather can be dried in a thin layer in a solar or cabinet drier. Spreading of the pulp can be repeated again on previously dried pulp and drying continued until the thickness reaches 2–3 cm. Finally, the leather is fumigated in a sulfur chamber and stored by wrapping tightly in butter paper: mango leather, tomato leather, papaya leather, custard leather, jackfruit leather, palm leather, jamun leather, etc.

2. *Flake:* Drying of a single layer of pulp until fully dried and taken from an aluminum plate as dry thin pieces is known as flake: tomato flake, papaya flake, corn flake, etc.

(c) *Preserves:* A preserve is made from properly matured fruit pieces or mature, tender vegetable pieces by cooking in heavy syrup until tender and transparent. In its preparation not less than 40 lb of fruits is used for every 55 lb of sugar; cooking is continued until a concentration of at least 60 % of soluble solid is obtained: Bael preserves, Karonda preserve.

(d) *Candy:* Fruit or vegetable pieces when impregnated with heavy sugar and subsequently drained and dried are called candied fruits. The total sugar content of the impregnated fruits or vegetables is kept at 75 % sugar to prevent fermentation. Candied fruits covered with a thin, transparent coating of sugar that imparts a glossy appearance are called glaced fruits. Candied fruit is coated with crystals of sugar by rolling on a mild steel plate having 0.64-cm-diameter holes to allow air to enter the box for evaporation of moisture. Examples: karonda, cherry, amla, ber, jackfruit

(e) *Pickles:* Fruit or vegetable pieces are mixed with 8–10 % salt and 3 % acid, then kept for 6–8 days in the sun at about 29.4–30 °C for fermentation with the help of lactic acid bacteria, and finally mixed with coarsely powdered spices with or without vinegar and sealed by covering fully with moisture-free edible oil to produce pickle. Pickles should be kept another 3 weeks in the sun: lime, mango, cauliflower, ber.

2.7 Preparation and Preservation of Unfermented Beverages

Fruit juices that do not undergo alcoholic fermentation, termed unfermented beverages, include natural and sweetened juices, RTS, nectar, cordial, squash, crush, syrup, fruit juice concentrate, and fruit juice powder. Barley waters and carbonated beverages are also included in this group (Girdhari et al. 2010).

1. *Selection and preparation of fruit:* Not all fruit juices are suitable for making fruit juice, either because of difficulties in the extraction of juice or because the juice obtained is poor in quality. Even some of the juicy fruits are not quite suitable as they do not yield juice of good beverage quality. The best juice is extracted from freshly picked, sound and suitable varieties at the optimal stage of maturity. Fully ripe, mid-season fruits, particularly citrus fruits, generally yield juice superior to that of fruits picked early or late in the season.

2. *Sorting and washing:* Decayed or damaged fruits do not yield juice. Small cull fruits, such as undersized, oversized, malformed, or blemished fruits, which do not fetch a good price in the fresh fruit market, are rejected. The fruits should be washed thoroughly with water, and in some cases scrubbed also while washing to remove any adhering dust and other extraneous matter. Residues of sprays of arsenic and lead should be removed: dilute HCl (23 l HCl in 455 l water) is adequate for this purpose.

3. *Juice extraction:* Juice from fresh fruits is extracted by crushing and pressing them. During extraction, the juice should not be unduly exposed to air, as oxygen in the air will adversely affect the colour, taste, and aroma and also reduce the vitamin content of the juice. Citrus juices, tomato juice, and even the more stable juices such as those of apples and grapes, deteriorate rapidly in quality when they are extracted by methods that expose them to air for unduly long periods. For products such as tomato juice, special extraction equipment has been designed recently to reduce incorporation of air to a minimum.

4. *Deaeration:* Fruit juice contains some air. Most of the air is present on the surface of the fruit particles and some is dissolved in the juice. In the case of citrus juices, particularly orange juice, which is highly susceptible to the adverse action of residual air, immediately after extraction the juice is subjected to a high vacuum whereby most of the air as well as other gases are removed. This process is known as deaeration. The equipment employed is fairly expensive. It is, however, necessary for large-scale production of orange and other pure fruit juices.

5. *Straining, filtration:* Fruit juices after extraction always contain varying amounts of suspended matter, which consists of broken fruit tissue, seed, and skin, and also various gums, pectic substances, and proteins in colloidal suspension. Usually coarse particles of pulp, seeds, and pieces of skin are removed by means of screens from almost all juices as their presence generally causes deterioration in the quality of the product. In the early years of the fruit juice industry, it was a common practice to completely remove all the suspended matter, including colloidal suspensions, before packing the juice in containers. Although this method no doubt improved the appearance of the product, it quite often resulted in lack of fruit character and flavour. The present trend is to let fruit juices and fruit beverages remain reasonably cloudy or pulpy in appearance. The recent comminuted fruit beverages, employing the whole fruit for extraction, are based on this concept and are claimed to be more nutritive than the clear juices.

6. *Clarification:* Complete removal of all suspended material from juice, as in lime juice cordial, is known as clarification, which is closely related to the quality of appearance and flavour of the juice.

7. *Addition of sugar:* All juices except those of grape and apple are sweetened by adding sugar. Sugar also acts as a preservative for flavour and colour and prolongs the keeping quality. Sugar-based products can be divided into three groups on the basis of sugar content: low (30 %), medium (30–50 %), and high (>50 %).

Sugar can be added directly to the juice or as syrup made by dissolving it in water, then clarifying by addition of a small quantity of citric acid or a few drops of lime juice and filtering.

8. *Fortification:* Juices, squashes, syrups, etc. are sometimes fortified with vitamins to enhance their nutritive value, to improve taste, texture, or colour, and to replace nutrients lost in processing. Usually ascorbic acid and beta carotene are added at the rate of 250–500 mg and 7–10 mg/l, respectively. Ascorbic acid acts as an antioxidant and beta carotene imparts an attractive orange colour. For a balanced taste some acids are added. Citric acid is often used for all types of beverages and phosphoric acid for the cola type of drinks.

9. *Preservation:* Fruit juices, RTS, and nectars are preserved by pasteurization, but sometimes chemical preservatives are used. Squashes, crushes, and cordials are preserved only by adding chemicals. The sugar concentration of syrup is sufficient to prevent its spoilage.

2.8 Preparation and Preservation of Fermented Beverages

Fermented beverages have been known to mankind from times immemorial. Grape wine is the most important among these. Wines made from fruits are named after the particular fruit employed. Thus, we have apple cider from apples, perry from pears, and orange wine from orange. Starch and sugar also are fermented to produce special types of liquors. In India, such liquors are known as neera juice of palm.

2.8.1 Alcoholic Fermentation

2.8.1.1 Preparation of Wine
1. *Selection of raw material:* Fruits intended for wine making are sorted to remove moldy and diseased fruits, then crushed between fluted rollers. Fruits are fermented slightly before pressing the juice, which helps in the extraction of colour and also facilitates pressing of the juice. Generally, the yield of juice is 60–70 %.

2. *Crushing:* Before crushing the fruit stems and stalks are removed. Crushing is done with a basket press.

3. *Addition of sugar:* Cane sugar is added to maintain at least 20 % total soluble. solids but not more than 24 %. If the fruits are sour, 70 g sugar is added for each kilogram of fruit, at the rate of 1.5 g for every 10 kg fruits, then mixed and allow to stand for 2–4 h. The preservative potassium magnesium sulfate (KMS) (1.5 g/10 kg fruit) inhibits growth of wild yeasts and spoilage organisms (Srivastava and Kumar 1994).

4. *Addition of wine yeast:* Wine yeast, such as *Saccharomyces ellipsoideus* inoculum, is added at 20 ml for every 5 kg fruits, about an hour after the addition of preservative. If the yeast is not available then potassium metabisulfite is not added. The yeast present in the skin of fruits can also ferment and produce wine but it is not of good quality.

5. *Fermentation:* Fruits are allowed to ferment for 2 days in a cool place, that is, at 22–28 °C. The mouth of the jar is covered with cloth during fermentation.

6. *Filtration:* The contents are filtered through a thin muslin cloth or a filter aid on the third day and the filtrate is again allowed to ferment in a cool place for another 10 days without any disturbance. During this period yeast cells and other solids settle at the bottom.

7. *Racking:* Siphoning off the fermented wine to separate it from the solid deposits is known as racking.

8. *Fining and filtration:* The newly prepared wine is sometimes not clear and requires fining and filtration. A suitable fining agent, such as bentonite, is added. All the colloidal material settles with the bentonite. The clear wine is siphoned off and filtered if necessary. Alternatively, the wine is stored in a refrigerator for about 2 weeks and thereafter the clear wine is siphoned off.

9. *Maturation:* When fermentation is complete, the clear wine is siphoned; the containers are filled completely and sealed airtight to exclude all air. In the course of

time, the wine matures. During this maturing process, which takes from 6 to 12 months, the wine loses its raw and harsh flavour and mellows considerably, acquiring a smooth flavour and characteristic bouquet and aroma. Barrels of oak wood are generally preferred for maturing as they impart a finer bouquet to the wine.

During the maturation process, there is natural clarification of the wine. Filter aids, such as egg white, can also be employed to bring about the clarification.

10. *Packing:* The volatile acidity of wine, which is mainly the result of acetic acid, should be low. High volatile acidity of 0.09–0.20 g/100 ml, expressed in terms of acetic acid, indicates the activity of acetic acid bacteria during the fermentation: this is not desirable. It is desirable to pasteurize the wine to destroy spoilage organisms and coagulate the colloidal materials which cause cloudiness in the wine. Wines are generally pasteurized for 1–2 min at 82–88 °C and then bottled. The bottles are closed with bark corks of good quality (Girdhari et al. 2010). Many wines are made from fruits having medicinal value (Tapsell et al. 2006). The following are well-known wines produced in various countries.

Champagne: A sparkling wine, made chiefly in France, from certain varieties of grapes such as Chardonnay and Pinot Noir: it is made in other countries as well. The fermentation is allowed to proceed to completion in bottles that are specially made to withstand the high pressure of gas produced during fermentation.

Neera: A wine made from juice of the palm tree. Varieties of palm wines or toddy are consumed all over the world (Steinkraus 1996). In India, the fermented palm sap is known as toddy or kallu.

Port: A fortified, sweet red wine made originally in Portugal but now produced in other countries also.

Sherry: A Spanish wine, matured by placing the barrels in sunlight for 3 to 4 months, where the temperature is as high as 54–60 °C.

Jack fruit wine: An alcoholic beverage made by fermentation of jack fruit pulp (Dahiya and Prabhu 1997).

Tokay: A very famous fortified wine made in Hungary.

Muscat: Prepared from Muscat grapes in Italy, California, Spain, and Australia.

Perry: Wine made from pears is known as perry. Its method of preparation is similar to thoat of apple cider. Wastes, culled fruits, and trimmings left over from canning may also be used for making perry.

Soor: A type of alcoholic beverage made by fermentation of fruits such as apricot, peach, pear, and apple, it contains 35–40 % alcohol (Rana et al. 2004).

Orange wine: Orange juice is sweetened by adding sugar and then allowed to ferment. The method of preparation is similar to that of grape wine. Orange oil should not be added to the juice as it hinders and sometimes stops fermentation.

Berry wine: Wines prepared from berries such as strawberry and blackberry are known as 'Berry wines.' These products are generally popular in other countries but are not common in India.

Feni: Cashew apple juice is fermented into a strong liquor known as 'feni.' It is registered, according to geographic indication, in places around Goa (Serkar and Mariappan 2007).

2.8.2 Lactic Acid Fermentation

2.8.2.1 Preparation of Pickles

The preservation of food in common salt or in vinegar is known as pickling. It is one of the most ancient methods of preserving fruits and vegetables. Pickles are good appetizers and add to the palatability of a meal. They stimulate the flow of gastric juice and thus help in digestion.

Several kinds of pickles are sold in the Indian market. Mango pickle ranks first, followed by cauliflower, onion, turnip, and lime pickles. These pickles are commonly made in homes as well as being commercially manufactured and exported. Fruits are usually preserved in sweetened and spiced vinegar, whereas vegetables are preserved in salt.

Pickling is the result of fermentation by lactic acid-forming bacteria, which are generally present in large numbers on the surface of fresh vegetables and fruits. These bacteria can grow in an acid medium and in the presence of 8–10 % salt solution, whereas the growth of a majority of undesirable organisms is inhibited. Lactic acid bacteria are most active at 30 °C, so this temperature must be maintained as much as possible in the early stage of pickle making. When vegetables are placed in brine, it penetrates into their tissues, and soluble material present in them diffuses in the brine by osmosis. The soluble material includes fermentable sugars and mineral. The sugars serve as food for lactic acid bacteria, which convert them into lactic and other acids. The acid brine thus formed acts upon the vegetable tissues to produce the characteristic taste and aroma of pickle.

In the dry salting method, several alternating layers of vegetables and salt (20–30 g dry salt/kg vegetables) are kept in a vessel that is covered with a cloth and a wooden board and allowed to stand for about 24 h. During this period, by osmosis, sufficient juice comes out from the vegetables to form brine. Vegetables that do not contain enough juice (e.g., cucumber) to dissolve the added salt are covered with brine (steeping in a concentrated salt solution is known as brining). The amount of brine required is usually equal to half the volume of vegetables. Brining is the most important step in pickling. The growth of the majority of spoilage organisms is inhibited by brine containing 15 % salt. Lactic acid bacteria, which are salt tolerant, can thrive in 8–10 % brine although fermentation takes place fairly well even in 5 % brine. In a brine containing 10 % salt, fermentation proceeds somewhat slowly (Pederson 1963). Fermentation takes place to some extent up to 15 % brine, but stops at 20 % strength. It is, therefore, advisable to place the vegetables in 10 % salt solution for vigorous lactic acid fermentation.

As soon as the brine is formed, the fermentation process starts and carbon dioxide begins to evolve. The salt content is now increased gradually, so that by the time the pickle is ready, salt concentration reaches 15 %.

Fermentation is completed in 7–10 days. When sufficient lactic acid has been formed, lactic acid bacteria cease to grow and no further change takes place in the vegetables. However, precautions should be taken against spoilage by aerobic microorganisms, because in the presence of air a pickle scum is formed that brings about putrefaction and destroys the lactic acid. Properly brined vegetables keep well in vinegar for a long time.

At present, pickles are prepared with salt, vinegar, or oil or with a mixture of salt, oil, spices, and vinegar. These methods are discussed below.

1. *Preservation with salt:* Salt improves the taste and flavor, hardens the tissues of vegetables, and controls fermentation. Salt content of 15 % or more prevents microbial spoilage. This method of preservation is generally used only for vegetables that contain very little sugar, and hence sufficient lactic acid cannot be formed by fermentation to act as a preservative. However, some fruits, such as lime, green chili, mango, etc. are also preserved with salt.

2. *Preservation with vinegar:* A number of fruits and vegetables are preserved in vinegar whose final concentration, in terms of acetic acid, in the finished pickle should not be less than 2 %. To prevent dilution of vinegar below this strength by the water liberated from the tissues, the vegetables or fruits are generally placed in strong vinegar of about 10 % for several days before pickling. This treatment helps to expel the gases present in the intercellular spaces of the vegetable tissue.

   Vinegar pickles are the most important pickles consumed in other countries. Mango, garlic, chilies, papaya, etc., are preserved as such in vinegar.

3. *Preservation with oil:* The fruits or vegetables should be completely immersed in edible oil. Cauliflower, lime mango, and turnip pickles are the most important oil pickles. Methods of preparation of some oil pickles are given next.

   Mango pickle: Use 1 kg mango pieces, 150 g salt, 25 g fenugreek (powdered), 15 g turmeric (powdered), 15 g nigella seeds,

10 g red chili power, eight cloves (headless), 15 g each black pepper, cumin, cardamom (large), and aniseed (powdered), 2 g asafoetida, and 350 ml mustard oil (just sufficient to cover pieces).

4. *Preservation with mixture of salt, oil, spices, and vinegar:*
   Cauliflower pickle: Use 1 kg cauliflower pieces, 150 g salt, 25 g ginger (chopped), 50 g onion (chopped), 10 g garlic (chopped), 15 g each red chili, turmeric, cinnamon, black pepper, cardamom (large), cumin, and aniseed (powdered), six cloves (headless), 50 g tamarind pulp, 50 g mustard (ground), 150 ml vinegar, and 400 ml mustard oil.

2.8.2.2 Problems in Pickle Making

1. *Bitter taste:* Use of strong vinegar or excess spices or prolonged cooking of spices imparts a bitter taste to the pickle.
2. *Dull and faded product:* Caused by use of inferior quality materials or insufficient curing.
3. *Shrivelling:* Occurs when vegetables (e.g., cucumber) are placed directly in a very strong solution of salt or sugar or vinegar. Hence, a dilute solution should be used initially and its strength gradually increased.
4. *Scum formation:* When vegetables are cured in brine, a white scum always forms on the surface because of the growth of wild yeast. This scum delays the formation of lactic acid and also encourages the growth of putrefactive bacteria, causing softness and slipperiness. Hence, it is advisable to remove scum as soon as it is formed. The addition of 1 % acetic acid helps to prevent the growth of wild yeast in brine without affecting lactic acid formation.
5. *Softness and slipperiness:* A very common problem, caused by inadequate covering with brine or the use of weak brine. The problem can be solved by using a brine of proper strength and keeping the pickles well below the surface of the brine.
6. *Cloudiness:* When the structure of the vegetable used in pickling, such as onion, is such that the acetic acid (vinegar) cannot penetrate deep enough into its tissues to inhibit the activity of bacteria and other microorganisms present in them, fermentation starts from inside the tissues, rendering the vinegar cloudy. This microbial activity can only be checked by proper brining. Cloudiness may also be caused by using of inferior quality vinegar, causing a chemical reaction between vinegar and minerals.
7. *Blackening:* Caused by the iron in the brine or in the process equipment reacting with the ingredients used in pickling, blackening is also caused by certain microorganisms.

2.8.3 Acetic Acid Fermentation and Vinegar Fermentation

Vinegar is a liquid substance consisting mainly of acetic acid in water, varying amounts of fixed fruit acids, colouring matter, salt, and a few other fermentation products which impart a characteristic flavour and aroma to the product. In the trade, vinegar is labelled according to the material used in its manufacture. For instance, vinegar made from barley or other cereals is called 'malt vinegar,' that made from grape juice is 'grape vinegar,' and so on.

2.8.3.1 Quality Standards

Vinegar is a liquid derived by alcoholic and acetic fermentation. It contains about 5 % acetic acid and has germicidal and antiseptic properties. It should not contain arsenic in amounts exceeding 0.0143 mg/100 ml nor any mineral acid, or lead except caramel. The amount of acid in the vinegar is expressed as 'grain strength,' which is ten times the percentage of the acetic acid present in it; a vinegar containing 5 % acetic acid is spoken of as vinegar of '50 grain strength.'

2.8.3.2 Types of Vinegar

Vinegar is made from various fruits and also from sugars. Two important types of vinegar are described here.

(A) Breved vinegars and, (B) Artificial vinegar

(A) *Breved vinegars:* Breved vinegars can be made from various fruits and sugar containing substances (molasses, honey) by alcoholic and subsequent acetic fermentation. Examples are fruit vinegar, potato vinegar, malt vinegar, molasses vinegar, and honey vinegar.

(B) *Artificial vinegars:* Artificial vinegars are prepared by diluting synthetic acetic acid or glacial acetic acid to a standard of 4 % and are coloured with caramel.

2.8.3.3 Steps Involved in Vinegar Production

Two important steps are involved in the preparation of vinegar: (1) transformation of the sugary substances of fruits etc., into alcohol by yeast (alcoholic fermentation), and (2) changing of the alcohol into vinegar by acetic acid bacteria (acetification) (Girdhari et al. 2010). The chemical reactions involved in these two processes can be represented as follows:

1. Fermentable sugar $(C_6H_{12}O_6) \rightarrow$ Ethyl alcohol $(2C_2H_5OH)$ + Carbon dioxide $(2CO_2)$ (under the presence of yeast)

2. Ethyl alcohol $(2C_2H_5OH)$ + Oxygen $(2O_2) \rightarrow$ Acetic acid $(2CH_3COOH)$ + Water (H_2O) (under the presence of vinegar bacteria)

2.8.3.4 Preparation of Vinegar

Vinegar is prepared by the following methods.

(A) Slow process

(B) Orleans Slow process

(C) Quick process

(A) *Slow process:* The slow process, covering a prolonged period, is generally adopted in India. The fruit juice or sugar solution is filled into barrels and allowed to undergo alcoholic and acetic fermentation slowly for at least 5–6 months in a warm, damp room. No special care is taken. To screen off dust and flies, the bung hole of the barrel is covered with a piece of cloth. The main drawbacks of this method follow:

1. Yield is low
2. Alcoholic fermentation is often incomplete
3. Acetic fermentation is very slow
4. Quality of the vinegar is inferior

(B) *Orleans slow process:* In this process, about three-fourths of the barrel is filled with the juice, inoculated with mother vinegar, and the barrel placed on its side. Two holes, each about 2.5 cm in diameter, are made on either side of the barrel just above the level of the juice in addition to the bung hole. These holes are screened with wire gauze or cheesecloth to exclude insects.

The barrels are kept in a warm place at 21–27 °C, and fermentation is allowed to proceed until the acid reaches its maximum strength. Under favourable conditions, it usually takes about 3 months for the complete conversion of the liquid into vinegar. About three-fourths of the vinegar is then withdrawn, and an equal quantity of fermented alcoholic juice is added for further vinegar fermentation. Vinegar produced by the Orleans slow process ages during the process of fermentation and is clear and of superior quality.

(C) *Quick process:* In this process an additional supply of oxygen is made available for the bacteria and the surface of the bacterial culture is also increased, resulting in rapid fermentation. The equipment used, known as an "upright generator," is a cylinder 3.6–4.2 m high and 1.2–1.5 m in diameter (Srivastava and Kumar 1994). The details of the three compartments are as follows:

1. *Distributing compartment:* This is about 30 cm above the central compartment and is separated from it by a partition that is perforated with a number of small holes. In this compartment, a revolving sparkler is fitted to allow the liquid to

trickle slowly over the material in the central compartment.

2. *Central compartment:* This is filled with beech wood shavings, pumice stone, rattan shavings, or straw to increase the surface area. This chamber is fitted with an adjustable opening near the bottom for admission of air.

3. *Receiving compartment:* This is the bottom chamber of the generator in which vinegar is collected. It is separated from the central compartment by a perforated partition about 1.5 m from the bottom of the generator.

The material in the central compartment is sprinkled and wetted with unpasteurized vinegar containing acetic bacteria. A mixture of the alcoholic fermentation product and vinegar (2:1) is then slowly trickled through the generator to promote the growth of vinegar bacteria. Within a few days the bacterial growth is enough for efficient functioning of the generator. The alcoholic fermentation liquid is now mixed with mother vinegar in the ratio of 1:2 to increase its acidity from 3 to 3.5 % and passed through the generator, where it is converted into acetic acid in a single passage.

2.8.3.5 Problems in Vinegar Production

1. *Insects and worms:* Vinegar flies (*Drosophila cellaris*), vinegar eels (*Anguillula* spp.), and the vinegar louse are important; these contaminants can only be avoided by maintaining proper sanitary conditions.

2. *Wine flowers:* A kind of film yeast; if the fermented juice is unnecessarily exposed to air, wine flowers grow on the surface of the liquid. This growth can be avoided by spreading a neutral oil such as liquid paraffin over the surface of the fermented liquid or adding 20–25 % of unpasteurized vinegar.

3. *Lactic acid bacteria:* Lactic acid bacteria are generally found in fermented juice. The bacteria interfere with acetic acid fermentation, cause cloudiness, and decrease the quality of the vinegar. This infestation can be avoided by using 20–25 % unpasteurized vinegar.

2.9 Preparation and Preservation of Other than Beverages

2.9.1 Selection of Fruits

For this product preparation, only plant-ripened and fully red tomatoes should be used. No green, blemished, and overripe fruits should be used. All green, blemished, and overripe fruits should be rejected as these adversely affect the quality of the product.

2.9.2 Washing and Trimming

Mere rinsing of fruits in water is not enough, because mold filaments and other microorganisms found in the cracks, wrinkles, folds, and stem cavities are not easily dislodged by gentle washing alone. For through cleaning, fruits should be washed in plenty of running water. For commercial production, rotary washers or through washers and soft roller brushes are generally used.

2.9.3 Pulping

Fruits can be pulped by the hot pulping process or cold pulping process.

2.9.3.1 Preparation of Puree and Paste

Fruit/vegetable pulp without skin or seeds, with or without added salt, and containing not less than 9.0 % of salt-free fruit/vegetable solids, is known as 'medium puree.' It can be concentrated further to 'heavy puree,' which contains not less than 12 % solids. If this is further concentrated so that it contains not less than 25 % solids, it is known as paste. On further concentration to 33 % or more of solids, it is called concentrated paste.

Cooking for concentration of the pulp can be done either in an open cooker or a vacuum pan. In an open cooker most of the vitamins are destroyed and the product become brown. Use of the vacuum pans, which are expensive, helps preserve the nutrients and also reduces the browning to a great extent. In vacuum pans the juice is boiled at about 71 °C only. Ordinarily fruit/vegetable juice can be concentrated to 14–15 % solids in an open cooker, but for obtaining higher concentrations a vacuum pan is required. Moreover, sterilization of the product is also possible in a vacuum pan. While cooking in an open cooker, a little butter or edible oil is added to prevent foaming, burning, and sticking. If, after cooking, the total solids content of the juice is higher than required, more juice is added to lower it; if it is lower, cooking is continued until the desired concentration is reached.

2.9.3.1.1 Judging the End Point

The total solids in the juice, in the beginning, during boiling, and at the finishing point can be determined with either a specific gravity hydrometer or refractometer or by drying the juice in vacuum at 70 °C or by measuring the volume (a known volume of juice is concentrated to a known volume of final product) with the help of a measuring stick.

2.9.3.1.2 Packing

The product can be packed in plain as well as lacquered cans. The prepared product is poured into the cans, scalding hot, at 82–88 °C, and the cans closed.

2.9.3.2 Preparation of Sauces and Ketchups

There is no essential difference between sauce and ketchup. However, sauces are generally thinner and contain more total solids (minimum, 30 %) than ketchups (minimum, 28 %). Tomato, apple, papaya, walnut, soybean, mushroom, etc., are used for making sauces.

Sauces are of two kinds: (1) *thin* sauces of low viscosity consisting mainly of a vinegar extract of flavouring materials such as herbs and spices, and (2) *thick sauces* that are highly viscous.

Sauces/ketchups are prepared from more or less the same ingredients and in the same manner as chutney, except that the fruit or vegetable pulp or juice used is sieved after cooking to remove the skin, seeds, and stalks of the fruits, vegetables, and spices and to give a smooth consistency to the final product. However, cooking takes longer because fine pulp or juice is used. Some sauces develop a characteristic flavour and aroma on storing in wooden barrels. Freshly prepared products often have a raw and harsh taste and must, therefore, be matured by storage. High-quality sauces are prepared by maceration of spices, herbs, fruits, and vegetables in cold vinegar or by boiling them in vinegar. The usual commercial practice is to prepare cold or hot vinegar extracts of each kind of spice and fruit separately, and then blend these extracts suitably to obtain the sauces, which are then matured. Thickening agents are also added to the sauce to prevent sedimentation of solid particles. Apple pulp is commonly used for this purpose in India, but starch from potato, maize, arrowroot (cassava), and sago is also used. A fruit sauce should be cooked to such a consistency that it can be freely poured without the fruit tissues separating out in the bottle. The sauce should be bright in colour. Sauces usually thicken slightly on cooling. By using a funnel, hot ketchup is filled in bottles, leaving a 2-cm head space at the top, and the bottles are sealed or corked at once. The necks of the bottles, when cold, are dipped in paraffin wax for airtight sealing. It is advisable to pasteurize sauces after bottling because there is always a danger of fermentation, especially in tomato- and mushroom-based sauces. Other sauces are more acidic and less likely to ferment but should be pasteurized nonetheless. For this purpose the bottles are kept in boiling water for about 30 min (Srivastava and Kumar 1994).

2.9.4 Problems of Making Other than Beverages

2.9.4.1 Black Neck

Formation of a black ring in the neck of bottles is known as black neck. It is caused by iron that gets into the product from the metal of the equipment and the cap/crown/cork through the action of acetic acid. This iron coming into contact with tannins in spice forms ferrous tannate, which is oxidized to black ferric tannate. This problem can be prevented by the following:

1. Filling hot product at a temperature not less than 85 °C
2. Leaving very little head space in bottles (the more the air, the greater is the blackening)
3. Reducing contamination by iron, sources of iron being salt and metal equipment
4. Partial replacement of sugar by corn syrup or glucose syrup, which contain sulfur and prevent blackening
5. Addition of 100 ppm sulfur dioxide or 100 mg ascorbic acid
6. Storing bottles in horizontal or inverted position to diffuse the entrapped air (O_2) throughout the bottle, thus reducing its concentration in the neck sufficiently to prevent blackening
7. Using cloves only after removing the flower/head

2.10 Preparation and Preservation of Jelly, Jam, and Marmalade

2.10.1 Jelly

2.10.1.1 Principles of Jell (Gel) Formation

Sugar, acid, pectin, and water are the four important constituents of jam, jelly, and marmalade and must be present approximately in the following proportions: pectin, 1.0 %, acid, 1.0 %, sugar, 60–65 %, and water, 33–38 %. Pectin is a complex organic compound of the CHO class present in varying amounts in practically all fruits and possessing the property of entering into a sort of equilibrium in presence of the other three ingredients either added to or present in a fruit pulp, juice, or the water extract of fruit. At a definite range of equilibrium of these ingredients, the compound formed has the consistency of a gel, which is the basic characteristic of jam, jelly, and marmalade. If any one of these is more or less than the required amounts, good gel will not form.

2.10.1.1.1 Pectin

Pectic substances, present in the form of calcium pectate, are responsible for the firmness of fruits. Pectin is a commercial term for water-soluble pectinic acid. In the early stage of development of fruits (unripe), the pectic substance is a water-insoluble proto-pectin that is converted into pectin by the enzyme protopectinase during ripening of fruit. In overripe fruits, because of the presence of the enzyme pectic methyl esterase (PME), the pectin is largely converted to pectic acid, which is water insoluble; this is one of the reasons both immature and overripe fruits are not suitable for making jelly and only ripe fruits are used.

Usually, about 0.5–1.0 % pectin in the extract is sufficient to produce a good jelly. If the pectin content is in excess, a firm and tough jelly is formed, and if it is less, the jelly may fail to set.

2.10.1.1.2 Sugar

This essential constituent of jelly imparts to it sweetness as well as body. If the concentration of sugar is high, the jelly retains less water, resulting in a stiff jelly, probably because of dehydration.

When sugar (sucrose) is boiled with an acid, it is hydrolyzed into dextrose and fructose, the degree of inversion depending on the pH and duration of boiling. Because of partial inversion of the sucrose, a mixture of sucrose, glucose, and fructose is found in the jelly. This mixture is more soluble in water than sucrose alone and hence the jelly can hold more sugar in solution without crystallization.

2.10.1.1.3 Acid

The finished product should contain at least 0.5 % but not more than 1.0 % total acids because a larger quantity of acid may cause syneresis. Jelly strength increases with the increase in pH until an optimum is reached. In general, the optimal pH value for jelly is 3.2.

2.10.1.2 Preparation of Jelly
2.10.1.2.1 Selection of Fruit

The fruit should be fresh, just ripe, and firm. It is often useful to use one-fourth ripe fruits and three-fourths just-ripe fruits for imparting flavour as well as for good setting of the jelly. Guava, sour apple, plum, karonda, wood apple, loquat, papaya, and gooseberry are generally used for preparation of jelly. Apricot, pineapple, strawberry, raspberry, etc., can be used but only after addition of pectin powder because these fruits have low pectin content.

2.10.1.2.2 Grading of Fruit
Grading should be done according to variety, size, and colour to make a uniform-quality product.

2.10.1.2.3 Preliminary Treatments
Fruits are to be washed thoroughly in running tap water. It is better to wash the fruits in diluted HCl (1:20) followed by washing in running tap water to remove dirt and residual spray materials.

2.10.1.2.4 Preparation of Fruits
Fruits are cut into small, 1.5-cm pieces along with edible seeds and skin with a sharp stainless steel knife. During preparation, the fruit pieces are placed in an aluminium container containing the required quantity of water to check surface oxidation. Usually no water is required to be added for very soft and juicy fruits (grape, orange, tomato, etc.). However, for soft, semi-hard, and hard fruits, 3/4 to equal, equal to 1.5 times, and 1.5–2.0 times water, respectively, is required during pectin extraction.

2.10.1.2.5 Extraction of Pectin
These fruit pieces along with water and half the total requirement of acid are cooked to soften and break down the fruit tissues and to release pectin. Lime juice can be used instead of citric acid. Addition of acid helps to release pectin from the fruit pieces, acts as a preservative, and improves the colour. The remaining amount of citric acid is added before reaching the end point.

2.10.1.2.6 Pectin Test
A pectin test is essential to calculate the quantity of sugar to be added per litre of pectin water extract, which can be done by the following methods.

1. *Gelmetre test or viscosity test*

   This test uses a 28-cm-long pipette-like glass apparatus with both sides open having downward graduations from 1.25 to 0.5. The clear pectin water extract is poured into the gelmeter up to the brim at a temperature ranging between 21 and 26 °C by blocking the lower end. The extract is then allowed to drop for only 60 s by removing the finger from the lower end. The graduation that the upper surface of the extract reaches is recorded. Dropping of extract depends on the concentration. If higher pectin is present in the extract, the higher will be the viscosity, resulting in slow dropping and a higher reading, that is, a higher amount of sugar is needed. Similarly, a lesser amount of sugar is needed if less pectin is present in the extract.

2. *Spirit test*

   A teaspoonful of strained pectin extract is placed in a test tube, three teaspoonfuls of methylated or rectified spirit added: the pectin present in the extract will be precipitated. If the precipitate formed consists of a big clot, the extract is rich in pectin; if the clot breaks up into two to three small pieces, the pectin content is medium; and if the precipitate consists of several small clots, the pectin content is very poor. In case of poor pectin content, the pectin extract may be concentrated further until it indicates medium pectin content or mixed 10–15 g pectin powder/l extract.

2.10.1.2.7 Addition of Sugar
Half the calculated amount of sugar is added when the mixture starts boiling. Addition of the full amount of sugar will unnecessarily increase the volume, resulting in charring of sugar and consumption of much heat energy. The first addition of sugar will partially inverted and form glucose and fructose (both are monosaccharides) and which do not crystallize at such a concentration when glucose as crystal recrystallizes on the surface of the jelly. The remaining half of the sugar is added at 102 °C during the frothing period. Inversion of sugar takes place in presence of enzyme (invertase) present in the fruit.

2.10.1.2.8 Boiling of Mixture

High temperature is to be applied to complete the processes within 45 min so that the pectin does not lose its jelling property.

2.10.1.2.9 Addition of Pectin Powder

When it is necessary, pectin powder is added at 10–15 g/l pectin extract with eight to ten times the amount of sugar. Both sugar and pectin are dissolved in a sufficient quantity of water at 60–70 °C and mixed with the product during boiling before addition of the second dose of acid.

2.10.1.2.10 End Point Test

(a) Temperature test: 105.6 °C at mean sea level. With every 150 m rise in altitude, the end point temperature drops by 0.6 °C.
(b) Sheet test or flake test
(c) Ball test
(d) Refractometer test
(e) Cold plate test
(f) Volume test
(g) Eye estimation or colour test

2.10.1.2.11 Filling Bottles

Fill the prepared jelly with the help of a glass rod or spoon into hot sterilized wide-mouthed bottles up to the brim, cover with a thin cloth, and leave undisturbed for 8–10 h.

2.10.1.2.12 Sealing

When the product in the bottles has settled and cooled, a thin layer of molten paraffin wax about 0.5 cm thick is to be poured over the surface of the jelly to seal the bottle airtight.

2.10.1.2.13 Labelling and Storing

Bottles are labelled properly with the date of preparation, name of product, recipe, and name of manufacturer. The bottles are then stored in a cool, dry, shaded place.

2.10.1.3 Problems in Jelly Making

2.10.1.3.1 Failure of Jelly to Set

1. Lack of acid or pectin or both in the finished product.
2. Addition of lesser amount of sugar, resulting in syneresis or weeping jelly.
3. Addition of excess amount of sugar, resulting in crystalline jelly.
4. Cooking below the end point, resulting in soft or syrupy jelly.
5. Cooking beyond the end point, resulting in tough or gummy or sticky jelly.
6. Cooking slowly for a longer period results in syrupy jelly because of destruction of the elasticity and jelling property of pectin and much inversion of sugar.
7. Addition of sugar at a time resulting in blackening of the product and complete inversion of sugar.
8. Disturbing filled container during setting of jelly may cause failure of jelly to set.
9. Using immature or overripe fruits that have no jelling property.
10. Addition of ingredients at an improper ratio.

2.10.1.3.2 Cloudy Jelly

When a jelly shows a cloud-like appearance and is not quite transparent, it is known as cloudy jelly.

Causes
1. Filling of jelly without the help of a glass rod, resulting in incorporation of air bubbles.
2. Use of immature fruits having insoluble starchy matter.
3. Crushing of fruits pieces either during extraction of pectin or during squeezing, resulting in mixing of fine insoluble fruit particles with the product.
4. Pectin extract when not properly clarified by sedimentation and by siphoning, resulting in cloudy jelly.
5. Scum if not removed properly during cooking (frothing time).
6. Premature gel formation: in extract with excess pectin content, lesser amounts of both sugar and acid are to be added and preparation should be completed promptly by adding more water.
7. Overcooking causes gel action to start before filling.

2.10.1.3.3 Crystalline Jelly

When on the surface or throughout the jelly sugar crystals appear because of recrystallization of sugar or when sugar does not have sufficient time to dissolve, this is known as crystalline jelly.

Causes

1. Addition of excess sugar
2. Overcooking or overconcentrating the product
3. Concentrating the pectin extract before addition of sugar so it does not have sufficient time to dissolve
4. Late addition of sugar and in one lot causing noninversion of sugar
5. Excess concentration of pectin extract where acid is less

2.10.1.3.4 Weeping or Syneresis Jelly

Spontaneous exudation of fluid from upper surface of jelly.

Causes

1. Addition of excess acid, which breaks down the pectin and thereby the jelly structure
2. Formation of tough jelly, causing premature gelation
3. Presence of less pectin in the finished product
4. Addition of less sugar

Precautions During Preparation of Jelly

1. Fruits used for jelly preparation should have 0.5–1.0 % pectin.
2. Just-ripe fruits should be selected.
3. Fruits should always be washed in running cold water before trimming.
4. Peeling should be avoided.
5. Utensils or knives made from iron or copper should not be used.
6. Ingredients should be used in proper amounts and at the proper time.
7. Boiling should be completed within 45 min after addition of first half quantity of sugar.
8. End point should be judged properly.
9. Bottles along with lids should be sterilized properly
10. Hot product should be filled into hot bottles with the help of a glass rod or spoon.

2.10.2 Preparation of Jam

The procedure of making jam is similar to jelly making, except that after the pectin test it is necessary to boil the pectin extract without straining of edible seed and skin. Apple, pear, sapota, apricot, loquat, peach, papaya, karonda, carrot, plum, strawberry, mango, tomato, and grape are used for the preparation of jam.

2.10.2.1   Problems in Jam Production

1. *Crystallization:* The final product should contain 30–50 % invert sugar. If the percentage is less than 30 %, cane sugar may crystallize out on storage, and if it is more than 50 % the jam will become a honey-like mass because of the formation of small crystals of glucose. Corn syrup or glucose may be added along with cane sugar to avoid crystallization.
2. *Sticky or gummy jam:* Jams tend to become sticky or gummy as a consequence of a high percentage of total soluble solids. This problem can be solved by addition of pectin or citric acid or both.
3. *Premature setting:* This is caused by low TSS and high pectin content in the jam and can be prevented by adding more sugar. If this cannot be done, a small quantity of sodium bicarbonate is added to reduce the acidity and thus prevent precoagulation.
4. *Surface graining and shrinkage:* This is caused by evaporation of moisture during storage of jam. Storing in a cool place can reduce it.
5. *Microbial spoilage:* Sometimes molds may spoil the jam during storage, but they are destroyed if exposed to less than 90 % RH. Hence, jams should be stored at 80 % RH. Mold growth can also be prevented by not sealing the filled jar and covering the surface of jam with a disc of waxed paper because mold does not grow under open conditions as rapidly as in a closed space. It is also advisable to add 40 ppm sulfur dioxide in the form of KMS.

2.10.3 Marmalade

Marmalade is a fruit jelly in which slices of the fruit or its peel remain in suspended condition. Marmalade is generally prepared from citrus fruits such as oranges and lemons in which shredded peel is used as suspended material. Citrus marmalades are classified into jelly marmalade and jam marmalade.

2.10.3.1 Problems in Marmalade Making

Browning during storage, which is very common, can be prevented by addition of 0.09 g KMS/kg marmalade and not using tin containers. KMS, dissolved in a small quantity of water, is added to the marmalade while it is cooling. KMS also eliminates the possibility of spoilage caused by molds.

2.11 Preparation and Preservation of Dried/Dehydrated Products

Drying is one of the earliest methods employed in the preservation of food. There is a minimum level of moisture necessary for each kind of spoilage organism to feed, grow, and develop and thus cause decomposition of food. In drying, the moisture content of a food is reduced to such a level that no organisms can grow in it; those present originally either remain dormant or die from lack of nourishment. Further reduction of moisture in certain types of foods is sometimes necessary to avoid other undesirable changes during storage.

2.11.1 Methods and Equipment for Drying

Foods can be dried either in the sun, called sun drying, or by artificial heat under controlled temperature and humidity conditions in specially constructed chambers called dehydrators, known as dehydration.

Dehydration, in general, has the following advantages over sun drying:
1. Drying is quicker and much time is saved
2. Deteriorative changes caused by the action of enzymes, etc., are much reduced
3. It can be practised in all places and under all climatic conditions
4. Product remains free from dust and attack of insects

5. The products will remain in better condition and will be more uniform in quality

2.11.1.1 Sun Drying

The only equipment required in sun drying is some suitable type of drying trays. These may be made of wire gauze, wooden strips, or bamboo strips fitted to wooden frames. The last two types, commonly known as slat bottom trays, are suitable for almost all kinds of fruits but not generally convenient and useful for vegetables, whereas galvanized wire gauze trays are suitable for practically all kinds of vegetables. The convenient size of trays for drying in the home is 53.34 cm × 80 cm × 3 cm. The strips are 3–4 cm wide and are fixed to the frame so as to provide space between two strips. Wire gauze of 8–10 mesh may be used.

2.11.1.2 Dehydration

The simplest form of equipment for dehydration on a small scale or in a home is called a home drier (Biaugeaud 1994). It consists of a galvanized iron sheet box 90 cm × 60 cm × 90 cm of which the sides and top are enclosed in a wooden frame. It is fixed on an angled iron stand about 40–45 cm high. The bottom of the box is made of mild steel plate 0.32 cm thick, having 0.64-cm-diameter holes all round to allow air to enter the box. About 10 cm below the top on both sides, there are metallic flaps 60 cm × 7 cm along the sides, to control the flow of moisture going out. Inside the box along both sides are suitable runways 3 cm wide, 3 cm thick, and 90 cm long; the lowermost is fitted at about 18 cm from the bottom of the box. The box has a suitable shutter and a hole for placing the thermometer in the top of the box. The dryer can hold seven trays. It may be heated by a stove, coal furnace, or any other suitable means.

2.11.2 General Procedure for Sun Drying and Dehydration

1. *Preparation of the materials:* The fruits and vegetables are, in general, selected and prepared by washing, peeling, trimming, etc.

Green mangoes are peeled with a knife, sliced; outer leaves and cores of cabbages are removed, shredded longitudinally into 0.50-cm-thick shreds; stalks, covering leaves, and stems of cauliflowers are removed, flowers broken into pieces of suitable size; potatoes are peeled and cut into 0.50-to 0.75-cm-thick slices; peas are collected from their pods. The materials should be washed thoroughly in running water.

2. *Pretreatments:* Fruits and vegetables are living tissues in which changes caused by the action of enzymes continue unless the activity of agents responsible for these changes is checked. These changes manifest themselves by affecting adversely the colour, flavour, etc. of the product. In many cases, after peeling and cutting the materials rapidly undergo discolouration. A variety of nonenzymatic chemical changes taking place in certain materials during drying also affect adversely the quality of the finished product. This type of deterioration during drying can be minimised or checked by one or more of the following treatments.

   (a) *Blanching:* Most vegetables are blanched to destroy the activity of enzymes, although fruits are not generally blanched because of danger of the loss of the soluble constituents of the fruit. Blanching of vegetables fixes the natural colour and imparts a bright appearance to the product. Blanching means dipping of vegetable pieces (tied loosely in a cloth) in boiling water (100 °C) for 2–3 min and then cooling them in water. It accelerates drying by softening their texture by partial cooking and thereby facilitating easy escape of moisture. Blanching also done in steam. Blanching in steam has a definite advantage over blanching in boiling water; the loss of soluble elements from the material is much less, resulting in better retention of flavour and nutrients. But for green peas, cabbage, and cauliflower, blanching in water is more useful.

   (b) *Lye peeling:* Certain fruits and vegetables such as peaches, apricots, guava, orange segments, sweet potatoes, and carrots can be more conveniently peeled by the action of boiling hot lye solution (1–2 % caustic soda solution; sodium hydroxide) for about 30 s to 4 min, depending on the thickness of the skin. Dipping of fruits in lye solution cracks their skins or peels, thus permitting a higher rate of drying, and also helps in better retention of sulfur dioxide.

   (c) *Sulfuring:* Sulfuring is the treatment of materials with sulfur dioxide, for fruits only. It bleaches the colour of the product, minimises enzymatic and oxidative changes leading to discolouration and off-flavour development and loss of nutritive value, and also acts as a preservative and helps against insect infestation. Fruits and vegetables may be sulfured by dipping into solutions of sulfurous acid or of salts such as sodium metabisulfite or potassium metabisulfite (0.2–1.0 % for 20–30 min) or by exposing the prepared material to the fumes of sulfur dioxide obtained by burning of sulfur (40–80 g/10 kg fruit slices for 30 min to 2 h) in a closed chamber or room (airtight). A typical sulfur box-like home dryer is also usable for sulfur fumigation. For fruits, fumigation with sulfur dioxide fumes is the most satisfactory method of sulfuring.

3. *Drying:* After preparation and suitable pretreatment as necessary, the fruits and vegetables are spread on trays and dried in the sun or a dehydrator. In sun drying, the trays must keep sufficiently raised from the ground level and their bottoms left open for free circulation of air. For protection against flies and other insects, the trays may be covered with galvanized wire gauze trays inverted over them or with a net of nylon or thin cloth. At night, the trays must be kept under a roof in a dry situation (Ingegno 1992). Generally, 4–7 days are required for complete drying.

   In dehydration, the temperature of the dehydrator during drying may be maintained between 60 and 70 °C according to the need of the product, and the trays should be inspected and interchanged occasionally.

4. *Testing of dryness:* Fruits when properly dried should be such that no moisture can be pressed out of any freshly cut ends and no natural grain of the fruit (as when fresh) will be seen if a cross section is made. Fruit should not also be so dried that it breaks off when bent or acquires a brown colour and burnt flavour. It should not be brittle and crisp. There should be no trace of moisture in the centre of larger vegetables when cut open.

5. *Sweating (conditioning):* After drying, fruits and vegetables must be conditioned for equalization of moisture by storing for a few days in a box or friction top tins, in a cool place, shaking the product occasionally.

6. *Packing and storing:* The dried fruits and vegetables for use in the home can be packed in sterile airtight jars or tins. For sale, they may be conveniently packed in a polythene bag sealed by a heated metal rod and further packed in large-sized containers (Colin 1992).

Name of fruits	Preparation	Sulfuring	Drying temperature
Banana (green)	Blanched in boiling water after peeling and slicing to 12-mm-thick slices, during slicing kept in 0.1 % ICMS solution to avoid discolouration	Sulfured in box for 2 h in fumes	60 °C for 20 h, ratio 5:1
Mango (green)	Peeled and cut into pieces of 12-mm-thick slices, sulfured for 2 h in fumes and then dried	Do	45–50 °C for 30 h or sun drying
Cabbage	Outer leaves and cores removed, shredded longitudinally into 0.5-cm-thick shreds	Blanched for 2–3 min in 1.0 % boiling sodium bicarbonate solution	60–65 °C for 12 h or sun drying, ratio 18:1
Cauliflower	Stalks, covering leaves, and stems removed, flowers broken into small pieces	Blanched 4–5 min in boiling water, steeped 45 min in 1 % KMS solution, and washed	55–60 °C for 12 h, ratio 35:1
Pea	Pods shelled to get grains (1 kg pod = 500 g seed = 135 g dried pea)	Blanched 1–2 min in boiling water	60–65 °C for 20 h, ratio 7.4:1
Potato	Peeled and cut into 0.25-cm-thick slice for fries and 0.5–1 cm thick pieces for curry	Blanched for 3–5 min in boiling water	60–65 °C for 8 h, ratio 7:1

Source: Magee and Wilkinson 1992; Matz 1984; Torres 1974; Thompson 1989

2.12 Preparation and Preservation of Canned Products

2.12.1 Definition

The process of storing fruit pieces in 33–50 % sugar solution (known as syrup) and vegetable pieces after blanching in 2–3 % salt solution (known as brine) together with 0.5 % citric acid, and if required ascorbic acid, is termed canning or bottling. The nutritive value and quality of the product remain unchanged.

2.12.2 Selection of Material

Fruits should be ripe, firm, and evenly matured, free from malformation, cracks, or insect damage. Overripe fruits are generally infected with microorganisms, but underripe fruits are shriveled and tough and would yield a pack of inferior quality on canning.

Similarly, vegetables should be tender, fully matured, firm, and deep in colour. Freshness in vegetables is one of the important factors (an hour from the field to canning factory is considered ideal).

2.12.3 Sorting and Grading

Sorting and grading are necessary to obtain a pack of uniform quality in variety, shape, size, flavor, and colour. After preliminary sorting, grading may be done by either hand or machine. A mechanical grader may be the screw or roller type.

2.12.4 Washing of Raw Materials

Microorganisms harbour mostly on the surface of damaged fruits and vegetables. Vegetables are produced near the soil and they contain less acid than fruits, so it is obvious that more infecting microorganisms will be on them. Nearly 80–100 % of the micro-organisms can be washed off by only steeping the material in hot water for a few minutes.

Washing may be done in different ways. Washing in a bucket of cold water is a common practice, which is not at all a proper method as dust and microorganisms from infected fruits or vegetables may spread to a fresh bucket. Washing in running tap water is a better method. Washing in hot water supplemented by compressed air through a jet is the best method. Fruits are sometime washed with a dilute solution of HCl at 1:20 ratio to remove residual spray materials. In all cases washing should be done separately by rubbing the produce thoroughly, especially for vegetables.

2.12.5 Blanching

Blanching is done in case of vegetables only. Treatment of vegetables by dipping for about 2–5 min in boiling water followed by rapid cooling before canning is known as blanching.

In special cases, sodium chloride (NaCl) or sodium carbonate (Na_2CO_3) is used in the water during blanching.

2.12.6 Preparation of Materials

The fruit and vegetables after washing are prepared for canning by the use of different types of knives such as a plain stainless steel knife with a sharp edge; peeling is done in different ways:

1. Hand peeling with knife
2. Machine peeling
3. Heat treatment (potato)
4. Dipping in lye solution (peach, apricot, guava)
5. Flame peeling (onion, garlic)

A peeling knife has a curved blade and guard to regulate the depth of peeling. It can be used for many fruits and vegetables. A pitting knife is used to remove the stone in litchi. A coring knife is used to remove fine pieces of fruit flesh. A coring loop is used to remove the seeds of oranges. Fruits or vegetables are then cut into pieces of 0.25–0.50 cm.

2.12.7 Different Types of Cans

(A) *Can*

1. Handmade can.
2. Hole and cap can have small holes for exhausting air.
3. Open-top sanitary can, slightly soldered at one side only. Cap has the same diameter as the can.
4. Black steel plate cans introduced during the war period when tin was scarce.
5. A string-opening composite can with a ripcord strip of tin that can open the can by a ring-type cord as developed by Metal Box of Calcutta.
6. Self-heating can: just by opening the cans the temperature of the product inside can be raised to 50 °C within 4 min, which is made possible by the use of calcium silicate and Fe_2O_3, Fe_3O_4. This mixture is placed at the centre of can, and just by opening it, the temperature is increased by the reaction of the chemicals.

(B) *Glass container*

1. Narrow-neck glass bottle with ordinary or velvet cork, crown cork. or pilfer-proof (pp cap) having a cork or polythene sheet inside the crown or pp cap for better sealing.
2. Wide-mouth glass jar with metallic or polythene (polyethylene) screwed lid or

cap. The cap has a rubber or polythene ring inside it to provide a good seal.

(C) *Plywood container*

This type of can is mainly used for a product that is consumed daily, but now it is used for temporary preservation of cordial and guava juice for future processing and known as a carboy.

(D) Shellac-laminated polythene container generally used for drugs and being used in preservation. In the future it may replace tin containers because of its low cost.

(E) Polyvinyl chloride container.

(F) Flexible container.

(G) Expanded polythene container.

(H) Silver-lined aluminum container.

(I) Laminated pouch: aluminum-lined paper or polythene pouch.

(J) Polythene tube.

(K) Metallic tube (aluminum, flexible).

(L) Lacquered can.

It is difficult to coat steel plate uniformly with tin during the process of manufacture. It is necessary, therefore, to coat the inside of can with some material such as lacquer that would prevent discolouration but would not impart its own flavour or injure the wholesomeness of the contents; this is achieved by lacquering the tin plate.

There are two types of lacquers:

(a) Acid resistant

(b) Sulfur resistant

The acid-resistant lacquer is ordinary gold colour enamel, and cans treated with it are called R-enamel cans. These cans are used for packing fruits of the acid group with soluble colouring mater. Acid fruits are of two kinds:

(a) Those in which the colouring matter is insoluble in water.

(b) Those in which the colour is water soluble. The first group, including peach, pineapple, apricot, grapefruits, etc., can be placed in plain cans. The second group, including raspberry, strawberry, red plum, and coloured grapes, are packed in lacquered cans.

(c) The sulfur-resistant lacquer also is of golden colour and the cans coated with it are called C-enamel or S.R. cans. These cans are used for non-acid products such as pea, corn, lima

bean, and red kidney bean to prevent discolouration of the contents and staining of the inside of the container.

2.12.8 Can Filling

Cans are washed and sterilized by 2–3 min in boiling water or in hot air vacuum. Water is then drained. Ordinary tin cans are generally used for vegetables and for noncolour fruits and lacquer cans are used for coloured fruits. The cans are filled with the material by a filling machine.

2.12.9 Pouring of Covering Liquid: Syrup and Brine

Sugar/salt is then mixed with water according to requirement, heated, strained through a thick muslin cloth, and poured on the material when still hot, mixing with 0.5 % citric acid or lemon juice for vegetables.

2.12.9.1 Objectives of Pouring Covering Liquid

(a) To remove air from spaces within the material and from unfilled space present within slices of the packed material

(b) To check oxidation, thereby preventing discolouration and loss of flavour

(c) An aid in further processing because it is a carrier of heat by which processing can be done rapidly and uniformly

(d) Reduces exhaust time

(e) Acts as preservatives

(f) Acts as shock absorber

(g) Improves the taste

(h) Checks enzymatic action

(i) Acts as soaking medium

(j) Prevents further entry of microorganisms

Apple, peach, pea, guava, and banana should be stored in acidified covering liquid, otherwise discolouration may start. Cans are filled with 175–180 °F hot covering liquid. Further high temperature will result in shrinkage of the pieces. A suitable head space should be left, so that when the lid is fitted, the space left inside the can between the surface of the covering liquid and inner surface of the lid is between 1/8 and 3/16 inch.

2.12.10 Lidding and Clinching

After being filled the cans are covered with a lid that is clinched by the first roller action of a double-roller can sealer. The lid remains sufficiently loose to permit air to be expelled. Exhaustion of air from the container is done by keeping the clinched can in a hot water bath. After clinching, the counting coding device is also incorporated in this machine.

2.12.11 Exhausting

The removal of air from the head space of a filled container is known as exhausting. It is important in canning: a good exhausting ensures a satisfactory vacuum in the can and thereby minimises oxidative changes that lead to discolouration of the product and loss of vitamins. Exhausting also releases the strain on the cans during processing and storing. Cans are exhausted either by heat treatment or a mechanical method.

In the first method, the cans are placed in 180 °F hot water contained in a tank for 10 min. The tops of the cans are kept ½–1½ inches above the water level. The air will exhaust from the clinched portion. At the end of exhaustion the temperature at the centre of the can should be 175 °F. The partial vacuum created inside the can preserves the taste and flavour of the material and keeps the product in hygenic condition for a longer period.

2.12.12 Sealing

After exhausting, the cans are sealed by the second roller action of the can sealer while the product inside the can is hot.

2.12.13 Processing

The term processing in canning means heating of the canned food to a certain extent that would

sufficient to eliminate all possibilities of fungal and bacterial growth. Sealed cans are placed in the tank containing water. A false bottom of thick cloth is placed, and the tank is then boiled until the temperature at the lowest heating point inside the centre of a can reaches 80 °C. Acid fruit requires less heat for processing, whereas nonacid vegetables require a much higher temperature. The dividing line between acid and nonacid foods is regulated at pH 4.5. So, a canned product of less than 4.5 pH can be processed at 10–15 lb pressure at 240 °F for 40 min. Processed cans are cooled rapidly to 100 °F in running water, or hot cans are passed through a tank containing cold water to stop further cooking, and thereby central browning of the product inside the can should be checked.

Can may be cooled

(a) By spraying of cold water through a water jet
(b) By exposing cans to air, which is time consuming
(c) Immersing hot cans in a tank containing cold water

Bottles are cooled as quickly as possible by keeping them in a well-ventilated space; this facilitates quick drying of the can's outer surface, thereby avoiding rust.

2.12.14 Testing of Defects

Finished cans are finally tested for leakage or improper sealing in the factory in batches.

2.12.15 Labelling and Storing

The cans are then dried, labeled, and stored in cool, dry, dark places.

2.12.15.1 Canning of Some Fruits and Vegetables
2.12.15.1.1 Mango
Juicy and fibrous varieties are not preferred. Suitable varieties are Langra, Bombai, or Himsagar. Firm but fully ripe mangoes are picked and ripened in straw. Mangoes are washed, peeled, and the fleshy portion is cut into six to eight longitudinal slices or into 2-cm cubical

pieces that are kept in 2 % brine (2 % sodium chloride) solution to check oxidation.

Pieces are then filled into a can. The covering liquid contains 35–40 % hot syrup (sugar) after mixing with 0.25 % citric acid. Clinched cans are then exhausted at 175 °F for 10 min by keeping them in water. The sealed cans are processed for 20 min. Plain cans are recommended for filling, and from 1.5 kg of raw materials 1 kg finished product can be prepared (Khurdiya and Roy 1986).

2.12.15.1.2 Litchi

Tree-ripened fruits are taken: freshness of fruit is the most important criterion for litchi canning. Bedana is a suitable variety. Outer shells are cracked, the fleshy portion is cut longitudinally at one side, and the stone is removed. Covering liquid contains 40 % sugar syrup, 0.5 citric acid, and 2 % ascorbic acid to inactivate the enzymatic action that causes browning. Prompt cooling of the can is essential as the fleshy portion of the fruit is white and tender. Plain cans are used.

2.12.15.1.3 Pineapple

The varieties Giant Kew and Kew are most suitable for canning as the fruits are larger in size with flat eyes. The Queen variety is also good for a quality product for its colour and flavour. Slicing is done at first with a pineapple slicer, then slices are made round shaped by using a pineapple puncher. Fruits also may be cut in cubes, triangles, etc. Different types of cans and other containers may be used in preservation. Covering liquid and other methods are same as for mango.

(a) Tin containers	1. Handmade can
(b) Glass container	2. Hole-and-cap type can
	3. Open-top sanitary can

2.12.15.1.4 Garden Pea

Uniformly matured, good-texture peas are suitable. Large-sized peas such as telephone, phenomenon, and bonnevel are suitable for canning. They are graded after shelling using sieves with hole diameter ranging between 9/32 and 13/32 of an inch. Peas may be graded by dipping in 1.04–1.07 specific gravity brine solution and discarding the

floating ones. The peas are then blanched in boiling water for 3–5 min. Plain cans are used with covering liquid containing 57 g sugar and 35 g salt in 2.8 l water (or 1.25 kg sugar with 750 g salt in 45 l water). Sometimes edible green colour and mint flavour are added with the covering liquid.

The cans are filled, exhausted, sealed, and processed under 10 lb pressure for 40–50 min. Canned fresh peas are commonly known as garden peas. Dried peas when soaked in water overnight and then canned are known as processed peas. For dried peas, the covering liquid is prepared by mixing 625 g salt with 2.5 kg sugar/l water. A 0.2 % solution of edible green colour is used with that covering liquid; 150 g is necessary for 45 l of covering solution.

2.12.15.1.5 Bean

Tender stringless beans are cut into 1-in. pieces, blanched, and preserved in brine solution. Plain cans are used.

2.12.15.1.6 Potato

Potatoes can be canned as whole or as slices. They should be sufficiently firm, blanched for 5–6 min, then peeled by hand or on a rotary vegetable peeler known as a potato baller. Slices are made 0.5 cm thick and placed in 2 % brine solution to prevent discolouration. The softer varieties of potato before canning are placed in 2.5 % calcium chloride solution for 1 h for firming. They are then washed well before filling into cans. Plain cans are used and covered with 2 % brine solution.

2.13 Spoilage in Preserved Food Products

In storage, food products are liable to spoilage for various reasons. After spoilage, the products may become unsuitable for consumption, or the product may have lost its normal colour yet may remain fit for consumption. Spoiled canned food exhibits characteristic differences in appearances, taste, and odour from normal canned food. It usually possesses a flat sour taste and offensive flavour, odour, and smell. Spoilage of product may be caused either by the activities of the microorganisms that contaminate the product or the bodies of microorganisms after their decay, causing poisoning of the food content.

Type of Spoilage

(A) *Internal spoilage*

(B) *External spoilage*

(A) *Internal spoilage may be of the following types:*
　1. *Internal microbial spoilage*
　　(a) *Gaseous spoilage*
　　　1. *Nonpoisoning*
　　　2. *Poisoning*
　　(b) *Nongaseous spoilage*
　2. *Internal physicochemical spoilage*
　3. *Internal biological spoilage*

2.13.1 Internal Microbial Gaseous Spoilage

2.13.1.1 Internal Microbial Nonpoisoning Gaseous Spoilage

Spoilage may be further classified as aerobic and anaerobic in type. The presence of yeast cells in fruit products are evidence of this type of spoilage with the production of alcohol and carbon dioxide. The rapid production of CO_2 accounts for frequent bursting of cans: this is an anaerobic type. Vegetables usually undergo anaerobic types of decomposition caused by low oxygen content. In canned vegetables of low acidic media, the spore-bearing gas formers *Bacillus sporogenes* and *Bacillus welchii* produce a very disagreeable putrid odour. Tomatoes are spoiled by anaerobic spore-bearing acid-tolerant bacteria such as *Clostridium pasteurianum* that produces a butyric odour. Tomato juice is occasionally spoiled by acid-tolerant, heat-resistant, flat–sour-producing thermophilic bacteria known as *Bacillus thermoacidurans*.

2.13.1.2 Internal Microbial Poisoning: Gaseous Spoilage

This spoilage is generally caused by sporophytic bacteria and produces a heavy amount of gas. *Bacillus botulinus* (i.e., *Clostridium botulinum*) became a grave concern to the canning industry, and even caused a 70 % death rate, because an outbreak of botulism means poisoning, for both commercial and home-canned products. In recent years, community canned foods are not affected by botulism (poisoning) but home-canned foods continue to do so. Usually canned foods are so badly decomposed as to be unfit for consumption. In most media, this spoilage produces a very characteristic penetrating odour resembling that of rancid butter: this is most pronounced in peas and least noticeable in fruits.

2.13.2 Internal Microbial Nongaseous Spoilage

Most of the flat sour spoilage in canned products is caused by the growth of thermophilic bacteria without production of gas, which develops only above 38 °C when canned products are not properly cooled and are allowed to stand in large stacks, which prevents rapid radiation of heat. Thus, the cans were provided with a favourable incubating temperature for the growth of bacteria. H^+ ion concentration also has a marked influence on the thermal death point of the thermophiles.

The usual source of contamination for thermophilic spoilage are the tanks of hot water used for blanching, hot brine, or cooling tanks. Sugar might be another source of contamination as in some stage of sugar manufacture it provided an opportunity for the growth and sporulation of thermophilies.

The growth of these bacteria is found even in 65 % sugar or in 10 % salt concentration. The bacteria grow more rapidly on glucose media. These organisms prefer media that are neutral or faintly alkaline and are readily inhibited by 1 % sodium benzoate at a pH value below 4.5. When pH is above 5.0, it is impractical to preserve the food. Spoiled food with this bacteria should never be tested, should not even be touched by the tongue, as this is very poisonous. If any food product is infected by this bacteria, prolonged boiling is necessary to destroy it. In the presence of oil, spores greatly increase their resistant power against heat.

Spoilage of canned foods caused by heat-resistant mold has also been found. The organism is *Byssochlamys fulva*, which can withstand 86–88 °C for 30 min in some fruit syrup. The content of the can is partially liquidified by these organisms. Occasionally, jam, jelly, and bottled juice are spoiled by *Penicillium* mold, whose spores survive the pasteurization temperature, although most of them are killed at 82 °C.

2.13.3 Internal Physicochemical Spoilage

1. *Defective tin plate:* Swelling in cans is caused by formation of H_2 gases inside the container, resulting from the acid content of covering liquid or fruit pieces contacting the corrosive tin plates. Plain cans are less susceptible to H_2 swell than the lacquered cans. Cane sugar promotes the reaction and production of H_2 gases. H_2 swelling is of two types:
   (a) *Flipper:* A mild positive pressure or sometimes a mild swelling where the lid can be brought down to its original position with finger pressure and the lid remains in that place even after removal of the finger pressure.
   (b) *Springer:* Also a mild swelling which can be replaced with finger pressure but the lid will return to its swollen position after lifting the finger from the lid. Corrosion may form pinholes and the can may leak. The content remains usable and fit for consumption.
2. *Improper washing:* This may cause residual spray material to remain, thereby poisoning.
3. *Low acidity of covering liquid:* Products when preserved in low acidity can produce H_2 gas. At low acidity up to pH 4.5, a chemical preservative such as sodium benzoate can check the growth of such bacteria as *Clostridium botulinum*. At pH 5, however, this preservative has no action and microorganisms may cause rapid production of H_2 gas and thereby swelling.

4. *Following improper processing technique:* Insufficient curing of the product will result in dull to faded colour.
5. *Using strong covering liquid:* Pieces may be toughened and shrivelled when placed in strong covering liquid.
6. *Production of bitter taste:* Results from use of stronger vinegar, prolonged cooking, or high amount of spices which also released tannins.
7. *Insufficient head space:* If the container is overfilled, any gas evolved inside the sealed container caused by a chemical or organic reaction will not have sufficient space for expansion and may cause flipper swell.
8. *Providing improper exhausting period and temperatures:* When exhausted for a longer period or in higher temperature than is recommended, a product may become browned. When a product is not fully exhausted, this may caused aid microorganisms to get an O_2 supply and thus survive.
9. *Improper sealing:* Faulty sealing may cause development of yeast spores inside the container with the production of CO_2. Bacteria such as *Bacillus coagulans* survive in a leaky can, but because they do not form spores they rarely survive the processing temperature. Similarly, improper sealing and insufficient oil may sometimes cause pickles to spoil by action of aerobes.
10. *Covering with insufficient or weak liquid:* Bacteria may survive, resulting in softening of the product. The product may become partially solidly packed so that heat penetration during processing is not adequate to kill the microorganisms.
11. *Insufficient processing temperature:* Spores of yeast may survive if the processing temperature is too low or too short in duration.
12. *Improper cooling:* When the product is not cooled promptly and allowed to stand in at temperatures that allow incubation of very harmful thermophilic bacteria, their rapid multiplication can occur rapidly. This temperature is 54 °C, resulting in the production of flat sour, and the product should never be tasted.

13. *Improper storage temperature:* The higher the storage temperatures, the more rapid will be the formation of H_2 gases, resulting in much corrosion of the tin plate.
14. *Metallic reaction with spices:* When iron of a corroded tin plate reacts with tannin of spices, this may produce ferrous tannate which by oxidation produces ferric tannate, causing blackening of the product. Similarly, with brass, iron, and copper, the content may spoiled by oxidation, causing a greenish or blackening colour of the product.

2.13.4 Internal Biological Spoilage

(a) *Blackening caused by oxidation:* Cut fruit and vegetable surfaces release enzymes which by oxidation turn the food brown. The action is accelerated in higher light and temperature. Browning of the cut surface is very often observed in apple, banana, guava, and mango.
(b) Browning may be caused by high exhausting and processing temperature because of partial charring of the cut surfaces.

The brown colouration of a cut surface may also caused by other enzymes or charring, including (1) reaction between nitrogenous matter and sugar, (2) reaction between nitrogenous matter and organic acid, (3) reaction between sugar and organic acid, and (4) reaction within different organic acids, which is known as the Maillard reaction.

(B) *External spoilage:* Mainly caused by inferior quality of steel used in cans, its covering by tin, and by lacquering. The quality of the can depends mainly on the porosity of its surface when it has not been covered properly.
  (a) *Quality of the tin plate:* Rusting for any reason from oxite; pinholes may occur, producing a leaky can and thereby allowing yeast to ferment.
  (b) *Improper drying:* If cans are not dried properly, corrosion of the tin plate may result.

(c) *Use of improper cooling water:* If chemicals such as sodium chromate are present in cooling water, leaky cans may result. Cooling water is generally infected with spores of microorganisms that may cause initial infestation in leaky cans.
(d) *Improper storing:* Coloured products when preserved in transparent glass or polythene containers and stored in direct or diffused sunlight or in an open place readily loose their normal colour. Tomato products, especially ketchup, should not be kept in light even if diffused.

References

Amerine MA, Pangborn RM, Roessler EB (1965) Principles of sensory evaluation of food. Academic, New York
Anonymous (2013) National horticulture board. In: Kumar B (ed) Indian horticulture database. Ministry of Agriculture, Government of India 85, Institutional Area, Gurgaon
Biaugeaud H (1994) Food processing equipment. Technical bulletin. Henri Biaugeaud, S.A., Arcueil
Colin D (1992) Recent trends in fruit and vegetable processing. Food Sci Technol Today 7:111–116
Dahiya DS, Prabhu KA (1997) Indian jack fruit wine. In: Symposium on indigenous fermented food, Bangkok, Thailand, 1977
Girdhari L, Siddappa GS, Tandon GL (2010) Prevention of fruits and vegetables. ICAR Publishers, New Delhi
Ingegno C (1992) A fresh look at dried fruit. Food Product Design, pp 48–50
Jood S, Ketarpaul N (2002) Food preservation. Agrotech Publishing Academy, Udaipur
Khurdiya DS, Roy SK (1986) Studies on canning of mango slices in covering syrup containing mango pulp. Indian Food Packer 40:50–54
Lal G, Siddappa GS, Tandon GL (1959) Preservation of fruits and vegetables. Indian Council of Agricultural Research, New Delhi
Magee TRA, Wilkinson CPD (1992) Influence of process variables on the drying of potato slices. Int J Food Sci Technol 27:541–549
Matz SA (1984) Potato chips. In: Matz SA (ed) Snack food technology. AVI Publishing Co., Inc., Westport
Pederson CS (1963) Processing by fermentation. In: Joslyn MA, Heid JL (eds) Food processing operations, vol 2. AVI Publishing Co., Westport
Rana TS, Datta B, Rao RR (2004) Soor, a traditional alcoholic beverage in Tons Valley, Garhwal, Himalaya. Indian J Tradit Know 3:59–65

Serkar S, Mariappan S (2007) Usage of traditional fermented products by Indian rural folks and IPR. Indian J Tradit Knowl 6(1):111–120

Sharma KD, Sethi V, Maini SB (1988) Studies on incorporation of concentrate in covering syrup for processing of apple rings in flexible pouches. J Food Sci Technol 35:371–374

Srivastava RP, Kumar S (1994) Fruit and vegetable preservation principles and practices. International Book Distributing Co. (Publishing Division) Chaman Studio Building and Floor, Charbagh, Lucknow-226004, UP, India

Steinkraus KH (1996) Handbook of indigenous fermented foods. Marcel Dekker, New York

Tapsell LC, Hemphill I, Cobiac L, Patch CS, Sullivan DR, Fenech M, Roodenrys S, Keogh JB, Clifton PM, Williams PG, Fazio VA, Inge KE (2006) Health benefits of herbs and spices: the past, the present, the future. Med J Aust 185:4–24

Thompson AK (1989) Recent advances in post-harvest technology of fresh fruits and vegetables. Private communication. Cranfield Institute of Technology, Silsoe College, UK

Torres M (1974) Dehydration of fruits and vegetables. Noyes Data Corporation, London

Value Addition in Vegetable Crops

3

S. Datta, R. Chatterjee, and J.C. Jana

Abstract

India is the second largest producer of vegetable next to China. Most of the vegetables are highly perishable and have limited life, which need to be marketed immediately or processed into varied value-added products. Diversification of vegetable products for export and domestic consumption reduces the risk of loss due to price fluctuation. This increased demand for value-added products is due to the change in the market behaviour, changes in consumer preferences and emergences of supermarkets, which have resulted in the usages of more value-added, ready-to-use vegetable products and vegetable in consumer packs. These products also have a high demand in the defence sector of the different country. Among the different vegetable value-added products, dehydrated potato, peas, carrot, cauliflower, tomato-based processed products, pickles and chutney from different vegetables are the most important. Apart from these lycopenes from tomato and watermelon, tomato seed oil, frozen vegetables, organic vegetables, minimal processed vegetables, consumer-packed vegetables, etc., are also important.

3.1 Introduction

India, the second largest producer of vegetables in the world, produces 162.12 million tons of vegetables from 9.21 million hectare areas in 2012–2013. The adaption of hybrid varieties and improved production technologies have helped in achieving the quantum jump in crop yield. But the real concern is that most of the vegetable crops are highly perishable and deteriorate rapidly after harvest, as a result 25–30 % of the fresh harvest goes waste during the process of postharvest operation before reaching to consumers. The loss of fresh produce can be minimized by adopting different processing and preservation techniques to convert the fresh vegetables into suitable value-added (Table 3.1) and diversified products which will help to reduce the market glut during the harvest season. The preparation of

S. Datta (✉) • R. Chatterjee • J.C. Jana
Uttar Banga Krishi Viswavidyalaya,
Pundibari, Cooch Behar, West Bengal, India
e-mail: suchanddatta@rediffmail.com

A.B. Sharangi and S. Datta (eds.), *Value Addition of Horticultural Crops: Recent Trends and Future Directions*, DOI 10.1007/978-81-322-2262-0_3, © Springer India 2015

43

Table 3.1 Different vegetable-based value-added products

Crop	Value-added products
Potato	Fried chips (chips), French fries, frozen products (potato patties, potato puffs, potato cakes, defrozen products, packed frozen dishes), dehydrated products (like potato flour, granules and flacks), wine, canned potatoes, etc.
Tomato	Tomato paste, ketchup, paste, chutney, sauce, tomato chilli sauce, tomato seed oil, canned tomato (in the form of fresh tomato, tomato juices, tomato-vegetable juice blend, tomato sauce and tomato ketchup), tomatine alkaloid, lycopene pigment, soup powder, etc.
Cabbage	Sauerkraut
Cauliflower	Dried cauliflower, frozen cauliflower, cauliflower pickle, etc.
Carrot	Carrot shred, frozen carrot, carrot powder, soup powder
Pea	Dehydrated peas, frozen pea
Sweet potato	Jam, pickle, soft drinks, etc.
Cassava	Fried chips, hot fries, crisps, nutrichips
Amaranth	Dry powder
Beet	Anthocyanin pigment
Watermelon	Lycopene pigment
Bitter gourd	Pickles
Pumpkin	Sauce
Amaranth	Powder
French bean	Canned, frozen products
Gherkin	Canned, brine solution, pickle
Cucumber	Canned, brine solution, pickle
Baby corn	Dehydrated corn, brine solution
Brinjal	Pickle

processed product will provide more varieties to the consumer, improve taste and appetite and enhance sensory properties of food. This will also help in the fortification of nutrients which is lacking in fresh produce. By adopting suitable methods for processing and value addition, the shelf life of fresh produce can be increased many-fold and that will enable the availability of the vegetables round the year to a wider spectrum of consumer in both domestic and international markets. A number of employment can be generated and better remuneration can be ensured to the farmers. For a successful processing of vegetables, the selection of right varieties with desirable qualities and good postharvest practices and the use of suitable preservation method need to be adopted to ensure the produce quality and wider acceptability among the consumers.

3.2 Global Market of Vegetable Products

This increased demand for value-added products is due to the change in the market behaviour, changes in consumer preferences and emergences of supermarkets, which have resulted in the usages of more value-added, ready-to-use vegetable and spice products and vegetable and spices in consumer packs. Value products of vegetable earn a sustainable amount through export earnings (Table 3.2) by exporting in the different countries (Table 3.3).

Table 3.2 Export value of vegetable products

	2010–2011		2011–2012		2012–2013	
Products	Quantity (MT)	Value (Rs. Crores)	Quantity (MT)	Value (Rs. Crores)	Quantity (MT)	Value (Rs. Crores)
Dried and preserved vegetables	49,009	373.33	64,794	526.78	68,520.25	637.96
Other processed fruits and vegetables	1,999,868	997.04	274,807	1,577.60	269,217	1,733.06

Source: Anonymous (2014)

3.3 Value Addition in Different Vegetable Crops

3.3.1 Potato

Potatoes are processed into different products, namely, chips, French fries, dehydrated product and canned potatoes, depending upon the dry matter content. Tubers with a high dry matter, high amylase-to-amylopectin ratio and low sugar content are most suitable for processed products. Quality attributes for the preparation of different value-added products of potato have been given in Table 3.4.

3.3.1.1 Potato Chips

Fresh potatoes with high dry matter content are preferred for potato chips, and these chips can be converted into potato flour. The chip colour is the result of the browning reaction between sugars and other constituents, mainly amino acids. The sugar causes browning because reducing the sugar imparts brown colour to the heat-treated

product through caramelization and interaction with amino acids. Chips from high-sugared potatoes become brown or black on frying, and discoloration also develops in boiled and peeled potatoes. The specific gravity or dry weight of potatoes determines the yield of chips. The specific gravity of potatoes is in turn influenced by various factors such as variety, maturity, management practices, etc. Dexter and Salunkhe (1952) reported that hot water and/or weak chemical treatments of potato slices prior to deep-frying can improve and make uniform light golden yellow potato chips. Pandey et al. (2005) reported that Kufri Chipsona-1 (Indian) and cv. Atlantic (American) produced high dry matter (>20 %), low reducing sugars and acceptable chips and showed good storability under ordinary storage at ambient temperatures. However, cv. Kufri Chipsona-1 having high resistance to late blight with better total yields would be preferred by the farmers. The commercial cultivation of cv. Kufri Chipsona-1 in west-central plains not only would ensure a steady supply of raw material to the processing units located in north-western plains but will provide the much needed raw material at a cheaper transportation cost (Table 3.5). At present in India Kufri Chipsonna-3 and Kufri Chipsona-4 are the suitable cultivars for making potato chips.

Joshi and Nath (2002) studied the effect of different pretreatments on the drying oil content and overall acceptability of fried chips from fresh and sprouted potatoes (Table 3.6).

3.3.1.2 French Fries

The process involves washing of fully mature potatoes and removal of small-sized, defective,

Table 3.3 Export market of different value-added vegetable products

Products	Countries
Dried and preserved vegetables	Sri Lanka, USA, UAE, Germany, France, the Netherlands
Pickle and chutney from vegetables and fruits	UK, USA, UAE, Germany, Canada, the Netherlands, Saudi Arabia
Other processed fruits and vegetables	USA, the Netherlands, UK, UAE, Indonesia, Philippines

Source: Khurdiya (2003)

Table 3.4 Quality attributes of potato for processing

| Characters | Type of potato products | | | |
	Chips	French fries	Dehydrated	Canned
Specific gravity	1.085	1.080	1.080	1.080
Dry matter (%)	22–25	20–23	22–25	18–20
Starch	15–18	14–16	15–19	12–24
Reducing sugar (%)	0.25	0.50	0.50	0.50
Shape/size preferred	Round to oval	Long oval shaped	Medium to large sized	Small sized

Source: http://agmarknet.nic.in/profile-potato.pdf

Table 3.5 Total and processing grade yield, reducing sugar, dry matter and chip colour of potato varieties

Variety	Total tuber yield (MT/ha)	Processing grade tuber yield (MT/ha)	Reducing sugars (mg/100 g fresh weight)	Percent dry matter (per cent)	Chip colour
Arinda	28.69	25.64	147.39	17.67	3.75
Raja	27.90	23.35	203.38	19.95	5.50
SL/85–482	21.59	17.74	89.63	21.39	3.50
K. Bahar	36.06	32.65	251.80	18.28	7.50
K. Chipsona-1	34.71	29.18	44.37	22.62	1.75
K. Chipsona-2	28.81	24.68	57.20	23.19	2.50
K. Anand	41.17	37.19	244.76	19.02	5.75
K. Jyoti	29.61	26.77	137.28	18.34	5.10
K. Sutlej	37.06	34.31	207.05	18.03	5.50
Atlantic	29.62	28.25	27.40	20.24	1.50

Source: Pandey et al. (2005)

Table 3.6 Effect of pretreatments and drying oil content and overall acceptability of fried chips from fresh/sprouted potatoes

	Partially dehydrated fried chips			Dehydrated fried chips		
	Fried chips[a]			Fried chips[a]		
Pretreatment	Moisture before drying	Oil %	Overall acceptability score (max 100)	Moisture before drying	Oil %	Overall acceptability score (max 100)
Fresh unpeeled potato chips						
Water soak[b]	37.6	36.4	62.5	4.8	35.5	57.5
Acid dip for 1 min[c]	37.4	30.6	81.3	4.2	30.0	86.0
Sulphite dip[d]	34.8	69.7	69.7	3.5	27.2	69.8
Citrate dip[e]	41.6	25.0	79.8	3.9	24.6	75.6
		SEm	2.2		SEm ±	0.7
		CD at 5 %	5.5		CD at 5 %	2.1
Fresh peeled potato chips						
Water soak[b]	44.9	35.1	53.6	4.7	34.8	50.2
Acid dip for 1 min[c]	43.3	29.5	81.1	4.0	29.1	79.4
Sulphite dip[d]	44.3	26.9	63.3	3.5	25.6	62.7
Citrate dip[e]	45.5	24.1	65.9	3.8	23.4	64.2
		SEm ±	2.1		SEm±	0.8
		CD at 5 %	6.0		CD at 5 %	2.2
Sprouted unpeeled potato chips						
Water soak[b]				4.5	41.2	48.3
Water blanch[f]				3.7	38.4	81.3
Acid dip for 2 min[c]				4.0	28.8	85.0
Acid dip for 3 min[c]				3.4	28.6	87.3
Acid dip for 4 min[c]				3.8	28.4	81.2
					SEm ±	1.2
					CD at 5 %	3.6

Source: Joshi and Nath (2002)

SEm standard error of mean, *CD* critical differences

[a]Moisture range in fried sample: 2.3–4.5 %

[b]Room temperature for 60 min

[c]0.5 % citric acid + 5 % NaCl + 0.25 % $CaCl_2$ at 85 °C

[d]Solution (3) + 0.25 % Na_2SO_3 at 85 °C for 1 min

[e]0.75 % sodium citrate + 0.15 % Na_2SO_3 + 0.1 % phosphoric acid at 80 °C for 1 min

[f]Boiling water at 99.6 °C for 2 min and chip to water ratio 1:6

degreened and discoloured area followed by peeling and cutting into rectangular pieces, blanching of pieces in boiled water at 70–90 °C for 5 min and draining out the excess water and then placing potato pieces for frying; once fried, the pieces are defatted by passing the product over a vibrating screen; after that the pieces are cooled to room temperature and placed in airtight polythene pouch or PET jar and frozen at 0 °C for 2–4 months. Kufri Chipsonna-3 is a suitable cultivar for making potato French fries.

3.3.1.3 Dehydrated Products (Like Potato Flour, Granules and Flakes)

When the dry matter content is high, lesser energy is required for removing the relatively lower amount of moisture from raw materials, but the product output is high. The stored potatoes are low in moisture and high in sugar and protein compared to fresh potatoes. To avoid the browning, generally, fresh potatoes are used to prepare potato flour.

3.3.1.4 Canned Potatoes

Potatoes with low dry matter contents are used when a firm piece is required for canning and curries. Whole or chopped potatoes either alone or mixed with other vegetables are canned. These canned potatoes have a good texture, flavour and crispiness. Potato patties, potato puffs and potato cakes are the important frozen product of potato and are made of mashed potato blended with baking soda and other ingredients and are eaten as dessert.

3.3.2 Tomato

Tomato is an indispensable item for various culinary preparations and mostly utilized for the preparation of a wide range of processed products such as juice, puree, paste, sauce, ketchup, etc. For processing purposes the tomato varieties should have higher total soluble solids (above 5.5 °Bx). Processed tomato is a major trade item in the world, but India does not figure anywhere among top exporters of the world. India is exporting processed tomato in the form of tomato paste and ketchup. Tomato processed products are exported to the USA, Saudi Arabia and Japan and in small quantities to Kuwait, Sri Lanka and the UAE.

3.3.2.1 Tomato Puree and Paste

Tomato puree is a processed product usually consisted of concentrated tomato juice with or without skin or seeds, with or without added salt. The juice concentration should not be less than 9 % for medium tomato puree, and for heavy tomato puree, it should not be less than 12 % solids. The juice concentration can be increased further to produce tomato paste, where it contains not less than 25 % tomato solids. On further concentration to 33 % or more of solids, it is called concentrated tomato paste.

The steps for the preparation of tomato puree/paste are extraction and staining of tomato juice from ripe fruits and then cooking for juice concentration as required, 12–14 °Bx for puree and up to 44 °Bx for paste, then filling the hot juice into sterilized bottles or canes followed by sealing and pasteurization in boiling water for 20 min and finally cooling completely and storing in a cool and dry place.

3.3.2.2 Tomato Sauce/Ketchup

Tomato sauce and ketchup are semisolid processed product of tomato made of strained tomato juice or pulp along with spices, salt, sugar, vinegar—with or without onion—and garlic paste. Sauces have smoother consistency and can be sieved. The sauce contains not less than 12 % tomato solids and 25 % total solids. Ketchup is a tomato sauce of a thick consistency with added spices.

The steps involved for the preparation of tomato sauce/ketchup are the selection of fully red ripe fruits; washing them properly and chopping into small pieces followed by boiling for 5 min at 70–90 °C to soften the tissues and then mixing a one-third quantity of sugar along with onion and garlic pastes and spices like red chillies, black pepper, clove cinnamon, vinegar, etc.; cooking till the volume reduces; then adding remaining sugar and cooking further; judging the end point by measuring the volume which will be one-third of the original volume of pulp or measuring the TSS (25 °Bx) with a hand refractometer; adding preservatives like sodium benzoate at 250 ppm and filling hot product into sterilized bottles followed by sealing and pasteurization in boiling water for 20 min; and finally cooling completely and storing in a cool and dry place.

3.3.2.3 Tomato Chutney

Chutney is a mixture of tomato pieces, spices, salt and sugar with chopped ginger, onion and garlic along with vinegar and/or sodium benzoate as preservatives. A good-quality chutney should be smooth, palatable and appetizing with a true flavour of tomato. The processing steps for tomato chutney are the selection of fully red ripe tomatoes; washing properly and blanching in hot water for 2 min; peeling and crushing and cooking gently to the desired consistency; mixing of required spices powder like red chilli, black pepper, large cardamom and cumin; and cooking further. Next, add salt and vinegar and mix the preservative sodium benzoate at 0.5 g/kg of the final product followed by filling hot product into sterilized bottles followed by sealing and pasteurization in boiling water for 20 min, and finally, cool completely and store in a cool and dry place.

3.3.2.4 Tomato Chilli Sauce

It is a highly spiced product made from ripe, peeled and crushed tomatoes and with salt, sugar, spices and vinegar with or without onion and garlic. The method of preparation is similar to that for tomato sauce except that the total unstrained pulp is used and seeds are not removed. The hot product is filled in bottles or cans and processed in water at 85–90 °C for 30 min.

3.3.2.5 Lycopene

Lycopene is the pigment principally responsible for the characteristic deep-red colour of ripe tomato fruits and tomato products. It has attracted attention due to its biological and physicochemical properties, especially related to its effects as a natural antioxidant. Although it has no provitamin A activity, lycopene does exhibit a physical quenching rate constant with singlet oxygen almost twice as high as that of β-carotene. This makes its presence in the diet of considerable interest. Increasing clinical evidence supports the role of lycopene as a micronutrient with important health benefits, because it appears to provide protection against a broad range of epithelial cancers. Tomatoes and related tomato products are the major source of lycopene compounds and are also considered an important source of carotenoids in the human diet (John 2006). The experimental results of high hydrostatic pressure (HHP) showed that more lycopene can be extracted from tomato paste waste in only 1 min at room temperature without any heating process. The highest recovery (92 %) was obtained by performing the extractions at 500 MPa pressure, 1 min duration, 75 % ethanol concentration and 1:6 (g/ml) solid/liquid ratio. From the viewpoints of extraction time, the extraction yield and the extraction efficiency, extraction by HHP shows a bright prospect for extracting lycopene from tomato paste waste (Jun 2006).

3.3.2.6 Tomatine

The steroidal glucoalkaloid alpha tomatine is a natural toxin found in all species of tomato. Alpha tomatine is extracted by a solvent extraction method by using methanol as a solvent (Erik et al. 1994).

3.3.2.7 Tomato Seed Oil

Tomato seed, a nonconventional source, has a promising potential source of edible oil. Tomato paste manufacturing unit generates 7.0–7.5 % solid waste of raw material and 71–72 % of which is pomace (Sogi and Bawa 1998; Sogi 2001). Seeds, the major constituents of pomace, contain 20.5–29.6 % lipid (Geeisman 1981; Latilef and Knorr 1983). The physical, chemical and sensory characteristics of the tomato seed oil (TSO) have been found comparable with sunflower oil. The refining process of the TSO has been standardized to obtain a colourless and odourless product similar to commercial refined oil. The TSO has been used successfully in the preparation of the cake and traditional breakfast item. For the preparation of TSO, tomato seeds are prepared from the pomace collected from the tomato paste manufacturing plant by sedimentation techniques and dehydrated at 70 °C for 5 min for 5 h in a cabinet dryer. The seed is expelled through an expeller, and oil was filtered through Whatman No. 1 filter paper to obtain crude oil. The crude oil is degummed, neutralized, blanched and deodorized according to Bhullar and Sogi (2000) to obtain refined oil.

Margarine is prepared by mixing TSO and hydrogenated fat at the ratio 70:30 and added to cream, inoculated for 8 h at 30 °C, churned to get white margarine, kept at 0 °C for 2 h and mixed with 3 % salt and 0.05 % annatto. The physiochemical as well as sensory analysis revealed that the margarine containing 30–46 % TSO could be prepared having characteristics comparable to the market sample. TSO was found suitable for the preparation of margarine like other vegetable oils with additional advantage of colour due to lycopene. Sensory characteristics of margarine

prepared from tomato seed oil (TSO) have been presented in Table 3.7.

3.3.3 Garden Pea

Fresh green peas are highly perishable due to high rate of respiration and cannot be stored for more than 2 days under room temperature. The tender green peas can be stored for a long period after processed into a different product for the year-round availability. The important processed products are dehydrated, frozen, canned and dried peas. However, pea cultivar differs for processed products.

3.3.3.1 Dehydrated Peas
Dehydrated peas are gaining popularity because they offer the advantage of greater shelf life, palatability and convenience during transport and

Table 3.7 Sensory characteristics of margarine prepared from tomato seed oil (TSO)

Margarine	Cream %	Colour	Appearance	Consistency	Odour	Taste
Crude TSO	60	6.22±0.67	6.10±0.57	6.10±0.53	6.60±0.52	6.50±0.53
	70	6.40±0.70	6.20±0.79	6.20±0.63	6.80±0.42	6.60±0.52
	80	6.50±0.85	6.50±0.53	6.40±0.52	6.90±0.32	6.80±0.42
Refined TSO	60	6.70±0.67	6.60±0.70	6.50±0.53	6.70±0.48	6.80±0.42
	70	6.80±0.63	6.50±0.53	6.40±0.52	7.00±0.45	6.90±0.57
	80	7.10±1.10	7.20±0.79	7.10±0.57	7.10±0.32	7.10±0.32
Market sample	–	1780±0.42	7.50±0.53	7.70±0.48	7.50±0.71	7.90±0.74

Source: Sogi and Kaur (2003)
Mean ± SD; $n = 15$

Table 3.8 Sensory scores for colour, flavour, texture, appearance and overall acceptability (OAA) of retreated pea samples fluidized bed dried at different temperatures

Sample/temperature	Colour	Flavour	Texture	Appearance	OAA
Unblanched					
100 °C	4.11	6.9	5.8	5.88	6.46
120 °C	3.66	6.4	5.9	5.81	6.38
140 °C	3.55	6.1	5.7	5.90	7.00
Water blanched					
100 °C	6.11	6.0	5.80	5.96	6.30
120 °C	5.77	5.8	5.79	6.20	6.37
140 °C	6.00	5.9	5.83	6.26	6.35
Alkali blanched					
100 °C	5.50	5.92	5.85	6.21	6.32
120 °C	6.40	5.92	5.86	6.20	6.30
140 °C	7.60	6.00	5.92	6.53	6.33

Source: Kar and Bawa (2002)

handling. The drying process should allow effective retention of colour, texture, flavour, taste and nutritive value, comparable to fresh peas (Table 3.8). The different steps for dehydration are washing of shelled peas and blanching in boiling water for 2 min by adding additives and preservatives followed by air-cooling and then drying in cabinet driers at 55–60 °C and packing in polythene bags and storing in a cool and dry place. Fluidized bed drying is gaining importance for pea drying.

3.3.3.2 Frozen Pea

Small dark green-seeded peas are preferred for freezing. Freezing of green peas is done at −18 °C by using polythene sheets which are resistant to low temperature and impermeable to water vapour. Matured peas are judged by dipping the peas in 10 % brine solution and the sinker is collected. The steps for freezing are washing of shelled peas and blanching for 2 min in boiling water followed by cooling by quick-freezing at −40 °C and packing in polythene bags and storing at −18 °C.

3.3.3.3 Canned Peas

For canning, tender, large-sized pods having sweet bold seeds are selected. Smooth- and wrinkle-seeded cultivars can be used for canning. However, cultivars growing from freezing are not suitable for canning because of a tendency for their skin to develop a bronze colour during processing or storage. The alcohol soluble solid content (11–16 %) of peas determined the quality of canned peas. The steps are washing of shelled peas and hot water blanching for 30 min at 90 °C and filling in cans with brine solution containing sugar and salt and followed by sealing and sterilization in boiling water at 121 °C for 15 min and finally cooling completely and storing in a cool and dry place.

3.3.4 Cabbage

3.3.4.1 Sauerkraut

Sauerkraut is highly popular in some countries of Europe and the USA. Sauerkraut means acid cab-

bage. It is a clean, wholesome product with a characteristic flavour, obtained by complete fermentation of shredded cabbage in the presence of 2–3 % salt. Good-quality sauerkraut should contain not less than 1.5 % acid, expressed as lactic acid, along with 0.5–2.5 % salt. Fully matured cabbage should be used for sauerkraut as the use of green cabbage may lead to defective colour and texture. The processing steps for sauerkraut are the selection of mature, white and sound headed cabbage, removal of core and outer leaves, slicing into fine shreds, addition of 2–3 % salt and fermentation by natural flora at 16–22 °C for a week; after that, the scum is removed, and the juice is drained out, and the fermented shreds are pasteurized at 75–80 °C for 10 min, and finally, the hot product is filled into sterilized jars or wide-mouthed bottles, sealed and stored.

3.3.5 Cauliflower

A fresh cauliflower is delicate in handling and ordinarily not stored for a long time due to its poor shelf life. Cauliflower can be stored converting into value-added products like dehydrated cauliflower, frozen cauliflower and cauliflower pickles. However, browning or blackening of florets, off-flavour development and rotting of florets are the problems for long-term storage. Adoption of suitable varieties, proper temperature management and effective pretreatment measure can maintain the quality of florets during long-term storage.

3.3.5.1 Dehydrated Cauliflower

Dehydrated cauliflower is used as an important ingredient in several food products including instant soup. Thakur and Jain (2006) reported that rehydration characteristics of cauliflower dried at 50 °C showed comparatively better reconstitution than others (Table 3.9). The water assimilates after rehydration were 74.2–76.3 %. Sensory attributes indicated that drying of cauliflower florets at 60 °C could yield a satisfactory dried product. Low-temperature drying maintains the good colour of the dried products in addition to lower taste and flavour scores. The

Table 3.9 Rehydration ratio and quality aspect of dehydrated cauliflower under different temperature regimes

Drying temperature °C	Reconstitution after 5 min		Reconstitution after 10 min	
	RR	PW	RR	PW
50	3.2	70.6	3.8	76.3
60	3.0	69.9	3.5	74.2
70	2.9	69.0	3.4	74.3
	Drying temperature °C			
	50	60	70	
Colour and appearance	3.4	3.0	2.5	
Texture and consistency	2.8	3.3	3.0	
Taste and flavour	2.4	3.2	2.4	
Overall acceptability	2.5	3.4	3.0	

Source: Thakur and Jain (2006)
RR rehydration ratio, *PW* % of water assimilates
Score level: *5* excellent, *4* good, *3* average, *2* fair and *1* poor

steps for dried cauliflower are washing of fresh-cut florets (2.5 cm thick) and blanching in boiled water at 95 °C for 5 min, pretreatments of the florets by potassium metabisulphite (KMS) and drying at 50–60 °C to reduce the moisture levels at 10 % followed by packing in airtight polythene pouch or PET jar.

3.3.5.2 Frozen Cauliflower

For frozen cauliflower, the fresh-cut florets (2.5 cm thick) are washed and blanched in boiled water at 95 °C for 5 min followed by cooling by quick-freezing at −40 °C, then dipping in 0.05 % KMS solution for 5 min to avoid browning and packing in polythene bags and storing at −18 °C. The packing material should be moisture proof and sustain low temperature to avoid dehydration during storage.

3.3.6 Carrot

Carrot is one of the most popular root vegetables and highly valued for its carotene content (2,400–3,000 µg/g). The crop is highly perishable and has a very short storage life with a limited period of availability. Hence, there is a need to preserve carrots for use during off-season. The value-added products are carrot shred, frozen carrot, fermented carrot and carrot juice. Drying and dehydration of carrots are carried out either by sun-drying or dehydration methods like cabinet drying, contact plate drying, freeze-drying, etc.

3.3.6.1 Carrot Shred

Dried shreds are used for *halwa*, instant mixes or noodle mixes after reconstitution. The steps for dehydration are washing of fresh-cut slices (5 cm) and blanching in low temperature for a long time by mixing with sugar followed by drying at 50–60 °C to reduce the moisture levels at 10 % and packing in airtight polythene pouch or PET jar. Shreds dried in open air had less reconstitution ratio. Sensory evaluation score of carrot *halwa* from reconstituted dried carrot shreds indicated its potential to use as base material for the preparation of carrot *halwa*. Machwad et al. (2003) studied on the chemical characteristics of carrots indicated their suitability for drying and feasibility of carrot shreds for further processing. Greater leaching losses were observed in reducing sugars and total sugars during pretreatments and had adverse effect on β-carotene content in all the samples. The reconstitution ratio of dried carrot shreds was higher in pretreated samples than untreated (Table 3.10). Shreds dried in open air had less reconstitution ratio.

3.3.6.2 Frozen Carrot

Carrot is frozen in the form of whole, sliced or diced carrot. The steps are washing, slicing (2 cm), blanching in steam or water, cooling and freezing at −40 °C, packing in polythene bags and storing at −18 °C.

3.3.6.3 Carrot Powder

Dehydration and powder making are highly effective methods for the preservation of carrot. Singh and Kulsherstha (2006) reported that blanching and boiling were the major pretreatments employed for the preparation of carrot powder. They recorded higher levels of beta-carotene and total dietary fibre content in carrot powder prepared by boiling method. The steps

Table 3.10 Effect of pretreatments and drying methods on chemical characteristics and reconstitution ratio of carrot shreds

Parameters	Control	Mechanically peeled and blanched	Mechanically peeled and sulphited	Mechanically peeled, blanched and sulphited	Lye peeled and blanched	Mechanically peeled and sulphited	Lye peeled, blanched and sulphited
Open air-drying							
Reducing sugar %	6.75	6.65	6.60	6.50	6.70	6.60	6.45
Nonreducing sugar %	3.15	2.85	2.75	2.65	2.75	2.70	2.35
Total sugar %	9.90	9.50	9.35	9.15	9.45	9.30	8.80
Beta-carotene µg/100 g	10.00	9.20	9.00	8.25	9.25	9.08	8.88
Reconstitution ratio	2.95	4.50	4.60	4.45	3.75	3.80	4.35
Cabinet drying							
Reducing sugar %	6.85	6.80	6.75	6.70	6.80	6.75	6.65
Nonreducing sugar %	3.80	3.45	3.35	2.75	3.35	3.25	2.80
Total sugar %	10.65	10.25	10.10	9.45	10.05	10.00	9.45
Beta-carotene µg/100 g	10.25	9.50	9.25	9.10	10.00	9.89	9.80
Reconstitution ratio	3.95	7.30	7.40	7.25	6.45	6.55	6.35
SEm	0.11	0.11	0.27	0.15	0.13	0.19	0.16
C.D. at 5 %	0.36	0.36	0.84	0.46	0.40	0.61	0.52

Source: Machwad et al. (2003)

are washing and cutting into 22 cm pieces, then cooking for 10 min followed by mashing and cooling at room temperature, then drying at 50 °C for 18 h and finally grinding into powder.

3.3.7 Wax/Ash Gourd

3.3.7.1 Candy/*Petha*
The famous Agra *petha* is a rectangular or cylindrical translucent soft candy made from ash gourd. It is a popular sweet dish among Indians due to the sublime flavour and good medicinal properties. The steps for the preparation of *petha* are as follows: peel and remove the ash gourd skin and seeds, cut into rectangular blocks and soak in lime water for 30 min, then drain the water, wash and clean, put in a bowl, sprinkle alum water (1 tsp for 1 kg gourd) and boil in a large pan for 10–15 min, drain the water and wash the pieces under running water, make a syrup of sugar and water (750 g sugar for 1 kg gourd), put drained pieces in syrup, boil till *petha* gets softened and sugar syrup is of three-thread consistency, remove *pethas*, drain out excess syrup, sprinkle essence and rose water and cool completely.

3.3.8 Pumpkin

3.3.8.1 Pumpkin Sauce
Pumpkin sauce is often used as a substitute of tomato sauce. The sauce is prepared from deep-orange flesh pumpkin cultivars. The steps are as follows: peel and chop into small pieces by discarding the seeds and stringy pith. Take a large pan and heat oil, add garlic paste, sauté till rich brown, add red chilli powder, mix powdered spices, then add pumpkin pieces and salt, cook in simmer till pumpkin is soft, add sugar and cook further till a soft lump is formed, cool to room temperature, add vinegar and blend in mixer, strain and add colour, mix properly and fill in sterilized bottles followed by sealing and pasteurization in boiling water for 20 min and store in a cool and dry place.

3.3.9 Cucumber and Gherkin

Cucumber and gherkin contain more than 90 % moisture in their fruits; hence, the freshness deteriorates rapidly immediately after harvest. To supply high-quality products during off-season, these crops can be preserved in a brine solution through canning or can be used for preparing pickles.

3.3.9.1 Canned
The steps for canning are washing and grading of fresh fruits, blanching in hot water followed by cooling, filling in can using 2–3 % salt solution, sealing and sterilization in boiling water at 121 °C for 15 min and finally cooling completely and storing in a cool and dry place.

3.3.10 Watermelon

Watermelon flesh is sweet and refreshing. The flesh can be utilized for preparing juice or extraction of lycopene pigment.

3.3.10.1 Watermelon Juice
Watermelon cultivars bright red in colour and high in TSS are selected for juice extraction. The edible red portion is separated, and the juice is extracted with the help of a mixer or juice extractor. The extracted juice is then strained with sieve and mixed with sugar, ginger powder or spices to improve the taste and acceptability. The juice can be stored by filling into sterilized bottles followed by sealing and pasteurization in boiling water for 20 min and finally cooling completely and storing in a cool and dry place.

3.3.10.2 Lycopene
Lycopene, a carotenoid, has antioxidant properties that may reduce the incidence of certain cancers. Watermelon is a natural source of lycopene. The lycopene content in commercial watermelon has been reported to be 45.1–53.2 µg/g fresh weight (Heinonen et al. 1989). These values are 60 % higher than those reported for fresh tomatoes (mean value of 30 µg/g fresh weight). Veazie

Table 3.11 Soluble solids, lycopene, hue and chroma values of different watermelon cultivars

Cultivars	Soluble solids	Lycopene μg/g fresh weight	Hue	Chroma
Seedless types				
King of heart	10.8	57.4	25.2	28.6
Tri-X 313	10.9	65.7	25.8	29.0
Summer sweet 5244	11.3	66.4	24.2	29.9
Crimson trio	11.0	71.2	22.6	29.6
Hybrid seeded type				
Royal sweet	10.5	38.6	30.3	24.1
Dumara	10.8	45.6	25.5	25.7
Sangria	11.0	46.1	25.3	26.9
Open-pollinated seeded type				
Black diamond	9.7	37.9	27.1	23.4
Calhoun grey	9.7	40.0	27.1	23.0
Crimson sweet	10.6	36.5	25.4	22.1
Dixie Lee	10.2	69.2	23.3	28.1

Source: Veazie et al. (2001)

et al. (2001) reported that lycopene content varied widely among the cultivars. They reported that the four cultivars have a mean value of 65.0 μg/g fresh weight. Seedless-type sample tended to have high amount of lycopene (>50.0 μg/g fresh weight) than seeded type (Table 3.11). The lycopene content can also vary on growing season.

3.3.11 Baby Corn

3.3.11.1 Soup Powder

Good-quality cobs are selected, husk and silk are removed and then cobs are blanched in boiled water at 95 °C for 5 min followed by dipping in a potassium metabisulphite (KMS) solution to maintain the colour and then dried in an oven or cabinet drier at 50–60 °C for 12–18 h to reduce the moisture levels at 6–8 % followed by milling in a disc mill to produce flour or powder.

3.3.11.2 Canned Baby Corn

The dehusked cobs are dipped in clean water, the floated cobs are removed and sinkers are dewatered and soaked with 0.05 % calcium chloride

solution for 10 min, then filled in a clean pasteurized can filled with a brine solution consisted of 2.5 % salt and 7 % sucrose solution and sterilized in boiling water for 10 min and finally cooled completely and stored in a cool and dry place. Vacuum packaging is another popular method where cobs are packed almost free of liquids by removing the air and sealed.

3.3.12 Cassava

3.3.12.1 Cassava Fried Chips

Fried cassava chips presently available in the market are often too hard to be bitten and bear no comparison with potato chips. This leads to poor acceptability of the product and lower price. Nowadays, excellent quality fried chips can be made from cassava tubers; by subjecting chips to mild acid treatment, this facilitates in the removal of excess starch and sugars from cassava slices, with the result that light-yellow crisp chips can be obtained, having a soft mouthfeel and good texture. Overmatured tubers having high starch content are less preferred for making fried chips. About 8–9-month-old tubers are ideal for making chips. After slicing to either longitudinal (4 cm × 1.5 cm) or round pieces (0.2 cm thick), tubers are soaked for 1 h (overmatured tubers can be soaked for 2 h) in 0.1 % acetic acid + 0.5 % brine (salt) solution (vinegar diluted 40 times yields 0.1 % acetic acid) (Padmaja and Moorthy 2006).

3.3.12.2 Cassava Starch

For extraction of food grade starch, the tuber is first washed and then rasped into a fine pulp for releasing most of the starch granules.

3.3.12.3 Cassava Fried Snack

A number of fried snack foods, pakkavada, hot fries, crisps, nutrichips, etc., have excellent taste and shelf life.

3.3.12.4 Sago

Sago is normally manufactured by using partially dried starch cake of cassava. Partially dried cake is globulated into special beads in a special type of shaker, and globules are graded by size, partially gelatinized by roasting on heated metal pans and then dried. Thereafter, dried humps are separated by passing through a spike beater and finally polished before packing. It is used as an infant and invalid food, preparation of pudding and sago wafer (Nanda 2008).

3.3.12.5 Other Snack Foods

Cassava and sweet potato can be made into several teatime snack foods like cutlets, puff, samosas, stuffed parathas, etc.; sweet potato leaves a rich source of nutraceuticals like beta-carotene, anthocyanin, etc., and can be used for making leaf pakoras. The soft and mealy texture varieties of sweet potatoes can be used for making gulab jamun by mixing cooked paste with milk powder and maidas.

3.3.13 Sweet Potato

3.3.13.1 Sweet Potato Jam

Sweet potato-based jam can be prepared by mixing mashed sweet potato paste with fruit pulp (banana and apple) in 70:30 ratio. Orange flesher sweet potato rich in beta-carotene is good for jam of high nutritive value.

3.3.13.2 Sweet Potato Pickles

Sweet potato can be used for pickle making, slicing into 1 cm cubes and soaking pieces in 1 % acetic acid (four times diluted vinegar) for 1 h. Soaked pieces are drained and used for pickle making.

3.3.13.3 Sweet Potato-Based Soft Drinks

Fine pulp of juice made from mango, orange, pineapple, etc., is mixed with sweet potato (30:70) and used for making sweet potato-based soft drinks (Padmaja and Moorthy 2006).

3.3.13.4 Sweet Potato Noodles

It is commonly used in noodle preparation. Sweet potato noodles are gluten-free; therefore, a small

amount of partially gelatinized starch is added before mixing with ungelatinized starch in order to function as a binder and to facilitate extrusion or sheeting of starch to prepare noodles (Nanda 2008).

3.3.14 Dehydrated Leafy Vegetables

Dehydrated leafy vegetables have the great potential to be used throughout the year for the preparation of several types of mixed vegetables after rehydration even during the off-season. Drying of leafy vegetables and making them useful for the future open up new vistas in the field of food technology. The pretreated vegetables were spread on an aluminium tray (1.05×0.45 m) at 1.5 kg/sq.m for cabinet drier and solar drier and in stainless steel tray (50×40 cm) at 1.0 kg/sq.m for low-temperature drier. Drying was carried out with the hot air flow rate of 1.2–1.8 m/s at a temperature of 58–60 °C in a cabinet drier (Kilburn make, model 0248) and solar drier (40–50 °C, 60–80 % RH) and low-temperature drier (40^+ –2 °C and 25–40 % RH) to a moisture content of 4–5 % in the finished product. Dehydrated leafy vegetables have the potential to become an important product because they are relatively inexpensive, easily and quickly cooked and rich in several nutrients essential for human health. Singh et al. (2006) studied the effect of driers on the drying and rehydration ratio of some vegetables and spices (Table 3.12), and the effect of driers on chemical constituents of dehydrated leafy vegetables have been presented in Table 3.13.

3.4 Conclusion

Home scale production of value-added products from different vegetables and spices is an age-old practice. Diversification of vegetable and spice products for export and domestic consumption reduces the risk of loss due to price fluctuation. Such produce is processed just after harvesting; it adds value and stabilizes the processed products

Table 3.12 Effect of driers on the drying and rehydration ratio of some leafy vegetables (on dry weight basis)

Parameters	Vegetables	Driers		
		Cabinet driers	Solar driers	Low-temperature driers
Drying ratio				
	Amaranthus	1:12.9	1:13.8	1:16.3
	Curry leaves	1:8.2	1:10.3	1:10.8
	Drm stick	1:15.0	1:16.5	1:17.8
	Methi	1:8.5	1:9.5	1:10.8
	Palak	1:20.7	1:22.9	1:25.0
Rehydration ratio				
	Amaranthus	4.4:1	4.3:1	4.2:1
	Curry leaves	3.7:1	3.7:1	3.6:1
	Drm stick	4.6:1	3.5:1	3.5:1
	Methi	4.0:1	3.9:1	4.1:1
	Palak	4.2:1	4.1:1	4.0:1

Source: Singh et al. (2006)

Table 3.13 Effect of driers on chemical constituents of dehydrated leafy vegetables (on dry weight basis)

Driers	Vegetables	Moisture (%)	Beta-carotene µg/100 g	Chlorophyll mg/100 g	Non-enzyming browning (OD at 420 nm)
Solar					
	Amaranthus	5.0	2,140	78.2	0.35
	Curry leaves	2.9	4,215	112.2	0.39
	Drm stick	6.1	4,080	65.4	0.32
	Methi	3.2	1,670	91.2	0.24
	Palak	4.2	2,515	65.0	0.55
Cabinet					
	Amaranthus	5.2	3,185	85.3	0.29
	Curry leaves	2.5	4,850	130.3	0.32
	Drm stick	5.5	4,775	78.1	0.41
	Methi	2.5	2,685	100.8	0.19
	Palak	3.5	2,975	70.0	0.53
Low temperature					
	Amaranthus	5.8	2,726	83.2	0.31
	Curry leaves	3.2	4,715	122.7	0.33
	Drm stick	6.7	4,520	72.3	0.44
	Methi	7.3	2,442	61.9	0.25
	Palak	4.2	2,518	69.1	0.57
	SEm±	1.27	100.8	22.63	0.27
	CD ($P=0.05$)	2.59	2,062.4	46.22	0.60

Source: Singh et al. (2006)

for a long time. The processed products can be available round the year at a reasonable similar price, and the value-added products have a demand in internal and external market. At present India earns an appreciable amount of foreign exchange by exporting an appreciable quantity of vegetable- and spice-based value-added products. Apart from these, products are gaining importance in the internal market due to a change in consumer preference and market behaviour. The production of these products is an urgent need of the hour to satisfy the consumers, and the producer can earn extra income. So, constant research and development activities of the following aspects will help to bring the new direction of value-added vegetable and spice industry:

1. Development of an easy method of value addition to fresh produce and new processed product from vegetables and spices
2. Dissemination of new techniques of value addition in vegetables and spices
3. Assurance of quality of the value-added products
4. Proper truthful packaging of these products
5. Proper marketing of these products in the internal and external market of the country
6. Popularization of these products through extension activities

References

Anonymous (2014) Indian Horticulture Database, National Horticulture Board, Ministry of Agriculture, Government of India, Gurgaon, India

Bullar JK, Sogi DS (2000) Shelf life studies and retaining of tomato seed oil. J Food Sci Technol 37:542–549

Dexter ST, Salunkhe DK (1952) Improvement of potato chip color by hot water treatment of slices. Mich Agric Exp Stn Q Bull 35:165

Erik AJK, Margareth ECMH, Jongen WMF (1994) Rapid high performance liquid chromatography for the quantification of alpha tomatine in tomato. J Agric Food Chem 42:475–477

Geeisman JR (1981) Protein from tomato seed. Ohio Rep Res Dev 66:92–94

Heinonen MI, Ollilainen V, Linkola EK, Varo PT, Koivistoinen PE (1989) Carotenoid in finished foods: vegetables, fruits and berries. J Agril Food Chem 37:655–659

John LMM (2006) Lycopene in tomatoes: chemical and physical properties affected by food processing. Crit Rev Food Sci Nutr 40(1):1–42

Joshi S, Nath N (2002) Effect of pretreatments on quality and shelf life of fried chips from sprouted tubers of potato variety "Kufri Chandramukhi". J Food Sci Technol 39(3):251–257

Jun X (2006) Application of high hydrostatic pressure processing of food to extracting lycopene from tomato paste waste. Int J High Press Res 26(1):33–41

Kar H, Bawa AS (2002) Studies on fluidized bed drying of peas. J Food Sci Technol 39(3):272–275

Khurdiya DS (2003) Promoting export of value added products of horticultural produce. Indian Hortic 48(1):32–35

Latilef SJ, Knorr D (1983) Tomato seed protein concentrate: effect of methods recovery upon yield and composition characteristics. J Food Sci 48:1583–1586

Machwad GM, Kulkarni DN, Pawar VD, Surve VD (2003) Studies on dehydration of carrot (*Daucus carota* L.). J Food Sci Technol 40(4):406–408

Nanda SK (2008) Postharvest technologies and utilization. In: Palaniswami MS, Peter KV (eds) Tuber and root crops. New India Publishing Agency, New Delhi

Padmaja G, Moorthy SN (2006) Value added food and food products from tubers. Indian Hortic 51(4):13–15

Pandey SK, Paul Khurana SM, Singh SV, Kumar D, Kumar P (2005) Evaluation of Indian and exotic varieties for sustaining processing industries in North Western parts of India. Indian J Hortic 62(2):155–159

Singh P, Kulsherstha K (2006) Studies on the effect of pre-treatment on the preparation of carrot powder. J Food Sci Technol, 43:145–147

Singh U, Sagar VR, Bahera TK, Kumar P (2006) Effect of drying conditions on the quality of dehydrated selected leafy vegetables. J Food Sci Technol 43(6):579–582

Sogi DS (2001) Functional properties and characterization of tomato waste seed proteins. Ph.D thesis, Guru Nanak Dev University, Amritsar, India

Sogi DS, Bawa AS (1998) Dehydration of tomato processing waste. Indian Food Packer 52:26–29

Sogi GS, Kaur J (2003) Studies on the preparation of margarine from tomato seed oil. J Food Sci Technol 40(4):432–435

Thakur AK, Jain RK (2006) Studies on drying characteristics of cauliflower. J Food Sci Technol 43(2):182–185

Veazie PP, Collins JK, Pair DS, Roberts W (2001) Lycopene content differs among red-fleshed watermelon cultivars. J Sci Food Agric 81:983–987

Value Addition in Spice Crops

4

S. Datta, S. Guha, and A.B. Sharangi

Abstract

Spices are very important horticultural crops and a variety of spices are used in everyday cooking. Value addition is the highest recognition of the value of the product through processing, packaging and marketing. In other words, it is the process of changing or transforming a product from its original state to a more valuable state. Nowadays, many value-added spices are used and they impart a special taste to food preparations. Value addition has several plus points, viz. the value added products are simple to carry, having long-lasting flavours, with low bacterial contamination, having higher income from food industry, used as preservatives and also in pharmaceutical industry. There are many regulatory agencies which control the quality of spices like ASTA, ESA, BIS, etc. Some prominent value-added products accredited globally are black pepper powder, pepper oleoresin, cardamom oil, curcumin, turmeric oleoresin, bleached ginger, garlic paste, onion powder, coriander oleoresin, etc. Big entrepreneurship v to be developed in large scale, and year round production of the value-added product for meeting the international demand is feasible. New product should be developed from different minor and underutilized spice crops, especially from herbal spices. Integrated approach for development, production and marketing strategies should give the new direction towards the value addition in spice crops. This chapter will summarize the current initiatives about the above with scopes and future direction in India as well as the world.

4.1 Introduction

Indian spices have a glorifying past dating back to more than 7,000 years. Even today, it holds the same spell. India is having to her credit as the largest producer, consumer and exporter of spices in the world (Table 4.1). Apart from the fresh vegetables and spices, India also exports an

S. Datta
Uttar Banga Krishi Viswavidyalaya,
Pundibari, Cooch Behar, West Bengal, India

S. Guha • A.B. Sharangi (✉)
Bidhan Chandra Krishi Viswavidyalaya,
Mohanpur, Nadia, West Bengal, India
e-mail: dr_absharangi@yahoo.co.in

A.B. Sharangi and S. Datta (eds.), *Value Addition of Horticultural Crops: Recent Trends and Future Directions*, DOI 10.1007/978-81-322-2262-0_4, © Springer India 2015

Table 4.1 List of a few spices with their common and botanical name, family and useful part(s)

Common name	Botanical name	Family	Useful part(s)
Black pepper	*Piper nigrum*	Piperaceae	Berry/fruit
Long pepper	*Piper longum*	Piperaceae	Berry/fruit
Cardamom	*Elettaria cardamom*	Zingiberaceae	Capsule/fruit
Ginger	*Zingiber officinale*	Zingiberaceae	Rhizome
Garlic	*Allium sativum*	Amaryllidaceae	Bulb/cloves
Vanilla	*Vanilla fragrans*	Orchidaceae	Beans/pods
Nutmeg and mace	*Myristica fragrans*	Myristicaceae	Kernel (N), aril (M)
Turmeric	*Curcuma longa*	Zingiberaceae	Rhizome
Clove	*Eugenia caryophyllus*	Myrtaceae	Unopened flower bud
Cinnamon	*Cinnamomum zeylanicum*	Lauraceae	Bark
Asafoetida	*Ferula asafetida*	Apiaceae/Umbelliferae	Latex/gum
Saffron	*Crocus sativus*	Iridaceae	Stigma of flower
Bay leaves	*Laurus nobilis*	Lauraceae	Leaves
Cumin	*Cuminum cyminum*	Apiaceae	Seeds
Caraway	*Carum carvi*	Apiaceae	Seeds/fruit
Fennel	*Foeniculum vulgare*	Apiaceae	Seeds/fruit
Poppy seeds	*Papaver somniferum*	Papaveraceae	Seeds
Amchur	*Mangifera indica*	Anacardiaceae	Fruit
Mustard	*Brassica nigra*	Brassicaceae	Seed
Fenugreek	*Trigonella foenum-graecum*	Fabaceae	Seeds and leaves
Coriander	*Coriandrum sativum*	Apiaceae	Seeds and leaves
Tamarind	*Tamarindus indica*	Fabaceae	Pod/fruit
Chillies	*Capsicum annuum*	Solanaceae	Fruit
Onions	*Allium cepa*	Amaryllidaceae	Bulb

appreciable quantity of spice-based value-added products. Value-added processed products means product which can be obtained from main products and by-products after some sort of processing and marketed to increased margin of profit. Value-added products in general indicate that in the same volume of primary products, a higher price is realized by means of processing, packing, upgrading the quality or other such methods. Diversification of spice products for export and domestic consumption reduces the risk of loss due to price fluctuation. More than 150 spice-based value-added products are now available for export. The integrated approach from harvesting to the delivery into the hands of a consumer, if handled and cared properly, can add value to the fresh produce in the market. But most of the fresh produce has limited life, although it can be stored at appropriate temperature and relative humidity at the same time. If such produce is processed just after harvesting, it adds value and stabilizes the processed products for a long time. The processed products can be available round the year at a reasonable similar price as the value-added products have a demand in internal and external market (Khurdiya 2003). Constant research and development activities in this regard will help to bring the new direction of value-added spice industry.

4.2 Global Market of Spice Products

India is the leading exporter of spice and spice products. We had the prestige of exporting nearly 40 % of the world's requirement of spices at one time, but the figure comes down to about 15–20% in the recent years (Narayana et al. 1999). India is the leading producer of spice and spice products. At present, India is exporting raw spices as well as value-added spice products to around 70 countries.

Table 4.2 Export of value-added spice products during the last few years

Year	Value-added spice products		Spices (including value-added spice products)	
	Quantity ('000) MT	Value (Rs. in cores)	Quantity ('000) MT	Value (Rs. in cores)
1993–1994	6.18	114.09	182.33	571.44
1994–1995	7.39	144.77	155.01	620.10
1995–1996	7.51	180.27	203.39	804.43
1996–1997	9.37	293.71	225.3	1,230.71
1997–1998	9.92	336.61	218.75	1,352.15
1998–1999	12.17	459.26	231.39	1,758.02
1999–2000	12.00	461.01	236.14	2,025.09
2000–2001	13.70	530.47	230.00	1,612.07
2001–2002	22.11	908.37	243.20	1,940.55
2002–2003	26.92	1,035.45	264.11	2,086.71
2004–2005	22.65	936.62	335.49	2,200.00
2005–2006	26.23	1,512.99	350.36	2,627.62
2006–2007	–	2,093.00	373.50	3,575.75
2008–2009	40.60	2,304.50	470.52	5,300.25
2009–2010	40.05	2,087.58	502.75	5,560.50
2010–2011	40.30	2,817.91	525.75	6,840.71
2011–2012	39.60	3,780.18	575.27	9,783.42
2012–2013	47.00	5,804.47	726.61	12,112.76

Source: Anonymous (2006), (2007), and (2014)

The major markets of Indian spices in terms of value are the USA (30.2 %), the UK (6.3 %), Japan (5.3 %), the UAE (4.8 %) and Germany (4.7 %) (Anonymous 2000). During 2012–2013, a total of 699,170 t of spices and spice products valued Rs 11,171.16 crore (US$ 2,040.18 million) has been exported from the country registering an increase of 22 % in volume and 14 % in rupee terms of value over the previous year (Siddaramappa 2013). The spices are classified into straight and value-added spices depending on marketing. Straight spices are those spices which are marketed in their natural form, while value-added ones are spice powders, oils, oleoresins, etc. Organic ginger, encapsulated oil, ginger drops, green ginger oil, powdered pepper, organic turmeric and natural turmeric pigments are other novel spice-based value-added products during the twenty-first century (Subbulakshmi and Naik 2002). Recently, export market is rapidly shifting from spices to value-added spice products. This increased demand for value-added products is due to the change in the market behaviour, changes in consumer preferences and emergences of supermarkets. Export of the value-added spice products has shown an increase of 38 % in terms of value in 2006–2007 compared to 2005–2006. It was estimated that in 2006–2007, value-added spices earned 2,093 crores by exporting against 1,512 crores during 2005–2006 and 5,804 crores during 2012–2012 (Table 4.2). Value-added spices exported from India are spice oil, oleoresin, curry powder, mint products, spice powder, dehydrated green pepper, freeze-dried green pepper, dehydrated garlic flakes, etc.

4.3 Advantages of Value-Added Products

1. More volume can be handled per unit area.
2. Do not impart colour to the end product (especially in the case of spice oil).
3. Not affected by bacterial contamination.
4. Easier to store.
5. Shelf life is longer as compared to straight spice.
6. It increases the development and growth of the allied industry and thereby provides additional employment opportunity.

7. New product can be developed.
8. Custom-made blinds are also offered to suit the specific requirement of the customer.
9. Hygienic and free from microorganism.
10. Contain natural antioxidants and are free from enzymes.

4.4 Value Addition in Spices

During the last three decades, India witnessed a boom in product development and export of value-added spice product. A large number of value-added products are prepared from spice crops (Table 4.3). Among the different value-added spice products, spice oil and spice oleoresin command a lion's share portion. With respect to different spices, black pepper, chilli and paprika, small cardamom, turmeric, ginger, cinnamon, cassia and nutmeg enjoy the monopoly in oil and oleoresin export. Spice oil and oleoresins are produced from most of the spices. Among them, few are discussed crop-wise with

Table 4.3 Value-added products from different spices

Name of the spice	Value-added products
Black pepper	Oleoresin, green pepper in brine, dehydrated green peppers, canned green pepper, white pepper powder, etc.
Paprika	Colour, paprika flavour
Ginger	Powder, wines, dry ginger starch from spent ginger preserves, ginger essential oil and oleoresin
Turmeric	Natural pigments, curcuminoids, oleoresins
Coriander	Powder, oleoresins
Cumin	Powder, oleoresin
Fennel	Sugar-coated fennel, oleoresin, whole, etc.
Fenugreek	Powder, dried fenugreek leaves, etc.
Tree spices (cinnamon, cassia nutmeg, cloves)	Obesity regulators, stimulators, nutraceuticals
Chillies	Powder, pickles, paste, oleoresin, oil, brined chilli, sauces

other value-added products, and important spice essential oil and oleoresins are discussed in Tables 4.16 and 4.17. Essential oil is a complex mixture of odorous and steam volatile components which are deposited in the subcuticular space of granular hairs, cell organelles, excretory cavities, canal and especially in the heart of the roots (Arya 2001). It is extracted by (1) water distillation, (2) water and steam distillation, (3) steam distillation, (4) solvent extraction and (5) supercritical fluid extraction (SCFE) by using CO_2. Steam distillation is a more widely used method for the extraction of spice essential oil. Quality management is an important aspect of spice essential oil marketing. Specific gravity, refractive index, optical rotation and composition are the major quality determinants of spice essential oil. Oleoresin may be defined as the true essence of the spice without impairing any aroma and flavour profile. It represents the complete flavour profile of spice soluble in a particular solvent used. It is the concentrated viscous, resinous components (except cellular compounds) obtained by solvent extraction in which volatile components gives aroma profile nonvolatile component for pungent and other taste principles. The residual solvent by the oleoresin should be below 30 ppm. It is mainly extracted by using an organic solvent or solvent mixture such as ethylene chloride, ethylene dichloride, acetone, ethyl acetate, alcohol, ethanol, methylene bromide, methylene chloride, petroleum ether, etc. Normally, solid to solvent ratio varies from 1 to 3 (Narayanan et al. 2000).

The essential oil and oleoresin are manufactured in Tamil Nadu, Maharashtra, Karnataka, Madhya Pradesh, Rajasthan, Orissa, Andhra Pradesh, Kerala, etc. Generally, its production area is specific, e.g. cinnamon oil is produced in Tamil Nadu and Kerala, and mint oil is produced in Terai areas of Jammu Kashmir and Uttar Pradesh. Oil used in perfuming industry is developed in Gazipur, Jainpur, Kannauj and Barsana, district of Uttar Pradesh; Ganjam and Parlakimidi, district of Orissa; and Pune and Pandharpur, district of Maharashtra.

4.5 Regulatory Agencies on the Quality of Spices

4.5.1 ASTA (American Spice Trade Association)

This organization plays an important role for the cleanliness specification of spices and herbs. For purposes of this specifications, rigorous monitoring is done to make the product absolutely clean by removing all extraneous matters. Extraneous matter is defined as everything foreign to the product itself and includes, but is not restricted to, stones, dirt, wire, string, stems, sticks, nontoxic foreign seeds, excreta, manure and animal contamination.

4.5.2 ESA (European Spice Association)

It represents the interests of its members vis-a-vis the competent bodies and departments of the European Union, as well as international institution and organizations.

4.5.3 BIS (Bureau of Indian Standards)

BIS has prescribed quality standards for 16 spices, spice powders, concentrates and oleoresins. Spice boards actively interact with BIS in finalizing standards for other spices and to upgrade the existing one.

4.6 Some Important Value-Added Spice Products

4.6.1 Black Pepper

4.6.1.1 Dehydrated Green Pepper

Thomas and Gopalakrishnan (1992) developed a process involving blanching and sulphiting combined with controlled drying and reduction of moisture to stabilize the green colour of pepper. Slightly immature green pepper (20–30 days prior to full maturity) is preferred for producing dehydrated green pepper. The cleaned pepper berries were subjected to blanching in boiling water for 10–30 min till the enzymes responsible for blackening the pepper are inactivated and polyphenols washed out of the berries. The pepper thus obtained was immediately cooled in water and subjected to sulphiting in potassium metabisulphite solution to fix the green colour. The sulphited berries were then washed and dried in a cabinet dryer at 50–55 °C for 12–15 h to get uniform green-coloured berries. Good quality dehydrated green pepper should content less than 8 % moisture. Boiling time depends on the maturity of the berries. Total heat inactivation of the enzyme was obtained after 10 min of boiling (Mathew 1994). Potassium metabisulphite has a phenolase inhibiting property and an ability to deter nonenzymatic browning (Varghese 1991).

4.6.1.2 Green Pepper in Brine

Four- to five -month-old peppers are preferred for this item. Harvest the immature spikes and clean it thoroughly. Light pepper and pinhead or broken berries are considered as extraneous matter. Twelve to 14 % common salt and citric acid not exceeding 0.6 % by mass of the packing media are used for the preparation. It is also an important item of export which has been presented in Table 4.4.

Table 4.4 Export of some value-added products of black pepper

Item	2001–2002		2002–2003	
	Quantity (MT)	Value (crore)	Quantity (MT)	Value (crore)
Pepper in brine	1,050	7.52	943	5.09
Dehydrated green pepper	326	6.70	771	12.72
Freeze-dried green pepper	12	1.11	44	3.44
White pepper	55	1.99	40	1.36

Source: Vasantakumar (2006)

4.6.1.3 Canned Green Pepper

The despiked and cleaned berries are immersed in water containing 20 ppm residual chlorine for about an hour. The berries are then immersed in 2 % hot brine containing 0.2 % citric acid, exhausted at 80 °C, sealed properly and processed in boiling water for 20 min. Canned pepper is then cooled immediately in a stream of running cold water. It is reported that pepper harvested 1 month prior to maturity is ideal for the manufacture of canned green pepper (Narayanan et al. 2000). Canned green pepper has greater demand in Western countries.

4.6.1.4 Bottled Green Pepper

The manufacturing process consists of despiking the fresh green pepper berries of uniform size and maturity immediately after harvest followed by cleaning, washing and steeping in 20 % brine solution containing citric acid. This is allowed to cure for 3–4 weeks. The liquid is drained off and fresh brine of 16 % concentration together with 100 ppm sulphur dioxide and 0.2 % citric acid are added. The resulting product is stored in containers protected from sunlight (Pruthi 1997). The product should possess the characteristic odour and flavour of green pepper with a colour range of pale green to green.

4.6.1.5 Freeze-Dried Green Pepper

India is one of the producers of freeze-dried black pepper. It is produced by vacuum drying at sub-freezing temperatures ranging from −12 to −40 °C. Freeze-dried green pepper retains the natural form of the green pepper, and it is considered far superior to dehydrated green pepper for its better colour, flavour and essential oil and piperine content. On rehydration, it retains the original green colour and shape of the green pepper. It finds a wide application in instant soups and dry meals for its special characters and subtle flavour. It is also used in a cheese industry.

4.6.1.6 Frozen Green Pepper

This is a relatively new and simple innovation for the diversification in pepper exports. Frozen green pepper is considered superior to green pepper in brine or dehydrated green pepper because of the following reasons:

(a) It has much better flavour, colour and texture and natural appearance.
(b) The cost in packaging is much less due to cheaper container than of cans, and thereby it offsets the extra cost of freezing to a great extent.
(c) It is a see-through container (unlike cans) and customer can see as to what they are buying.
(d) Although its freezing is expensive, its cost is reduced through cheaper flexible plastic pouches and is gaining popularity.

For producing frozen green pepper, the desired maturity green pepper is blanched. Freezing is done at 40 °C by covering the green pepper in a 2 % sodium chloride, 0.25 % citric acid and 0.1 % ascorbic acid. Frozen green pepper is used in fresh salad and frozen meals.

4.6.1.7 Cured Green Pepper

Collect 3–4-month-old pepper spike for the preparation of cured green pepper. After proper washing of spike, berries are separated from the spike and treated with concentrated sodium chloride solution containing dilute citric acid and preservatives. Thereafter, berries are drained from the salt solution and aseptically sealed in cellophane or flexible pouches. RRL, Trivandrum, developed this product.

4.6.1.8 Green Pepper Pickle (in Oil, Vinegar or Brine)

Green pepper pickle is very popular in many states, notably in Kerala, Karnataka, Tamil Nadu, Gujarat, Maharashtra, etc. People relish it with rich meals as an appetizer. Recipes are almost the same as for domestic mango pickles. Green pepper is also prepared in 15–16 % acidified brine and in vinegar similar to other vegetables.

4.6.1.9 Mixed Green Pepper Pickle

Green pepper berries are mixed with lime pickles, mango pickles, mixed cauliflower and carrot pickles, brinjal pickles and bitter gourd pickle with or without green chillies and sliced ginger (fresh). They are quite popular all over India. However, their preparation is very limited. These pickles can be preserved in good condition for 6–9 months in ambient environment.

4.6.1.10 White Pepper

White pepper is the white inner corn obtained after removing the outer skin or pericarp of the pepper berries. It is preferred over black pepper in light-coloured preparations such as sauces, cream soups, etc., where dark-coloured particles are undesirable. It imparts pungency and a modified natural flavour to the foodstuff (Sudharshan 2000). Varieties like Balankotta and Panniyur 1, with large-sized berries, are ideal for making white pepper. Moisture level of white pepper is brought down below 11 % by drying. On an average, the recovery from fresh pepper to dry white pepper is 28 %. The world demand for white pepper is about 50,000 MT, which is about 25 % of the black pepper produced worldwide. White pepper should fetch at 40 % higher price than the black pepper. Major producers of white pepper are Malaysia, Indonesia and Brazil. Chief consumers of white pepper are west European countries, the USA and Japan.

White pepper can be prepared using by any one of the following techniques (Pruthi 1980):
1. Water steeping and rotting technique (retting)
    (a) From fresh ripe berries
    (b) From dried berries
2. Steaming or boiling technique
3. Chemical technique
4. Decortications technique
5. Pit method

4.6.1.10.1 Water Steeping Technique
4.6.1.10.1.1 Water Steeping Technique by
            Using Fresh Berry
The harvested ripe spikes or berries are packed in gunny bags and steeped in water tanks or under running water for 7–10 days. Underdeveloped and light berries are separated. Outer rind of the berries is then removed by rubbing them in hands or trampled. The deskinned berries are then washed, drained and put into galvanized iron vessel containing a solution of bleaching powder for

2 days. Then they are drained, dried under the sun and cleaned.

4.6.1.10.1.2 Water Steeping Technique by
            Using Dried Berries
The dried black pepper berries are steeped in water for 10–15 days, after which they are removed, rubbed, washed thoroughly, steeped again in bleaching solution for 2–3 days, drained, dried under the sun and sold as white pepper. However, the white pepper is more easily prepared from using fresh ripen berries than from dried black pepper.

4.6.1.10.2 Steaming or Boiling or CFTRI
            Method
This process consists of the steaming or boiling of fresh berries for about 15 min. The boiled or soften berries are then passed through a motorized fruit pulping machine for removing outer skin. The deskinned berries are then washed, bleached and dried under the sun to get white pepper. The skin of the berries are immediately utilized for steam distillation for the recovery of valuable pepper oil. But the recovery and quality is comparatively lower than the different water steeping methods.

4.6.1.10.3 Pit Method
In this method, fully matured ripe berries are packed in gunny bag, and place the bag in the pit and cover with soil. Watering should be done above the pit. As a result, moisture and microbes come in contact with the berries. Pericarp is rotten and then berries are washed, bleached and dried properly. This method requires less time and water as compared to water steeping method.

4.6.1.10.4 Microbial Removal of Skin
Dry black pepper is deskinned by using nonpathogenic bacterium *Bacillus subtilis*. Retting should be carried out at P^H 6.8–7.0 in a minimal nutrient medium with dry black pepper. Within 2–4 days, the whole black pepper is deskinned. Thankamony et al. (1999) studied about the quality parameters of black pepper and white pepper from fresh and dry berries which have been presented in Table 4.5.

Table 4.5 Quality parameters of black pepper and white pepper from fresh and dry berries

Sample	Moisture per cent	Volatile oil per cent	Piperine per cent
Dry black pepper	10.0	3.6	4.2
White pepper form black pepper	10.0	3.2	4.5
Fresh green pepper	75.0	3.6	4.0
White pepper from fresh green pepper	10.9	3.6	4.4
Commercial white pepper	9.0	1.0	2.0

Source: Thankamony et al. (1999)

Table 4.6 Some high-quality varieties of black pepper

Variety	Essential oil per cent	Oleoresin per cent	Piperine per cent
Kottanadan	4.0	17.0	6.0
Kumarakodi	4.0	14.0	5.0
Panniyur 1	2.5	9.5	3.0
Sreekara	5.0	13.0	5.0
Subhakara	5.0	12.4	3.4
Malabar excel	5.4	14.6	3.5

Source: Zachariah (2005)

4.6.1.10.5 Chemical Technique

Joshi (1962) developed a chemical technique for the production of white pepper. According to them, the whole dried black pepper was steeped in five times in its weight of water for 4 days and treating with 4 % NaOH solution and boiling the mixture.

4.6.1.11 Ground Pepper

Ground pepper is obtained by grinding pepper without adding foreign matter. Grinding can be accomplished by employing equipment like hammer mill, pin mill or plate mill. The ground product is further sieved to the required size and packed. In order to reduce the flavour loss in ground spices due to excessive heat produced during grinding, cooling arrangements are required. TNAU has developed a low-temperature device in which dry ice is used to cool the grinding zone of the mill (Anonymous 2001). Cryogenic grinding by controlled injection of liquid nitrogen directly into the grinding zone helps in retaining more of volatile oils, improving fitness and minimizing distortion in the natural composition of the powder (Pruthi 1980).

4.6.1.12 Pepper Essential Oil

The characteristic aroma of black pepper is due to the presence of volatile oil, which can be recovered by steam distillation or water distillation (Pruthi 1980). But steam distillation is the common one. The essential oil contains mainly a mixture of terpenic hydrocarbons separator. Volatile oil of black pepper is slightly greenish with a mild non-pungent taste. Generally, black pepper berry contains 2–6 % essential oil, and the main components of the essential oil are phellandrene, sabinene, α-pinene and β-pinene.

4.6.1.13 Oleoresin

Oleoresins are concentrated products obtained by the extraction of ground pepper using solvents like hexane, ethanol, acetone, ethylene dichloride, ethyl acetate, etc. The production of oleoresin is greatly influenced by the solubility of the solvent used. Normally, solid to solvent ratio varies from 1 to 3 (Narayanan et al. 2000). The extraction temperature should be 55–60 °C. The choice of solvent is an important aspect to getting higher recovery and better quality of products, e.g. hexane gives very low yield because of low solubility of piperine in hexane and therefore is not used for extraction pepper oleoresin. High yields of oleoresin (12–14 %) containing 19–35 % volatile oil and 40–60 % piperine are obtained with good quality pepper (Lakshmanachar 1993). Solvent-free pepper oleoresin can be obtained by supercritical CO_2 extraction. Pepper oleoresin is a dark green viscous liquid with a strong aroma and pungent taste. The list of some black pepper cultivar having high oleoresin and essential oil has been presented in Table 4.6.

4.6.1.14 Dehydrated Salted Green Pepper

Dehydrated salted green pepper is a product developed by Pepper India Corporation; it is a 100 % substitute for green pepper in brine and is much more convenient as it is easier for transportation and storing as it does not involve any brine solution. It is a product which can be used instead of pepper in brine as it contains both pepper and salt in the same proportion and at the same time maintains the natural green colour.

4.6.1.15 Piperine
Piperine is the main principle component of black pepper. Piperine content varied from 3 to 6 %. It is sparingly soluble in water but readily soluble in alcohol. Its concentrated form can be produced by centrifuging the black pepper oleoresin in a basket centrifuge. It is mainly used in food, pharmaceutical and nutraceuticals industries.

4.6.1.16 Other Novel Products
Pepper concrete, pepper absolute, pepper essence, pepper paste, pepper emulsion, etc., have also been made and marketed abroad.

4.6.1.17 Pepper by Products
There are different by-products available in the market, viz. the paper rejection or waste or in the unfertilized bud and the stem and inflorescence stalks. In order to economize the use of pepper as a condiment and to replace it in times of scarcity, many products having the characteristic taste and pungency of pepper have been prepared.

4.6.1.17.1 Pep-Sal or Pepper Sal
A new flavouring substance named pepper sal from waste black pepper and common salt pepper sal had found acceptance as a flavouring agent for salads, drinks and meat dishes.

4.6.1.17.2 Pepper Hulls
Pepper hulls or shells removed during the preparation of white pepper are sold separately as a light- to dark-brown powder with a very pungent odour and taste and had been found useful for flavouring tinned foods.

4.6.1.17.3 Light Powder
The underdeveloped berries where in photosynthesis were not complete, lacking in starch synthesis, are therefore very light but rich in oil,

piperine and oleoresin content. They are being used in manufacturing oil, oleoresin concentrate and paste in various meat products.

4.6.1.17.4 Pepper Pinheads
Pepper pinhead (berries with the size of 'pinheads') is fairly rich in oil and oleoresin, but its quality is somewhat inferior.

4.6.1.17.5 Pepper in Curry Powder Spice Blends
Many spice mixtures or blends like 'curry powder', 'garam masala', and a host of other masalas, which are flooding the markets in India and abroad, have pepper as one of the important components. They also find use in soups, pickles, sauces, chutneys, etc.

4.6.2 Garlic

4.6.2.1 Garlic Powder
The export of garlic in the form of powder instead of as raw bulbs can realize a higher unit value. Converting garlic bulb into powder adopting CFTRI technology can also reduce the wastage occurring during storage. Garlic powder is being preferred in Europe and the USA, primarily due to standardization of flavour levels possible in finished product. About 20 parties have been given process know-how for the production of dehydrated garlic powder. Some of them are exporting the product. Various products of garlic, viz. garlic oil, dehydrated garlic powder, garlic juice and extract, pickled garlic, etc., can be prepared. Of these products, garlic powder is perhaps the most important. However, it suffers from the disadvantage of browning and caking on storage besides losing the flavour. This was obviated by the development of 'stabilized' garlic powder.

4.6.2.2 Garlic Essential Oil
Garlic oil, recovered by steam distillation of fleshy ground cloves, is a reddish-brown overpowering liquid. The pungency of the product makes it difficult to use directly, and it is commonly diluted in vegetable oil or microencapsu-

lated dextrin. The chief constituents of the oil are diallyl disulphide (60 %), diallyl trisulphide (20 %), allyl propyl sulphide (6 %) and probably a small quantity of diethyl sulphide. Diallyl polysulphide and diallyl disulphide are said to posses the true garlic odour. One gram oil is equivalent in flavouring terms to 900 g fresh garlic or 200 g dehydrated powder. It is being used in beverages, ice cream and ices, confectionary, baked goods, chewing gum and condiments.

4.6.2.3 Garlic Juices

Garlic juice is obtained by repeated extrusion of garlic tissue, flash heating (140–160 °C) and then cooling to 40 °C. The juice contains both flavour and aroma active compounds and their precursors as well as sugars. This kind of pressed juice is used in the food industry for seasoning.

4.6.2.4 Dehydrated Garlic and Garlic Flakes

Processed garlic is becoming a product of considerable importance in garlic industry and also reduced the transportation cost during transit and avoidance of losses of bulb during storage. It is also an important item of trade (Table 4.7). Processed garlic is a highly competitive commodity in the international market. Dehydrated products such as flakes, granules, powder and other products from the important by-product are being prepared and marketed worldwide. Garlic dehydration is carried out by using tunnel dehydrator, sliced garlic are dried in series of hot air tunnels, and the dried sliced are prepared in the form of flakes and powder. Garlic is commercially dried up to 6.5 % moisture content. To produce 1 kg of dried product, approximately 3.2–4.5 kg of fresh garlic are required.

Table 4.7 Export of dehydrated garlic flakes/powder from India

Year	Quantity (tonnes)	Value (Rs. in lakhs)
1994–1995	162	69
1995–1996	332	142
1996–1997	978	333
1997–1998	360	181
1998–1999	440	228
1999–2000	598	261

Source: Sankar et al. (2005)

Flow chart for the preparation of garlic flakes

Garlic bulb – storing and washing
↓
Slicing
↓
drying of the sliced garlic (6.5 per cent moisture)
↓
Garlic flakes
↓
Further processing, packing and marketed.

4.6.2.5 Garlic Salt

The garlic salt content should not be more than 81 % and the moisture-free white garlic powder should be between 18 and 19%. Garlic salt has much wider culinary potential than the powder, and tablespoon is considered to be equivalent to a clove of fresh garlic.

4.6.2.6 Garlic Paste

In a market, garlic paste is available as in pure form or mixed with ginger or onion. Selected, peeled fresh garlic cloves are cut into chunks. Put ingredients in a blender adding little quantity of water, just to help in grinding, and pack in airtight container and store in refrigerated condition. Instead of water, edible oil is also added during blending process to enhance the shelf life of the produce. Garlic paste was successfully exported first to London (Pruthi 1980).

4.6.2.7 Preparation of Paste

Garlic bulbs (variety: 'Punjab Selection 1') procured from Fatehgarh Churian, in the state of Punjab. The garlic bulbs were subjected to mild pressure by hand to separate into cloves. Cloves were dried in a tray drier at 40 °C for 30 min to facilitate peeling. Peeling was done manually. After peeling, cloves were blanched at 90 °C for 15 min in hot water followed by grinding in a laboratory grinder. This ground material was passed through a 14-mesh sieve to obtain the product of uniform consistency; 10 % sodium chloride (w/w) was added to increase the total soluble solids (TSS). The final pH was maintained to about 4 by addition of 30 % citric acid (w/v) solution. The prepared paste was filled in glass bottles and thermally processed (Ahmed et al. 2003).

4.6.2.8 Garlic Pickles

For garlic pickles, cloves fermented in brine solution and taste-ground spices, oil, salt, citric acid/lactic acid and other permitted preservatives are added to enhance the sour taste of pickles. The product is packed in a sterilized bottle, and after thorough mixing of all ingredients, bottle is topped with a lot of edible oil to plug any air gaps among the cloves.

4.6.3   Onion

4.6.3.1 Dehydrated Onion

Desirable quality traits needed in white onions for dehydration or manufacture of onion powder are (1) white-coloured flesh; (2) full globe to tall globe shape of bulbs with 5–6 cm diameter; (3) high TSS content, above 15 %, preferably 20 %; (4) high degree of pungency; (5) high yield good; and (6) high quality. Dehydrated onion is produced by removing water from the raw onions to a maximum level of 4.25 % and then milling it to a specific particle size. Prior to drying, onion is cleaned and peeled and top removed; the peeled onion is washed and sliced. Onions are dehydrated without blanching or sulphating in order to protect enzyme system which develops onion flavour when onion cells are cut or broken. Onions are dried on a stainless steel belt, which passes then three to four stages, following a carefully controlled time/temperature programmed by circulating hot air. The conditions can be controlled to deliver onion with a maximum of 5 % moisture, without any heat damage to the product.

4.6.3.2 Onion Powder

Onion powder is prepared by grinding dehydrated onion slices in a hammer mill to a suitable mesh. It is highly hygroscopic, and hence the important precaution regarding its storage is to keep airtight containers in a cool, dark places to avoid absorbing moisture; becoming granular, caky and pasty; and ultimately getting mould attack. International quality standard for dehydrated onion and onion powder has been described in Table 4.8.

Table 4.8 International quality standard for dehydrated onion and onion powder

Physicochemical characteristics	Dried slices and flakes	Onion powder
Moisture content, per cent, maximum	8.0	6.0
Total ash (per cent) (dry basis), maximum	5.5	5.5
Acid insoluble (per cent) (dry basis), maximum	0.5	0.5
Extraneous matter (rots, papery skin) (per cent) (dry basis), maximum	0.5	0.5

Source: ISO 5559 – 1983: International standard for dried onion/onion powder

4.6.3.3 Onion Salt

It is prepared by mixing 19–20 % of onion powder with 78 % free-flowing, pulverized and refined iodized table salt and anti-caking agent like anhydrous sodium sulphate which prevents absorption and caking of onion powder during storage.

4.6.4   Small Cardamom

4.6.4.1 Essential Oil

Cardamom oil is produced commercially by steam distillation of crushed fruits. The oil content varied from 3.1 to 10 %, and the main components of the essential oil are 1,8-cineole (25–45 %), α-terpinyl acetate (28–34 %), linalyl acetate (1–8 %), sabinene (2 %), limonene (2–12 %) and linalool (1–4 %). In general, spices from the recent harvest, which have not suffered excessive volatile oil loss, should be employed for in order to obtain good yield. The best oil yield is obtained with Alleppey Green cultivar. The distillation time required for the extraction of essential oil is 4 h.

4.6.4.2 Oleoresin

On solvent extraction, small cardamom produces 8–9 % oleoresin. It is dark green in colour and pungent in nature and replacement strength is 1:20.

Table 4.9 Comparison report of different ginger drying methods

Sample no.	Method	Particulars	Temperature (°C)	Time taken (hour)	Recovery per cent	Volatile oil per cent	Moisture
1	Scrapped w/o cutting	Sun drying	34	104	28.8	1.24	12.3
2	W/o scrapping and cutting	Sun drying	34	Could not be assessed			73
3	W/o scrapping and ½ sliced	Sun drying	34	50	24.8	2.00	11.2
4	W/o scrapping and ½ sliced	Vacuum drying	60	48	22.4	1.08	11.0
5	W/o scrapping and cutting	Vacuum drying	60	48	23.4	1.7	13.1
6	Scrapped w/o cutting	Vacuum drying	60	48	22.4	1.20	11.2
7	W/o scrapping and cutting	Hot air oven	60	48	24.8	1.97	13.0

Source: Mani et al. (2000)
W/o Without

4.6.4.3 Decorticated Seeds and Seed Powder

Decorticated cardamom seed generally commands a disproportionately lower price than whole cardamoms due to the fairly rapid loss of volatile oil during storage and transportation. Seed powder is also sold to some extent.

4.6.5 Ginger

4.6.5.1 Ginger Oil

On steam distillation dried, cracked ginger yields 0.5–2.0 % of pale-yellow, viscid volatile oil. Oil is generally extracted from unscraped dried ginger or from ginger scrapping. It is economical and convenient to recover oil and oleoresin from dried ginger than fresh ginger. Ginger oil is greenish to yellowish in colour and mobile with the characteristics warm and aromatic odour. The main component of essential oil is α-zingiberene. The oil is sparingly soluble in 95 % alcohol, but it is soluble in 90 % alcohol. The oil and oleoresin content varies from cultivar to cultivar (Nybe et al. 1980) which has been presented in Table 4.9.

4.6.5.2 Ginger Oleoresin

Ginger oleoresin is obtained by extraction of powder ginger with suitable solvent like alcohol, acetone or any other solvent like ethylene dichloride. It contains gingerol, zingerone, shogaol, volatile oil, resin, phenols, etc. The amount of oleoresin extracted by alcohol was much higher than that extracted by acetone or ethylene dichloride (Natarajan et al. 1972). The oleoresin content of different ginger cultivars varies from 4 to 10 %, and it is viscous dark brown in colour. The amount of essential oil in oleoresin varies from 7 to 28 %; non-pungent substances may amount to 30 %. The amount of essential oil is an important factor in the evaluation of oleoresins, and the Essential Oil Association of America has specified a content of 18–35 ml volatile oil per 100 g of oleoresin.

4.6.5.3 Dehydrated Ginger

The ginger is sun dried or dehydrated mechanically. Sun-dried ginger is available in two forms: unscraped or scraped. Green ginger is placed in a wire gauge cage and dipped in a boiling solution of 20 %, 25 % or 50 % sodium hydroxide solution for 5, 1, 0.5 min. The cage is pulled out of the lye bath, and rhizomes are placed in 4 % citric acid for 2 h, and then they are washed and dried. The loss in weight due to lye peeling is about 12.5 %; mechanical peeling of 60 s gives a product of equal in essential oil content to hand-peeled ginger.

However, hand peeling was found to be superior to mechanical peeling in giving a product which is uniform in appearance, size and colour. Peeling facilities, subsequent drying of ginger and drying in a through-flow drier at a temperature not exceeding 60 °C give a satisfactory product. Sreekumar et al. (1980) recorded the dry recovery of ginger ranged from 17.7 % in China to 28.0 % in Tura. They reported the cultivars Maran, Jugigan, Ernad, Manjeri, Nadia, Himachal Pradesh, Tura and Arippa having dry recovery of

Table 4.10 Quality parameters of dry ginger

Parameters	Quantity in per cent
Ash	6.80
Acid-insoluble ash	0.93
Alcohol-soluble extract	7.50
Water-soluble extract	11.25
Extraneous matter	1.20
Volatile oil	1.28
Moisture	13.40
UV fluorescence	Yellow

Source: Mani et al. (2000)

above 22 % were quite suited for the conversion of dry ginger. The correlation between dry ginger percentage and crude fibre was negatively significant. Ratnambal et al. (1987) studied about the correlation matrix of different quality matrices of ginger and comparison and quality of dry ginger report of different ginger drying methods have been presented in Tables 4.9 and 4.10, respectively.

4.6.5.4 Bleached Ginger

For the preparation of bleached ginger, proper peeling of the fresh rhizome is done. The peeled rhizome is washed and then repeatedly immersed in milk of lime and allowed to dry in the sun dry until the ginger receives a uniform coating of lime (calcium hydroxide) and assumes a bright colour; the final drying takes about 10 days. Finally, the product is well rubbed with gunny cloth to remove bits of skin and to provide a smooth finish.

4.6.5.5 Ginger in Brine

Fresh ginger aged for five to 6 months should be properly harvested and washed in fresh water. After peeling, it is preserved in 16 % brine and 1 % citric acid solution. Sometimes, 0.5 % sulphur dioxide may be added with this solution. Ginger in brine is exported in the foreign market. It is mainly used for the preparation of salted ginger, pickles and for some special Japanese dishes.

4.6.5.6 Salted Ginger

Fresh ginger with low fibre, harvested at immature stage, is cleaned and soaked in 30 % salt solution containing 1 % citric acid. After about 14 days, remove the rhizomes from salt solution, clean and preserve in cold condition. This can be stored in 1–2 % brine containing citric acid.

4.6.5.7 Ginger Powder

For the preparation of ginger, proper peeling of the fresh rhizome is done. The peeled rhizome is washed and then immersed in boiling water for 10 min. Then dry it properly and thereafter powder is made with special type of machine. The grain size of the ginger powder should be 50–60 mesh.

4.6.5.8 Ginger Paste

It is a novel product which can meet regular requirement of fresh ginger where it is not readily available. It is also a convenient preparation compared to fresh ginger which requires peeling and cutting or crushing before domestic use. This product therefore may compete with fresh ginger as it can be prepared in a peak harvesting season when the price of the fresh ginger is very low (George 1996).

4.6.5.9 Other Value-Added Products of Ginger

1. Ginger preserve
2. Ginger candy
3. Soft drinks like ginger cocktail
4. Ginger pickles, salted in vinegar or in vinegar mixed with other materials like lime, green chillies, etc.
5. Alcoholic beverage like ginger wine, ginger brandy and ginger beer

4.6.5.9.1 Ginger Candy

After proper washing of the fresh ginger, peeling is done. Then pieces are made in appropriate sizes. This is prepared by crystallizing soaked fresh ginger pieces in sugar solution (60:40). A layer of sugar is spread over the ginger and then dried under the sun for 6–8 h. After proper packaging, it is sent to the market. Ginger candy is mainly used for chocolate preparation.

4.6.5.9.2 Ginger Beer

It is prepared by fermenting ginger extract. For this, six part dry ginger, three part hops and other spices are mixed and boiled for 20–30 min.

After filtering this syrup, it is mixed with sugar and cooled up to −100 °C. Then fermentation is done with yeast and citric acid if required.

4.6.5.9.3 Ginger Juice

Flow sheet for the extraction of ginger juice:

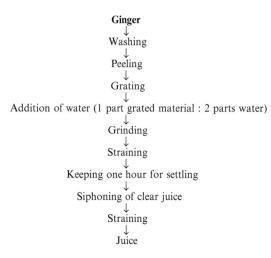

```
                    Ginger
                      ↓
                   Washing
                      ↓
                   Peeling
                      ↓
                   Grating
                      ↓
Addition of water (1 part grated material : 2 parts water)
                      ↓
                   Grinding
                      ↓
                   Straining
                      ↓
       Keeping one hour for settling
                      ↓
         Siphoning of clear juice
                      ↓
                   Straining
                      ↓
                    Juice
```

4.6.6　Cumin

4.6.6.1 Cumin Oil

Ground cumin on steam distillation yields 2.5–4.5 % of colourless or pale yellow essential oil. The chief constituents of the oil is cuminaldehyde (20–40 %), which is chiefly used in perfumery. The yield of oil depends upon the quality and age of the seed, variety and methods of extraction. The highest oil recovery was recorded when 35–48 mesh and 48–65 mesh for ground cumin were used to extract the essential oil (Sagani et al. 2005).

4.6.6.2 Cumin Powder

Cumin powder is an important item of value-added product gaining commercial importance in world trade.

4.6.7　Turmeric

4.6.7.1 Cured Turmeric

It is an important item of the value-added product of turmeric. This product is popular in some parts of the world.

4.6.7.2 Turmeric Powder

Turmeric powder of commerce in the form of hard fingers is powdered in two stages to get a fine powder of about 60 mesh. The colouring matter is stable to heat but sensitive to sunlight and hence does not pose much problem during grinding even though heat is generated. In view of this, 'cryogenic grinding' may not be necessary for powdering turmeric.

4.6.7.3 Volatile Oil

The volatile oil derived from steam distillation of crushed turmeric is an orange-yellow liquid occasionally slightly fluorescent with an odour of turmeric. The dried rhizomes yield 5–6 % of oil and fresh ones give 0.24 % essential oil. About 58 % of the oil is composed of turmerones (sesquiterpene ketones) and 9 % tertiary alcohol.

4.6.7.4 Oleoresin

The CFTRI, Mysore, has standardized the techniques for the manufacture of oleoresin from ground turmeric by solvent extraction followed by vacuum concentration. On an average turmeric cultivar contains 15 % oleoresin. The semi-liquid viscous stuff contains both volatile aromatic principles and nonvolatile acrid fraction covering the overall aroma and flavour in a concentrated form. Acetone, alcohol and ethylene dichloride were found suitable for extraction of oleoresin from turmeric. Oleoresin is highly viscous orange-brown product containing 30–35 % curcumin and 15–20 % volatile oil and has characteristic turmeric aroma.

4.6.8　Chilli

Chillies are valued for its quality, which includes oleoresin, capsaicin and capsanthin.

4.6.8.1 Oleoresin

In the food and beverage industry, chilli has acquired a great importance in the form of oleoresin, which permits better distribution of colour and flavour in food as compared to chilli powder. The food industry generally prefers to use large, highly colour and less pungent chillies for the preparation of oleoresin. Export share of different oleoresins has been presented in Table 4.11. Chilli oleoresin, which is prepared from dried chilli powder by solvent extraction, represents the complete flavour. Oleoresin content varied from variety to variety (Table 4.12). According to Singh et al. (2001), acetone is the best solvent for the extraction of chilli oleoresin (Table 4.13) which gave the maximum chilli oleoresin. The capsaicin and colouring matter was also more in acetone-extracted chilli oleoresin. The optimum time for the extraction of chilli oleoresin is 5 h at 50 °C. The main product, high pungent oleoresin can be used for pharmaceutical purposes utilizing medical properties of capsaicin. Chilli oleoresin is red in colour and replacement strength is $1 \cong 10$.

4.6.8.2 Capsanthin

There is a great demand for natural chilli fruit colour, which is used in processing food in place of synthetic colours. Basically, the colouring matter of chillies is a mixture of carotenoids, yellow and red pigments, which encompass carotenes and xanthophylls. Capsanthin and capsorubin are the red pigments, and yellow include beta-carotene, cryptoxanthin and zeaxanthin. Considerable variability exists among the genotypes with respect to colour value.

4.6.8.3 Capsaicin

The hot flavour of chilli is caused by capsaicin and allied constituents, which have good counter-irritant function. It is used in pharmaceuticals and cosmetics due to this property and perhaps helps

Table 4.11 Export share of different oleoresins

Spice oleoresin	Export amount (Rs. in crore)	Per cent share of export
Paprika oleoresin	132.90	42.13
Capsicum oleoresin	37.41	11.86
Chilli oleoresin	1.78	0.56
Pepper oleoresin	70.30	22.28
Turmeric oleoresin	27.51	8.72
Ginger oleoresin	13.63	4.32
Nutmeg oleoresin	8.60	2.73
Cassia oleoresin	4.76	1.51
Celery oleoresin	3.40	1.08
Cardamom oleoresin	1.95	0.62
Cumin oleoresin	1.47	0.47
Other spice oleoresins	1.15	0.36

Source: Deshpande (2005)

Table 4.12 Oleoresin content (in per cent) of different genotypes of chilli

Genotypes	Oleoresin	Genotypes	Oleoresin	Genotypes	Oleoresin	Genotypes	Oleoresin
Surya	16.6	CH-1	16.2	ELS-11-2	16.5	MY-12(3)	16.5
S-2529	17.2	S-2528	12.5	S233-2	11.1	I-16	13.0
Lorai	13.5	S20-1	11.2	CH-3	17.0	ELS-82	12.5
ELS-13	15.2	H-44	15.6	S-2539	15.3	LT-4	12.7
Tiwari	14.3	Kiran	16.8	Perennial	9.9	Pepsi	14.8
LLS	16.2	Pusa Sadabar	12.0	S-2545	14.0	Ooty round	13.9
S-2530	14.3	Pusa Jawala	16.1	MF-41-2	16.1	CD ($P=0.05$)	1.02
BG-1	16.2	Punjab Guchhedar	14.3	RHCH UP	11.5		
Punjab Surkh	16.7	Punjab Lal	15.3	LS-111	16.4		
MS-12	17.0	S-2319	16.1	CA-87-1	15.8		

Source: Singh and Hundal (2004)

Table 4.13 Effects of different solvents and extraction times on oleoresin yield (per cent) and also the effect of different solvents on capsaicin and colouring matter of chilli oleoresin

Solvent	Oleoresin yield (per cent)	Extraction time (hours)	Oleoresin yield (per cent)	Capsaicin (per cent)	Colouring matter (ASTA units)
Acetone	16.25	4	14.56	2.15	421
Ethyl acetate	14.21	5	16.02	1.75	364
Ethyl alcohol	16.18	6	16.06	2.06	310
C.D. at 5 %	0.56	–	0.56	0.12	11.8

Source: Singh et al. (2001)

Table 4.14 Performance of *Capsicum chinense* for quality characters under rainy and summer season

Genotypes	Capsaicin (per cent)		Oleoresin (per cent)		Carotenoids (per cent)	
	Rainy	Summer	Rainy	Summer	Rainy	Summer
CC 23	1.32	1.1.58	8.47	10.07	0.22	0.25
CC 13	2.73	3.03	11.53	14.97	0.36	0.35
CC 7	2.52	2.72	13.37	17.57	0.41	0.36
CC 2	2.68	2.42	19.43	18.43	0.32	0.37
CC 15	1.61	1.98	9.07	11.53	0.32	0.35
CC 30	2.40	2.32	9.43	9.03	0.22	0.27
CC 28	2.57	2.56	17.93	21.53	0.33	0.34
CC 31	2.24	2.54	11.57	12.93	0.18	0.19
CC 3	3.18	3.58	8.53	12.03	0.29	0.30
CC 11	2.86	3.01	7.53	13.97	0.26	0.38
CD ($P=0.05$)	0.07	0.06	0.49	0.42	0.009	0.007

Source: Robi and Sreelathakumary (2006)

in absorption and movement of bowels. Capsaicin has many medicinal properties, specially as an anticancerous agent and instant pain reliever and used in preparation of pain balm, vapour rubs. Capsaicin and its allied principle have both antimicrobial and antioxidant property. Pushpakumari and Pramod (2009) reported that capsaicin in mild concentration (0.025–0.075 %) is used as good cream for temporary relief of minor aches and pains of muscles and joints associated with arthritis, sprains and strains. Clinical studies revealed that natural capsaicin directly inhibits the growth of leukaemic cells. Considerable variability in capsaicin content of sun-dried fruits of chilli varieties was reported by several workers and has found to range from 0.2723 to 1.1267 mg per 100 g of dry chilli powder (Table 4.14). Among the different chillies, *bhut jolokia* is the hottest chilli of the world, and it was officially declared on 9 September, 2006, by Gunnies World Records Ltd. that *bhut jolokia* is the hottest of the all spices.

4.6.9 Tamarind

4.6.9.1 Tamarind Pulp Powder
Half-matured tamarind is suitable for the preparation of tamarind powder. It is generally prepared by drying the fruit pulp in the dryer especially in a cabinet dryer at 50–60 °C till it reaches a moisture content of 5 %. After drying, powder is prepared from the pulp. The colour of tamarind is due to the presence of anthocyanin. It is highly soluble in water and used to impart the raspberry red colour in the different food products. Tamarind powder is mainly used as red colourant in rasam, sambar, jam, jelly, etc.

4.6.9.2 Tamarind Concentrate
For the preparation of tamarind concentrate, the pulp is heated with boiling water, and soluble solids are extracted to the extent of 20 % using countercurrent principle. It is then sieved and concentrated under vacuum in a forced circulation evaporator and directly filled in cans or bottles. The method is developed by CFTRI, Mysore. Tamarind concentrate sets like jam after cooling. The soluble solid content is about 68 %. It has a higher self life than tamarind pulp. Freshly harvested fruit pulp (not older than 5–6 months) gives good quality products. The composition of the concentrate is tartaric acid (13 %), invert sugar (50 %), pectin (2 %), protein (3 %), cellulosic material (2 %) and moisture (30 %) (Chempakam and Peter 2000).

4.6.9.3 Tamarind Kernel Powder
It is prepared by decorticating seed and pulverizing the kernels. Decortication is either done by soaking the seed in water or by roasting the seed. To avoid enzymatic reaction for better storage life,

sodium metabisulphite is added as preservative. The nutritive value of tamarind kernel protein is more or less equivalent to that of cereal proteins. Its total carbohydrate and mineral contents are favourably well compared with oats and gram. It is much cheaper than other starches and can be used as thickener in foods for human consumption. Tamarind kernel powder (TKP) is mostly useful in textile and paper industries as it contains starch. Jellose is the major polysaccharide present in the kernel powder which can be effectively used as a remedy against diarrhoea, dysentery and colitis. Bharadwaj et al. (2007) reported that genotype AKT-14 gave maximum TKP; however, TKP prepared from AKT-10 was found better with high content of starch. Among the different treatments studied for TKP preparation, significantly maximum TKP recovery was obtained when the seed was roasted in the microwave oven at 180 °C for 3 min followed by sand roasting treatment. However, sand roasting treatment was found very easy and convenient for tamarind kernel powder preparation.

4.6.10 Kokum

4.6.10.1 Kokum Syrup

Fruits, after washing with clean water, are cut in two halves and all seeds and pulp are removed. This is followed by pouring of sugar into the opened fruit halves forming one layer. In this way, alternate layers of kokum rind halves and sugar are put in cleaned food grade plastic drums for 7 days. When the sugar gets dissolved and the kokum rind gets extracted by osmosis, the syrup is strained through 1 mm sieve or muslin.

4.6.10.2 Kokum RTS

Kokum juice is evaluated for TSS and acidity. Then the required quantity of citric acid and sugar is added to raise the brix and acidity levels to 200 °Bx and 0.3 % acidity, respectively. The normal parameters of kokum RTS are juice, 20 %; acidity, 0.3 %; water, quite sufficient; and sodium benzoate (preservative), 140 mg/kg of the final product. During preparation, the mixture is

boiled till all ingredients including sugar, citric acid and water are dissolved. At the end, preservative is added; the product is filtered, bottled and sealed followed by pasteurization and cooling of filled bottles.

4.6.10.3 Kokum Squash

It is prepared by maintaining juice, 25 %; TSS, 45 %; acidity, 1.2 %; brix, 450; and sufficient water and sugar where all the ingredients are dissolved through boiling. Sodium benzoate at 610 mg/kg of the final product is added as preservative at the end before filling of bottles. Then they are sealed, pasteurized and cooled.

4.6.10.4 Kokum Rind Powder

Dried kokum fruits without any seed and extraneous mater are cut into bits, loaded in percolator and extracted with hot water to obtain kokum concentrate. The concentrate is mixed with a suitable carrier and dried under controlled conditions to obtain kokum powder. It may also be obtained through spray-drying technique.

4.6.10.5 Kokum Honey and Kokum Wine

Kokum honey may be produced by establishing apiculture unit within the plantation of kokum trees. It has, of course, not been successfully implemented and under trial. However, kokum wine is being produced by fermenting kokum juice (containing 4 % sugar, approx.) largely at Goa.

4.6.11 Coriander

4.6.11.1 Coriander Essential Oil

Coriander essential oil is prepared by steam distillation method. The oil content varied from 0.1 to 1.7 %. Generally, large fruited types produced lower oil yield as compared to small fruited types. The main composition of the oil is d-linalool (45–70 %), α-pinene (6.5 %), β-terpene (10 %), camphor (5 %), limonene (1.7 %) and geranyl acetate (2.6 %). Coriander essential oil is a colourless or pale-yellow liquid. The aroma is pleasant, sweet and somewhat woody and spicy.

4.6.11.2 Coriander Oleoresin

Coriander oleoresin is prepared with the help of solvent extraction method. It is brownish yellow in colour and the replacement strength of the oleoresins is 1:33. In commercial oleoresin, essential oil content varies from 5 to 40 %.

4.6.11.3 Coriander Dal

Coriander dal (dhania dal) is mainly used as an adjunct in supari or pan masala. Coriander dal is obtained from the seeds of coriander. The coriander seeds are dehusked, flaked and given a mild heat treatment. Then, it is salted. The treated seeds are highly flavoured and consumed as a digestive chew. The traditional method of preparing dhania dal from the whole seeds of coriander is quite laborious and time consuming, besides being less efficient with high amount of brokens. In this method, the coriander dal is soaked overnight in salt water and partially dehydrated and its husk is removed manually. CFTRI has developed a process to get superior quality coriander dal with less brokens. In this process, indigenous milling equipments are used along with improved conditioning technique.

4.6.11.4 Coriander Fatty Oil

Coriander seed contains 19–21 % fatty oil, and it is brownish green in colour with a similar odour to that of coriander essential oil. Fatty oil has soft consistency having leathering properties. It is mainly used as a flavouring agent for spiritual liquors, chocolate and cocoa industries.

4.6.11.5 Coriander Powder

Coriander powder is prepared by grinding matured dried coriander. Ground spices are highly hygroscopic in nature. Various changes like loss of volatile oil, caking of the products, microbial spoilage and infestation of mould and insect occur during storage and transportation due to absorption of moisture from atmosphere. Proper packaging is essential for the protection of the above-mentioned hazard. It is generally packed in polythene pouches for better packaging; it should be packed in aluminium foil laminated LDPE and polycell-laminated pouch.

4.6.12 Fenugreek

4.6.12.1 Fixed Oil

Fenugreek seed contains about 7 % fixed oil. It has marked drying properties. The dried oil is golden yellow in colour and is insoluble in ether. It is recently used in perfumery industries.

4.6.12.2 Volatile Oil

Seed contains less than 0.02 % essential oil.

4.6.13 Fennel

4.6.13.1 Essential Oil

On distillation of crushed fennel seed, essential oil (1–6 %) is obtained. It is a colourless or pale-yellow liquid with characteristic odour and taste. It is mostly used in different food and flavouring industries and also used in pharmaceutical industries. The main component of the essential oil is anethole (70 %) and fenchone (6 %).

4.6.14 Ajowan

4.6.14.1 Essential Oil

Steam distillation of crushed ajowan seed produces 2–4 % essential oil. It is mainly used for preparations of different medicines. The main component of the essential oil is thymol (53–80 %).

4.6.14.2 Thymol

It is prepared by treating ajowan essential oil with aqueous alkaline solution and regenerating thymol from it by extracting with ether or steam distillation. It is also mainly used in pharmaceutical industries.

4.6.15 Celery

4.6.15.1 Volatile Oil

The seed contains 2–3 % volatile oil. It is yellowish to dark greenish in colour, and the main components are d-limonene (19.5–68 %), B-selinene (8 %)

and n-butyl lidene phthalide (8 %). It is mostly used as fixative and as an ingredient of novel perfumes, pharmaceutical preparations and medicines.

4.6.15.2 Seed Oleoresin

It is prepared by extraction of crushed dried celery seeds with suitable solvents like food grade hexane or ethylene dichloride. It is used as food flavourant and also used in pharmaceutical and perfumery industries.

4.6.15.3 Celery Fatty Oil

The fruit also yields 7 % of fatty oil. This oil is used as antispasmodic and nerve stimulant.

4.6.15.4 Celery Salt

It contains about 25 % of celery seed powder or ground celery and 75 % of common salt.

4.6.16 Some Other Spice Value-Added Products

In spite of the above spice products, some value-added spice products are also prepared from the different spices. These are curry powder, organic spice and consumer-packed spice, encapsulated spice essence, etc.

4.6.16.1 Curry Powder

Because of varied physical forms, aroma and flavour, an individual spice does not produce the preferred flavour. Indian curry powder is the answer to it, i.e. Indian curry power can be able to provide preferred and desirable flavour aroma and taste. Generally, curry powder is made from a blend of several spices; these numbers vary from a minimum 5 to 20 spices depending upon the end use. Spices are ground mixed in definite proportions to produce curry powder. Curry powder generally contains about 85 % spices, 10 % farina (i.e. flour materials) and 5 % salt. It is believed that farina acts as a binder and salt acts as natural preservative. Some of the important ingredients in the curry powder are coriander, cumin, turmeric, chillies, black pepper, cumin, clove, nutmeg, cardamom, ginger, turmeric, cassia, etc.

Table 4.15 Range of different spices in typical formulation of curry powder

Ingredients	Typical range (%)
Ground coriander	10–50
Ground cumin	5–20
Ground turmeric	10–35
Ground fenugreek	5–20
Ground ginger	5–20
Ground celery	0–15
Ground black pepper	0–10
Ground red pepper	0–10
Ground cinnamon	0–5
Ground nutmeg	0–5
Ground cloves	0–5
Ground caraway	0–5
Ground fennel	0–5
Ground cardamom	0–5

Source: Tainter and Grenis (1993)

A wide range of products are covered under the categories 'curry' like mixed powder, curry powder, curry mixture, vegetable masala, chicken masala, tandoori masala, fish masala, etc. Typical formulation of curry powder has been presented in Table 4.15. Apart from their traditional use in oriental cooking, curry powder is finding place in industrial food such as instant snacks and curry powder-flavoured sauces and soups.

4.6.16.2 Consumer-Packed Spice

The packaging has gained considerable importance as it increases the shelf life of spices. The popular packaging materials are glass bottles, rigid plastic containers and flexible aluminium foil pouches. Exporting consumer-packaged spices can earn higher unit value for the same quantity. The use of flexible attractive consumer packs for ground spices and spice products has gained popularity in the internal market of India, and this will help establish the brand image in the international market. Consumer package must assure the unadulterated and good quality spices and spice products. The value addition ranges from 40 to 380 % as compared to straight spices depending upon the spices and spice products, packaging media, weight of the material, market, etc. Generally, the weight of the institutional package ranges from 0.3 to 1.0 Kg, whereas retail package varies 30–200 g. It is important to note that with the stiff competition that India is facing in spice market, building up brand image is essential in case of packed spices. Quality upgradation as per ISO-9002 standard is necessary because recently India is facing stiff competition in the international market from Guatemala, Malaysia, Indonesia and Brazil.

4.6.16.3 Encapsulated Spices

Encapsulated spices are prepared with essential oil or standardized oleoresins with gum Arabic or modified starch as encapsulent. Such products have long shelf life and the flavour is only released when they are mixed with water. Sharma and Arya (1995) found that gum acacia is most suitable for micro-encapsulation of cardamom oil and cumin oleoresin. Eighty to 90 % organic volatiles were found to retain in this process. These have five to ten times more strength than the ground spices. The encapsulated spice extractives are used for high temperature application such as baked or retorted products. These spice flavours are slowly released into the products at the appropriate processing temperature.

4.6.16.4 Organic Spices

The spices from organic farming can also be considered as value-added products. They fetch premium price in the international market. The price may be higher by 20–50 % and in some cases even 100 % than the spices grown from conventional farming (Arya 2000). India has the tremendous scope for the production of organic spices. Organic spice production requires a lot of precaution, viz. selection of land, fertilizer management, disease and pest management, storage, packaging, etc. Registration and certification is an important task in organic agriculture. In the 4th World Spice Congress held at Madras, emphasis was given for the identification of compact organic block and registration of organic spice growers. At present, Spice Board provides the guidelines for the production and export of organic spices. During the recent times, organic spices share 1–1.5 % of the global spice markets.

Organic spice trade increases day by day all over the world due to increasing health conscious among the people. The major importers of the organic spices are the United States, Europe and Japan. Organic certification is a growing world demand for organic food, and it is an important task for export of organic spices. Certification is an accreditation process for the producer of organic food and other agricultural products. The requirement of quality standard, production inputs, storage, processing, packaging, etc., for organic spices varies from country to country (Sajindranath 2005). Products may also display logo of the certification body that approved them. Some of the organic spices certifying agencies in India are Spice Board, APEDA, etc. Certification is done on the basis of different standard fixed national and international bodies.

CAC Codex Alimentarius Commission
IFOAM International Federations of Organic Agriculture Movements
USDA United States Department of Agriculture (US)
NOP National Organic Programme (US)
QAI Quality Assurance International
UKROFS UK Register of Organic Food Standard
JAS Japanese Agricultural Standard
NPOP National Programme for Organic Production (India)
NAPP National Accreditation Policy and Programme (India)

4.6.16.5 Spice Essence

Spice oil and oleoresins are highly concentrated in nature. For this reason, they are not directly used in the food products. Spice essence is prepared by mixing oil and oleoresin with acceptable food grade solvent, which make the produce an acceptable diluted form. These solvents are harmless in food stuffs. The common solvents used for this purpose are glycerol, propylene, and polypropylene alcohol. Sometimes in commercial spice essence, synthetic flavour components are also mixed. Some common traded spice essence are small cardamom, vanilla, clove, nut meg, all spice, mint, coriander, fennel, rosemary, etc.

Table 4.16 Some essential oil with their characteristics

Sl. no.	Name	Oil per cent	Colour	Constituents
1	*Acorus calamus*	1.5–3.0 %	Yellow colour	Asarone (82 %)
2	*Allium cepa*	6.01–0.015 %	Dark brown	d-n-Propyl disulphide, methyl n-propyl disulphide, vinyl disulphide
3	*Alpinia galanga*	0.5–1.5 %	Pale yellow to olive yellow	Methyl cinnamate (50 %), 1,8-cineol (20–25 %), α-pinene, d-pinene, sinigrin
4	*Amomum subulatum*	2.5 %	–	α-Terpinyl acetate
5	*Apium graveolens*	2–3 %	Yellow to greenish brown	d-Limonene (19.5–68 %), B-selinene (8 %), n-butyl lidene phthalide (8 %)
6	*Armoracia rusticana*	0.2–1.0 %	–	Sinigrin derived allyl isothiocyanate diallyl sulphide, phenyl propyl thiocyanate
7	*Artemisia dracunculus*	0.5–0.7 %	Pale yellow to amber	Methyl chavicol (45–75 %), anethole (10 %) α- and β-pinenes, phellandrene, camphene
8	*Carum carvi*	4.5–7.5 %	–	50–85 % d-carvone, α- and B-pinene and sabinene
9	Cassia leaf oil	1.0–2 %	Yellowish	Eugenol (70–80 %), cinnamic aldehyde
10	*Cinnamomum tamala*	2 %	–	Eugenol (80–85 %) and cinnamic aldehyde
11	*Cinnamomum zeylinicum*	0.5–3.0 %	Fresh light yellow, stored red	Cinnamaldehyde (65–75 %), eugenol, benzyl aldehyde
12	*Crocus sativus*	0.6–1.4	–	Safranal-aldehyde, pinenes and cineol
13	Curcuma longa	5–6 %	Orange yellow	Turmerones (58 %) and 9 % tertiary alcohols
14	*Eugenia caryophyllus*	15–17 %	Yellowish green	Eugenol (81–83 %), eugenol acetate (7 %), B-caryophyllene (9 %), humulene
15	*Illicium verum*	3–3.5 %	Colourless to pale yellow	Anethole (85–90 %), α-pinene, phellandrene, p-cymene, limonene
16	*Laurus nobilis*	1–3 % (bark)	–	1, 8 cineol (35 %), methyl eugenol (4 %), α-pinene (12 %) B-pinene (11 %), α-terpin (6 %), limonene (4 %), α-terpinyl acetate (6 %) sabinene (5 %) and eugenol (2 %)
17	*Mentha arvensis* (Japanese mint)	1.5–4.0 %	Golden yellow	Menthol (70–90 %), α-pinene, α-1 limonene, caryophyllene
18	*Mentha citrata*	0.2–0.5 %	Colourless yellow or yellowish green	Carvone (50–70 %), dihydrocarvone, α-limonene, phellandrene, menthol, 1,8-dineol
19	*Mentha piperita*	0.5–4.0 %	Pale yellow	Menthol (30–55 %), menthone (15–35 %) methyl acetate (4–7 %), menthofuran, isomethane, limonene
20	*Mentha spicata*	0.2–0.55 %	–	Linalool acetate (45–50 %) and linalool (7 %)
21	*Murraya koenigii*	2–2.5 %	Dark coloured	1-Sabinene (35 %), 1-α-dinene (27 %), diterpene (16 %), 1-terpenel, (7.7 %), 1-caryophyllene (6.7 %), cadinene (5.2 %)
22	*Myristica fragrans*	6–16 % 4–15 %	Pale-yellow colour	Sabine (15–50 %, α-pinene (10–22 %) and B-pinene (7–18 %), limonene (3–4 %), 1,8-cineol (1.5–3.5 %), myristicin (1–13.5 %)

(continued)

Table 4.16 (continued)

Sl. no.	Name	Oil per cent	Colour	Constituents
23	*Nigella sativa*	0.5–2.0 %	Yellowish brown	Carvone (50–60 %), d-limonene, cymene
24	*Ocimum bassilicum*	0.4–1.0 %	–	Linalool (40–65.3 %) and methyl chavicol (25 %)
25	*Petroselinum crispum*	2.0–5.0 %	–	Apiol (36 %), myristicin (13 %), α-pinene
26	*Pimpinella anisum*	1.5–6 %	Colourless to pale yellowish	Trans anethole (80–90 %), methyl chavicol (10–15 %) anethole (2 %) and anisaldehyde (12 %)
27	*Rosmarinus officinalis*	1.5–2.5 %	Pale yellow to almost colourless	α-Pinene (25 %), camphoo (15–25 %), boneol (15–20 %), 1,8-cineol, bornyl acetate
28	*Salvia officinalis*	1.5–2.5 %	Pale yellow to greenish yellow	Thujone (25–45 %), linalyl acetate (10 %) α-pinene (2 %), 1,8-cineol, β-caryophyllene, limonene
29	*Thymus vulgaris*	2–2.5 %	Colourless to yellowish reddish	Thymol (45 %)

Source: Datta (2007)

Table 4.17 Colour and replacement of strength of some oleoresins

Sl. no.	Name	Colour	Replacement of strength
1	*Cuminum cyminum*	Brownish to yellow green	1 ≅ 20
2	*Anethum sowa*	Greenish to pale amber in colour	1 ≅ 20
3	*Cinnamomum zeylinicum*	Dark brown liquid	–
4	*Eugenia caryophyllus*	Brownish green to greenish brown	1 ≅ 20
5	*Allium sativum*	Brownish yellow	1 ≅ 50
6	*Majorana hortensis*	Dark green	1 ≅ 40–50
7	*Apium graveolens*	Greenish	1 ≅ 21
8	*Petroselinum crispum*	Deep green	1 ≅ 33
9	*Rosmarinus officinalis*	Greenish brown	1 ≅ 20

Source: Datta (2007)

4.7   Future Strategies

Indian spices enjoyed a glorious history from very ancient times. India also exports a lion's share portion of value-added spice products to the world market like Indian spices. In the last four decades, India developed and exported large value-added products. Among these different spices, essential oils and oleoresins are the major ones. These are mainly produced from, namely, black pepper, ginger, chill and paprika, turmeric, cardamom, cinnamon, cassia, nutmeg, coriander, fennel, cumin, celery, etc. In the last decade, India is exporting different dehydrated spices, frozen spices, freeze-dried spices, spice essence, emulsions, liquid extract, paste, etc. In the future, attention should be given to the development of new products like instant spice, different spice blends, dry soluble spice, canned spices, bottled spice, spice decoction, etc. With the production point of view, the value-added product should be produced in a hygienic manner using modern techniques (supercritical fluid extraction for the production of essential oil) and maintain the quality rigid standard as prescribed by the different importing countries and agencies. Organic spices should be popularized among the farmers from the export point of view. Big entrepreneurship should be developed for the large-scale and year-round production of the value-added product to meet the international demand. New product should be developed from different minor and underutilized spice crops, especially from herbal spices.

Integrated approach for development, production and marketing strategies should give the new direction towards the value addition in spice crops.

References

Ahmed J, Pawan P, Shivhare US, Kumar S (2003) Effect of processing temperature and storage on colour of garlic paste. J Food Sci Technol 39(3):266–267

Anonymous (2006) Spices statistics. Spice India 19(8):32

Anonymous (2007) Spices exports peak. Spice India 20(6):11

Anonymous (2014) Review of export performance of spices during 2012–2013. www.indianspices.com/pdf/Major item wise export-2013.pdf

Anonymous (2000) Data bank. Indian Food Ind 19(1):66

Anonymous (2001) Report of the NATO Task Force on spices, plantation crops including cashew. ICAR, New Delhi

Arya PS (2000) Spice crops of India. Kalyani Publishers, Ludhiana

Arya PS (2001) Ginger production technology. Kalyani Publishers, Ludhiana

Bharadwaj S, Mahorkar VK, Panchbhai DM, Jogdande ND (2007) Preparation of tamarind kernel powder for value addition. Agric Sci Dig 27(3):194–197

Chempekam B, Peter KV (2000) Processing and chemistry of tamarind (Tamarindus indicus Linn). Spice India 13(4):9–12

Datta S (2007) Value addition of spices with special reference to quality. In: Sharanghi AB, Acharya SK (eds) Quality management in horticulture. Agrotech Publishing Academy (ATPA), Udaipur, pp 1–360. ISBN 81-8321-090-2

Deshpande AA (2005) Paprika improvement for exports and value addition. In: Souvenir of national symposium on current trends in onion, garlic, chillies and seed spices-production, marketing and utilization, November 2005. NRC for Onion and Garlic, Pune, pp 25–27

George CK (1996) Ginger: quality improvement at farm level. In: Sivadasan CR, Kurupu PM (eds) Quality improvement of ginger. Spices Board (Government of India), Kochi, pp 7–15

Joshi D (1962) White pepper. Indian Patent 70:349

Khurdiya DS (2003) Promoting export of value added products of horticultural produce. Indian Hortic 48(1):32–35

Lakshmanachar MS (1993) Spices oil and oleoresin. Spice India 6(8):8–10

Mani B, Paikada J, Varma P (2000) Different drying methods of ginger. Spice India 13(6):13–15

Mathew AG (1994) Blackening of pepper. Int Pepper News Bull 18(1):9–12

Narayana DBA, Brindavanam NM, Dodriyal RM, Katiyar KC (1999) Value addition in spices: an overview. Indian Spices 36(2):13–16

Narayanan CS, Sreekumar MM, Sankarikutty B (2000) Industrial processing. In: Ravindran PN (ed) Black pepper (Piper nigrum). Harwood Academic Publishers, Amsterdam, pp 367–379

Natarajan CP, PabmaBai R, Krishnamurthy MN, Raghawan B, Shankaracharya NB, Kuppuswamy S, Govinbarajan VS, Lewis YS (1972) Chemical composition of ginger varieties and dehydration studies on ginger. J Food Sci Technol 9:120–124

Nybe EV, Sivaraman Nair PC, Mohanakumaran N (1980) Assessment of yield and quality components in ginger. National seminar on ginger and turmeric. Calicut, pp 24–29

Pruthi JS (1980) Spices and condiments: chemistry, microbiology and technology. Academic Press Inc, New York, p 450

Pruthi JS (1997) Diversification in pepper utilization. Int Pepper News Bull 15(8):5–9

Pushpakumari KN, Pramod S (2009) Health benefits of spices. Spice India 22(4):16–22

Ratnambal MJ, Gopalan A, Nair MK (1987) Quality evaluation of ginger in relation to maturity. J Plant Crops 15(2):108–117

Robi R, Sreelathakumary I (2006) Seasonal influence on oleoresin, capsaicin, carotenoids and ascorbic acid contents in hot chilli. Indian J Hort 63(4):458–459

Sagani VP, Patel NC, Golakia BA (2005) Studies on extraction of essential oil of cumin. J Food Sci Technol 42(1):92–95

Sajindranath AK (2005) Accreditation and certification of organic products. Spice India 18(1):44–50

Sankar V, Mahajan V, Lawande KE (2005) Prospect of garlic processing industry. Spice India 18(10):23–26

Sharma GK, Arya SS (1995) Microencapsulation of spice oleoresin and essential oil. Beverage Food World 22(1):29–31

Siddaramappa S (2013) Development schemes of spices board. In: Proceedings of the National Symposium on Spices and Aromatic crops (SYMSAC VII), Madikeri, Karnataka, 27–29 November 2013, pp 162–172

Singh R, Hundal JS (2004) Evaluation of chilli genotypes for oleoresin. Spice India 17(4):31–32

Singh R, Gill BS, Hundal JS (2001) Studies on extraction of chilli oleoresin. Spice India 14(5):13–15

Sreekumar V, Indrasenan G, Mammen MK (1980) Studies on quantitative and qualitative attributes of ginger cultivars. In: Proceedings of the National Seminar on ginger and turmeric, Calicut, 8–9 April 1980, pp 47–49

Subbulakshmi G, Naik M (2002) Nutritive value and technology of spices: current status and future perspectives. J Food Sci Technol 39(4):319–344

Sudarshan MR (2000) White pepper: a simple value addition. Spice India 13(6):5–7

Tainter DR, Grenis AT (1993) Spices and seasonings. NCH Publication, New York

Thankamony V, Menon NA, Amna OA, Sreedaharan VP, Narayanan CS (1999) Bacterial removal of skin from white pepper. Spice India 12(9):10–11

Thomas PP, Gopalkrishnan N (1992) A process for the unwrinkled dry ball shaped green pepper. Indian Spices 29(4):9–10

Varghese (1991) Add a touch of green. Pepper News 15(4):9–11

Vasantakumr K (2006) Processing and product development of spices subsidiary and minor products of black pepper. Spice India 19(11):6–10

Zachariah TZ (2005) Value added products from black pepper. Spice India 14(5):16–21

M.S. Mebakerlin and S. Chakravorty

Abstract

Floriculture is presently considered as the most lucrative agro-enterprise in terms of profit making. As the flowers are the utmost perishable horticultural farm produce, there remains some hindrance in proper marketing following standard postharvest management practices by the common farmers. Hence, value addition by the agro-industries is another important arena for proper utilisation of fresh ornamentals in either garden-fresh or processed form. Different kinds of value-added products are nowadays formulated and marketed by the companies which include essential oils, flavours, fragrance, pharmaceutical and nutraceutical compounds, insecticidal and nematicidal compounds, pigments and natural dye, vanilla-based products, gulkand, rose water, etc. Besides floristry items, flower arrangements and floral ornaments are important value addition with the fresh flowers and other ornamentals. Another important arena is the production of dry/dehydrated flowers in which India has achieved a considerable success regarding in-house production and worldwide marketing. The different methods for the production of value-added products from flowers and other ornamentals are discussed in this chapter.

5.1 Introduction

One of the most rapidly growing sectors of this era is the floriculture industry that has significantly increased the economy of the country through the profits gained by marketing and export of not only the cut flowers and ornamentals but also the value-added products that can also be derived from these plants. Value-added floriculture is a process of increasing the economic value and appeal of any floricultural commodity through changes in genetics, processing or diversification (Verma et al. 2012). By application of such innovative techniques, the grower is able to receive an additional income by converting an indistinctive raw material into a unique

M.S. Mebakerlin • S. Chakravorty (✉)
Palli Siksha Bhavana, Visva-Bharati, Sriniketan,
Birbhum, India
e-mail: snehasish.chakravorty@gmail.com

A.B. Sharangi and S. Datta (eds.), *Value Addition of Horticultural Crops: Recent Trends and Future Directions*, DOI 10.1007/978-81-322-2262-0_5, © Springer India 2015

and alluring product. However, before the producer can make such an enterprise a successful one, he/she has to be able to exert a substantial amount of time, labour and skill, more than that which is usually required for other farming operations.

5.2   Types of Value-Added Products

Value addition has become a magnanimous incentive towards the expansion of floricultural trade by the art of preservation of ornamental plants and the creation of novel products that appeal to the tastes and preferences of the customers. The procreative skills such as flower arrangements, artificial colouring of flowers, aqua packing for better presentation, three-dimensional window packing of flowers for greater visibility, garlands, venis, bouquets, greeting cards using petal-embedded craft papers, dry flowers, potpourris, etc. are some of the value-added products that have gained an escalating acceptance in the global market. Other value-added products obtained from flower crops are essential oils, flavours, fragrance, pharmaceutical and nutraceutical compounds, insecticidal and nematicidal compounds, pigments and natural dye, gulkand, rose water, vanilla products, etc.

5.2.1   Floral Ornaments

The innovative nature of mankind with the innate desire for beautification and decoration has led to the use of flowers and flowering parts for indoor decoration, hair decoration, making garlands, bouquets and for worship.

5.2.1.1 Garlands
Floral garlands are possibly one of the most ancient methods of using flowers for decoration with special preference for sweet-scented flowers. The highly demanded flowers for this purpose are jasmine and tuberose and other flowers include marigold, chrysanthemum, crape jasmine (*Tabernaemontana divaricata*) and rose. Red

hibiscus is particularly used for making garlands that are offered to the Goddess Kali who is worshipped among the Hindus of West Bengal region. Garlands can be made from a single type of flower or different flowers in combination. By using needle and thread of cotton, nylon or silk material, the flowers are held together, and in the case of heavy garlands, fine wire strings are used.

5.2.1.2 Floral Bangles and Floral Crowns
These are popular among Indian dancers who also wear them with garlands. The scented flowers used for making such bangles and crowns are jasmine and tuberose, while marigold and *Tabernaemontana* which are non-scented are also used. Earrings and *Bajubandhs* which are also used in ceremonial functions are also made from flowers. For hair decoration, flowers such as *Crossandra* and *Barleria* are widely used in southern part of India for decorating the hair as 'gajra', while other flowers like tuberose, jasmine or *Michelia champaca*, singly or in combination, can also be used. 'Veni', a style of hair decoration by attaching the flowers along the plait of the hair during marriage ceremonies or while performing cultural dances.

5.2.1.3 Rangoli
The creation of beautiful patterns and diverse forms on the floor by using colourful flowers of varying kinds is another contriving innovation among the Hindus in India. This custom of 'rangoli' also known as 'kolam' has a religious overtone and is done during puja in front of temples or in social functions such as marriages and festivals such as Diwali, Onam and Pongal. Petals are mostly used for the purpose of decoration but other plant parts like attractive leaves, creepers and flowers as a whole can also be used.

5.2.1.4 Buttonholes
Buttonholes are also fascinating items for floral decoration by which males wear them in their coat collars at special occasions like weddings and other grand ceremonies. Almost any flower can be used but usually roses and orchids are preferred.

5.2.1.5 Flower Bouquets

One of the many ways of honouring a person in celebration of a special occasion is the presentation of flowers that have been beautifully arranged along with foliage plants tied together to form a flower bouquet. Bouquets are of many types and have different shapes from flat to round, and when arranged in a basket, it is considered to be of higher taste and beauty.

Flat bouquets are made using a hard poster paper in white or any other pleasing colour, and the stem length of each flower is 45–75 cm. Over the poster paper, an aluminium foil or silver paper can be laid to make the bouquets more attractive. The flowers are then laid flat over the paper and held in position using cellotape. Ornamental foliages like thuja, ferns, baby eucalyptus, ivy and asparagus can be used as filler material. Different colours of satin ribbons can be used to tie and decorate the bouquet. For preparing round-shaped bouquets, the flowers are arranged in a whorl to form a cone shape with the stem end becoming tapering, while the flower end comes in a round whorl. Along with the flowers, ornamental foliages may be added to increase its decorative appeal. The base of the bouquet is firmly tied with gunny twine, which is camouflaged with a silk ribbon.

5.2.2 Flower Arrangement

Flowers are a symbol of beauty, love, peace, affection and tranquillity. They are eminently present in ceremonies and gatherings as a requisite for decoration and thus provide an aesthetic appeal to the surrounding. By using various flowers with their different colours to suit the occasion, flowers can be arranged in attractive forms to add variety and excitement to the surrounding. Flower arrangements are also commonly seen as part of home decoration to give out a friendly and cordial atmosphere. A good flower arrangement should have an attractive colour combination and contrast, size, quality, firmness, harmony and distinction.

There are two main styles of flower arrangements:

1. European style: in this style, importance is given on mass arrangement in the form of art only.
2. Japanese style: in this style, every branch, flower and line is well described based on principles taught by philosophers and has deep spiritual significance (Bhattacharjee and De 2005).

The Japanese style of flower arrangement follows three basic lines of traditional ikebana, namely, 'heaven' called shin, 'man' called soe and 'earth' called hikae. Shin is the tallest followed by soe and hikae. A basic style is the moribana, meaning 'piled flowers', in which arrangements are made in shallow containers. This arrangement looks very natural and is often referred to as the natural style of ikebana. Another style called nageire, meaning 'thrown in', needs tall upright vases for making the arrangements. Here also, flowers are arranged as naturally as possible.

There is another method called jiyu-bana (free flowers) which can be arranged in both moribana and nageire styles. While in the latter two cases, generally only those materials that are available in the season are used. In jiyu-bana, one can use combinations of wood, metal or any other material. In zenei-ka (abstract style), the arrangement does not simulate nature. Plant materials are used in combination with any other material that a person can conceive. However, the arrangement should be pleasing and in agreement with the surrounding. There is still yet another style, zenei-bana (avant-garde), which is a piece of art where a beautiful sculpture is created using wood, stone, rock or metals and may depict any natural scenery. Lastly, there is the combination of fruit and flower arrangement which the Japanese call morimono (Randhawa and Mukhopadhyay 2012).

Besides Japanese styles, there can be an all-foliage arrangement, using all foliage types in combination. Presently, interior decorators and flower arrangers have developed finer skills in the art of flower arrangement.

5.2.2.1 Centre Table Arrangement

This kind of flower arrangement requires one to make sure that the floral design holds equal beauty on all sides from any point where it is viewed. It should also be low enough to provide a clear view to any person who is either sitting or standing. Such arrangements are placed on the centre of tables found in the living room, dining room and passages for the incoming and outgoing people. While designing flower arrangements for centre table, the designer first makes a mental division of two or four sections so the floral material can be divided accordingly. In this way, a uniform design is created that can be observed from all sides. The long-stemmed flower buds are placed at the lower level, while the open blooms are at the top. The same principle is also followed for arranging flowers on the office desk or on the reception table.

5.2.2.2 Side Table Arrangement

Side table arrangements are created with the objective of viewing only from one side of the table, and therefore, the beauty of the floral design is focused on the frontal side of the arrangement. However, before the preparation of the design, other factors should be considered. The size of the decoration should be in proportion with the display table. In order to serve a particular theme, the colour scheme should be determined with caution, also the necessary materials should be properly selected along with the flowers and foliage to express the theme accurately. Such kind of flower arrangement can follow the Western, Indian or ikebana style of flower arrangement.

Flowers with long stems such as tuberose, gladiolus, larkspur and goldenrod are suitable. Leaves of gladiolus, ferns and palms can also be used. The arrangement of flowers and foliage from the boundary towards the centre is usually followed leaving the attractive blooms for the focal point (Bose et al. 1999).

5.2.3 Dried Flowers

With the reality that all living beings perish, including plants and ornamentals that wither away and lose their natural beauty, it has set in the mind of a few to develop a method to retain the appearance of such species and thus the art of flower drying followed. Anything from botani-

cals such as stems, twigs, branches, bark, leaves/foliage, flowers, thorns/spines, fruits, cones, seeds, roots and minor forest products like lichens, fleshy fungi, mosses, selaginella, ferns, etc. in a dried form comes under the domain of dry flowers (Verma et al. 2012). The varieties of flowering plants that can be employed for drying treatment include globe amaranth, celosia, marigold, etc. Dried flower industry has been identified as a potential area for export, and it constitutes 15 % of global floral business (De 2011). It has been found that the Netherlands ranks first in export of dried flowers followed by the USA, Mexico, India, Columbia and Israel. The export of dried flowers and plants from India is about Rs. 100 crore per year. The USA is the largest consumer of dried and artificial flowers estimated at US$2.4 million annually, followed by Germany and the UK (Bhattacharjee and De 2005).

Somani (2010) mentioned some of the ornamental flowers that are dried for their colour quality:
- *Red*: cockscomb, peony, pomegranate, sumac, zinnia
- *Pink*: gladiolus, larkspur, peony, snapdragon, statice
- *Yellow*: acacia, goldenrod, marigold, strawflower, yarrow, zinnia
- *Blue*: cornflower, delphinium, globe amaranth, thistle, hydrangea, larkspur
- *Green*: foliage, grasses, seedpods
- *Orange*: marigold, strawflower, zinnia
- *Violet*: gladiolus, heather, lilac, statice, stock

5.2.3.1 Time of Harvesting for Drying

The time of harvest varies from flower to flower, and generally, the flowers that are processed for drying are cut when they have reached maturity but before the colour is deteriorated. An experiment was conducted at TNAU, Coimbatore, and it was found that half bloom (22 days from bud appearance) and full bloom (39 days from bud appearance) are ideal for drying, bleaching and dyeing in *Gomphrena*. Half bloom (18 and 23 days from bud appearance) and full bloom (21 and 29 days from bud appearance) are ideal for drying in French marigold and zinnia, respectively (Lourdsamy 1998).

Roses are preferred at half-bud stage for drying; foliages, when the colour is fresh and clear, are harvested at young age; grasses are cut during their maturity stage and seedpods and cones have to be collected at full maturity before they burst. Vegetables and fruits to be dried should be firm and free from blemishes and infections (Somani 2010).

5.2.3.2 Uses

Dehydrated flowers and plant parts have been used for designing distinctive and artistic greeting cards, landscapes and interior decorative items with dry flowers sealed in glass containers. It can also be used in the preparation of potpourri, flower baskets, twig baskets, front-facing arrangements, mirror frames and table centres. Dry flowers and floral crafts have an everlasting value that can be cherished for long periods if they are protected from moisture and dust. With the fact that they can be maintained and cherished for years without being affected by the harsh winter and the hot summer, the beauty and worth of dried flowers are highly valued.

5.2.3.3 Dried Floral Arrangements

Floral arrangements, using dried flowers and leaves, are known as dry decoration, and they provide more flexibility than the fresh flowers. They are economical as they can be used over and over again, and there is no problem of wilting. The length of the stalk varies between 15 and 40 cm, and it can also be extended by using wire (preferably canes) stalks that are attached with hot glue guns. Dried flower arrangements are classified as 'main blooms', 'liners' and 'exotics':

Main blooms: these constitute the main bulk and play a key role in flower arrangements because of their shape, size and/or colours. Most common species used as main blooms include statice, strawflower, nigella, larkspur and roses.

Liners: the ornamental grasses that are mainly used to provide a linear accent to a flower arrangement are called as liners. Such grasses include *Avena* (animated oats), *Halaris* (canary grass), *Phleum* (timothy) and *Triticale* (ornamental wheat). Maize, sorghum, spiked millet, dried branches and twigs of trees and shrubs can also be used in addition to the grasses.

Exotics: from the name itself, we can understand the type of plants that fall under this category. These plants include lotus heads, palm spears and okra pods.

Fillers: these constitute the group of products that are used for adding mass to flower arrangements (Somani 2010).

5.2.3.4 Process of Dehydration of Flowers and Plant Parts

Dehydration is the removal of moisture from any material under artificially produced heat and controlled temperature, humidity and air flow. In the case of living materials, the drying process checks any chemical changes and prevents microbial growth (Bhattacharjee and De 2010). For the preservation of ornamentals and foliage plants, the following methods are mentioned:

5.2.3.4.1 Air Drying

It has often been referred to as 'hang and dry' method, and it is considered to be one of the simplest, easiest and cheapest methods of drying. Flowers like *Helichrysum* (strawflower), *Acrolinum* (paper daisy) and *Limonium* (statice), having a crisp texture, are hung in an inverted position or kept in a container by which they can easily be dried, taking into account that the room should be properly ventilated to avoid rotting of the flowers before they are dried. Other than garden flowers and wild plants, seedpods as well as grasses can also be dried in the same manner.

5.2.3.4.2 Sun Drying

To facilitate quick dehydration, the plant material is embedded in a drying medium in any container after which it is kept under the sun. Sun drying is a common practice for flowers like cornflower, eucalyptus, poppy pods, lotus pods, palm leaves, etc. Also, flowers like marigold, pansy, zinnia and pompon chrysanthemum are dried by embedding them in sand in an upside down manner before they are exposed to the sun. Solar dryers have also been utilised to fulfil the same purpose for most flowers and herbs.

5.2.3.4.3 Press Drying

In this method of drying, the plant parts and foliage are placed in folds or unglazed sheets of newspaper or blotting paper sacrificing the original shape of the plant material. Unless the water vapour completely escapes from the plants, the risk of microbial attack can hardly be prevented resulting in failure of drying and loss of material. Verma et al. (2012) observed that flowers like roses, carnation and helichrysum are press dried at 40–45 °C in a an electric hot air oven for 120, 133 and 72 h, respectively, whereas 24 h is optimum for leaves of thuja, *Adiantum* and *Nephrolepis* and flowers of hibiscus, marigold and *Calliandra*. In water drying, ornamentals like hydrangeas, yarrow, bells-of-Ireland and celosia have been reported to dry well by placing the stems initially in a few centimetres of water in a warm, dry and dark location, and the water is allowed to evaporate.

5.2.3.4.4 Embedded Drying

By this method, the original shape, size and colour of the ornamentals are maintained, and the problem of petal shrinking is avoided. The materials used for embedding and drying flowers and foliage should be very fine (0.02–0.2 mm). The flowers and foliage are carefully embedded either in sand or silica gel or borax or in metallic or plastic or earthen containers at room temperature in a well-ventilated room. Solely dependent on weather conditions, drying may prove to be successful, but it is a time-consuming method. Misra et al. (2009) conducted an experiment by which zinnia had been dried by embedding method using electrical oven and microwave oven. It was observed that the size and colour of the flowers remained unaltered both in sand and silica embedding after dehydration with significant loss in flower weight due to loss of water.

5.2.3.4.5 Hot Air Oven Drying

Flowers are placed in an electrically operated hot air oven at a controlled temperature (40–50 °C) and specified duration, depending on the plant size, structure and moisture content of the material, in an embedded condition. This is, undoubtedly, the fastest method of dehydration; however, the disadvantage is that the colour of the plant material is lost.

5.2.3.4.6 Microwave Oven Drying

This is a rapid method of drying plants like snapdragon, China aster, chrysanthemum, etc. with less heat being generated. It is based on the principle of liberating moisture by agitating molecules in the organic substances with the help of electronically produced microwaves (Verma et al. 2012).

5.2.3.4.7 Glycerine Drying

Fresh and fairly matured foliage is treated with hygroscopic chemical with the objective to retain the suppleness of the plant material. The glycerinising chemical is considered to substitute moisture present in the plants responsible for maintaining the foliage form, texture and sometimes the colour.

5.2.3.4.8 Skeletonising

This refers to the removal of all plant tissues with only the veins of leaves intact on the plant. It was found that heavy textured leaves respond better to this kind of technique of preservation.

5.2.3.4.9 Freeze Drying

By the application of heat from solid state (ice) by sublimation, moisture is removed from the plants, which requires high vacuum and low temperature. With a minimum temperature of 12 h, the flowers are dried in a special freeze-drying machine at a temperature starting from −10 °C.

5.2.3.4.10 Predrying Treatment

The retention of texture, shape and overall acceptability of dried flowers depends on the speed of dehydration. A common floral preservative that can be used to improve flower shape and colour is citric acid. It acts as an acidifying agent, thereby lowers the pH of the solution and also prevents the blockages of xylem vessels and helps to improve flower size, shape and colour.

5.2.3.4.11 Care After Drying and Dehydration

Dried flowers and foliage require proper care and maintenance as they are very delicate, brittle and fragile. After press drying, the dried flowers are stored either in a dry location or in desiccators till they can be used. In the case of embedded drying, the containers are tilted to remove the desiccants, and the dried flowers are carefully picked and dusted with a fine hair brush to remove the remaining desiccants. To prevent microbial attack on the dried products, a small quantity of silica gel should be placed inside the container to absorb moisture. Also, storage containers should be dust-free. Mothballs should be kept in storage containers to protect them from insect attack.

5.2.3.5 Colouring of Flowers
5.2.3.5.1 Tinting

In cases where the flower pigment is absent, light or dull in colour, an eminent technique has been developed by artificial colouring of flowers using edible dyes and stains such as food colours, Feulgen stain, bromocresol blue, bromocresol green, eosin yellow, ammonium purpurate and phenol red. Red-, blue-, green- and yellow-coloured flowers can be obtained from white flowers of tuberose, spider lily, candytuft and white ixora. Other flowers with the potential for tinting are white gladiolus, different orchid species, loose flowers of jasmine, crossandra, chandni, etc. (De 2011).

5.2.3.5.2 Bleaching

According to Somani (2010), bleaching and dyeing are the two important processes related to commercial dry flower production. It involves chemical processes that can change the ability of colour bodies to absorb light by changing their degree of saturation (Verma et al. 2012). In bleaching, both oxidative and reductive bleaches are available, of which the level of efficiency depends on the measure of accessibility of bleach to lignin. Some of the commonly used oxidative bleaches are sodium chloride, hypochlorite and peroxidise, and reductive bleaches are sodium sulphide, hydrosulphite and dioxide. Sodium chlorite (10 %) is an effective bleaching agent for Gomphrena followed by hydrogen peroxide (30 %) (Lourdusamy 1998).

5.2.3.5.3 Sulphuring

In sulphuring, sulphur dioxide acts to bleach coloured plant material and, below a certain concentration, to fix red colour in some flowers. Colour fixation is associated with acidification of the tissues.

5.2.3.5.4 Dyeing

For increasing the aesthetic value, dyes can be applied to the fresh, dried or bleached ornamental plant parts, as per the changing seasons and fashions. The most widely applied technique of dyeing is the immersion of the plant material into the suitable dye. Immersion dyeing is improved by removing the waxy cuticle with NaOH from the dried plant material and by adding surfactants to improve the contact between the dye bath solution and the plant material. In addition to dyeing (staining), the plant material can also be painted and commonly silver or gold paint is used (Bhattacharjee 2006).

5.2.3.6 Potpourri

Potpourri is a mixture of dehydrated flowers, berries and leaves, seeds, stems and roots (De 2011). Dried flowers, herbs, grasses and to a lesser extent seedheads are used by florists to design the semi-permanent, maintenance-free beautiful decorative arrangements and the potpourri mixes (Bose et al. 1999). The basis of a potpourri is the aromatic oils located within the plants (De 2011). Thus, the design and making of potpourri is based on the principle that these aromatic oils are slowly released into the atmosphere which gives a pleasing aroma to the surrounding. The two types of potpourri that are available are wet and moist, both of which require a fixative to absorb the aromatic oils, after which they are gradually discharged. The materials that are selected for the preparation of this special kind of dry flower arrangement should be resistant to mould, nontoxic, free from noxious odours and sufficiently robust to withstand mechanical bending.

Some of the common fixatives that are used in the preparation of potpourri are finely ground non-iodised salt, orris root, sweet flag, gum, benzoin, storax and ambergris. In our country, the major flowers like marigold, rose and bougainvillea are used with globe amaranth and cockscomb. These flowers and flower parts are contained in a beautifully designed glass bowl with different colours and shapes or they are packed in a colourful satin or muslin sachet. Potpourris have been prepared not only for the purpose of creating sweet fragrance but also to repel moths and protect woollen clothes in storage.

5.2.4 Perfume and the Components Constituting the Perfumery Industry

Perfume is a mixture of fragrant essential oils, aroma compounds, fixatives and solvents used to give the human body, animals, objects and living species a pleasant scent (Verma et al. 2012). It is a man-made product that contains odoriferous compounds that give out a pleasing aroma.

5.2.4.1 Essential Oils

An essential oil is a concentrated, hydrophobic liquid containing volatile aromatic compounds from plants and its parts (Verma et al. 2012). These are chemical compounds of an odoriferous nature, which are highly volatile, insoluble in water but soluble in organic solvents. They contain mixtures of organic compounds belonging to different classes of compounds such as terpenes, phenols, phenyl terpenoids, aliphatic compounds, etc. (De 2011). The chemical constituents by which essential oils are made of have been classified as non-oxygenated and oxygenated hydrocarbons. The non-oxygenated hydrocarbons are hydrocarbons of terpene class and are of little importance in perfumeries. Alcohols, aldehydes, ketones and ethers form the group of oxygenated hydrocarbons (Bhatacharjee et al. 2005).

Essential oils are derived from natural raw materials that are the various parts of a plant, e.g. leaves, flowers, bark, roots and seeds (Verma et al. 2012). Many essential oils constitute one, two or three major compounds that are responsible for the particular fragrance used in the perfumery industry such as citral in lemongrass and *Litsea cubeba* oil, citronella in Eucalyptus citriodora and Java citronella oils. The major flowers that are used for the extraction of essential oils in India are jasmine, rose, chamomile and clary sage. In jasmine, *J. sambac* and *J. grandiflorum* yield 0.15 % concrete and 55–62 % absolute. Among the rose species, *Rosa damascena, R. bourboniana, R. centifolia, R. moschata, R. gallica* and *R. alba* are used for the extraction of oils on a commercial scale. The best quality essential oil in the world is obtained from *R. trigintipetala* (0.03 %). Other important sources for perfumery are sandalwood, vetiver, palmarosa, cypress, etc. (De 2011).

For the extraction of essential oil, the flowers are harvested at the stage when they are fully open, in the early hours of the day. Delay in picking would result in a loss of concrete yield. Furthermore, to prevent any discolouration and yield loss, the flowers should be carefully handled after harvest.

The methods by which the essential oils are extracted from these are mentioned below:

5.2.4.1.1 Steam Distillation

In this method, pressurised steam is allowed to pass through the fresh flowers that are kept in a plant chamber, and the heat produced forces out the volatile oil from the plant cells which also get evaporated along with the steam. After condensation, the oil is separated while the distillate can be used as floral water.

5.2.4.1.2 Solvent Extraction

The extraction of essential oils with solvents such as petroleum ether or benzene is practised whenever the oil with its natural flavour is required. Here, the plant material is saturated with the solvent that releases the aromatic compounds. After the extraction process, the solvent is filtered and concentrated by heating at a constant temperature of 75 °C till it gets evaporated off leaving behind the perfumery compounds.

5.2.4.1.3 Enfleurage or Cold Fat Extraction

This is an, comparatively, older method where flowers like jasmine, tuberose, violets, etc., that usually produce essential oils even after they are picked, are placed over a blend of fats, which absorbs the oils and fragrances of the flowers. After a number of days, the flowers are replaced by fresh ones until the fat is saturated. The volatile oils are then separated from the fat by treatment with alcohol.

5.2.4.1.4 Maceration
In this method, the plant material is chopped and digested with hot oil at 45–80 °C for several hours, during which successive batches of the new flowers are added to the filtrate up to 20 times. Finally, the essential oil is collected separately by treatment with alcohol.

5.2.4.1.5 Expression
Here the fruits are compressed or squeezed in claw-shaped bowls where the juice is sucked out of the fruit through a cannula inserted in the pulp, while the oil released is rinsed with water and separated by centrifugation.

5.2.4.1.6 Adsorption
In this process, hot air or inert gas is passed over the aromatic plant material which is then led through the activated carbon from which the essential oils are recovered by solvents (Bhattacharjee and De 2005).

5.2.4.1.7 Uses of Oils and Their Share (Verma et al. 2012)

Uses of oils	% share
Flavours in food industry	55–60 %
Fragrance in perfumes or cosmetics	15–20 %
Pharmaceutical preparations as starting material	15–20 %
Natural products	5–8 %
Isolation of aroma	15–20 %

5.2.4.1.8 Applications

Essential oil	Application
Geranium, ylang-ylang	Perfumery compounds
Orange, peppermint	Flavouring industry, essences
Clove, aniseed	Antiseptic oils
Cinnamon, ginger, pepper	Snack food industry
Rose, *kewra*, sandalwood, peppermint	Chewing tobacco industry
Lavender, rose, jasmine	Aromatherapy and *agarbatti* industries

5.2.4.2 Flavours and Fragrances
In an effort to affect one's mood and promote health, certain fragrances are derived from the petals of selected flowers which produce oil-based compounds of alcohols and sugars surrounded with glucose which causes the formation of scentless glucosides. A distinct aroma is detected when the glucoside is hydrolysed by enzymes, and the scent increases with additional moisture. Chemical constituents responsible for fragrance are rhodinol in rose, geraniol in geranium, nerol in magnolia and eugenol in clove. Compounds that contribute to the essence are terpenes, esters, aldehydes, ketones, alcohols and phenols, each of which possesses antimicrobial effects and medicinal properties. Esters are fungicidal and sedative; aldehydes are lemon scented and are sedative; ketones help in congestion and respiratory complaints. Among alcohols, linalool is the most common, and these are antiseptic and antiviral in nature. Eugenol, carvacrol, anethol and estragole are the major phenolic compounds and are bactericidal in nature.

Fragrances are used in the manufacture of soaps, detergents, creams, lotions, hair oil and other cosmetics. Fragrances are also used in leather and rubber textiles, as well as in plastics and craft papers.

Flavour is actually a combination of odour and taste. Characteristic flavours can be obtained from different cultivars or species of a flower crop from which the essential oil can be extracted. For example, among the rose cultivars, apple flavour is obtained from *Rosa eglanteria* (fragrant sweet briar rose) and *Rosa wichuraiana* (memorial rose); balsam flavour from *Rosa rugosa rubra* (red Japanese rose), *Rosa gallica* (French rose) and *Rosa damascena* (damask rose) and honey and musk from *Rosa moschata* (musk rose), *R. multiflora* (baby rose or Japanese rose), *R. arvensis* (field rose) and *R. sempervirens* (evergreen rose). For flavouring biscuits, mint, rosemary, thyme and sage are used and rose petals, sage, rhubarb, mint, bay leaf and lavender for jams, jellies and puddings.

5.2.5 Phytochemicals

Xanthophylls, carotenes and flavonoids, namely, anthocyanin, flavonols and flavanones are secondary metabolites specially known as phytochemicals as they impart taste, aroma and colour to the food as well as having pharmaceutical and nutraceutical properties that can be used in medicine and agriculture. They are strong antioxidants and are mostly involved in photoreception and photoprotection.

Marigold is utilised for the extraction of important phytochemicals like lutein and zeaxanthins which are effective against cancer, heart disease, cataract and age-related nuclear degeneration. Lycorine and tazettine from dried bulbs of *Narcissus tazetta* (narcissus) have antiviral and antileukemic properties. Dahlia tubers are rich in insulin and fructose, phytin and benzoic acid and are effective against diabetes. *Calendula officinalis* contains celandine, flavonoids, polysaccharides and pectic substances and is used for the treatment of fever, ulcers, burns and wounds.

From orchids, loroglossin is the most common glucoside that has been isolated from different species. Dendrobine from *Dendrobium nobile*, malaxine from *Malaxis congesta* and phalaenopsine from *Phalaenopsis mannii* are other phytochemicals obtained from orchid species (De 2011).

5.2.6 Plant Pigments

Pigments responsible for colours of flowers are carotenoids, chlorophylls, anthocyanins and anthoxanthins. The carotenoids are a group of orange, yellow and orange-red fat-soluble pigments. These contain hydroxyl groups called xanthophylls and occur as esters of fatty acids. Chlorophylls are also fat-soluble pigments like carotenoids. They are of two types, chlorophyll a and chlorophyll b, and occur in plants in the ratio of 3:1.

Flavonoids are a group of compounds distributed in the plant kingdom which are water soluble and consist of anthocyanins that are red, blue and purple pigments. The anthoxanthins are orange to yellow in colour and are water soluble.

Some of the different pigments isolated from their respective ornamental plants are as follows:
- Pelargonidin: dahlia, geranium, verbena, tulip, petunia
- Cyanidin: dahlia, petunia, lily, chrysanthemum, dendrobium, rose, petunia
- Petunidin: petunia, parash (*Thespesia populnea*)
- Lutein: marigold, *Hemerocallis* (day lily)
- Anthocyanin: sweet pea, tulip, zinnia, euphorbia
- Delphinidin: tulip, lupine, petunia, rhododendron
- Quercetin: rose, waterlily, sweet pea, camellia

5.2.7 Natural Dyes

Dyes are colourants or colouring substances that are added to something to change its hue or colour. Chemically, dyes are benzophenones, glycosides, quinines and flavones. Apart from the application in the food industry, dyes are used as colouring agents in paints, ink, leather, wood, paper, fur, cosmetics, medicines and toothpastes. They can also be applied to fresh, dried or bleached ornamental plant parts in order to increase their aesthetic value as per the changing seasons and fashions.

Yellow-coloured dyes extracted from chrysanthemum are used in cosmetic and food industry. They are also derived from flowers of golden rod (*Solidago canadensis*) which have also been used for the preparation of cosmetics. The dye extracted from the leaves, bark and fruits of *Cassia auriculata* is used in the leather-tanning industry. An orange-red dye used in cosmetics and medicine has been derived from the arils of *Bixa orellana*. Marigold yields a yellow- and orange-coloured dye that can be used in food industry and poultry feed. Henna (*Lawsonia inermis*) leaves produce orange dyes that are used in cosmetic and leather-tanning industry.

5.2.7.1 Extraction of Orange-Red Dye from *Bixa orellana* (Annatto)

A fine suspension of colouring matter is obtained by soaking the seeds along with the adhering pulp in hot water for several hours. The seeds are removed and the liquid is allowed to ferment for a week until the dye is settled at the bottom. The sediment that has thickened is dried and cut into small cakes. For procuring the pure form, the red pulp is dissolved in an organic solvent like chloroform and evaporating the solvent to get a paste-like matter. This type of dye is used in soaps, pomades, fabrics, paints, varnishes and burns.

5.2.7.2 Extraction of Red Carthamine and Safflower Yellow from *Carthamus tinctorius* (Safflower)

The dried florets of safflower are repeatedly washed with acidulated water to get the water-soluble safflower yellow. Carthamine, the water-insoluble orange-red dye, is extracted by treating the leftover residual mass with sodium carbonate which is then precipitated out using dilute acids. The dye can be used as a substitute for saffron, and it produces different shades of colours, cherry red, rose pink, crimson or scarlet used in cotton and silks. It is also used for colouring cakes and biscuits.

5.2.7.3 Extraction of Blue Dye from *Indigofera* sp. (Indigo)

The dye, indigotin, is extracted by crushing and immersing the freshly cut plants in water for 10–15 h in 'indigo vats'. Woad leaves (*Isatis tinctoria*) are added to enhance the fermentation process for the conversion of soluble indigotin to indigo white. The liquid is then agitated continuously by passing through beating or oxidising vats equipped with paddle wheels, and the operation is stopped by heating as soon as the blue colour develops. The blue dye settles at the bottom and the liquid is drained off. The bluish mass is further granulated by boiling with water. The filtered sludge is pressed and cut into small tubes after which they are air dried and graded. Indigo

is utilised for dyeing and printing cotton, rayon and also for dyeing wool.

5.2.7.4 Extraction of Orange Dye 'Henna' from *Lawsonia inermis* (Henna)

The dried leaves are the primary source for the extraction of the dye which is obtained in the form of a powder, after which it is mixed with water and applied to the parts needed for dyeing. It is harmless and does not cause any irritation to the skin. It is also used for colouring leather, and the oil, obtained by steam distillation, has been used for making perfumes since ancient times (Bhattacharjee and De 2005).

5.3 Value Addition in Commercial Flowers

5.3.1 Rose

Rose is one of the most beautiful perennial flowering plants having different growth habits (shrub, climber or trailing plant) displaying a diversified range of colours and different shades of white, yellow, pink, orange and red. These are perennial in nature and are native to the Himalayan regions of Asia.

5.3.1.1 Value-Added Products

1. *Rose water*: it is prepared by boiling the flower petals with water and collecting the condensed steam which is used as sherbets or as cleansing lotions, eye drops, etc.
2. *Rose oil or attar of rose*: this constitutes the essential oil extracted from the rose along with the wax that has been collected from the petals. It is transparent pale yellow or yellow grey in colour and is generally obtained from *Rosa centifolia* and *Rosa damascena*. Rose oil has its use in the perfumery and cosmetic industry.
3. *Concrete*: essential oils that have been extracted by solvent method using petroleum spirit or hexane and evaporated at low

temperature and under reduced pressure yield a reddish brown waxy solid known as concrete. The range of concrete recovery is 0.18–0.30 % and is highly valued in the perfumery industry.

4. *Gulkand*: rose petals and sugar are mashed together in the ratio of 1:2 and dried in the sun. The gulkand that has been prepared is used as a laxative and also as a flavouring agent.

5. *Pankhuri*: rose petals that have been dried under a shade are used in preparation of cool summer drinks and incense.

6. *Rose hips*: the fruits of this plant are called as rose hips, and they are a rich source of vitamin C, pectin (11 %) and (3 %) a mixture of malic acid and citric acid. They contain laxative and antidiuretic properties and also help prevent cancer and cardiovascular diseases as they contain important phytochemicals like carotenoids, plant sterols and tocotrienols and a high level of anthocyanins, catechins and other polyphenols. Rose hips are used for preparing apple sauce, soups, puddings, jam, jellies, etc.

7. *Rose tea*: dried and crushed rose hips are used for making herbal tea that promotes health by preventing digestive tract infection and skin diseases.

8. *Medicinal roses*: *Rosa centifolia* and *Rosa damascena* have medicinal properties in which the roots are used for treating intestinal disorders, rickets, diarrhoea and haemorrhages. The leaves are used to treat wounds, ophthalmia, hepatopathy and haemorrhoids. The flowers are emollient, expectorant, cardiotonic, anti-inflammatory, digestive, carminative and antidiarrhoeal.

5.3.2 Chrysanthemum

Chrysanthemums are herbaceous annuals that occur as single or spray types with large flower heads having a wide range of colours from white, yellow to pink. They are commonly known as 'mums' and are native to Asia and North East Europe.

5.3.2.1 Value-Added Products

1. *Garlands*: varieties that are specially used for the purpose of making garlands are 'Indira', 'IIHR-Sel-5', 'IIHR-Sel-6', 'Jaya', 'Shanti', 'Red Gold' and 'Meera'.

2. *Potpourri*: dried flowers of yellow or white mums are used for providing a soft aroma and decorative purpose.

3. *Edible chrysanthemums*: sweet drinks can be prepared by boiling the yellow or white flowers of chrysanthemums and are common in Asia. Tea made from these flowers is effective against influenza. A rice wine that has been flavoured with chrysanthemum flavours is popular in Korea and is known as 'gukhwaju'.

4. *Chrysanthemum insecticides*: pyrethrin, an active component that can be extracted in the form of oleoresin by pulverising the flowers of *Chrysanthemum cinerariifolium* (pyrethrum), is contained in the seeds and applied as a suspension in water or oil or as a powder. These are used as safe insecticides as they are biodegradable and nonpersistent.

5. *Medicinal chrysanthemum*: chrysanthemum extracts are rich in alkaloids, volatile oils, sesquiterpene lactones, flavonoids, choline, chrysanthemin and vitamin B1 and have proven to contain antibacterial and antimycotic properties and have the potential to combat HIV. They are also used for treatment against eye infections, fever, headaches, bad breath, etc.

5.3.3 Carnation

It is an herbaceous perennial. The plant grows to a height of 80 cm producing glaucous greyish green to blue-green slender leaves, and the flowers occur singly or as a cluster in a cyme which are sweet scented and are red, white, yellow and green in colour.

5.3.3.1 Value-Added Products

1. *Carnation concrete and absolutes*: the concrete recovered from the petals of these flowers ranges from 0.2 to 0.3 %, and when treated

with alcohol, it yields carnation absolute, which are both used for perfumery.

2. *Dry flowers*: carnation flowers are quick to dry (3–5 h) and are commonly used for floral decoration and dried flower arrangements. They are also used as potpourri, scented sachets and cosmetic products.

3. *Medicinal carnations*: the carnation flowers are utilised in the medicinal field to treat kidney and bladder problems, skin infections and constipation. It has been reported as an alexiteric, antispasmodic, cardiotonic, diaphoretic and nervine. They have also been effective against poisoning, muscle spasms, heart diseases and nervous breakdown.

4. *Edible carnations:* the flower petals of carnation are clove scented and candied and are generally used for garnishing and flavouring in salads.

5.3.4 Anthurium

Anthuriums are herbaceous perennial plants and are characterised by the presence of colourful spathe and hundreds of small flowers that form the spadix. They are commonly known as flamingo flower and belong to Araceae family of the plant kingdom.

5.3.4.1 Value-Added Products

Generally grown as potted plants or cut flowers, they are found to make beautiful flower arrangements and bouquets and for other decorative purposes. The spathes are available in varying colours from green, white, cream, orange to pink:

1. Standard anthuriums: they are characterised by their heart-shaped spathes.
2. Obake anthuriums: they are bicoloured, generally green with a distinct spathe colour.
3. Tulip anthuriums: these are upright cup-shaped spathes with an erect spadix.

5.3.5 Gerbera

Commonly known as 'African daisy' or 'Transvaal daisy', these plants are herbaceous

perennial, grow to a height of 60–70 cm bearing a large capitulum with striking, two-lipped ray florets in yellow, orange, white, pink and red colour.

5.3.5.1 Value-Added Products

These flowers are excellent for flower arrangements and preparing attractive bouquets as they have a longer keeping quality as compared to other cut flowers. They are generally grown as border plants, in flower beds and containers or window boxes. They also possess medicinal value as they act as effective antispasmodic, anodyne and antitussive.

5.3.5.1.1 Gladiolus

These are perennial bulbous flowering plants which produce sword-shaped longitudinal grooved leaves enclosed in a sheath with flower spikes that are large and one-sided; hence, they are known as 'sword lily'. The flowers are bisexual, each subtended by green bracts and have a diversity of colours from pink to reddish or light purple with white markings or white to cream or orange to red.

5.3.5.2 Value-Added Products

1. *Bouquets*: the flower spikes used for this purpose should be harvested in the morning or at night when only two to three flowers have opened.
2. *Flower arrangement*: they make spectacular flower arrangements, but one should keep in mind to add floral preservatives to the vase water before arranging the flowers.
3. *Medicine*: it is commonly used in Africa for treating dysentery, constipation and diarrhoea. It is also used as an energy booster and prescribed for hypochondriacs.
4. *Edible gladiolus*: the corms of *G. edulis* and *G. dalenii* can be consumed as food by boiling them in water.
5. *Scented gladiolus*: *G. callianthus* having irregularly petalled white flowers, with a dark purple patch at the centre, is delicate and highly scented species of this genus.

5.3.6 Tuberose

The perennial bulbous flowering plants produce clusters of white waxy fragrant flowers that bloom at the top of the spike. The bright green leaves grow in a cluster at the base and also as clasping leaves along the stem. They are commercially cultivated as cut and loose flowers in the tropical and subtropical regions.

5.3.6.1 Value-Added Products
1. *Floral ornaments*: tuberose is commonly used for preparing garlands, venis, floral bangles, earring and floral crowns. They are also used for decorative purposes in wedding ceremonies and traditional rituals.
2. *Essential oils*: essential oil of tuberose, extracted from the petals through solvent extraction method, contains eugenol, benzyl alcohol, farnesol, butyric acid, methyl benzoate, nerol, geraniol and methyl anthranilate. Usually, 1,150 kg flowers yield 1 kg absolute and concrete recovery ranges from 0.08 to 0.11 %.
3. *Medicines*: the fragrance of tuberose is known to relax the mind and it enhances the blood circulation. The oil obtained from it is used to treat skin infections, nausea, vomiting, etc.
4. *Edible tuberose*: in Indonesia, the flowers are cooked and used in preparation of vegetable soups and sauce.

5.3.7 Jasmine

Jasmine is a woody shrub, climbing or trailing vine which can grow up to a height of 3–4.5 m producing highly fragrant white or yellow flowers.

5.3.7.1 Value-Added Products
Essential oils: the scented oil obtained by steam distillation from the flowers of *J. occidentale* and *J. grandiflorum* is non-toxic and nonirritant and is rich in linalool, eugenol, benzyl benzoate, benzyl acetate, methyl anthranilate, indole and others. These are valuable in the cosmetic and perfumery industry for making perfumes, incense, soaps, shampoos and creams.

Herbal medicines: the different plant parts such as flower, leaves, stems, seeds and roots are useful in the pharmaceutical industry. The flowers are effective against jaundice and other venereal diseases; the flower buds are treatment against ulcers, vesicles, boils, skin diseases and eye disorders. The leaves can be used to treat breast cancer and mouth ulceration. The oils are antidepressant, antispasmodic, antiseptic, sedative anti-daturine, etc.

Jasmine tea: the flowers of *J. sambac* are used for the preparation of tea in China and Japan.

Jasmine syrup: jasmine scones and marshmallow (*Althaea officinalis*) are popular among the French which are prepared from the extracts of jasmine flowers.

Hair decoration: the fragrant flowers are used in decorating the hair of women as floral crowns or venis.

5.3.8 Marigold

Marigold is an herbaceous annual or perennial and produces flowers having both disc and ray florets in *varying shades of yellow and orange*.

5.3.8.1 Value-Added Products
1. *Phytochemicals*: marigold flowers are a rich source of terpenoids, flavonoids, carotenoids and thiophenes that have found varied uses in preparation of medicines and insecticides.
2. *Natural dyes*: a yellow cloth dye known as 'egandai' or 'gendia' is extracted from the flowers of *Tagetes erecta* and produces bright colour with mordants.
3. *Industrial application*: the essential oils are widely used in the preparation of soaps, detergents, disinfectants, mosquito repellents, flavourings of food, etc.
4. *Edible products*: *T. lucida* are used in salads and as an aromatic herb which is added to soups, sauces, chicken dishes, etc.; the leaves are also used in the preparation of herbal tea.

5.4   Value Addition in Aquatic Plants

Aquatic plants are indispensible components of natural water bodies that absorb minerals and carbon dioxide in the water. They are classified as oxygenators, marginals, floaters and deep aquatics. Oxygenators release large quantities of oxygen in the water, e.g. buttercup and water violet, while marginals are those ornamentals that are planted in pots or baskets at the edge of the ponds, e.g. *Primula*. Floaters like water hyacinth help prevent growth of algae and provide enough cover for tadpoles and fish and deep aquatics like water hawthorn or waterlilies, aponogeton, *Nuphar*, etc.

There are also aquatic plants that act as vegetative filters that take up ammonia, nitrate, phosphate and even toxins through their roots and keep the water free from solids and chemicals. Such plants include water hyacinth, water lettuce, watercress, waterlily, cattail, sweet flag, etc.

Edible aquatic plants: the leaves of aquatic mint, bacopa, Vietnamese cilantro and water spinach, the tubers of arrowhead, duckweed, lotus and taro and all parts of perennial rice, water celery and watercress are eaten.

Medicinal aquatic plants: the leaves of pennywort (*Hydrocotyle* spp.) are used for treating arthritis, and marshmallow acts as a cough suppressant and wound healer.

Aquatics for decoration: plants like *Nelumbo nucifera* are offered to the gods in the temples.

Ornamental aquatics: *N. nucifera*, *N. pubescence* and *Eichhornia crassipes*.

5.5   Value Addition in Cacti and Succulents

The group of plants belonging to the family Cactaceae and many other plants belonging to the genera *Agave*, *Aloe*, *Cotyledon*, *Echeveria*, *Euphorbia*, *Haworthia*, *Kalanchoe*, *Sansevieria*, *Sedum* and *Sempervivum* comprise the family of succulent plants as they have the characteristic ability to store water and thus help them to survive for extended period, especially during droughts. While all cacti are succulents, not all succulents are cacti (Bose et al. 1999).

The white-, cream-, yellow-, dark orange-, vermillion-, dark red-, bright purple- and blood red-coloured flowers of cacti are hermaphrodite, stalked or sessile and bell shaped and develop from the upper portion of the areoles. Some may bloom during the day while some bloom at night.

5.5.1   Importance and Uses

Cacti are used in beautifying the landscape and are suitable as potted plants, for hanging baskets, miniature gardens, window garden, rock garden, trough garden, dish garden, bowl garden, tray garden and desert garden:

Fragrant cacti: *Astrophytum myriostigma*, *Echinocereus luteus*, *Echinopsis campylacantha*, *Epiphyllum darahii*, *Hylocereus extensus*, *Nyctocereus serpentinus*.

Cacti for indoors: *Chamaecereus silvestrii*, *Gymnocalycium mihanovichii*, *Echinocereus procumbens*, *Mammillaria* spp., *Opuntia microdasys*.

Cacti for bowl garden: *Cleistocactus strausii*, *Chamaecereus silvestrii*, *Echinocereus grusonii*, *Mammillaria elongata*, *Opuntia pilifera*.

Edible cactus: edible cactus is characterised by its fleshy leaves called pads and is commonly known as nopales, nopalitos, cactus paddles or cactus pods. These are rich in beta-carotene, iron, vitamin B, vitamin C and calcium. They can be eaten raw or cooked. The leaves of *Pereskia aculeata*, the pulp of *Echinocactus horizonthalonius*, the stem and pulp of *Ferocactus wislizeni* and the stem, flower buds and flowers of *Opuntia basilaris*, *O. dejecta* and *O. clavata* are consumed as food. Cacti that can be used for the preparation of beverages are *Carnegiea gigantea* and *Opuntia megacantha*.

Among succulents, there are those that are edible such as *Adansonia gregorii*, *Bombax ceiba* var. *leiocarpum*, *Brachychiton populneus*, *Calandrinia balonensis*, *Dioscorea bulbifera*,

Doryanthes excelsa, *Portulaca bicolor*, *Portulaca oleracea*, *Sarcozona*, *Tetragonia decumbens*, etc. Some of the succulents are also found to have medicinal properties, e.g. the boiled leaves and roots of *Aloe ferox* (cape aloe) are used as laxative, arthritis, eczema, conjunctivitis, hypertension and stress; the pulp obtained from the fruit of *Adansonia digitata* is effective against fever, diarrhoea and haemoptysis; and the tubers of *Dioscorea* spp. are used to treat hysteria, convulsion and epilepsy.

5.6 Value Addition in Some Ornamental Orchids

Orchids are perennial herbs and unlike most plants, they do not grow in soil but on rocks (lithophytes), trees (epiphytes) and even on decaying organic matter (saprophytes). They possess aerial roots that have a special covering of loose and spongy tissue called 'velamen' which absorbs moisture and nutrients that wash over them. Orchids are grown mostly for their flowers, but the seedpods of the vanilla orchid provide the characteristic flavour and aroma in foods and cosmetics.

Orchids are grouped into two major categories – monopodial and sympodial orchids. Monopodials are those in which the stem has an unlimited apical growth, and the roots are not restricted to its basal portion, e.g. *Phalaenopsis* and Vanda, while sympodials do not have a single upright stem but rather, they have a more or less horizontal growth habit that consists of a pseudobulbs that can store water and thus survive without water for a long duration, e.g. cattleya, cymbidium, oncidium and dendrobium.

5.6.1 Value-Added Products

Orchids are commonly used in decorations for various occasions such as wedding ceremonies, and they also make beautiful flower arrangements. The attractive flowers and foliages have been used for making attractive floral ornaments like corsages and boutonnieres. Corsages can be

creatively and distinctively be worn on wrist, ankle or neck or decorate shoes, handbags, hair, hats, etc. Boutonnieres are worn by men on the left lapel of suits. Orchids like Cymbidium, Dendrobium, Oncidium, *Paphiopedilum* and Odontoglossum make beautiful corsages and boutonnieres.

Some orchids are also used as medicine due to their healing properties. The uses of dried orchids range from immune system build-up, cancer treatment, eyesight improvement, fever, etc. and have been widely exploited in China. The parts mostly used are the stems and bulbs as they contain nutritive substances. For example, *Bulbophyllum odoratissimum* is used to treat tuberculosis, chronic inflammations and fractures. *B. inconspicuum* is used as an expectorant and is effective against stomach cancers.

Vanilla is a popular species that have been utilised in the flavouring industry to prepare foods and beverages. In Bhutan, *Cymbidium hookerianum* has been used for flavouring curries, giving the food a slight bitter flavour. The fragrant leaves of Dendrobium have been used as a condiment for rice in Malaya.

Dendrobium catenatum canes are boiled and drank as tea for rejuvenation of strength after an illness. The leaves of *Dendrobium chrysotoxum* are dried and also drank as a herbal tea.

Some of the orchids are also consumed as vegetables such as, Dendrobium, whose flower is consumed as a fried dish in Thailand and is used for garnishing cakes and desserts in European countries.

5.7 Conclusion

Value addition has become an important component of the floricultural industry. New products and processes, including innovative methods, are continually being devised. Standardisation of techniques for the production of pigments, essential oils and natural dyes from flowers and the potential crops for these purposes should be carefully evaluated. The principal objective of value addition is to increase the value of raw materials and deliver to the customers unique products as per the needs and demands of the market. Some of the factors that require consideration, when one has undertaken such a complex enterprise, are the methods of storage, packaging and transport to the local or distant market to preserve the quality of the product and thus promote profitability. With proper planning and management skills, value-added products can generate higher return, open new markets, create brand recognition and add variety to farm operation.

References

Bhattacharjee SK (2006) Advances in ornamental horticulture. Pointer Publishers, Jaipur

Bhattacharjee SK, De LC (2005) Post Harvest Technology of flowers and ornamental plants. Pointer Publishers, Jaipur

Bhattacharjee SK, De LC (2010) Advanced commercial floriculture. Aavishkar Publishers, Distributors, Jaipur

Bose TK, Maiti RG, Dhua RS, Das P (1999) Floriculture and landscaping. Naya Prokash, Calcutta

De LC (2011) Value addition in flowers and orchids. New India Publishing Agency, New Delhi

Lourdusamy W (1998) Flowers and vegetable breeding. Grower Books, London

Misra P, Banerji BK, Dwivedi AK (2009) Dehydration of Zinnia in an oven for value addition through embedding technique. J Ornament Hort 12(3)

Randhawa GS, Mukhopadhyay A (2012) Floriculture in India. Allied Publishers Private Limited, New Delhi

Somani LL (2010) Floriculture and landscaping- at a glance. Agrotech Publishing Academy, Udaipur

Verma AK, Gupta A, Kumar D, Dhiman MR (2012) Post Harvest Technologies for Commercial floriculture. New India Publishing Agency, New Delhi

Flower Arrangement Toward Value Addition

6

S. Das, T. Mandal, and S. Sultana

Abstract

Flower arrangement is the art of using flowers and plant materials to create a pleasing and balanced composition. Professionally designed flower arrangements incorporate the elements (line, form, space, texture, and color) and the principles of flower arrangement (balance, proportion, rhythm, contrast, harmony, and unity). There are two main styles of flower arrangement: Western style and *ikebana*/Japanese style. *Ikebana* incorporates the three main line placements of heaven, man, and earth. In contrast, the Western style emphasizes color and variety of botanical materials not limited to just blooming flowers, in mass gatherings of multiple flowers, and is characterized by symmetrical, asymmetrical, horizontal, and vertical style of arrangements. Three general styles – line arrangements, mass arrangements, and line-mass arrangements – are in use today. There is no end to the many possible variations of the three basic styles of flower arranging. The designer can use their own imagination to create new arrangements that will express their ideas and personality.

6.1 Introduction

Arranging flowers gives a chance to participate in an art activity, to express creativity, and to make home or room more livable and attractive. Flowers, when casually placed in a vase, are attractive because of their beautiful color and shape, while they are even more appealing when arranged in a stylish way. Most flower arrangements are made for a certain place or purpose, and a good flower arrangement should be expressive of some theme or idea.

A flower arranger participates in an art form which is different from other arts with its medium because unlike other arts here, the medium is alive and will die quickly unless given proper care. In selecting plant materials, their visual characteristics or design elements like color,

S. Das (✉) • T. Mandal
Department of Floriculture & Landscaping,
Bidhan Chandra Krishi Viswavidyalaya,
Mohanpur, Nadia 741252, West Bengal, India
e-mail: srabani.das.1989@gmail.com

S. Sultana
Department of Spices & Plantation Crops,
Bidhan Chandra Krishi Viswavidyalaya,
Mohanpur, Nadia 741252, West Bengal, India

shape, texture, size, space, and expressiveness must be considered.

6.2 Flower Arrangement

Flower arrangement is an art of organizing the design elements inherent in plant materials and accessories according to the design principles.

6.2.1 Principles of Flower Arrangement

Principles of flower arrangement are the guidelines which are used by the floral designer to create a beautiful composition. The principles of flower arrangement are as follows:

Proportion is the relationship between size and shape among objects and parts of objects. In floral design, there are three aspects which determine proportion: the location of the arrangement, the height of the arrangement, and the materials used in the arrangement. In regard to form or space, proportion deals with the relative area or volume. The design must be in proportion to the place of its display which could involve a room, a table, or sometimes a person wearing the design. The design must be suitable for and in harmony with its location. The height of an arrangement should be at least 11/2 times the height of its container's greatest dimension. Four aspects of the container (container's physical dimensions, its color, its material and texture, and its shape) are the primary factors in determining appropriate height. A heavy, dark container supports taller arrangements, whereas a glass container provides an excellent base for shorter arrangements. Proportion in regard to shape would describe the number of round forms in relation to spike forms. Proportion of size dictates the number of small flowers in relation to large flowers. Using flowers and foliage of equal or nearly equal size helps in achieving proper proportion.

Balance in a flower arrangement refers to the equilibrium and equality of the arrangement both in physical and visual appearance.

Physical balance is the actual stability of plant materials within a container, which ensures that the arrangement will not fall over and can stand freely on its own.

Visual balance is the perception of an arrangement being balanced that can be reached by putting equal weight on both sides of the central axis. Visual balance must be evident from side to side, top to bottom, and front to back view. Two types of visual balance are symmetrical and asymmetrical.

Symmetrical balance is present when a design has equal material placements and weight on both sides of its central axis. Oval, round, fan, triangle, inverted-T, and vertical arrangements are examples of symmetrical designs.

Asymmetrical balance is present when a design has unequal material placements and weight on both sides of its central axis. Asymmetrical balance is achieved through compensation and counterbalancing. Crescent, Hogarthian curve, fan, diagonal, vertical, ikebana, scalene, and right triangle are examples of asymmetrical designs.

6.2.1.1 Designers Use Three Main Techniques to Help Achieve Balance

Toward the outside of an arrangement, lighter materials and colors are used whereas toward the center heavier materials are used. Dominant plant material is placed along a design's central vertical axis, allowing heavier plant materials to be placed higher in a design. This technique is called as centering. The practice of balancing plant materials on one side of a design with visually equal materials on the opposite side is called counterbalancing which is essential in asymmetrical design and can be used to avoid strict mirror imaging in symmetrical work.

Rhythm: Rhythm refers to the orderly organization of design elements to create a visual pathway. The two types of rhythm are regular, repeated rhythm and free, variable rhythm. Regular, repeated rhythm is created when materials are repeated at regular intervals from the top to the bottom of an arrangement. Free, variable rhythm is an unstructured style with suitable flowing movements which is created with unstructured materials.

Five techniques, namely, radiation, repetition, transition, variety, and contrast, are used to achieve both regular and free rhythm. Stems that radiate naturally from the container are referred to as radiation. The stems appear as spokes on a wheel from top view. Crossing stems are avoided. Repeating design elements throughout an arrangement to create unity in the work is repetition. It includes color, shape, space, and lines. Transition is referred to the smooth and gradual change from one material to another. Variety referred to the use of diverse assortment and different components in one design. It focuses attention and stimulates interest and imagination. Contrast describes objects with striking differences beyond mere variety or diversity.

Focal Point: Focal point is the center of interest of an arrangement. It can be created by using large flowers; using of special-form flowers such as orchids; using dark shades, various concentration of plant material; using framing materials; using strong color contrast, radiating the plant materials to a particular area; using accessories; etc. It is usually placed low in the arrangement near the center, just above the container to break the horizontal line of the container. The focal point immediately attracts the attention of the viewer, has weight, and gives stability to the arrangement.

Dominance, Accent, or Emphasis: Dominance is the greater force of a design element, such as more round forms or more curved lines. Dominance can be achieved through the use of a dominant color, size, shape, or texture or by using larger forms or stronger colors. Accent is accomplished by introducing sharp contrast in form, size, or color and is used to first attract attention to the focal point and from that point to every detail in order of its importance.

Scale: Scale is the size relationship between flower and flower or between flower and container. Better scale balance can be obtained by grouping the smaller flowers, thus giving them more individuality. Placing small and large elements together accentuates the contrast between the elements. In miniature arrangements, scale requires special attention.

6.2.2 Design Elements

6.2.2.1 Color

Effective use of color is important in floral arrangements. Reds, oranges, and yellows are generally bright and stimulating and are considered warm colors which tend to be advancing colors (to the eye). Blues, greens, and violets are usually tranquil, peaceful, and restful and are considered cool colors which tend to be receding and have less visual weight than warm colors.

6.2.2.2 Twelve-Hue Color Wheel

The twelve-hue color wheel is helpful in determining which colors will work well together in a design.

Primary Colors

Red, blue, and yellow are considered as primary colors. These three primary colors can be combined to create all other colors.

Secondary Colors

Orange, green, and violet are considered as secondary colors. These are the results of combining two primary colors in equal proportion.

Intermediate Colors

The results of combining primary and secondary colors are considered as intermediate colors. Blue-green, blue-violet, red-violet, red-orange, yellow-orange, and yellow-green are in this category.

6.2.2.3 Color Harmonies

The following color harmonies or color combinations are generally pleasing to the eye.

6.2.2.4 Related Color Harmonies

Monochromatic color harmony refers to one hue and its tones, tints, and/or shades.
Example: light orange through orange to brown
Analogous color harmony refers to colors near each other on the color wheel.

Examples: yellow, yellow-orange, and orange-green, yellow-green, and yellow.

6.2.2.5 Contrasting Color Harmonies

Direct complements refer to the colors lying opposite one another on the color wheel. Strong contrast can be achieved by using direct complements.

Examples: red and green-yellow and violet

Split complements refer to a key hue combined with two hues on either side of its direct complement on the color wheel.

Examples: yellow with red-violet and blue-violet red with yellow-green and blue-green.

Triad refers to three hues equally spaced on the color wheel. It is eye-catching.

Examples: red, yellow and blue-red-orange, yellow-green, and blue-violet.

6.2.2.6 Light

Light affects design in many ways like it may change apparent colors of materials and enhance form, texture, or depth. Lighting will be limited to artificial lighting in the form of overhead fluorescent lights. If the display area is in open-air condition, indirect natural light may be present during the day.

6.2.2.7 Space

Space refers to the open area in and around the design or to the spaces within individual plant materials selected for use in the design.

6.2.2.8 Line

Line is the primary foundation of a design. It creates a visual path for the eye to follow through a design and also function to establish the structural framework or skeleton of a design. It may be long or short, straight or curved, weak or strong, etc.

6.2.2.9 Form

Form refers to the three-dimensional aspect of a design. It may either be closed (compact, massed, with few open spaces) typical of a mass design or open (with spreading parts which produce spaces between the parts) typical of a line or line-mass design. A closed form appears to be heavier than an open form of the same size. It also applies to the shape of individual components of a design. These are generally put in to three main groups according to their shape:

Spiky/linear shapes are useful for line and line-mass arrangements and are often used to form the skeleton of an arrangement.

Examples: gladiolus, iris, ornamental grasses, twigs, branches, etc.

Mass/rounded shapes are useful for line-mass or mass arrangements and it may be used to develop the focal point in line-mass arrangements or make up almost all of a mass arrangement.

Examples: chrysanthemum, marigold, rose, zinnia, etc.

Spray or filler shapes are useful for mass and line-mass arrangements and are also used as background fillers and as space fillers in mass arrangements.

Examples: baby's breath, ageratum, ferns, asparagus leaves, parsley, etc.

6.2.2.10 Pattern

Pattern refers to the design formed by solids and spaces. Individual components in a design have a pattern of their own, and the individual components are combined into an overall pattern.

6.2.2.11 Texture

Textural contrast and variety can add interest to a design. Plant texture refers to the surface quality of the plant material. The texture of a plant material may be fuzzy, glossy, smooth, rough, etc. It also applies to the overall effect of the arrangement of the petals or florets.

6.2.2.12 Size

Flowers and leaves of different sizes are selected. Usually buds and small flowers are used at the top and edges of an arrangement, while large leaves or fully opened flowers are placed low in an arrangement, and a large element is often used as a focal point. In floral design, size also applies to apparent or visual size. Size of a flower may be expressed as delicate, heavy, or bold.

6.2.3 Attributes of Design

6.2.3.1 Expression

Expression is an attribute of floral design. Through the artistic and creative selection of components, a mood, feeling, or idea may be expressed and communicated to the viewer.

Example: daffodils and pussy willows suggest spring; cattails make one think of a swamp; sunflowers are casual and suggest summer; white roses are formal and elegant.

6.2.3.2 Harmony

Harmony is the principle that produces a feeling of unity throughout the entire composition. To obtain harmony, all components must have something in common like size, shape, texture, idea, time of flowering, color, etc.

6.2.3.3 Unity

Unity is that quality of a design which expresses congeniality, cooperation, and a spirit of working together as a unit. Unity is the fitting or bringing together of all of those principles and elements which go into making up the design or composition.

6.2.4 Materials of Design

6.2.4.1 Cutting Flowers and Foliage

Successful floral arrangement begins with fresh plant materials that have been properly handled and prepared. Beauty and good composition of an arrangement is determined by the way they are selected, cared for, and arranged and not by the cost or rarity of plant materials used. Flowers and foliages are cut at the evening before making the arrangement to allow time for proper conditioning. Plant materials are handled carefully only by the stems. Most flowers can be kept at its best if cut when nearly fully open as many tight buds or young leaves wilt rapidly and do not take up water. Whereas some flowers will keep best if cut in the bud stage or when they are just beginning to open. Fully open or faded flowers are past prime and usually will not keep well. Flowers and foliages are cut on a slant with a sharp knife or flower shears to enable flowers to absorb more water. Stems are cut longer than required for arrangement. Extra stems are cut in case of damage. Flowers and foliages are cut in a bucket of cool water as they are cut from the garden and are placed indoors in a cool and dark place. Recut stems indoors at a 45-degree angle. Stems are cut underwater to prevent air from entering the stem and interfering with water uptake. Only an inch or so of the stem is placed under the water to cut it. Flowers and foliages are stored in a cool, humid place, out of the sun and away from drafts. These should not be stored with fruits, vegetables, or decaying flowers or leaves as these give off ethylene which shortens the life of many cut flowers.

6.2.4.2 Conditioning Plant Materials

Conditioning is an important factor in successfully arranging and exhibiting cut plant materials as its purpose is to allow the cut plant material to absorb as much water as possible. Plant materials which are not conditioned properly appear dried out or wilted. Proper conditioning prolongs the life of the arrangement. Plant materials are stand in lukewarm water to a depth of half their length overnight in a cool, dark place to allow the stem, leaves, and blossoms to absorb water to enable them to retain optimum beauty. The next morning, stems are again cut underwater at a 45-degree angle. All foliages are removed below the water level of the container as the submerged foliage decays, creating foul water. The cut plant materials are left in water until they feel stiff and can be used in arrangements. Some flowers with fleshy, fibrous stems such as cockscombs or sunflowers last longer if about ½in. of their cut ends is dipped in boiling water before being placed in a container. Woody stems are peeled back and split to allow the stem to absorb more water. Plants that exude a milky substance are sealed by searing the cut end, using a flame, or by dipping it into powdered alum. Hollow-stemmed flowers are filled with water before being placed in a container. Cut flowers are hold upside down, the stems are filled with water, and the flower holding is inverted a finger over the cut end of the stem and placed in water. Some plant material (generally foliage) gets benefit from overnight complete submersion in lukewarm water.

6.2.4.3 Care of Arrangement

The arrangement must be placed out of the sun and away from drafts, hot air ducts, and radiators. At night, it should be put in a cool place to prolong the life of the flowers. The container must be full of water and it should be checked daily. The water is changed occasionally, and at the time of rearranging, at least 1 in. from each stem must be removed.

6.2.4.4 Containers

Almost anything that can hold water may be used as a container for an arrangement. The important thing is to select a proper size, shape, color, and material that will harmonize with the plant materials and the place where the arrangement is to be placed.

The container should be in scale with the table on which it is displayed. The larger the container, the more flowers are required. Small- or medium-sized containers are most useful. About two-fifths of the overall size of the arrangement is allowed for the container.

The shape of the container must suit the design of the arrangement. Simple shapes with clean lines are best. Those in the form of animals, heads, pianos, etc. are avoided, except for special occasions.

Tans, browns, grays, and greens are useful colors which can be harmonized easily with plant materials and with most backgrounds. White containers call attention to them and are often difficult to use effectively.

Pottery containers, ranging from rough bean pots to delicate china vases, are the most frequently used ones. Many vegetable dishes, cereal bowls, and sugar bowls also make good flower containers. In using clear glass, it should be remembered hat stems and the holder will be seen. Metals such as copper, bronze, brass, etc.

are the most versatile and suggest strength. Aluminum and stainless steel are modern, while iron suggests weight. Baskets with liners to hold water come in many shapes and are excellent for mass or naturalistic arrangements. Most plastic containers are very light in weight.

The spirit of the plant material and container should go in same way. Pussy willows and daffodils are good in brown or green pottery but not in an elegant silver bowl.

6.2.4.5 Holders

A good holder provides the freedom to position stems at desired position and hold them in place securely. A suitable holder is chosen for the style of arrangement planned, the plant materials to be used, and the container to be used.

Pinpoint Holders

Pinpoint holders are best suited for line and line-mass arrangements in low bowls or shallow containers. The holders are fastened securely to the clean, dry container with waterproof floral clay such as Posey clay or cling. Stems are stuck directly onto the pins or are wedged between them. Several thin stems are bind to each other or to a larger stem with a rubber band, string, or floral tape to put on a pinpoint holder. Thick woody stems are easier to insert by splitting the ends.

Floral Foams

Wet floral foams hold stems in place and supply water to the plant materials. Floral foams are available both in wet or dry forms; wet foams are used for arranging fresh plant materials, while dry floral foams are used for dried or silk floral arrangements. Wet floral foams should not be reused due to the existing holes in the foam that will not supply water to the plant materials. Floral foam is inexpensive and easily available. Floral foam is best for line-mass or mass arrangements. The foam is soaked before use, in a pail of water until it barely floats. A clear floral preservative is added to the water to extend the freshness of the floral arrangement. A piece of foam is cut, is fit in the container tightly, and forces it into place; the container must be mostly filled with the foam. A small piece of the corner of the foam is cut off so that water may be added to the arrangement later as needed. Floral tape can be used to secure the foam. The container is then filled with water and the stems are inserted to the bottom of the foam; it is more important with heavy or large flowers that the stem be placed all the way to the bottom of the foam for added support. After placing in foam, a stem must not be pulled out of it, as this may remove the stem end from contact with the water or foam, causing the flower to wilt.

Chicken Wire

Chicken wire or floral netting of 1-in. mesh, preferably enameled green, is best for mass arrangements in vases or deep bowls. The entire container is filled with chicken wire and it is extended to an inch or so above the top. Stems are passed through at least three layers of meshes or wire. A pinpoint can be used beneath the chicken wire for holding vertical stems precisely in place.

6.3 Western Flower Arrangement

6.3.1 Line Arrangements

Line arrangements which develop a dynamic feeling of action, movement, and life are the adaptations of Japanese styles and have three lines or placements. A clean-cut, sparse look is produced by giving emphasis on the linear quality of a few branches, leaves, or flowers. The spaces between the plant materials are as important as the materials used. Color is of less concern than line, shape, space, expressiveness, etc. Line arrangements encourage creative experimentation and originality due to the natural lines and spaces of the plant material. Thicker and fuller materials are selected to develop a strong, sturdy effect. Most linear arrangements have asymmet-

rical balance and are to be viewed only from the front. These may be tall in relatively small containers and still have good balance since so little material is used. Low, flat bowls, compotes, or pedestal bowls using pinpoint holders are best for it.

6.3.2 Mass Arrangements

Traditional mass arrangements having thick, full look and requiring much plant material are adapted from European floral designs. Rather than the individual flowers, leaves, or branches, the whole colorful mass of flowers and foliages is emphasized, and these may be of many shapes, such as circles, domes, crescents, ovals, and triangles. Color is the most important element in mass arrangements. Though round or spray-shaped flowers generally dominate, spiky shapes are also good for triangular arrangements. Good development of depth is important. A block of wet floral foam is generally used in these arrangements.

6.3.3 Line-Mass Arrangements

Line-mass arrangements which are the combination of the strong line of Japanese styles with the massed effect of European floral designs, were developed in the United States, are often called Contemporary American. These arrangements have a neat, uncluttered look with definite line, a well-defined mass, and plenty of open spaces. Strong lines are established by spikes or spikelike flowers or leaves or by leafy or bare branches in the upper part. The designer decides whether to emphasize the line or the mass. Whichever portion between the line and mass dominates must blend together. Line-mass arrangements are often triangular shaped, mostly asymmetrically balanced, and meant to be seen from the front only.

6.3.4   Types of Flower Arrangement Designs (Figs. 6.1, 6.2, 6.3, 6.4, and 6.5)

Circular Arrangement

In the circular arrangement, flowers are arranged in round shape. This is one of the easiest forms of arrangement and common to view in conferences and executive meetings; the flowers are arranged in a circular form, thereby presenting a soothing appearance to the viewers.

Fig. 6.1   Circular arrangement

Triangular Arrangement

In the triangular arrangement, flowers are arranged in specific shapes and up to a proportionate height. This can be a formal or an informal arrangement.

Fig. 6.2   Triangular arrangement

Fan Arrangement

This arrangement is a horizontal flower arrangement form which does not possess height. The concentration, in this kind of arrangement, lies on one single big flower and is generally arranged in shallow containers.

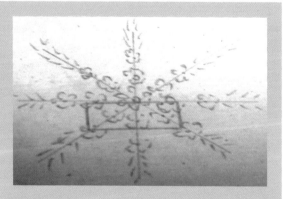

Fig. 6.3   Fan arrangement

Crescent Arrangement

It is an informal pattern of arrangement where flowers are arranged in a curvaceous moon-shaped form. The arrangement is refreshing and mostly seen on coffee tables.

Fig. 6.4   Crescent arrangement

Vertical Arrangement

It is a form where flowers with small stems are a major concern. Tulips, roses, and carnations are the most appropriate ones for this kind of arrangement, though small fillers are also used to create a symmetrical appearance.

Hogarthian Curve

The "S"-shaped sophisticated asymmetrical design was named after an English scientist, William Hogarth. Tall, stemmed cam pots, raised containers are best suited for this design as a portion of the floral line extends below the rim.

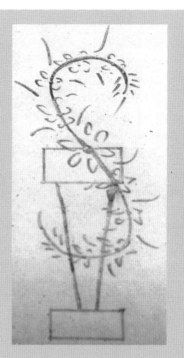

Fig. 6.5   Hogarthian curve

6.4   Ikebana: Japanese Flower Arrangement

The word *ikebana* means "the Japanese art of flower arrangement," though it includes freshly cut branches, vines, leaves, grasses, berries, fruit, seeds, flowers, wilted and dried plants, as well as glass, metal, and plastic materials.

According to legend a monk Ikenobo Semu originated the art form when he made an offing of flowers on a Buddhist altar statue of Buddha at Rokkakudo Temple in Kyoto. The Ikenobo school of flower arranging was founded in 1462 and is named after him.

The practice of ikebana, which was also called *kado* or the way of flowers, has been pursued as a form of meditation on the passage of the seasons, time, and change. Its religious origins and strong connection to the natural cycle of birth, growth, decay, and rebirth give it a deep spiritual resonance. The main idea behind ikebana is to arrange the flowers to heighten the appeal of a vase in a tea room and use flowers to represent heaven, earth, and humanity. Ikebana creates a harmony of linear construction, rhythm, and color in which the vase, the stems, leaves, and branches are part of the art form as well as the flowers. Unlike Western appreciation of flowers, which emphasizes quantity and color, Japanese flower arrangement is based on three main lines that symbolize heaven, earth, and humankind.

According to Sogetsu School, there are three main lines in a Japanese flower arrangement: *shin* (heaven), *soe* (man), and *hikae* (earth). *Hikae* is placed at the bottom with an angle of 75° to the vertical line. *Soe* is placed after *hikae* at 45° angle to vertical line at other side. At the same side, *shin* is placed in between *soe* and *hikae* with an angle of 15° to the vertical line.

6.4.1 Types of Ikebana

Moribana
It is a natural design where piled up flowers are used in shallow vases.

Jiyubana
It is a free flower arrangement.

Morimono
In this kind of arrangement, fruits and vegetables are used along with flowers.

Nageire
In this Japanese flower arrangement, tall vases are used.

Zeneika
In *zeneika* type of arrangement, straight materials with uneven height are used.

Zeneibana
Beautiful sculpture is created in this kind of arrangement by using wood, stone, rocks, etc.

6.4.2 Some Terms Used in Flower Arrangements

Basing
Covering the area at the base of a floral arrangement is generally known as basing. Basing establishes focal emphasis at the lower portion or at the base of the arrangement and adds texture, color, shape, dimension, and visual balance to an arrangement.

Layering
Layering or stacking is a basing technique in which leaves or other thin, flat materials are placed one on top of the other for visual effect. Leaf edges get most attention in this technique.

Terracing
It is similar to layering except that the materials used are thicker and more three dimensional, with which more depth can be achieved. Lotus pods and sponge mushroom are two dried materials that are suitable for this purpose.

Pave
It is a basing technique in which practically identical small flowers or pods are placed close to each other and close to the floral foam. Each placement retains its own identity which is important to make this technique different from other basing techniques.

Clustering
In this kind of basing method, materials are placed so closely together that they lose their individual identity. The quantity and shape of each individual flower is obscured to the point that the cluster acts as one unit. Carnations are an excellent choice for clustering because all of the petals are almost the same; placing closely together, they all blend into one another and in the process lose their individual identity.

Pillowing
A basing method that is a variation on clustering. Pillowing is used at the edge of a container where a rounded shape would make a good transition from the container to the design.

Grouping

It refers to placement of identical floral materials in an arrangement and separating them from each other with clearly defined negative space. There must be space between individual groups of elements. Negative space which makes each group distinct is an open space deliberately included in an arrangement. It is called "negative" because it separates the individual elements.

Banding

Banding is mainly used as a decorative addition to an arrangement, in which some type of material, such as metallic wire or ribbon, is wrapped around one or several stems, which creates a strong visual statement.

Binding

It involves tying a relatively large quantities of material that are bound together with string, twine, or raffia (not with floral tape) and worked into an arrangement. Raffia, a natural material obtained from palm trees, can be used for this purpose.

Shadowing

In the shadowing technique, two pieces of identical materials are placed in an arrangement, one immediately behind the other. Shadowing is most effective in arrangements with only a few flowers and foliage elements.

Sequencing

In sequencing, floral materials are arranged in a progressively changing pattern. Sequencing is most powerful when the change in the pattern of color, size, or texture is gradual.

Framing

Framing involves enclosing specific floral materials within an area. Materials for the frame must have some relevance to the materials used in the design and the design as a whole. Frames are commonly made by placing straight or curved twigs at the right and left of the design, which isolates the focal materials of a floral composition by demanding the viewer's attention.

Zoning

Zoning works in almost same as grouping, but the space between the groups of floral materials is much more distinguishable.

Parallelism

Parallelism is one of the classical floral design styles and is considered a generic design technique and an extension of zoning. It involves positioning materials in separate vertical groupings and each group is kept separate and distinct.

Skeletonizing

It is a way of trimming foliage to give added emphasis to the branch or stem. Sword fern can be used for this purpose.

References

Culbert JR (1978) Flower arranging (Circular 1154). Urbana-Champaign University of Illinois. http://www.archive.org/details/coursescat19982000univ

Honeywell ER (1958) Principles of flower arrangement. Purdue University Agricultural Extension Service, Lafayette

Roy Chowdhury N, Misra HP (2001) Text book on floriculture and landscaping, Kalyani Publisher, (1):231–242

Value Added Products from Medicinal Plants

7

S. Maitra and P. Satya

Abstract

Medicinal plants are considered as nature's antidote against human ailments and disorders having considerable agribusiness potential especially for the biodiversity-rich countries. The sane utilization of such flora lies in the development of value-added products. Value addition can be done directly through minor processing or indirectly through maintaining quality standard. The harvesting season and index for industrial utilization of some medicinal plants are summarized here. The processing and storing techniques after harvesting to maintain the quality standards along with sensory, macroscopic or microscopic evaluations and physico-chemical analysis are also discussed here. GACP and GMP are the two terms closely associated to collection and value addition in medicinal plants, the gist of which is also presented here. The different forms of value-added products like dust, pill, syrup, fortified food, cosmetics, tinctures, blended products, etc. are discussed in detail with some relevant pictures. The different processes like maceration, decoction, percolation, soxhleting, solvent extraction, SCFE, CCE, sonication, etc. are elaborated here. The biotechnological interventions for targeted raw material production, biopharming, characterizing and thereafter engineering the biosynthetic pathways to reorient the species to produce manifold active principles are having enormous potential in the field of drug development.

7.1 Introduction

Nature is a treasury with enormous range of flora of which one third are known to have miraculous healing properties. From the time immemorial human being utilized this natural wealth to cure their ailments. Medicinal plants are the generous aid from the nature, rich in secondary metabolites and obviously the potential source of medicine.

S. Maitra (✉)
Floriculture, Medicinal and Aromatic Plants,
Uttar Banga Krishi Viswavidyalaya,
Pundibari, Cooch Behar, West Bengal, India
e-mail: soumenmaitra@rediffmail.com

P. Satya
Crop Improvement Division, ICAR-Central Research
Institute for Jute and Allied Fibres, Nilganj,
Barrackpore, West Bengal, India

A.B. Sharangi and S. Datta (eds.), *Value Addition of Horticultural Crops: Recent Trends and Future Directions*, DOI 10.1007/978-81-322-2262-0_7, © Springer India 2015

India has a rich ancestry of traditional knowledge related to the use of medicinal plants in the health-care system which is perceived from the ancient literatures like Charaka Samhita and Sushruta Samhita. In the era of Vedas, 129 medicinal plants and in Kalpasutras about 519 medicinal plants were referred for their remedial traits (Kaushik and Dhiman 2000). Due to the increased use of herbal remedy throughout the world, the global market of medicinal plants would be five trillion USD by the end of 2050 (Kalia 2005) which is a potential money spinner for the biodiversity-rich developing countries. India is a biodiversity-rich country that possesses 15 different agroclimatic zones, 10 vegetation zones, 8 phytogeographic regions, 28 biotic provinces, 426 habitats for specific species and representing 1 of the 12 mega biodiversity centres of the world with two hotspots having 47,000 different plant species out of which *3,500 are medicinal* (Tyagi et al. 2005), while 1,100 species are used in ISM and 650–700 species are used in herbal industries, yet only 175–200 species are used commercially (Maitra 2012). Being a gene-rich country, India harbours about 320 species of wild relatives of cultivated plants, contributing a large section of species gene pool. Indian herbal flora belongs to 2,200 genera under 386 plant families (Maitra and Satya 2009). According to nature of plant species, Indian medicinal plants are largely trees (33 %), followed by herbs (32 %), shrubs (20 %), climbers (12 %) and others (3 %) (Tiwari 1999). The worldwide phyto-pharmaceutical industries poised to grow about 70 billion USD as compared to 16.50 billion USD during late 1990s. The yearly growth rate becomes nearly 15–20 % for European and 25–30 % for North American market (Chatterjee 2010). India plays a vital role in the world herbal drug industry with an 8.13 % share; still it is lagging behind against 28 % of China. The Indian herbal products showed 4.95 % growth to the world herbal industry from 1991 to 2002, but at the same time, the trend growth rate of China was 7.38 % (Singh and Vadera 2010).

7.2 Classification of Medicinal Plants

Medicinal plants are classified into several ways out of which the major methods are stated here.

7.2.1 Classification According to the Plant Type

In this process, the medicinal plants are classified according to plant types like medicinal herbs, medicinal shrubs, medicinal trees and climbing medicinal plants. Besides, some fungi, mosses, ferns, etc. are well known for their therapeutic values. All those come into a single group called 'others'. The examples of each group are presented in Table 7.1.

7.2.2 Classification According to Economic Plant Part

Medicinal plants are also classified according to the economic plant part like root, stem, whole plant, bark, fruit, seed, stem, flower, leaves, wood and bulbous organs, etc. The examples of each group are presented in Table 7.2.

7.2.3 Classification According to the Nature of Secondary Metabolites (Chemosystematic or Chemotaxonomic Approach)

Medicinal plants are recently being classified on the basis of chemical nature of their secondary metabolites or active principles. This approach has added advantage over the binomial nomenclature as the plants are classified botanically as well as biochemically at the same time. It has tremendous importance to the herbal drug developers. The examples of each group are presented in Table 7.3.

Table 7.1 Classification of medicinal plants according to type of plant

Class	Example Botanical name	Family
Trees	*Aegle marmelos*	Rutaceae
	Alstonia scholaris	Apocynaceae
	Azadirachta indica	Meliaceae
	Cassia fistula	Caesalpiniaceae
	Cinchona ledgeriana	Rubiaceae
	Phyllanthus emblica	Euphorbiaceae
	Terminalia arjuna	Combretaceae
	Terminalia bellerica	Combretaceae
	Terminalia chebula	Combretaceae
Herbs		
Annual	*Andrographis paniculata*	Acanthaceae
	Euphorbia hirta	Euphorbiaceae
	Papaver somniferum	Papaveraceae
	Plantago ovata	Plantaginaceae
	Solanum khasianum	Solanaceae
	Swertia chirayita	Gentianaceae
Biennial	*Atropa belladonna*	Solanaceae
	Digitalis purpurea	Scrophulariaceae
	Hyoscyamus niger	Solanaceae
Perennial	*Acorus calamus*	Araceae
	Alpinia galanga	Zingiberaceae
	Asteracantha longifolia	Acanthaceae
	Bacopa monnieri	Scrophulariaceae
	Cephaelis ipecacuanha	Rubiaceae
	Ferula assafoetida	Apiaceae
	Colchicum luteum	Liliaceae
	Nardostachys jatamansi	Valerianaceae
Shrubs	*Adhatoda vasica*	Acanthaceae
	Coscinium fenestratum	Menispermaceae
	Embelia ribes	Myrsinaceae
	Ephedra gerardiana	Gnetaceae
	Gaultheria fragrantissima	Ericaceae
	Phlogacanthus thyrsiflorus	Acanthaceae
	Sida cordifolia	Malvaceae
Climbers	*Aristolochia indica*	Aristolochiaceae
	Hemidesmus indicus	Periplocaceae
	Ipomoea paniculata	Convolvulaceae
	Piper longum	Piperaceae
	Tinospora cordifolia	Menispermaceae
	Tylophora indica	Asclepiadaceae
Others	*Claviceps purpurea*	Hypocreaceae
	Fucus serratus	Fucaceae
	Laminaria cloustoni	Laminariaceae
	Penicillium notatum	Aspergillaceae

Table 7.2 Classification of medicinal plants according to economic plant part

Class	Example Botanical name	Family
Root (29 %)	Rauwolfia Serpentina	Apocynaceae
	Withania somnifera	Solanaceae
Whole plant (16 %)	Catharanthus roseus	Apocynaceae
	Acalypha indica	Euphorbiaceae
	Bacopa monnieri	Scrophulariaceae
Bark (14 %)	Cinchona ledgeriana	Rubiaceae
	Alstonia scholaris	Apocynaceae
Fruit (10 %)	Aegle marmelos	Rutaceae
Seed (7 %)	Centratherum anthelminticum	Asteraceae
	Hydnocarpus kurzii	Flacourtiaceae
Stem (6 %)	Aristolochia indica	Aristolochiaceae
	Coscinium fenestratum	Menispermaceae
Leaves (6 %)	Adhatoda vasica	Acanthaceae
Flowers (5 %)	Madhuca indica	Sapotaceae
Wood (3 %)	Santalum album	Santalaceae
Bulbous organs (4 %)	Acorus calamus	Araceae
	Alpinia galanga	Zingiberaceae
	Colchicum luteum	Liliaceae
	Dioscorea composita	Dioscoreaceae
	Urginea indica	Liliaceae

7.3   Value Addition

Value-added product refers to the processing of raw herbage into a different state which can improve its efficacy, longevity and look for whom a manifold return would become realized over the additional expenditure due to processing. Value addition becomes nowadays a new attraction to the entrepreneurs of herbal products leading to development of several categories of produces which were only confined into limited items in the recent past. Value addition is very important in the new millennium where medicinal plants are used globally as home remedy, OTC drugs and crude drug for the manufacturers requiring proper quality standards as per worldwide recognized guidelines failing which the produce will consider as inferior and fetch a very low or no price. Value addition can be achieved in medicinal plants by direct minor processing of collected or cultivated raw herbage to improve the quality, indirectly by maintaining quality standards through some chemical test or by processing the raw herbage into different forms.

7.3.1   Harvesting

Direct value addition can be done by collecting the produce in the proper season. In majority of the cases, the produces are collected from the natural habitat as MFP/NTFP at any time of the year without considering the seasonal variation of active principal concentration of the particular species; e.g. Cinchona is a medicinal plant grown for its bark containing quinine – the concentration of which is in traces during the rainy season. Unlike other horticultural and field crops, medicinal plants must be harvested at the industrial maturity state of the economic plant parts. The harvesting season and index of some medicinal plants is presented in Table 7.4.

Table 7.3 Chemosystematic classification of medicinal plants

Class	Example
Alkaloid yielding	
Pyrrole	*Nicotiana tabacum*
Pyrrolidine	*Erythroxylum coca*
Alugelin/imidazole	*Pilocarpus grandiflorus*
Pyridine	*Ricinus communis*
Piperidine	*Piper longum*
Tropane	*Datura stramonium*
	Atropa belladonna
	Hyoscyamus niger
Quinoline	*Cinchona ledgeriana*
Isoquinoline	*Cephaelis ipecacuanha*
	Zizyphus mauritiana
Indole	*Strychnos nux-vomica*
Phenanthrene	*Papaver somniferum*
	Thalictrum hazarica
Purine	*Coffea arabica*
	Theobroma cacao
Steroidal	*Holarrhena antidysenterica*
	Solanum tuberosum
Glycoside yielding	
Cyanogenic/ cyanophoric	*Prunus serotina*
	Manihot esculenta
Cardiac	*Digitalis purpurea*
	Strophanthus caudatus
Saponin	*Dioscorea composita*
	Saponaria officinalis
Steroidal	*Gentiana kurroo*
	Cynanchum stauntonii
Triterpenoid	*Glycyrrhiza glabra*
	Adina rubella
	Cephalaria scoparia
	Bacopa monnieri
Phenolic	*Moringa oleifera*
	Pimenta dioica
	Populus davidiana
Simple phenolic	*Vanilla planifolia*
Anthracene	*Cassia angustifolia*
	Rhodomyrtus tomentosa
Coumarin	*Aegle marmelos*
	Seseli montanum
Flavonoid	*Acacia catechu*
	Phoenix dactylifera
Anthocynidin	*Rosa sp.*
	Vaccinium myrsinites
	Rubus occidentalis
	Ribes nigrum

Table 7.3 (continued)

Class	Example
Thiocyanin	*Brassica nigra*
Resinous	*Ipomoea maxima*
Tanin	*Saraca indica*
	Acacia mearnsii
Resin	
Oleoresin	*Copaifera lansdorffii*
Gum resin	*Commiphora molmol*
Oleo-gum-resin	*Ferula assafoetida*
Balsam	*Cinnamomum zeylanicum*
Organic acid	*Pyrus malus*
	Citrus sp.
	Tamarindus indicus
	Phyllanthus emblica
	Terminalia chebula
Fixed oil	*Guizotia abyssinica*
	Papaver somniferum
	Glycine max
	Linum usitatissimum
	Juglans regia
	Carthamus tinctorius
Volatile oil	
Alcoholic	*Pelargonium graveolens*
	Cymbopogon winterianus
	Cymbopogon martinii
	Rosa moschata
	Rosa damascena
	Rosa centifolia
Aldehyde	*Cymbopogon flexuosus*
	Coriandrum sativum
Ester	*Satureja montana*
	Thymus vulgaris
	Cananga odorata
	Populus balsamifera
	Gaultheria procumbens
Ketone	*Mentha piperita*
	Citrus reticulata
	Zingiber officinale
Phenolic	*Origanum vulgare*
	Thymus serpyllum
	Eucalyptus globulus
Phenol-ether	*Pimpinella anisum*
	Piper nigrum
	Myristica fragrans
	Anethum graveolens
Oily hydrocarbon	*Humulus lupulus*
	Ocimum basilium
	Litsea zeylanica

Table 7.4 The harvesting season and index for industrial utilization of some medicinal plants

Name of the plant	Harvesting season and index
Rauwolfia serpentina	It is a crop of 18 months. Harvesting can be done at leaf-shedding stage after completion of this period preferably in October–November
Withania somnifera	It is a crop of about 6 months. Harvesting can be done when the fruits become mature and take crimson red colour, and the plants are about to dry
Plantago ovata	It is a crop of 4 months. Harvesting can be done when the plant becomes yellowish, and the spikes turn brownish. It is harvested on a clear sunny day at daytime not in the morning
Cassia angustifolia	It is a crop of 4–5 months. Harvesting can be done at 50, 90 and 130 days. Bluish green leaves and young pods are preferred
Aloe barbadensis	It is a perennial plant. Individual leaves of at least 700 g are harvested
Andrographis paniculata	It is a crop of 5–6 months. Leaves are harvested after every 2 months (three pickings)
Gymnema sylvestre	It is a perennial plant. Periodical harvesting of vines is done except in peak winter months
Commiphora mukul	It is a perennial plant. Gum tapping can be done in plants of 8–10 years of age during November to March
Acorus calamus	It is a crop of 8 months. Creeping rhizomes are harvested in the month of November
Swertia chirayita	It is a crop of 4–5 months. One or two leaf stripping can be done during growth phase at 2-month intervals. Final harvesting by uprooting the plant is done during seed maturity in June–August
Mucuna pruriens	It is a crop of 5–6 months. Pods are harvested as and when mature
Piper longum	It is a perennial plant. Fully mature but unripe fruit stage is the ideal stage of harvesting of spikes. Thicken roots are harvested after every 3 years
Solanum khasianum	The greenish-yellow berries are harvested as and when available. To make it synchronous, it is harvested at 5–6 months age of crop
Glycyrrhiza glabra	It is a perennial plant. The roots are harvested in 2–3 years in winter season
Mentha arvensis var. *piperascens*	It is harvested at 3 months after planting for the first time, 2 months after first harvesting and finally 2 months after second harvesting when flowering occurs
Papaver somniferum	Lancing can be done in capsules of 15 days after flowering

7.3.2 Processing

After proper harvesting, the raw herbage needs to be dried thoroughly (in most cases) and stored in proper package or container checking admixture with inert materials and microbial decontamination through fungal growth and mould development. Fresh herbal materials are classified into woody (comparatively less prone to spoilage and may be stored at airy sacks), leafy, whole herbage and fleshy (more vulnerable to spoilage and require polyethylene-lined gunny bags), flower and delicate flower parts (highly susceptible to spoilage and mechanical damage and store in polyethylene-lined corrugated boxes), herbage having volatile oils and resins (highly susceptible to volatilization and store in airtight container) and herbal extracts and compound (store in airtight container). The storage condition may also determine the quality of pro-

duce. The humidity level should be such that no fungal growth and mould formation will occur. For this purpose, desiccants are used in stores. Nowadays, Silica gel is used commercially as a desiccant in storages due to its excellent capacity to absorb moisture from the air as well as it is reusable. The dried materials must be stored at low temperature than the ambient to protect denaturation of active principles. Some materials that need protection from light should be stored in opaque containers. Sometimes rodents and insects appear as potential problems in storage of produce reducing quality, and hence clean and hygienic situation must be maintained. Though the produce is graded according to the size relationship or through other grading methods as stipulated, species-wise, sensory, macroscopic or microscopic evaluations are essential to categorize the dried produce. In some cases, visual observations are enough to judge the quality of the produce, but in some other cases, they need to evaluate the surface qualities. Still in some valuable materials where the presence of adulterants in tiny quantity may reject the produce, physicochemical analysis related to moisture content, presence of foreign organic matter, total acid content, alcohol-soluble extractive, water-soluble extractive, total ash content, acid-insoluble ash content, pesticide residue and presence of microorganism/contamination through microorganism is essential which is called as microscopic examination of the produce or alternatively termed as indirect value addition. After physico-chemical analysis of the raw material, the crude drug will be semi-processed into different products like pills, powders or extracts which are finally processed through proper testing (accelerated ageing test) and packaging into the value-added herbal products.

7.4 GACP

Herbal medicines are directly related to health, and hence the quality assurance is utmost important. The quality of the finished product largely depends on the raw herbage or crude drugs which are collected both from natural vegetation and commercial cultivation units worldwide. Faulty selection of site for collection of raw material or improper cultivation and collection technique may lead to the presence of chemical residues in the raw materials which is not desirable. To ensure the quality assurance of the herbal medicines, the World Health Organization (WHO) developed the guidelines for good agricultural and collection practices (GACP) for medicinal plants in 2003 which was later on modified by the European Medicines Agency (EMEA) for the European Union in 2006 (Date of effect – 01.08.2006). The main objectives of the GACP are the following:

- Contribute to the quality assurance of medicinal plant materials used as the source for herbal medicines, which aims to improve the quality, safety and efficacy of finished herbal products.
- Guide the formulation of national and/or regional GACP guidelines and GACP monographs for medicinal plants and related standard operating procedures.
- Encourage and support the sustainable cultivation and collection of medicinal plants of good quality in ways that respect and support the conservation of medicinal plants and the environment in general.

This guideline contains the authentication of cultivated medicinal plants through selection, botanical identity and botanical specimen preparation, specification of seed and other propagating materials, cultivation through Conservation Agriculture technique following no tillage system as far as practicable which includes the stipulations related to site selection, ecological environment and social impact, climate, soil, irrigation and drainage and plant maintenance and protection. It also discusses about the standard harvesting technique, knowledge and skill of the person concerned. Good collection practices for medicinal plant part of the guideline contain the details about the necessary permission, technical planning, selection of plants, collection practices and expertise of the person concerned. The common technical aspect part holds the stipulations related to postharvest processing, bulk packaging and labelling, storage and transportation, equipment, quality assurance, documentation and skill development of the person concerned. Besides it also discusses about the ethical and legal considerations, IPR and benefit sharing, threatened and endangered species and ultimately the research needs.

GACP is a very important document for the global quality assurance of raw materials for herbal medicines.

7.5   GMP

Good manufacturing practices may be explained as the production of drugs from raw herbage or crude drugs following standard techniques under hygienic situation and quality assessment of drugs through stipulated testing procedures that ensures a quality finished product. The objective of this is to produce a quality medicine to safeguard the health of the user. There are standard stipulations for GMP certification which varies country-wise but the objective remains the same. The manufacturing unit or pharmaceutical or phyto-pharmaceutical house needs to be GMP certified. GMP fulfilment is a must prior to release of a new herbal drug. It is one of the weapons to the drug control department to check adulteration in drugs. GMP inspections in India are conducted by state FDA.

7.6   Products

Medicinal plants are traded both as raw herbage or crude drug and processed products. The demand for a wide range of species is increasing with the expansion of market and development of new end uses. The herbal products are nowadays classified as traditional and conventional herbal medicinal products, fortified foods, dietary supplements, foodstuffs and cosmetics. The crude drugs are used in pharmaceutical companies for isolation of single purified drugs, preparation of plant extracts, development of advanced extract or as starting material for the production of other semi-synthetic pharmacologically active substances.

The value-added products from medicinal plants include the following:

Production of quality seeds and other planting materials, which can be used for expansion of commercial cultivation of potential medicinal plants as dearth of quality planting material is an important bottleneck. This type of venture can be taken up by seed companies, commercial nurseries as well as progressive farmers under the supervision of NSC/NSCA/SSC/SSCA. Commercial cultivation as well as separation of economic plant parts, initial drying of the raw material (either in sun or in driers) and marketing of dried products which is a simple value addition approach can be taken up both by farmers and entrepreneurs. Sorting of economic plant parts into different grades like *Rauwolfia* and *Withania* roots, *Aloe* leaves, *Piper longum* fruits, *Cassia angustifolia* leaves and pod shells and *Hyoscyamus niger* leaves, *Catharanthus roseus* leaves, stem and roots separately, whole herbage of *Andrographis paniculata*, etc. is the next step of value addition after drying which needs knowledge on the industrial maturity, market needs and quality of crude drug. Preparation of product in powder form is another kind of value addition which can be taken up at cottage industry level using simple grinder and packing machines. Nowadays many kinds of such material like Kalmegh powder, Triphala Churna, Ashwagandha root powder, Chirayata Churna, Stevia powder, Centella powder, Brahmi powder, Gudmar Churna, Punarnava Churna, as well as dried fruit pieces of Aonla, etc. have been marketed by several entrepreneurs. Separation of rosy mucilaginous husk from dried seed of *Plantago ovata* is one of the important value addition approaches as Psyllium husk is one of the major traded item from India. It involves the installation of a machine having a number of shellers where the grinding pressure is so adjusted that only the husk is removed. Herbal teas are currently an important choice worldwide. This value addition technique involves mixing of crude drug in powdered form of specific mesh with tea and packed either in tea bags, in sachets or in containers. Several concerns are presently marketing arjuna tea, stevia tea, pomegranate tea, raspberry tea, strawberry tea, lemon tea, tulsi (Basil) tea, chamomile tea, etc. China is credited with utilizing more than 200 medicinal plants as herbal teas (Rao et al. 2012). Separation of *Aloe* gel from *Aloe* leaves is another type of value addition. It involves the installation of a plant in which the yellow-coloured gel is separated from the raw herbage and bottled. Development of galenicals as liquid or solid extract or in powder form or tinctures is an important arena for value addition of medicinal plants. Now in the health-care systems, the use of plant-based extracts is getting momentum. Solvent extracts of many medicinal plants having standard concentration are used throughout the world. The extracts of several plants like Aloe, Basil, Senna, Cinchona, Ipecac, Mints, Belladonna, Bacopa, Chilli, Aonla, Guggal, Rauwolfia, Ginseng, etc. are marketed by several industries. Some of those are marketed like modern drugs in the form of pills or capsules. Still some are marketed as powders like Withania root powder. Isolation of intermediate for the development of semi-synthetic drugs like diosgenin and hecogenin which can further be used for the development of steroidal drugs is also practiced by the pharmaceutical industries. Isolation of pure active principles from raw herbage and identification of newer compounds of plant origin are the major focus of the recent pharmaceutical research. The effective isolation and quality assessment technique at industrial scale of many medicinal plants have been developed by the houses. Several other value-added products like herbal juices and drinks and confectionaries are now developing by the related industries. Phytomedicine or botanical pesticides are nowadays widely used in the agricultural sector. Some easy recipe and techniques are now available through which the farmers can prepare it and use the home-made products like Dasagabya. Herbal disinfectant solutions are also important addition in this field. Several other formulations like neem oil, garlic oil and basil oil for phytomedicine or botanical pesticide purpose and neem oil, basil oil, vetiver oil, citronella oil etc. for health care purpose are coming on a regular basis. Cosmeceuticals are now becoming popular worldwide. This kind of product includes creams, lotions, soaps, perfumeries, hair oils, etc. The future significant value-added products from medicinal plants will be the nutraceuticals and fortified foods which are having enormous potentiality. The materials which are sources of high antioxidant – the scavengers of free radicals – are very much demanded today as it prevents ageing and cancer to some extent. These materials when incorporated in the foodstuffs or in prepared foods are working very efficiently.

7.7 Processes of Extraction

Extraction is the process of separation of active principles from the inactive component using standard techniques. There are several classes of extracts like powdered, fluid, tinctures, decoctions, semisolid and infusions which need further processing to develop traditional or modern herbal drugs. Hence, the extraction procedure has a pivotal role in the quality of the finished product. The raw herbage can be extracted through maceration which is a solvent-based extraction technique. On the other hand, decoction needs boiling of herbage in water for a specific period. Circulatory extraction, infusion and digestion are the different forms of maceration technique. Soaking, filtration and concentration are the techniques where agitation during soaking in alcohol (solvent) can be done in a simple mixer cum juicer followed by filtration through filter paper and finally concentration can be done using rotary vacuum evaporator and laboratory hot plate (Odey et al. 2012). The difference in root and stem extracts of three medicinal plants using this technique as reported by them is presented in Table 7.5.

Percolation is the method of extraction of active principles for preparation of tinctures. Use of soxhlet is another important solvent-based extraction technique for continuous extraction under boiling temperature of the solvent. Sometimes for better extraction of active principles, fermentation technique is used where soaking for a specific period is done for in situ generation of alcohol facilitates the extraction. Supercritical fluid extraction (SCFE) is another technique which involves low solvent use and develops high sample throughput that allows

Table 7.5 Root and stem extract of some medicinal plants

Plants	Percentage yield	
	Root extracts	Stem extracts
Vernonia amygdalina	13.6	11.96
Gongronema latifolia	10.75	10.01
Nauclea latifolia	19.90	18.30

greater recovery percentage with no solvent residue in the final product. Sonication means use of ultrasound having 20–2,000 KHz frequency that increases permeability of cell walls and facilitates extraction of active principles. Countercurrent extraction (CCE) by grinding the herbage and producing a slurry facilitates quick extraction without any risk of degeneration of active compounds due to high temperature. The new patented phytonics process is a health- and environment-friendly approach of extraction of active principle with superb quality which needs no further processing after extraction.

7.8 Biotechnological Approaches

The area of biotechnical research in medicinal plants has received considerable attention of public and private enterprises for the lucrative commercial markets of plant-derived drugs.

The major focuses of biotechnical researches in medicinal plants are the following:

7.8.1 In Vitro Propagation and Hairy Root Culture

The major problem arising from large-scale harvesting of produce from medicinal plants is the limited availability of source plant, which mostly exists in natural habitat. Indiscriminate harvesting is rapidly depleting the existing resource; moreover, it is not sustainable as in near future the source will be depleted. This leads to loss of germplasm, increased vulnerability of the ecosystem as well as the people involved in the enterprise. The list of threatened taxa of medicinal plants has already touched the 10,000 mark; thus, alternate method for large-scale cultivation of plants under in vitro condition is a primary emphasis of medicinal plant business sector.

Ideally, any plant part can be cultured in vitro for large-scale multiplication. Often, the active ingredient of interest may be confined to certain plant parts like alkaloids in roots; thus, mass multiplication of plant parts instead of the whole

plant is economical. Moreover, the in vitro plants are selected from best varieties and are disease-free and the production system can be automated in bioreactors reducing input cost. Instead of whole plant, specific organs can be cultured, economizing space and increasing productivity. In addition, these organs can be stored in specific minimal medium for a long period, which is advantageous for production systems which often need to be synchronized with the demand in the market. Organ culture is well established for hairy root culture, where the hairy root-inducing bacterium '*Agrobacterium rhizogenes*' is cocultivated with plant roots. Such cultures are genetically stable, have higher secondary metabolite production capacity and can be grown in hormone-free media for a long period, sometimes over 4 years. Hairy root culture systems are well established for *Atropa belladonna*, *Gingko biloba*, *Panax ginseng*, *Coleus forskohlii*, *Gmelina arborea*, *Papaver somniferum*, *Ammi majus*, *Camptotheca acuminata*, *Pueraria phaseoloides*, *Rauwolfia serpentina*, *Hyoscyamus albus*, *Withania somnifera* and some other species. Hairy root culture has also successfully been used to produce active compounds from nearly to be extinct medicinal plants such as devil's claw (*Harpagophytum procumbens*), which is used for treating arthritis. The hairy root culture system allows large-scale harvest of two active compounds harpagoside and harpagide.

Instead of culture of organs, individual cell culture has also been successful for harvesting of secondary metabolites from the medicinal plants. These cells are cultured in large bioreactors, and the production systems are automated for higher efficiency of the system. For example, Phyton Biotech, a US-based company, uses cell culture for production of paclitaxel, an intermediate for production of taxol, which is widely used for treatment of cancer patients worldwide.

7.8.2 Molecular Pharming/ Biopharming

Molecular pharming or biopharming is the method of large-scale production of pharmaceuticals in the plant system. The pharmaceuticals may be plant derived, animal derived or synthetic. The target plant is genetically engineered for production of the pharmaceutical compound. Under such condition, the genetically altered plant becomes a medicinal plant. The term molecular pharming also includes large-scale production of pharmaceuticals in the plant system either in the form of cell culture or organ culture. The genes encoding for high-value pharmaceutical production are engineered in the target plant for production of desired pharmaceutical compounds such as therapeutic proteins and monoclonal antibodies. Genetic transformation technology is well established in maize, a high storage tissue (kernel) containing crop. This crop has high potential for molecular pharming. It is estimated that about 500 kg of monoclonal antibodies can be produced from a good maize crop of about 40–100 ha area. Besides, tobacco, carrot and safflower are also being used for molecular pharming. The USA in 2012 has approved molecular pharming of a plant-based enzyme to treat a rare genetic disease, the Gaucher's disease, which has been produced from cultured carrot cells. The main challenge in the area of molecular pharming is to develop low-cost processes for large-scale industrial cultivation and extraction of the target compound.

7.8.3 Characterization of Biosynthetic Pathways

Using the new genomics and transcriptomics technologies, the pathways of secondary metabolite production are elucidated from important medicinal plants. *Catharanthus roseus* is considered as a very important medicinal plant for treatment of cancer, as two anticancer drugs, vinblastine and vincristine, along with numerous other alkaloids (terpenoid indole alkaloids) are obtained from this plant species. Using an Illumina HiSeq2000-based RNA sequencing strategy, 44.2×10^9 bases of RNA have been sequenced in *C. roseus*. The information after bioinformatics processing has helped to develop

a metabolic pathway database named *CathaCyc* (Van Moerkercke et al. 2013), which identified 390 pathways, 1,347 enzymes and 1,322 compounds. In another study, expression analysis of terpenoid biosynthesis genes in *Litsea cubeba* using Illumina paired-end sequencing revealed 68,648 genes, of which 16,130 unigenes were assigned to 297 pathways.

DNA microarray technologies are widely used in medicinal plant research for identification and authentication of medicinal plants, gene expression profiling and whole transcriptome characterization. Besides techniques like cDNA-AFLP, serial analysis of gene expression (SAGE) and its variants (superSAGE, longSAGE), massively parallel signature sequencing (MPSS), EST sequencing and whole-genome RNA sequencing are novel techniques for identification and characterization of biosynthetic pathways in medicinal plants. For example, next-generation massively parallel sequencing of *Podophyllum hexandrum* and *Podophyllum peltatum* transcriptomes has helped in identification of podophyllotoxin biosynthesis pathway. EST analysis was used in *Polygonum minus* to identify key genes involved in flavonoid biosynthesis, namely, chalcone synthase, flavonol synthase and leucoanthocyanidin dioxygenase.

Recently, gene clusters controlling the biosynthetic pathway of some alkaloids have been identified through transcriptome analysis. In opium poppy, a 10-gene cluster has been identified for expression of five enzymes involved in biosynthesis of noscapine, an antitumor alkaloid. Ten genes were also identified to control synthesis of steroidal glycoalkaloids (SGAs) like α-solanine in tomato and potato.

7.8.4 Engineering of Biosynthetic Pathways

Successful engineering of biosynthetic pathways to produce desirable protein or secondary metabolites is a long-term goal in transgenic research of medicinal plants. The engineered products have been synthesized either in microorganisms like *E. coli*, single-cell eukaryote like yeast or higher plants. Some examples are the following:

- Engineering of gene for hyoscyamine 6-β hydroxylase (*h6h*) from *Hyoscyamus niger* in *Atropa belladonna* has resulted in increase of production of scopolamine.
- Simultaneous introduction and over-expression of two genes putrescine N-methyltransferase (*pmt*) and *h6h* in transgenic *H. niger* hairy root cultures was achieved to produce significantly higher levels of scopolamine.
- Introduction of genes for TR-I (*trI*) and h6h from *H. niger* led to synthesis of novel-acetylated forms of tropine in *N. tabacum*.
- Over-expression of 3-hydroxy-3-methylglutaryl coenzyme A (HMG-CoA) reductase showed limiting enzyme in the mevalonate pathway doubled linalool production in yeast.
- A sesquiterpene synthase gene isolated from *Helianthus annuus* was engineered to synthesize high amount of farnesyl diphosphate in yeast.
- A limonene synthase gene from spearmint (*Mentha spicata*) was constitutively expressed in spike lavender (*Lavandula latifolia*) plants.

In conclusion, biotechnological application in medicinal plant research is a vibrant area of research with potential to deliver miracle solutions for diseases and ailments. Many established technologies are being commercially exploited to produce life-saving drugs, while new discoveries are expanding the horizon of current knowledge in this field.

References

Chatterjee SK (2010) Status of medicinal plants in global perspective: biodiversity destruction and conservation. In: Proceedings of Botanicals in integrated Health Care Convention 2010, 26–28 December 2010. The Agri Horticultural Society of India, Kolkata, India, pp 32–36

Kalia AN (2005) Worldwide trade scenario of medicinal and aromatic plants. In: Tyagi CS, Verma PK, Hooda JS, Yadav OP, Goyal RK (eds) Course compendium-winter school on advances in medicinal, aromatic and under-utilized plants research, 29 September–19 October 2005 at CCSHAU, Hisar, Haryana, India, pp 271–273

Kaushik P, Dhiman AK (2000) Medicinal plants and raw drugs of India. Bishen Singh Mahendra Pal Singh, Dehra Dun

Maitra S, Satya P (2009) Ethnomedicinal plants of India. In: Singh HP, Pandey ST, Singh B (eds) Dynamics of medicinal and aromatic plants. Indus Valley Publication, Jaipur, pp 74–99

Maitra S (2012) Conservation of biodiversity of medicinal plants in Northern part of West Bengal: instant potential approaches. In: Saha H, Ghosh ML, Gangopadhyay G, Saha D, Singh PK, Sarker S, Das SC (eds) Biodiversity conservation – fundamentals and applications. DumDum Motijheel College & Sarat Book Distributors, Kolkata, pp 69–74

Odey MO, Iwara IA, Udiba UU, Johnson JT, Inekwe UV, Asenye ME, Victor O (2012) Preparation of plant extracts from indigenous medicinal plants. Int J Sci Technol 1(12):688–692

Rajeswara Rao BR, Rajput DK, Nagaraju G, Adinarayana G (2012) Scope and potential of medicinal and aromatic plants products for small and medium enterprises. J Pharmacogn 3(2):112–114

Singh V, Vadera S (2010) Export potential of herbal and medicinal plants in India. SME WORLD (November, 2010). Web-address – http://www.smeworld.org/story/special-reports/export-potential-of-herbal-and-medicinal-plants-in-india.php (visited lastly on 20.02.2014)

Tiwari DN (1999) Medicinal plants for health care. Yojana, pp 8–17

Tyagi CS, Yadav OP, Verma PK (2005) Conservation of biodiversity of medicinal and aromatic plants. In: Tyagi CS, Verma PK, Hooda JS, Yadav OP, Goyal RK (eds) Course compendium- winter school on advances in medicinal, aromatic and under-utilized plants research, 29 September–19 October 2005 at CCSHAU, Hisar, Haryana, India, pp 8–11

Van Moerkercke A, Fabris M, Pollier J, Baart GJ, Rombauts S, Hasnain G, Rischer H, Memelink J, Oksman-Caldentey KM, Goossens A (2013) CathaCyc, a metabolic pathway database built from Catharanthus roseus RNA-Seq data. Plant Cell Physiol 54(5):673–685

Postharvest, Product Diversification and Value Addition in Coconut

8

D.K. Ghosh

Abstract

Coconut is a crop with a large potential for varied use and considered to be the most important and useful among the tropical palm. Almost all the parts of coconut are useful both for domestic purpose and in industrial units and it is called 'nature's supermarket'. Product diversification, adoption of stringent quality standards for coconut products and increased productivity are some of the ways to make this industry's fate bright. Coconut is a smallholder crop, and millions of rural people depend on it for survival. Its development particularly in postharvest activities could be the base for rural development in the coconut-producing countries. Copra, coconut oil, desiccated coconut (DC), coconut cream, coconut milk, virgin coconut oil, spray-dried coconut milk powder, coconut chips, cream, nata de coco, coconut jam and young tender coconut are the convenience coconut products, and some by-products are coconut fibre, e.g. coir and coir products, mats, matting, brushes, brooms and rubberised coir mattresses, and shell products, e.g. charcoal, activated carbon, etc. For coconut-based oleochemicals including fatty alcohol, fatty acids, methyl esters, tertiary amines, alkanolamides and glycerine, there has been a growing demand in the world market. In the present chapter, the aspects discussed are coconut, harvest, post harvest, products, by-products, processing industry, prospect, problems, health benefit of products, etc.

8.1 Introduction

The coconut is a crop with a large potential for varied use. It is considered to be the most important and useful among the tropical palm. Almost all the parts of coconut are useful both for domestic purpose and in industrial units. In beauty and utility, no other tree can surpass the coconut tree. It is the most extensively grown nut in the world,

D.K. Ghosh (✉)
Bidhan Chandra Krishi Viswavidyalaya,
Mohanpur, Nadia, West Bengal, India
e-mail: drdipakghosh08@gmail.com

A.B. Sharangi and S. Datta (eds.), *Value Addition of Horticultural Crops: Recent Trends and Future Directions*, DOI 10.1007/978-81-322-2262-0_8, © Springer India 2015

the most important palm. It provides people's basic needs such as food, drink, shelter, fuel, furniture, medicine, decorative materials and much more. They are a necessity and a luxury. It is the 'heavenly tree', 'tree of life', 'tree of abundance' and 'nature's supermarket'. Though India produces more than 13 billion nuts per annum, the postharvest processing is presently confined to the production of edible and milling quality of copra, coconut oil and coir and coir-based products. The growth of product development and by-product utilisation is considerably lower in comparison with other countries like the Philippines, Indonesia and Thailand. The postharvest processing and marketing sector of coconut in India is still at its infancy stage. Roughly 40 % of the coconut production in the country is made into copra of which about 80 % is converted into oil. More than 48 % of coconut is consumed as raw nuts in household and 10 % in the tender form for drinking purpose (Nanda Kumar 1998). Coconut industry in India at present survives mainly on coconut oil industry. The price of coconut and copra therefore move with the price of coconut oil which very often experiences yearly, monthly and daily variation. In this context product diversification and by-product utilisation are a must (Thampan 1996a). Technological research in the country has been successful in evolving appropriate processing technologies for the profitable utilisations of some of the product and by-product of the coconut palm. Moreover, there is a growing demand for different products of coconut both in the domestic and export markets. Hence, it has become the right time to encourage the development of coconut-based industry in India. However, development of nontraditional sector like coconut-based industry requires techno-economic support, market research and promotion.

Lack of awareness and actual skills on coconut postharvest technologies has caused significant losses starting from the harvesting of the nuts, seasoning, drying and storage. While wastage and losses occur at different stages, the copra drying stage or the efficiency of the drying process at the farm level is the most critical stage as this affects subsequent losses in terms of product quality and reduced prices. Harvesting of immature nuts causes the production of rubbery copra with high moisture content. If one allows the nuts to fall naturally, without harvesting or picking the nuts from the tree, the losses due to overripe nuts or germinated nuts are likely to occur. This could be as high as 10 % of the total harvest especially with varieties that are early germinating. As the growing embryo utilises the stored food in the endosperm, the copra produced from germinated nuts would be thinner, lighter and with lower oil content. Losses due to pilferage and losses due to nuts that are hidden or covered by thick weeds or shrubs could also range from 5 % to 10 % of the total harvest if one does not regularly harvest his or her coconuts. To avoid these losses, it is recommended that the 45-day cycle of coconut harvesting be adopted. Seasoning of unripe nuts for 2–4 weeks should also be practised. Farm sanitation, e.g. weeding of thick shrubs and grasses in the spaces between coconut palms, is highly recommended to prevent losses due to uncollected nuts. As mentioned earlier, major postharvest losses are caused by improper drying of copra as a result of a lack of know-how on the proper drying technology and the lack of incentives to adopt the recommended copra dryers and the appropriate copra drying methods. Improperly dried copra or copra with high moisture content are prone to aflatoxin contamination. Coconut researchers have also identified beetles, cockroach, a moth and an earwig to be associated with deteriorating copra and copra cake. Studies reveal that after 1 year of storage, copra weight loss due to pests will be as high as 5–10 %. Spraying of suitable insecticides may be done, but this is not practised due to its prohibitive cost. Sanitary practices in the copra warehouse are the best recommended alternative to control these pests. Generally, these pests are considered a minor problem when compared to the attack of aflatoxin-related moulds or fungi. Other factors cited to contribute to copra/copra cake deterioration are presence of wet or improperly dried copra, rubbery copra, delays in transport, long storage period and unsanitary conditions in the farms and warehouse. Long storage time also favours the breeding of copra pests or the proliferation of aflatoxin-related moulds.

8.1.1   Coconut Yield and Harvesting

Bearing coconut palms produce nuts throughout the year, although yields may vary with the season. A normal-bearing, adult palm produces at least one matured ready-to-harvest bunch of coconuts every month. Depending on the variety, the number of nuts per bunch can vary from 5 to 15. The theoretical number of bunches per palm that can be harvested annually is about 14 from tall coconut varieties and 16 from the dwarf species. It usually takes 12 months for a nut to mature from pollination to harvest. Husk colour is the best indicator of coconut maturity. To attain good quality products, it is advisable that coconuts be harvested at the right maturity. Thus, only nuts that are partially or completely brown should be harvested. Nuts harvested at the tenth month or colour-break stage should be stored or seasoned for some time to increase copra and oil yield. To obtain maximum copra and oil recovery, nuts must be harvested when fully ripe. At this age of maturity, the estimated age is from 11 to 12 months. Although this stage is ideal for copra production, in practice, green and immature nuts (about 10 months old) are sometimes included during harvest especially as harvesters are paid on a per nut basis in certain countries. Immature nuts when converted into copra will produce rubbery copra with low oil recovery. Rubbery copra is also susceptible to insect and mould attack due to its high moisture content. Immature nuts should therefore be segregated for seasoning for about 2–4 weeks. Seasoning is done under a shed, preferably with a concrete or wooden floor.

8.1.1.1   Harvesting

In practice, the harvesting cycle varies from 45- to 60- or 90-day periods. However, considering the hired labour cost, the recommended harvesting cycle is every 45 days for practical and economic reasons. Two to three bunches of coconuts could be harvested from each palm if this cycle is followed. This harvesting cycle has been found to yield a good number of mature nuts with high copra and oil recovery. The methods of harvesting coconuts vary among countries or even among provinces within the same country. Producers from certain countries, especially in the Pacific, do not harvest their coconuts. Mature nuts are just left to fall on the ground and gathered by the farmer or the members of the farming family at regular intervals. There are two common methods of harvesting coconuts. These are the pole and the climbing method. A third method is only practised in Thailand, Malaysia and Indonesia. This procedure involves harvesting of mature coconuts using of trained monkeys. The pole method of harvesting is common in many countries in the region. In this case a harvesting scythe attached to the end of a long bamboo pole is used. The palm-climbing device is useful and advantageous for harvesting operations in places where traditional palm climbers are not available. The device is more efficient than manual climbing. With its use around 80 trees are harvested a day. There is also no risk of falling from the tree. In research stations and seed farms, the gadget could be useful for breeding purposes. Although both the pole and the climbing method of harvesting require considerable experience and skill to be performed safely and efficiently, each has its own advantages and disadvantages. Harvesting using bamboo poles is generally faster, more efficient, less tedious and less dangerous when compared with climbing. With bamboo poles, a harvester could also harvest more nuts per unit of time from more trees. On the other hand, the advantage of harvesting by climbing is that the climber/harvester could clean and inspect the crown of the palm for pest and disease attack. However, the cuts made to construct steps in the trunk in certain countries to facilitate climbing make the trees less suitable for timber purposes and fractures serve as entry points for pests. Harvesting coconuts by using trained monkeys is considered efficient and cost-effective especially in areas where labour has become scarce.

8.1.1.2   Harvest Maturity Indices

Coconuts are harvested at two different stages of development, depending on the intended use. Coconuts intended to be consumed fresh for the water content and jelly-like meat should be harvested when the fruit have reached full size but at an immature stage with soft inner white meat (endosperm). Fruit intended to be harvested for copra and further processed into oil should be harvested at a mature stage, when the inner white meat has thickened and hardened. Several different indices can be used to determine coconut maturity. These include time from flowering, fruit size, external appearance, and amount and texture of the meat. Water coconuts should be harvested soon after the fruit has reached full size but while it is still immature. This coincides with maximum water content and occurs about 7 months after flowering. In immature coconuts, the skin surface around the cap on the top of the fruit is typically whitish yellow. Also, the short stem above the individual coconuts that originally contained the male flowers will have partially dried coconuts intended for copra production that should be harvested when fully mature, which requires about 12 months from flowering. The skin will have turned mostly brown. The stem on top of the coconut is also brown at full maturity. At this stage, the coconuts will have their maximum copra content and oil recovery. At the fully mature stage, the meat is firm and is eaten without processing or may be shredded, dried to produce copra and then squeezed to produce coconut milk, a water-oil emulsion. Although the fully mature stage is ideal for copra production, in practice, immature coconuts are sometimes included during harvest if the entire bunch is picked. Immature coconuts will produce rubbery copra with low oil recovery. Rubbery copra is also susceptible to insects and mould due to its high moisture content. External fruit appearance is an indicator of maturity. Depending on cultivar, coconut fruits are green, yellow or gold in colour when immature. Water coconuts should be harvested at one of these colour stages. The fruit will turn a brownish colour as they become mature. Fruit size is also indicative of maturity. The fruit should be fully developed in size before being harvested, either as a water coconut or for copra. The amount and texture of the meat is a destructive index of harvest maturity. Several randomly selected fruits of different sizes are cut open to determine the amount and firmness of the meat. Other fruits of similar size from the same cultivar are assumed

to be in the same stage of maturity. The meat of water coconuts should be thin, soft and jelly-like. The water content of immature fruit is high but is gradually absorbed into the meat with increasing fruit maturity. The meat of coconuts harvested for copra should be thick and firm with limited water content.

8.1.2 Harvest Methods

The fruit from shorter growing coconut trees may be harvested from the ground using a machete or knife to cut the stem just above the shoulder of the fruit. The fruit may also be twisted or snapped off the tree by hand. Fruit borne on mature tall trees may be harvested with the aid of a ladder or climbing device, by skilled climbers, or by using a sharp blade attached to a long pole. The coconuts are generally left to drop to the ground and collected after the entire tree has been harvested. If the majority of fruit on an entire bunch is ready for harvest, the coconuts can be harvested as a unit by severing the stem just above the first fruit in the cluster.

8.1.3 Preparation for Market

The coconuts should be gathered from the field soon after harvest and taken to a shaded collection site. Fruits which are unmarketable due to excessive insect damage, decay or undesirably small fruit size are discarded. The remaining fruits should be prepared for market by cleaning/dehusking, sorting and packing.

8.1.4 Cleaning/Dehusking

The surface of coconuts should be cleaned with a damp cloth or cotton gloves to remove excess dirt, dust or undesirable stains. Water coconuts marketed domestically are typically not dehusked. However, most coconuts intended for export must be dehusked to reduce the transport weight and volume. Coconuts should be dried for several days at ambient temperature before dehusking by hand. The outer coloured skin (exocarp) and the fibrous inner husk (mesocarp) are stripped away by striking the coconut against a sharp-pointed metal stake mounted on a platform. A few impaling strokes loosen the husk, making it easier to be removed. A machete can also be used to initiate the dehusking process. Dehusked coconuts are oval to round in shape with the eyes showing. To prevent browning, the dehusked coconuts can be dipped in a 1–3 % sodium metabisulphite solution for 2–5 min. This treatment prevents browning for a period of 5–7 days. A fungicide may be included in the sodium metabisulphite solution to inhibit the growth of surface mould. The discarded husks can be placed several layers deep over the dehusked coconuts to help reduce desiccation.

8.1.5 Grading

Water coconuts intended for export should be graded according to size, uniformity of shape and degree of skin blemishes if they are not dehusked. The fruit should be categorised into small, medium and large sizes. The fruit should be uniform in shape and free of noticeable blemishes or skin damage from insects, diseases or physical injury. Surface colour should be uniform and characteristic of the cultivar. The preferred skin colour of non-dehusked water coconuts in the export market is green. Dehusked coconuts should be free of stress cracks and not have deeply sunken eyes. The fruit should not have any protruding germination tubes, leakage of water around the eyes or surface mould. When shaken, the fruit should have a sloshing sound, indicating the presence of water in the coconut. Any fruit that does not have a sloshing sound when shaken should not be packed for market. The most common size for exported dehusked coconuts is between 750 and 850 g, although the acceptable sizes typically range from 600 g to 1 kg.

8.1.6 Waxing

The market life of water coconuts can be extended by waxing the fruit with paraffin. Waxing significantly reduces weight loss and is also very effec-

tive in reducing stress cracking of dehusked coconuts during transport. The fruits are waxed by rapidly dipping them in a tank of melted paraffin.

8.1.7 Packing

Coconuts are packed in various types of containers, depending on the market destination. Domestically marketed water coconuts may be sold in bulk or packed in large synthetic or mesh sacks of known fruit count per sack. If the coconuts are sold in bulk, the fruit are usually loaded onto the bed of a large truck and transported to the destination market. Considerable manual labour is required to load and unload the bed of a truck with loose coconuts. In large-scale operations, the loading process is made more efficient by packing the fruit in large wooden bins on top of pallets. A hand jack or forklift can be used to move the bins onto the bed of the truck. If domestic marketed fruit are dehusked, usually 40–50 coconuts are put in the sack. Wooden crates may also be used. Coconuts for export are usually dehusked and packed in strong well-ventilated fibreboard cartons, with minimum 275 psi test strength. The carton typically has a net weight of 18 kg. Uniform-sized fruit should be packed in each carton. Dividers may be used to separate individual fruit. Wrapping of husked coconuts in thin polyethylene film will significantly reduce weight loss. There is a lesser export market demand for coconuts that have not been dehusked. However, some demand exists in high-priced niche markets. For these markets, water coconuts are packed by count in 3 kg fibreboard cartons. Typically two or three fruits are packed in each carton.

8.1.8 Temperature Control

Although coconuts are of tropical origin, the ideal storage temperature range is 0–1 °C. They can be stored satisfactorily at this temperature range for 2 months. Therefore, it is possible to ship dehusked coconuts successfully by refrigerated sea container to any destination worldwide.

Moderately cool temperatures of 12–16 °C will allow up to a 3-week market life. In the absence of refrigeration, the market life of fresh coconuts is short. They can be held in a shaded location at ambient temperature for up to 2 weeks without a significant loss in quality.

Water coconuts that have not been dehusked store for a longer period than dehusked coconut. The sugar content in the water of dehusked fruit declines more than that in non-dehusked fruit. Also, the acidity of the water of dehusked coconuts increases to a higher level than in non-dehusked coconuts. Consequently, the taste of stored dehusked coconuts becomes less desirable than non-dehusked fruit. The husk helps to preserve fruit quality and increase the storage life of water coconuts.

8.1.8.1 Relative Humidity

The ideal storage relative humidity (RH) for water coconuts is between 80 % and 85 %. Coconuts are subject to weight loss and transpiration loss of the water at low RH storage. However, if the RH is above 90 %, the fruit is susceptible to surface mould.

8.1.9 Coconut Dehusking and Splitting

After harvesting, the succeeding operations are collection of the nuts, ripening and dehusking. Harvested nuts are usually gathered together on a single layer on the ground. If the soil is moist, there is always the tendency for the nuts to germinate. Hence, nuts are not allowed on the damp ground for a long time, but are moved to a drier place. As mentioned earlier, the nuts are kept for about a month to ripen on the ground. This practice promotes desirable changes in the greener or somewhat less mature nuts: the coconut meat is said to grow thicker and harder thus producing a better quality copra if copra is desired or a more suitable material for desiccated coconut production. Immature nuts tend to produce rubbery copra. Producers claim that seasoning or storage of 10–11-month-old green nuts for 1 month or so improves the coconut kernel. This reduces

the tendency to produce rubbery copra. Also dehusking is easier. Coconuts in the husk are very bulky. They are dehusked first before being transported in trucks or carts. Dehusking is a manual procedure. The principal part of the dehusker is a sharp-pointed shard of steel (a part of the native plough) positioned vertically with the point up and the broader part firmly placed on the ground. The farmer-operator impales the coconut on the sharp point with a strong determined downward movement. A few impaling strokes loosen the husk, making it come off (usually) in one piece. Impaling requires accuracy and nerve. Hence, it has been difficult to get dehuskers in countries that are still trying to set up coconut plantations. Since dehusked coconut is an important article of commerce locally, husking therefore becomes mandatory. The coconut husks are left with the farmer. In the places where there is a coir fibre industry, the husks may be sold to this industry. Most often the husks are not sold but are used as fuel for drying copra. If little or no copra is made, there is an accumulation of coconut husks. Since the coconut meat is found well inside the nut and is firmly attached to the shell, certain steps are necessary before drying the coconut kernel.

8.1.10 Nut Splitting

After the coconut is dehusked, the hard but brittle shell is exposed and can be split open into two halves using a machete. The coconut water is drained off leaving the cups ready for the drying stage. The meat is still attached to the shell. During the drying process, the meat shrinks and is easily detached or scooped out from the shell. These cups of coconut meat are then dried further. Some farmers also practise nut splitting using a heavy machete even without dehusking the coconut. After nut splitting, the water is allowed to drain off. With meat still attached to the shell and the shell to the husk, the halved nuts are dried under direct sunlight. During the drying process, the meat becomes detached or is scooped out from the shell with a scooping knife. The cups of meat are then further dried into copra.

8.2 Product Range of Coconut

Of the world production of coconut, more than 50 % is processed into copra. While a small portion is converted into desiccated coconut and other edible kernel products, the rest is consumed as fresh nuts. The coconut palm also provides a series of by-products such as fibre, charcoal, handicrafts, vinegar, alcohol, sugar, furniture, roofing and fuel, among others, which provide an additional source of income. Diversified local uses of the coconut palm number over 200. A wide range of coconut products are internationally traded. There are more than 50 unprocessed, semi-processed or processed coconut products entering the international markets in small and big quantities. Aside from copra and coconut oil, other exports which have a significant volume are desiccated coconut, copra meal, cocochemicals (fatty acids, fatty alcohol, methyl ether), shell charcoal and activated carbon, fibre products, coconut cream, milk, powder and nata de coco. Although global production continues to increase, the growth rate of domestic use was faster, therefore reducing the exportable supply of kernel products to about one third of the production. However, there has been a sizeable increase in the export of nontraditional products in recent years. Both copra and coconut oil are traded internationally. In the past, the main export was copra. In the 1960s over 1 million tons of copra was traded worldwide per annum. The volume declined to about 900,000 t a year in the 1970s further declining to an annual average of 350,000 t in the 1980s. This immense drop occurred when the producing countries established domestic copra processing plants in response to their desire to obtain more value-added products. In 1994 copra exports further dropped to 234,874 t with Papua New Guinea at 53,767 t followed by Indonesia, Vanuatu, the Philippines, Solomon Islands and Malaysia with exports below 40,000 t each. The downtrend in copra exports is likely to continue.

From the utilisation point of view, coconut products can be grouped as follows (Rajagopal 1998):

8.2.1 Food Products

(a) *Wet meat or kernel*: the kernel of 7–8 months old is soft and rich in sugar and protein and eaten as such. The kernel of 11–12 months old is used in household purpose and processed in different ways.
1. *Coconut milk (CM)* – it is used as substitute of dairy cream. Preserved form of milk, e.g. canned cream milk (40 % fat) or whole milk (12.5 % fat), is available in the market.
2. *Coconut jam (CJ) and coconut syrup (CS)* – CJ is an excellent bread spread and used in confectionery. CS with 44.5 % soluble solids gives a delicious instant milk white drink.
3. *Coconut skimmed milk (CSM)* – it is a good source of quality soluble protein in a place where animal protein is deficient.
4. *Coconut flour (CF)* – the fibrous residue obtained after expelling the milk.
5. *Desiccated coconut (DC)* – white kernel is desiccated to a moisture level of less than 3 %. It is an important commercial product used in confectionery and has worldwide demand.
6. Coconut chips, coconut shreds, etc.

(b) *Coconut water and its product*:
1. *Coconut water* – from 6 to 7 months old, tender coconut water is effective for gastroenteritis, etc. and rich in vit B minerals, sugar, etc.
2. *Nata de coco* – a gelatinous substance that can be added to soft drinks.
3. *Coconut vinegar* obtained by fermentation of coconut water.

(c) *Coconut toddy and its product* – toddy is obtained by tapping unopened spadix. Fermented toddy is also used. Coconut jaggery is produced from toddy with high % of sugar (88.3 %).

(d) *Copra* – dried endosperm is called copra. The two types of copra are edible copra and milling copra. An edible copra is available as a ball copra or cup copra.

(e) *Coconut oil* – it is extracted from copra with appropriate devices. Coconut oil was the target of attack for its high content of saturated fat (lauric and myristic acid). But they belong to medium- and short-chain fatty acid and so they are rapidly oxidised when consumed. A study conducted by Biochemistry Department of KAU reported that consumption of coconut oil did not elevate blood cholesterol level.

(f) *Virgin coconut oil*

(g) *Coconut cake* – the cake left after extraction of coconut oil from copra; it is a valuable cattle feed rich in fat.

8.2.2 Nonfood Products

1. Coconut husk (CH) – the husk usually forms 25–45 % of the weight of whole ripe nut.
2. Coir fibre and coir pith are the products derived from coconut husk. Of the husk 30 % constitute coir fire and the balance 70 % is pith and outer skin. Two types of fibres are white and brown.
3. Coir pith is a waste product and can be used to conserve moisture, as manure, to make hard board, etc.
4. Composted coir pith – a good manure.
5. Coconut shell.
6. Shell charcoal, activated carbon, etc.
7. Coconut shell handicrafts.

8.2.3 Miscellaneous Products

Coconut wood is currently used as substitute for conventional wood. Coconut leaves and midribs are used in different purposes for making broom, basket, handicraft, etc (Ghosh 2008).

8.3 New Application for Value-Added Products

1. Organic food – Bulk of coconut is raised without the application of inorganic fertiliser and pesticides. So there is scope for organic products.
2. Virgin coconut oil – Gaining popularity as healthy oil as well as oil widely use in

pharmaceuticals, nutraceuticals and cosmeceuticals. There are scope for large number of small-scale units. The price of an ordinary oil is 400–600 US \$/nut but for VCO the price range is US \$ 3,000–6,000/nut.

3. Functional foods – Designer foods, medical food, longevity food, super food, perspective food, therapeutic food, etc. are the used concept. Coconut food products with lauric acid and MCFA have a greater role to play in the fast-developing functional food particularly baby food, nutraceuticals, pharmaceuticals, etc. to build up resistance against viral, fungal and bacterial disease.
4. Functional drink – Sport drink, energy drink wellness drink markets are continuously growing. Tender nut water and matured nut water with various minerals and vitamins as well as coconut milk can have a wider market both in domestic and international.
5. Cosmoceuticals – Coconut oil rich in C12-C14 FA (lauryl and myristyl) is good for skin care and cleaning products.
6. Oleochemicals – There is a good scope for coconut-based oleochemicals.
7. Biofuel and bio-lubricants – Time is not far to consider coco biofuel in Andaman and Lakshadweep Island.
8. Premium grade monolaurin and HIV/AIDS – Coconut oil with 48.6 % lauric acid is potential source for producing monolaurin (Lauricidin) which has been experimentally found to reduce the virus. If this happens, then the coconut produced in the world may not be adequate.
9. High-value/eco-friendly coir products (Tables 8.1, 8.2 and 8.3).

The main coconut product, copra (dried kernel), and its derivatives coconut oil, copra cake and desiccated coconut represent a major source of foreign exchange. For several small nations, especially in the Pacific, copra is the principal source of foreign exchange. Coconut oil is the leading commercial product which contributes nearly 7 % of the total supplies of vegetable oils in the world. The world's production of coconut oil reached 3.1 million tons in 1996. Traditionally, the coconut is dried to produce copra, and the oil

Table 8.1 List of coconut products

Different coconut products		
Activated carbon	Coconut oil	Coconut vinegar
Ball copra	Coconut oil – medicated	Coconut water concentrate
Brooms	Coconut oil – virgin	Coconut water soda
Coconut de husked	Coconut oil solvent extracted	Coconut wood handicrafts
Coconut biscuit	Coconut oilcake	Coir pith briquette
Coconut chips	Coconut pickles	Coir products
Coconut chutney powder	Coconut seedlings	Copra
Coconut flakes	Coconut shell	Desiccated coconut
Coconut honey	Coconut shell charcoal	Nata de coco
Coconut husk	Coconut shell crockery cutlery	Ready to cook fried coconut meat gravy
Coconut husk handicrafts	Coconut shell handicrafts	Technology source
Coconut jaggery	Coconut shell ice cream cups	Tender coconut
Coconut jam	Coconut shell powder	Tender coconut cart
Coconut milk and cream	Coconut squash	Tender coconut minimal processed
Coconut milk powder spray dried	Coconut sweets	Tender coconut snowball
		Tender coconut water – packed

Table 8.2 Consumption pattern of coconut in India

Item	In lakh ton	In million nut	Percentage
1. Total production	–	12,355	100
2. Raw nuts	–	7,613	61.6
Mature including		6,013	48.7
Seed tender nut		1,300	10.5
D C		300	2.4
3. Copra	7.27	4,742	38.4
Milling	5.22	3,822	30.9
Edible	1.35	920	7.5
4. Oil	3.55	–	–
Edible	1.40	–	39.4
Toiletry	1.65	–	46.5
Industry	0.50	–	14.1

Indian Coconut J 29(4): 4–27

Table 8.3 Coconut-based food products and its coconut maturity requirement

Maturity	Product
6–7	Buko ball, leather-type product, instant buko juice
7–9	Canned buko in syrup, fresh young tender nut, dried buko chips/crisps, pie filling and other beverages
9–10	Nata de coco, sweetened meat bars (bukayo)
10–12	Desiccated coconut, creamed coconut, canned coconut milk, sweetened coconut milk (condensed, jam, honey, syrup, candy), unsweetened coconut milk, cheeses, coco yogurt, coco skimmed milk, powder edible coconut kernel, coco flour, ingredient for baked products

is then obtained from the copra by expression or prepress solvent extraction methods. The residual product after the oil extraction from coconut meat contains 18–25 % protein but is too fibrous for use in monogastric diets. Consequently its main use is ruminant feeding. Desiccated coconut is the dried, white, particulate or shredded food product manufactured from freshly peeled coconut kernels. The world production of desiccated coconut is around 200,000 mt a year, the bulk of which is exported and is the second next most significant coconut product in global trade. In 1996 the total volume of desiccated coconut imported was 174,000 mt tons mainly purchased by Europe and the USA.

8.4 Copra

After the coconut is dehusked, the hard but brittle shell is exposed and can be split open into two halves using a machete. The coconut water is drained off and the meat attached to the shell is dried. During the drying process, the meat shrinks and is easily detached or scooped out from the shell. Copra is produced from the dried coconut meat. The copra quality is influenced by the method and the manner of drying the meat. Improper drying may result in contamination of the meat with certain harmful aflatoxin-producing moulds, including the dangerous yellow-green mould, *Aspergillus flavus*. Aflatoxin is harmful to both humans and animals. It is therefore extremely important that the coconut meat be properly dried.

The three common methods of drying are sun drying, kiln drying and hot-air drying. Smallholders typically use sun-drying or kiln drying methods, whereas larger producers may use hot-air dryers. During drying, the moisture content of the coconut meat is reduced from about 50 % down to 6 %. For producers using sun drying, it is important to thoroughly clean the floor or pavement before spreading the cut coconut halves. Make sure that soil and other extraneous matter are not mixed with the coconuts. Plastic sheeting may be used under the coconut meat to avoid direct contact with the ground. Split the coconuts and expose the meat only when certain that drying can start immediately or within 4 h from splitting in order to prevent mould formation. When there is a threat of bad weather, defer nut splitting. A portable cover made of plastic sheeting should be available to protect the coconut meat from rain or dew. The covers are normally shaped like roofing (inverted V's) to allow aeration. Continuous drying for 4–5 days in good sunlight typically lowers the moisture content to 6 %. If the weather suddenly turns bad during the sun-drying period and is expected to remain so for some time, use of mould inhibitors is recommended. For producers using kiln dryers and hot-air dryers, a temperature of 35–50 °C (95–122 °F) should be maintained for the first 16 h of drying followed by 50 °C (122 °F) during the next phase until a final moisture content of 6 % is reached. It is important that drying begins within 4 h after the coconuts are split in order to prevent mould contamination. In drying copra using kiln dryers, it is important to use a clean source of fuel and minimise the amount of smoke that passes through the drying coconut meat. The colour of the dried copra will depend on the source of fuel and cleanliness of the smoke. The colour of copra obtained from hot-air dryers is typically whiter than from kiln dryers since the coconut meat is dried by uncontaminated hot air that passes

through the coconut meat. Since smoke does not come in contact with the meat, the copra produced from hot-air dryers is clean and white. The moisture content of the dried meat can be estimated by pressing it between the thumb and forefinger. If the dried meat does not stick to the thumb and readily drops when released, a moisture content of approximately 6 % has been obtained. In India standard contract terms for milling copra were specified in as early as 1949. Since then, these form the basis of transactions in the domestic market. The terms apply to sundried and smoke-dried copra, but the smoked copra cannot be tendered against a contract for sun-dried copra. The oil content, the colour and appearance and the moisture content are variable. These characteristics are demonstrated in the grades and standards used for copra. The following are the details of contract terms for milling copra.

Copra is produced after drying the coconut kernel. Copra and the coconut oil as well as the cake derived from it are a major source of foreign exchange for many coconut-growing countries in Asia, the Pacific and Africa. The quality of copra and copra cake is influenced by the method and the manner of drying the coconut kernel. Improperly dried copra gives rise to certain moulds, the most harmful of which is the yellow-green mould called *Aspergillus flavus*, and other aflatoxin-related moulds. Aflatoxin is harmful both for man and animals. It is therefore extremely important that the coconut kernel be properly dried to prevent the attack of aflatoxin-related moulds. Processing of mature nuts to copra has several problems. Improper processing results in low oil yield and incidence of aflatoxin. Proper postharvest practices, as well as proper drying and storage, can increase the oil yield to about 20 %. Proper drying of coconut results in copra with lower moisture content and lower incidence of aflatoxin. Copra is mainly produced by small coconut holders using sun-drying or smoke-kiln methods. Hot-air dryers are also used to a limited extent. Copra making involves different steps between harvesting and marketing of the product. Of these, drying the coconut kernel or reducing its moisture content from 50 % to 6 % most influences the quality of the product.

The following are ten guidelines for producing aflatoxin-free copra:

1. Harvest only fully matured (brown) nuts. These are the 12-month-old or older nuts.
2. Do not pay the harvester for immature (green) nuts; instead penalise them for picking green nuts.
3. For producers selling husked nuts to desiccated coconut factories, segregate the 'fouls' for processing into copra. Never mix the 'regular copra' with the copra from 'foul' nuts. They tend to have high mould growth.
4. When preparing copra, split the nuts and expose the meat only when certain that drying can start immediately or within 4 h from splitting (exposure) to prevent mould formation. When there is the threat of bad weather, defer nut splitting.
5. If the weather suddenly turns bad during the sun-drying period and is expected to remain so for some time, use of mould inhibitors is recommended.
6. For producers practising sun drying, maintain cleanliness in the drying area. Clean pavement or floors before spreading fresh coconut meat. Make sure that soil and other extraneous matter are not mixed with the meat. Plastic sheeting may be used under the coconut meat to avoid direct contact with the ground.
7. Have on hand a portable cover (plastic sheeting) to protect coconut meat from rain and dew. These are shaped like roofing (inverted Vs) to allow aeration. On extended downpours, heat and dry the copra within 24 h.
8. Continuous sun drying for 4–5 days (in good sunlight) shall achieve 6 % moisture content.
9. For producers using smoke-kiln dryers and other types of dryers, a drying temperature of 35–50 °C should be maintained for the first 16 h of drying followed by 50 °C during the next phase until a final moisture content of 6 % if reached. It is important that drying should begin 4 h after the nuts are split to prevent mould contamination.
10. Pressing the copra between the thumb and forefinger, the thumb against the white meat

is a quick test for 6 % moisture content. If the copra kernel (white portion) does not stick to the thumb and readily drops when released, the 6 % (approximately) moisture level has been achieved.

8.4.1 Drying

There are several methods and practices in drying the coconut kernel or in making copra. The methods vary from that which is considered primitive and traditional to one that adheres to certain scientific principles of drying. The three common methods of drying are (a) sun drying or solar drying, (b) kiln drying which is either direct or semi-direct drying and c) indirect drying using hot-air dryers.

8.4.1.1 Sun Drying

Where weather conditions permit, sun drying can produce good quality copra. This method is used only during the dry season and when drying only small quantities of nuts. Since sun drying requires no expenses for fuel, the overall drying cost is considerably cheap compared to other copra drying methods using fuel-fed dryers. Fuel saved could mean possible additional farm income when sold or transformed into high-value products like coconut shell charcoal, activated carbon, coir, etc., leading to the maximum utilisation of farm resources. Because the dryer is capable of producing clean, white and edible copra, copra produced should command a premium price. Moreover, its adoption could promote a high degree of consciousness in the production of superior quality export products.

8.4.1.2 Kiln Drying

There are two types of smoke dryers commonly used by coconut farmers: the direct and semi-direct types. Primarily, both types have the same heating principle but differ only in design and manner of firing or charging fuel. The direct dryer is designed in such a way that the fire bed is directly located below the copra bed. On the other hand, the design of semi-direct dryer is superior to the direct type. The hearth where fuel charging/feeding is done is located on one side of the dryer, connected to the drying bed by a tunnel-like flue.

8.4.1.3 Direct Smoke Copra Dryer

The direct smoke dryer is commonly used by coconut farmers in many coconut-producing countries in the world. The smoke dryer has a grill platform usually of split bamboo which constitutes the drying area. Halved nuts in the shell are placed on this grill. Underneath the platform is a fire hearth where coconut shells and husks are burned slowly to provide the heat for vaporising the water from the coconut meat. Generally, there is no chimney. The coconut meat shrinks upon drying and may be removed or scooped out from the shell. The meat is then further dried in the smoke dryer. The basic features which make the direct smoke dryer preferred by farmers are the high thermal efficiency of the dryer (the coconut meat is directly heated), the low cost of construction (the component parts are available on the farm), the simplicity of the design and the low cost of fuel. However, copra produced from this dryer are usually dark, sooty with smoke and at times scorched. Since the fuel is burned inside a pit underneath the drying bed, the dryer has to be attended when it is in operation to prevent the dryer from burning.

8.4.1.4 Semi-direct Copra Dryer

It is a simple structural design, cheap and easy to build. The dryer has a combustion pit located about 3 ft away from the drying bed. The hot combustion product is channelled to the drying bed via an underground tunnel. The dimension of the excavation pit is 6 ft in width, 12 ft in length and a depth of 4 ft. The pit floor of the firing chamber is slightly inclined upwards towards the end portion, which is designed to direct the flow of heated air. Dry coconut husks are used for fuel. It has a capacity of 2,000 nuts which are dried after 20–25 drying hours with resultant moisture of 6 %. Due to the ease of structural design and operation, needing only inexpensive and locally available construction materials, this dryer is deemed to be socially adaptable and economi-

cally ideal for small coconut farmers. Since the total construction cost is within the reach of small coconut farmers with minimal fuel costs, the overall production cost per kilo of copra would be much cheaper. Reflecting that fuel consumption per batch is approximately 50 % of nut capacity, the savings per coconut husk (50 %) plus coconut shell have a higher commercial value. This would mean additional financial benefits for the coconut farmers.

8.4.1.5 Copra Drying Using Hot-Air Dryers

In drying copra using hot-air dryers, the coconut meat is dried by means of uncontaminated hot air that passes through the copra bed. Since the smoke does not come in contact with the kernel, the copra produced is clean and white. If properly done, copra drying using hot-air dryers produces good quality copra with 6 % moisture content. There are quite a few hot-air dryer designs. The common ones are (a) the modified Kukum hot-air dryer and (b) the Cocopugon or the brick hot-air dryer.

8.4.1.6 The Modified Kukum Dryer

The modified Kukum dryer is an indirect natural draught dryer measuring 1.83 m in width, 3.66 m in length and 2.13 m in height. About 2,000 nuts (average size) can be accommodated (volume of drying bed: 2.8 m^3). Its heat exchange is made up of three standard oil drums welded together with five semicircular baffles installed alternately inside the drums at distance of 0.46 m. The furnace measures 3 ft in length and 2 ft in width and is made of steel plastered with 6 cm thick cement-ash mixture inside. The furnace is provided with a slanting grate and door to regulate air entry. A butterfly valve is also provided at the chimney to control the temperature. About 30 h are needed to dry one batch to 6 % moisture content. Based on a 10 h operation time per day, drying takes 3 days. About 8.7 min are needed to produce 1 kg copra with the modified Kukum dryer. The modified Kukum dryer produces good quality copra. However, maintenance and repair costs are high. The metal parts of the dryer, which start to corrode as soon as the dryer is being constructed. Frequent use of the dryer will reduce corrosion, but never stop it. Since copra is a low-price product, the use of stainless steel or even the application of primer is not economical. Exposure to high temperatures, aggressive fumes and water induces corrosion of the metal dryer components.

8.4.1.7 Cocopugon Hot-Air Brick Copra Dryer

The Cocopugon is a further improvement of the modified Kukum dryer. Instead of using metal drums as the heat exchanger common in Kukum dryers, the Cocopugon uses bricks. Bricks are known for their high strength, durability and dimensional stability. The proportions of the Cocopugon are 260 cm in width, 360 cm in length and 200 cm in height. Standard fire bricks and 2.5″ crown bricks are used for the chimney and the heat exchanger, respectively. The dryer can accommodate 2,000 average-sized nuts per batch (volume of drying bed: 3.33 m^3). To facilitate ease of loading and unloading, the right side of the drying bed wall is removable. A one-step stair and platform is also provided on the same location. Unlike dryers with metal heat exchangers, this dryer needs to be preheated. Firing should be done first before loading the split nuts. The burner can accommodate about 200–300 husks. Refuelling has to be done every 3–5 h. The heat stored in the bricks will be released slowly after the last firing on the first drying day, such that drying will continue for several hours without adding fuel (husks). After a preheating time of 3.5 h and a loading time of 2 h, the average temperature in the bottom layer is 66.3 °C. The burner then has to be fed five to seven times for the whole drying period. Formerly, this could only be accomplished in 1 day at a feeding interval of about 3–4 h assuming a constant fuel feed rate. Unloading could be done after the dryer has cooled down on the second day. If operated on a 2-day schedule, five firings are needed on the first day and another two to three firings on the second day. Unloading will be done the next morning to utilise the heat stored in the bricks. If the baffle in the chimney is closed during night-time, embers can still be found inside the burner on the following morning, making it easy to continue firing. The temperature curve for the burner has several small peaks indicating the maximum temperature per feeding interval. The effect on the drying bed temperature is minimal, thus having an almost constant drying temperature. Even if the burner is fully loaded, the resulting temperature in the drying bed does not exceed 90–95 °C, thus eliminating the risk of producing scorched copra. Since the heat exchanger or the burner covers almost the whole area inside the dryer body, the temperature distribution is very uniform. The difference in temperature between the highest and lowest value is less than 5 °Kelvin. A standard deviation of 3 °K indicates a very constant temperature. During operation, the dryer operator spends 2 h per batch at the dryer, meaning the labour requirements are cut down by more than 50 % to 4.1 min per kilogram copra compared with the modified Kukum dryer. The farmer can therefore leave the dryer in between fuel feedings and use his or her time for other activities.

8.5 Coconut Oil

Coconut oil is used in the country as a cooking fat, hair oil, body oil and industrial oil. Coconut oil is made from fully dried copra having maximum moisture content of 6 %. Steam cooking of copra is also practised by some millers to enhance the quality and aroma of oil. Coconut oil is marketed in bulk as well as in packs ranging from sachets containing 5 ml to 15 kg tins. The branded coconut oil in small packs is mainly marketed as hair oil and body oil. There are several brands known for their superior grade oil which have export market throughout the world. India has unbeatable quality advantage in this sector. Refined coconut oil is also manufactured in the country for industrial uses. Refined coconut oil is mainly used in the manufacture of biscuits, chocolates and other confectionery items, ice cream, pharmaceutical products and costly paints. Generally, filtered coconut oil is used for cooking and toiletry purposes.

8.5.1 Virgin Coconut Oil

Virgin coconut oil is also made from the milk extracted from raw kernel. This is done on a small scale by the traditional method which is now partially mechanised or on a large scale by adopting wet processing technology. Coconut milk is fermented, and then by mechanical process, water is separated from oil. No heating or application of sunlight or dryer is done for the process. Virgin coconut oil (VCO), extracted from fresh coconut meat without chemical processes, is said to be the 'mother of all oils'. It is rich in medium-chain fatty acids, particularly lauric acid, and is a treasure trove of minerals, vitamins and antioxidants and is an excellent nutraceutical. It has about 50 % lauric acids, having qualities similar to mother's milk, thus confirming its disease-fighting ability. When lauric acid enters human body, it gets converted to monolaurin, which has the ability to enhance immunity. Several studies have confirmed that this compound has the ability to kill viruses including herpes and numerous other bacteria. Its antiviral effect has the ability to considerably reduce the viral load of HIV patients. VCO is not subjected to high temperatures, solvents or refinement procedures and therefore retains the fresh scent and taste of coconuts. It is rich in vitamin E, is non-greasy and non-staining and is widely used in soaps, lotions, creams and lip balms. The health benefits of VCO are second to none, ranging from speeding up body metabolic system and providing immunity against a horde of commonly prevalent diseases. Coconut palms are grown widely in the coastal tracts of the country. Copra, the dried kernel, is the chief/commercial product from coconut, which is mainly used for oil extraction. Copra normally has an oil content varying from 65 % to 72 %. Coconut oil is an important cooking medium in southern parts of the country especially in Kerala State. Besides, the oil has various industrial applications. It is used in the manufacture of toilet soaps, laundry soaps, surface active agents and detergents, hair tonics, cosmetics, etc. It is used throughout the country as a hair oil as it helps in the growth of the hair. As a massage oil, it has a cooling effect on the body. Owing to these qualities, coconut oil has

a potential market in the country. Since the price of coconut oil in the international market is very much lower than the domestic price, the quality and attractiveness of consumer packs are important factors to compete in the world market. While the demand for coconut oil for cooking purpose is elastic, its demand as hair oil is inelastic. For the extraction of oil from copra, the common method still prevailing in our country is by using rotary chucks. But the efficient system of extraction of oil is by the use of expellers.

8.5.2 Process

The Coconut Development Board and Central Food Technology Research Institute (CFTRI), Mysore, have developed a technology for production of virgin coconut oil and dietary fibre. The technology will be transferred to entrepreneurs having a sound financial background and experience in production and marketing. Interested entrepreneurs may apply to the Board in the prescribed application form which is downloadable from the link given below. The selected entrepreneurs will be required to pay a technology transfer fee of Rs. 50,000. The Board reserves the right to reject all or any of the applications received.

Coconuts are deshelled followed by paring and dewatering. Pared coconuts are disintegrated by passing through rotary wedge-type coconut cutter having a sieve plate (3 mm hole) through which shredded coconut gratings are expressed in a screw/hydraulic press to extract fresh coconut milk. The coconut milk is filtered and passed through a high-speed centrifuge wherein the coconut oil gets separated from the coconut milk. The coconut oil is then packed in consumer packs in an automatic packing machine.

8.5.3 Qualities

Virgin coconut oil (VCO) is abundant in vitamins, minerals and antioxidants, thus making it the 'mother of all oils'. Extracted from fresh coconut kernel without any chemical processes, it is the purest form of coconut oil, water white in colour. Virgin coconut oil is a major source of lauric acid and vitamin E. The virgin coconut oil is free from trans fatty acid and high in medium-chain fats (MCFA) or medium-chain triglycerides (MCTs) known as lauric acid, which is identical to special group of fats found in human breast milk. VCO is widely consumed as MCT oil for weight loss treatment, etc. MCTs are more easily and rapidly digested than other types of fats, as they require lower amounts of enzymes and bile acids for intestinal absorption. MCTs are metabolised very quickly in the liver and are reported to encourage an increase in energy expenditure, while decreasing fat storage. Numerous studies suggest that substituting MCT oil for other fats in a healthy diet may therefore help to support healthy weight and body composition. High quality of this oil makes it an ideal massage oil for babies and also for skin and hair applications. It protects the skin from infections caused by bacteria, viruses and fungi and prevents dandruff and hair loss. It even eases muscular pain and supplements your body with antioxidants. Antioxidant is your body's natural defence against free radicals. It prevents the chain reaction of free radicals and mars sagging and unsightly wrinkles. Rich in vitamin C and vitamin E, virgin coconut oil slows down the ageing process and assures the best of life and beauty to your skin. It also helps in the absorption of fat-soluble vitamins A, D, E and K.

8.5.4 Product Specification

Free fatty	0.5 max	C 8:0	4.6–10 %
Moisture and volatile matter	0.25 max	C 10:0	5.5–8.0 %
Refractive index	1.4481–1.4491	C 12:0	45.1–50.3 %
Iodine value	4–11	C 14:0	16.8–21.0 %
Saponification number	250–260	C 16:0	7.5–10.2 %
Unsaponifiable matter	0.2–1.5 % max	C 18:0	2.0–4.0 %
Sp gravity	0.908–0.921	C 18:1	5.0–10.0 %
Polenske value	13.0 min	C 18:2	1.0–2.5 %
C 6:0	0.4–0.6 %	C 18:3	0–0.2 %

8.5.5 Coconut Oil Processing

Coconut can be processed into many products; the next part focuses on the processing of coconut oil. Two technologies, (a) the hot oil immersion drying technology and (b) the ram press coconut oil extraction technology, were selected as the most practical. Though their viability is site specific, women and members of the farming family may run the technology.

Coconut oil is one of the main traditional products derived from the meat or kernel. It is a mixture of chemical compounds called glycerides containing fatty acids and glycerol. The different fatty acids present in coconut range from C6 to C18 carbon atom chains. Coconut oil processing methods or technologies are classified into two (2) major types: the dry and the wet processes. The oil extraction technology which starts with copra as the raw material is termed as the dry process, while the method that uses fresh coconuts as starting material is generally called the wet process. Dry processing of meat for oil production involves the conversion of coconut meat into copra prior to expelling and refining. This process is however done off farm, in an oil mill. From the farm, the copra goes through a series of traders. Storage in warehouses ranges from 2 weeks to 2 months. At the mill, the copra undergoes the following steps:

1. Cleaning: Copra is transferred from the warehouse to a mill by a series of floor conveyors, rotor lift and overhead conveyors. Copra is cleaned of metals, dirt and other foreign matter manually by picking or through the use of shaking or revolving screens, magnetic separators and other similar devices.

2. Crushing: Copra is broken into fine particle sizes of about 1/16″–1/8″ by high-speed vertical hammer mills to facilitate oil extraction.

3. Cooking/conditioning: The crushed copra that has about 5–6 % moisture is passed through a steam-heated cooker. This brings the temperature of the copra to the conditioning temperature of about 104 °C (220 °F). At the conditioner, the copra is maintained at about 104–110 °C (220–230 °F) for about 30 min to ensure uniform heat penetration before oil extraction. Moderately high temperature

facilitates the expelling action. Oil is able to flow out more easily due to decrease in viscosity proteins and other substances present in the copra. Heating dries and shrinks these substances. Moisture content of copra is about 3 % when it leaves the conditioner.

4. Oil extraction: In the expeller, the milled copra is subjected to high-pressure oil extraction, first by a vertical screw and finally by a horizontal screw. To control the temperature during extraction, the main shaft is provided with water cooling, and cooled oil is sprayed over the screw cage bars. The temperature of the oil should be kept at about 93–102 °C (200–215 °F) to produce light-coloured oil and effect good extraction.

5. Screening: The oil extracted in the expeller flows into the screening tanks to remove the entrained parts from the oil. The foots settle at the bottom and are continuously scooped out by a series of chain-mounted scrapers which lift the foots to the screen on top of the tank. While travelling across the screen, oil is drained out of the foots. The filtered oil flows into a surge tank from where it is finally pumped to the coconut oil storage tank.

6. Filtration: The oil is passed through a plate and frame filter press to further remove the solids in the oil. Two filter presses are provided – one on duty while the other is being cleaned and dressed. Maximum filtering pressures reach about 60 psi. The filtered oil flows into a surge tank from where it is finally pumped to the coconut oil storage tank. Coconut oil produced from good quality copra is clear, is low in fatty acid and has good coconut aroma. However, crude coconut oil from bad quality copra is dark, turbid and high in free fatty acids (FFA), phosphatides and gums and has an unpleasant odour. To render this oil edible, it has to undergo a refining process. Typically, 5 % of the weight of the crude oils is lost in refining, but the loss can be as high as 7.5 %.

Refining consists of neutralisation, bleaching and deodorising. Neutralisation reduces the FFA to improve the taste and appearance of the oil. It is done by reacting sodium hydroxide with free fatty acid to form an oil-insoluble precipitate called soap stock. The amount of sodium hydroxide required to neutralise is 1.418 kg NaOH per 1 % FFA content per ton of crude coconut oil. In actual practice, it is reported that 10 % excess is added to ensure complete neutralisation. This amounts to 1.50 kg NaOH per ton of crude coconut oil per 1 % FFA. The soap stock is then removed once it settles out. This is converted either into acid oil by treatment with sulphuric acid or into soap by complete saponification. Phosphatides and gums are removed by physical refining in which the first stage involves treating the oil with phosphoric acid. These are then separated from the oil either by centrifugation or decantation. Bleaching takes out most of the dissolved or colloidal pigments responsible for the colour of crude oil. Either activated carbon or bleaching earth such as bentonite (1–2 %) or a combination of both is added to neutralise the oil under vacuum while heating it to 95–100 °C. Afterwards, the bleaching agents are removed by passing the oil through a filter press. Deodorisation removes volatile odours and flavours as well as peroxides that affect the stability of the oil. It is done by heating the oil to a temperature between 150 and 250 °C and contacting with live steam under vacuum conditions (29 psig pressure).

8.5.6 Coconut Oil Extraction: The Hot Oil Immersion Drying (HOID) Technology

The hot oil immersion drying (HOID) technology or the 'fry-dry' process is a method of extracting coconut oil from fresh coconut meat (wet process). The process involves grating and then drying the freshly cut coconut kernel by immersing it in hot oil. The dried residue is subsequently removed from the hot oil, drained and passed through a screw press where the oil is extracted under pressure leaving a dry cake. The HOID or 'fry-dry' process is indigenous to West and North Sumatra and North Slaws in Indonesia and is currently practised all over the country. It was reported that a few areas in the Philippines have used the technology but not many are fully knowledgeable of its application. It is believed that there are good prospects and a wider scope for the introduction and application of the HOID technology not only in other parts of Indonesia but also in many areas in Asia and the Pacific, especially in medium- and large-scale operations. The HOID technology is an alternative method of producing coconut oil. The oil is generally of a better quality and preferred by certain segments of the population, especially in Indonesia, because of its distinctive coconut flavour. The HOID oil can be used directly as cooking oil, without chemical refining. In certain parts of Indonesia, HOID oil is sold at prices higher than that of refined palm oil. The method of HOID oil production involves the following steps:

1. The fresh coconut meat is delivered to the processing plant where it is inspected, washed and cut into pieces with a hammer mill or a grater.
2. The grated kernel is then fried in the pan of hot coconut oil at approximately 120 °C for 20–45 min depending on the oil temperature and ratio of fresh meat to the coconut oil used. Care must be exercised not to add too much meat at once during the frying because the immersion of the cut coconut kernel results in a rapid evolution of steam which can result in oil spillage. Stirring of the grated coconut is occasionally done during the frying. The drying process is completed when there is no more steam produced, the coconut meat becomes yellowish to brown and the temperature of the coconut oil in the pan increases.
3. The fried particles are then taken out of the oil by means of perforated spoon affixed to the end of a long wooden handle. The meat is then dumped in a filter box, and the oil is allowed to drain through a meshed plate at the base of the container.
4. The drained, cooked brown coconut particles, rich in coconut oil, are then fed to the hopper whence it is fed to the screw press. The expelled oil is passed through a mesh plate and settled in a tank before it is pumped or poured into the main settling tank.

5. The oil is then clarified by settling the oil in the tank. Sometimes a filter press is used. Once clarified, the oil can be sold directly in the market as cooking oil without further chemical refining. The main equipments used in a small HOID processing plant are as follows:
   (a) Hammer mill or grater – this is used to cut the fresh coconut kernels. In some areas in Indonesia, the kernel is grated.
   (b) Drying pans – either circular or rectangular in shape. These pans are equipped with wooden stirrers and spoons for removing out the dried meat manually.
   (c) Furnace – this is used to heat the pans by burning the wood, coconut shell or husk in the combustion chamber.
   (d) Screw press – this is used to extract oil from cooked, brown coconut meat.
   (e) Filter press or setting tank.
   (f) A draining tank and other handling equipment such as scooper, tray, metal and rattan baskets.

The viability of the process is most sensitive to the price of the raw material, price of oil and the oil yield. Thus, it is important that the plant must be designed and operated to minimise oil losses and maximise returns from efficient operation of the whole system. The viability of the process is also dependent on the site.

8.5.6.1 Ram Press Coconut Oil Extraction

Ram press coconut oil extraction is a method of expelling oil from dried coconut either in the form of dried fresh coconut gratings, copra or dried residue from aqueous coconut processes. The ram press also called the Bielenberg press was developed by the Appropriate Technology International, a Washington-based NGO, in 1985 through its Village Oil Press Project in Tanzania. It is manually operated, a low-cost piece of equipment which was originally designed to be used by smallholder farmers to process soft-shelled sunflower seed to obtain scarce cooking oil. The original design of the ram press was arduous to use and took two men to operate.

Recently, the Natural Resources Institute (NRI) of the UK has carried out some work on improving small-scale coconut oil extraction methods using the participatory approach, particularly involving women in the rural areas in Asia, the Pacific and Africa. One of the design advancements of the ram press is a version that is smaller and easily operated by a woman. The newly designed ram press has a long, pivoted lever which moves a piston backwards and forwards inside a cylindrical cage constructed from metal bars spaced to allow the passage of oil. At the end of the piston's stroke, an entry port from the feed hopper is opened so that the oilseed or the squeezed coconut gratings (called chicha in Tanzania) can enter the cage. When the piston is moved forward, the entry port is closed and the chicha is compressed in the cage. The compressed chicha is pushed through a circular gap at the end of the cage. The width of this gap, which can be varied using an adjustable choke, controls the operating pressure of the press. The lever mechanism of the press is such that it can operate pressures greater than those obtained in most manually operated presses and as high as those in small-scale expellers. While the ram press has a low seed throughput, it has the advantage of continuous operation. Laboratory and field trials conducted by the NRI in Tanzania indicated that the ram press was suitable for pressing sun-dried squeezed coconut gratings or chicha. The oil extraction efficiency achieved was 60–70 %. Although more arduous to use when processing chicha than when processing sunflower seed, women users in the villages of Zanzibar found that the ram press is easy to operate, especially when several changes in operation were made.

8.5.7 Coconut Oilcake

Coconut cake is the residue left after the extraction of oil from copra which is mainly used as a cattle feed. Coconut cake contains 4–5 % oil which is extracted by solvent extraction process. This oil is generally used for industrial purpose, and de-oiled cake is used to make mixed cattle feed. There are a few such units in the country especially in Kerala.

8.6 Tender Coconut Water

The water of tender coconut, technically the liquid endosperm, is the most nutritious wholesome beverage that the nature has provided for the people of the tropics to fight the sultry heat. It has caloric value of 17.4 per 100 g. 'It is unctuous, sweet, increasing semen, promoting digestion and clearing the urinary path', says Ayurveda on tender coconut water (TWC). It's a natural isotonic beverage with the same level of electrolytic balance as we have in our blood. It's the fluid of life. The major chemical constituents of coconut water are sugars and minerals and minor ones are fat and nitrogenous substances.

Numerous medicinal properties of tender coconut water reported are as follows:

1. Good for feeding infants suffering from intestinal disturbances.
2. Oral rehydration medium.
3. Contains organic compounds possessing growth-promoting properties.
4. Keeps the body cool.
5. Application on the body prevents prickly heat and summer boils and subsides the rashes caused by small pox, chicken pox, measles, etc.
6. Kills intestinal worms.
7. Presence of saline and albumin makes it a good drink in cholera cases.
8. Checks urinary infections.
9. Excellent tonic for the old and sick.
10. Cures malnourishment.
11. Diuretic.
12. Effective in the treatment of kidney and urethral stones.
13. Can be injected intravenously in emergency case.
14. Found as a blood plasma substitute because it is sterile, does not produce heat, does not destroy red blood cells and is readily accepted by the body.
15. Aids the quick absorption of the drugs and makes their peak concentration in the blood easier by its electrolytic effect.
16. Urinary antiseptic and eliminates poisons in case of mineral poisoning (Table 8.4).

Table 8.4 Analysis of mature and tender coconut water

Content	Mature coconut water	Tender coconut water
Total solids %	5.4	6.5
Reducing sugars %	0.2	4.4
Minerals %	0.5	0.6
Protein %	0.1	0.01
Fat %	0.1	0.01
Acidity mg %	60.0	120.0
pH	5.2	4.5
Potassium mg %	247.0	290.0
Sodium mg %	48.0	42.0
Calcium mg %	40.0	44.0
Magnesium mg %	15.0	10.0
Phosphorous mg %	6.3	9.2
Iron mg %	79.0	106.0
Copper mg %	26.0	26.0

Table 8.5 Amino acid composition of coconut water (% of total protein)

Alanine	2.41	Leucine	1.95–4.18
Arginine	10.75	Lysine	1.95–4.57
Aspartic acid	3.60	Proline	1.21–4.12
Cystine	0.97–1.17	Phenylalanine	1.23
Glutamic acid	9.76–14.5	Serine	0.59–0.91
Histidine	1.95–2.05	Tyrosine	2.83–3.00

Table 8.6 B vitamins in coconut water

Nicotinic acid	0.64 µg/ml	Folic acid	0.003 µg/ml
Pantothenic acid	0.52 µg/ml	Thiamine	Trace µg/ml
Biotin	0.02 µg/ml	Pyridoxine	Trace µg/ml
Riboflavin	<0.01 µg/ml		

8.6.1 Sugars

Sugars in the forms of glucose and fructose form an important constituent of the tender nut water. The concentration of sugars in the nut water steadily increases from about 1.5 % to about 5–5.5 % in the early months of maturation and then slowly falls reaching about 2 % at the stage of the full maturity of the nut. In the early stages of maturity, sugars are in the form of glucose and fructose (reducing sugars), and sucrose (non-reducing sugar) appears only in later stages which increases with the maturity while the reducing sugars fall (Ghosh and Bandopadhyay 2010). In the fully mature nut approximately 90 % of the total sugars is sucrose.

8.6.2 Minerals

Tender coconut water contains most of the minerals such as potassium, sodium, calcium, phosphorous, iron, copper, sulphur and chlorides. Among the minerals more than half is potassium, the concentration of which is markedly influenced by potash manuring. Tender coconut water being rich in potassium and other minerals plays a major role to increase the urinary output.

8.6.3 Protein

Coconut water contains small amounts of protein. The percentage of arginine, alanine, cystine and serene in the protein of tender coconut water is higher than those in cow's milk. Since it does not contain any complex protein, the danger of producing shock to the patients is minimised (Table 8.5).

8.6.4 Vitamins

Tender coconut water contains both ascorbic acid and vitamins of B group. The concentration of ascorbic acid ranges from 2.2 to 3.7 mg/ml, which gradually diminishes as the kernel surrounding the water begins to harden (Table 8.6).

8.6.5 Minimal Processing of Tender Coconut Water

Perishability of tender coconut is relatively high, and once the tender coconuts are detached from the bunches, its natural freshness will get lost within 24–36 h even under refrigerated conditions unless treated scientifically. The bulkiness of tender coconut is due to the husk which accounts for two-thirds of the volume of tender nut. Handling of tender coconuts will be easy if a major part of the husk is removed. But when partial removal of husk is done, the colour of the nut will be changed to brown thereby reducing the attractiveness of the nut. Technologies for minimal processing of tender coconut have been developed for retaining the flavour and to prevent discolouration. The technology for minimal processing of tender coconut developed by Kerala Agricultural University (KAU) involves dipping partially dehusked tender coconut in a solution of 0.50 % citric acid and 0.50 % potassium metabisulphite for 3 min. The product can be stored up to 24 days in refrigerated condition at 5–7° centigrade. By using this process, tender coconut can be transported to distant place served chilled like any other soft drink. Optimised uniform size facilitates using of plastic crates and insulated chill boxes for transporting and storage.

8.6.6 Preservation and Packing Tender Coconut Water in Pouches/Aluminium Cans

The Coconut Development Board in collaboration with Defence Food Research Laboratory, Mysore, has developed the technology for packing tender coconut water in pouches/aluminium cans with shelf life of more than 6 months under normal ambience condition and 12 months under refrigerated condition. There are about half a dozen units in India that availed this technology and are doing commercial production. The technology is available to entrepreneurs at a total lump sum transfer fee of Rs. 3 lakhs.

8.6.7 Snowball Tender Nut

Snowball tender nut is a tender coconut without husk, shell and testa which is ball shaped and white in colour. Coconut of 8 months old is more suitable for making SBTN in which there is no decrease in quantity of tender water and the kernel is sufficiently soft. The process has been developed for making the SBTN. Important steps involved in the process are dehusking of the nut, making groove in the shell and scooping of the tender kernel in ball shape without breakage by using a scooping tool. The groove has to be made by using a machine which is under the progress of development. Two types of copra, namely, milling and edible, are made in India. Milling copra is used to extract oil, while edible grade of copra is consumed as a dry fruit and used for religious purposes. Milling copra is generally manufactured by adopting sun drying and artificial means. Substantial quantity of milling copra is manufactured using modern hot-air dryers resulting in the availability of superior quality copra which is required for the manufacture of best grade coconut oil. A good number of farmers' cooperative societies are also involved in the manufacture and marketing of milling copra. Milling copra is available in different grades. Edible copra is made in the form of balls and cups. Different grades of edible copra are available in the market according to the size, colour, etc.

8.7 Convenience Products from Raw Kernel

Desiccated coconut (DC), coconut cream, coconut milk, virgin coconut oil and spray-dried coconut milk powder are the convenience coconut products manufactured in the country. Desiccated coconut is used as a substitute to grated raw coconut in various food preparations. Desiccated coconut is marketed in bulk as well as in small packs. Defatted desiccated coconut is also available in the country. Processed coconut cream/coconut milk is used in various food preparations as a substitute to the milk extracted from raw kernel in the traditional method. They are available in cans and tetra packs. Spray drying is the best method for the preservation of coconut milk. The product has advantages such as less storage space, bulk packaging possible at low cost and long shelf life. Spray-dried coconut powder is manufactured by one unit in the country.

8.7.1 Coconut Chips

CPCRI has developed a process for producing the coconut chips by osmotic dehydration followed by hot-air drying. Fresh kernels of matured coconut containing reasonable amount of water are to be used for this process. The time of osmotic dehydration will be 1 h. The drying time in hot-air dryer is about 6 h. The conversion ratio from fresh kernel weight to sweet coconut ships is about 50 %. On an average about 120–150 g chips can be obtained from a coconut. After osmotic dehydration of the coconut slices, instant coconut ships can also be prepared by drying in the microwave oven. Coconut chips like sweet chips, spicy chips, medicated chips and sweet chips with different flavour can be prepared. The coconut chips are hygroscopic in nature. Hence, the chips must be packed in the aluminium foil laminated with LDPE pouches, which will maintain its flavour and crispness up to a 6-month period without affecting its microbial and biochemical qualities. To avoid the breakage of chips during transportation, it may be packed as billow packet. The coconut chips are crispy in nature and ready-to-eat form. No frying is required before the consumption. It is having its own good coconut flavour as no oil is used for frying. It can be used as snacks. After rehydration of the chips, it can also be used as fresh kernel.

8.7.2 Desiccated Coconut

Dehydrated coconut meat in the grated or shredded form is desiccated coconut. The process involves shelling, paring, disintegrating, drying, sieving and packing. The desiccated coconut is used in confectionery and other food industries. It could also be used in the households for culinary preparations as a substitute to grated fresh coconut. In India, manufacture of desiccated coconut is mainly concentrated in Karnataka state, and the production is mainly absorbed by the organised food industries in different parts of the country for various end uses. A study conducted by the Coconut Development Board has revealed that a growing consumer demand for desiccated coconut could be developed in the country by resorting to organised market promotion activities. The survey has also shown that desiccated coconut in consumer packs is acceptable not only in nonproducing states but also in South Indian coconut-growing states. From the survey it was revealed that sizeable section of the middle class and upper class families residing in cities and towns would prefer desiccated coconut, if quality product is readily available at reasonable rates. For project details for setting up of desiccated coconut unit, you may contact the technology section of the Board. Desiccated coconut is the dehydrated, shredded white kernel of the coconut. It is produced from fully ripe coconut kernel under strict hygienic conditions for human consumption. It is used both in household foods and processed foods particularly in ready-to-cook mixes and in packaged and canned foods. In the bakery and confectionery industries, desiccated coconut is a preferred product. Nutritionally desiccated coconut is not different from fresh coconut kernel. It retains all the characteristic features of the wet kernel including the original nutrients. Good desiccated coconut is crisp and snow white in colour with a sweet, pleasant and fresh taste of the wet kernel. The production of desiccated coconut involves dehiscing of fully matured coconut. This involves detailing without breaking the kernel, removing the brown test and slicing the pared kernel into two halves to release the nut water. Next comes washing and sterilising the kernel pieces either by passing them through boiling water or subjecting the pieces to live steam. Stabilising the sterilised kernel pieces is done by immersion in a solution of sulphur dioxide. Finally, the kernel is disintegrated or shredded into standard or fancy cuts. In the final stage, the kernel is dried, cooled, graded and packaged in moisture- and odour-proof containers. In the desiccating process, the wet kernel is shredded into nine different cuts. These cuts are grouped under three broad categories such as granular cuts, shred cuts and speciality cuts. The cuts are further processed at the destinations to satisfy specified end-use requirements.

Table 8.7 Composition of Neera

Total solids (g/100 ml)	15.2–19.7	Citric acid (g/100 ml)	0.50
pH	3.9–4.7	Alcohol in %	Nil
Specific gravity	1.058–1.077	Iron (g/100 ml)	0.15
Total sugars (g/100 ml)	14.40	Phosphorus (g/100 ml)	7.59
Original reducing sugars (g/100 ml)	5.58	Ascorbic acid (mg/100 ml)	16.30
Total reducing sugars (g/100 ml)	9.85	Total protein (g/100 ml)	0.23–0.32
Total ash (g/100 ml)	0.11–0.41		

The more common products so produced are (1) sweetened coconut, (2) toasted coconut, (3) coloured coconut and (4) creamed coconut.

8.8 Neera (Coconut Flower Sap) and Its Products

Neera, the non-alcoholic and nutritious drink from the immature inflorescence of coconut, can be promoted due to its potential for value addition, employment generation and better returns to the coconut farmers. The vascular sap collected from immature unopened coconut inflorescence is popularly known as 'Neera' in fresh form. It is a sugar-containing juice, a delicious health drink and a rich source of sugars, minerals and vitamins. It is sweet and oyster white in colour and translucent. It is tapped from the coconut inflorescence, filtered and pasteurised, and bio-preservatives are added to preserve the product. Treated Neera can be preserved in cans up to 2 months at room temperature.

8.8.1 Uses of Neera

Neera is popular as a delicious health drink. It is good for digestion, facilitates clear urination and prevents jaundice. The nutrient-rich 'sap' has low glycaemic index (GI of only 35) and hence diabetic-friendly since very low amounts of the sugar is absorbed into the blood. It is an abundant source of minerals, 17 amino acids, vitamin C and broad-spectrum B vitamins and has a nearly neutral pH. Diversified value-added products like coconut flower syrup, jaggery and coconut palm sugar are produced from Neera (Table 8.7).

8.9 Coconut Toddy

Toddy tapping is an organised industry in traditional coconut-growing tracts in the country. Toddy on fermentation becomes an alcoholic drink. Arrack and vinegar are also manufactured from coconut toddy. In Goa commercial arrack obtained by distillation of coconut toddy is known as coconut fenny.

8.10 Coconut Flower Syrup

This is a product similar to coconut palm jaggery with high content of minerals. It is a rich source of potassium. It has good content of sodium and is free from total fats and cholesterol. It is produced when fresh Neera is heated and concentrated into a syrup. The product can be used for healthy food preparations, as topping on wide variety of appetiser, dessert or beverage, for delicious serving and for a healthy lifestyle. It is also used as a health drink in connection with Ayurveda and other systems of medicine.

8.11 Coconut Palm Jaggery

The strained unfermented coconut sap is boiled, crystallised and transferred into suitable moulds to prepare jaggery. The semisolid jaggery solidifies gradually by the cooling effect of the atmosphere into a crystallised hard substance. The recovery of jaggery from syrup is 15 %. It is used as a sweetening agent for the preparation of dishes and is superior to cane jaggery. The nutritional composition of coconut palm jaggery (in 100 g) is as follows: thiamine 21.00 mg, riboflavin 432.00 mg, nicotinic acid 5.24 mg and ascorbic acid.

8.12 Coconut Palm Sugar

The coconut palm syrup or jaggery can be crystallised to produce fine granules of sugar. Transition of coconut jaggery into a ground granule sweetener is more accepted by global markets. The recovery of palm sugar from coconut palm jaggery is 15 %. The application of this sugar is tremendous and offers huge potential owing to its most important health attributes, the low glycaemic index and the high nutrient content. It can be the most suited alternative sweetener, especially when agave sugar is being rejected owing to the high fructose content. This alternative sugar industry is estimated to be a $1.3 billion industry and hence the market prospects are enormous. It is understood that in producing countries like Indonesia, around 50,000 mt of coconut sugar is produced per month and 6 lakh mt produced per year. The product has good local market in Indonesia (Tables 8.8, 8.9, 8.10 and 8.11).

Tapping of 25 % of inflorescences (three inflorescences) in a palm for production of Neera will yield additional income to the farmer and to the tapper, without much reduction in yield. The coconut farmer communities in the federations of

CPS can work out a viable proportion of production of Neera and jaggery so that the products can cater to defined segments of the society. Neera being a sweet and nutritious health drink can be promoted among all sections of the society. Coconut palm jaggery owing to their low glycaemic index can be positioned as a safe and healthy

Table 8.8 Nutritional composition of coconut palm sugar (in 100 g)

Component	Qty in mg	Component	Qty in mg
Moisture	0.06 %	Minerals	5.24 mg
Protein	432 mg	Calcium	18.9 mg
Carbohydrate	11.0 mg	Phosphorous	1.9 mg
		Iron	5.2 mg

Table 8.9 Economics of production of Neera[a] and value-added products from Neera

Assumptions	
Number of palms per hectare	175
Total number of palms tapped in a hectare	Limited to 80
Number of inflorescence tapped per tree per annum	3
Average productivity per ha in Kerala	7,365 nuts/year
Approximate cost of 1 l of Neera	Rs. 50
Conversion ratio from Neera to jaggery	6:1
Approximate cost of 1 kg of coconut palm jaggery	Rs. 250

[a]Commercialisation of Neera is possible only after amendment of Abkari Act

Table 8.10 Estimated returns from Neera

Item	Yield of Neera		
Yield of Neera in litres	At 1 l/tree/day	At 2 l/tree/day	At 3 l/tree/day
Yield of Neera from 80 palm tapped for 3 months	7,200 l	14,400 l	21,600 l
Returns from Neera at Rs. 50 per litre	3.6 lakhs	7.2 lakhs	10.8 lakhs
Returns per palm	Rs. 4,500	Rs. 9,000	Rs. 13,500
Returns to farmer in the proportion of 75 % of production	2.7 lakhs	5.4 lakhs	8.1 lakhs
Returns to tapper (25 % of production)	0.9 lakhs	1.8 lakhs	2.7 lakhs

Table 8.11 Estimated returns from coconut palm jaggery

Item	Production of Neera at		
	At 1 l/tree/day	At 2 l/tree/day	At 3 l/tree/day
Yield of jaggery in kg	1,200 kg	2,400 kg	3,600 kg
Returns from jaggery at Rs. 250 per kg	3.0 lakhs	6.0 lakhs	9.0 lakhs

alternative sweetener. If we convert jaggery further to coconut palm sugar, value addition is still better.

8.13 Secondary and Derived Product

In major coconut-producing countries, several products and by-products are processed for export. They are coconut fibre products (coir and coir products, mats, matting, brushes, brooms, rubberised coir mattresses), shell products (charcoal, activated carbon) and coconut-based food products (coconut milk, cream, nata de coco, coconut jam, young tender coconut). For coconut-based oleochemicals including fatty alcohol, fatty acids, methyl esters, tertiary amines, alkanolamides and glycerine, there has been a growing demand in the world market. Today, technologies exist for many other value-added products from the coconut tree, its fruit as well as the wastes generated. These technologies are not centred in any one country but are scattered across the major coconut-growing regions. Coconut food processing technologies that are adaptable by individuals or groups of coconut farmers will be featured here. These coconut-based technologies require very simple locally available materials, and their operation is quite easy to follow. The coconut-based products that are derived from these technologies may be consumed by the farming family or sold in the domestic market, thus adding value to the coconut and enabling the farming family to earn additional income. These technologies are (a) coconut vinegar making, (b) moulded coconut sugar, (c) coconut jam, (d) nata de coco, (e) soap making, (f) coconut shell charcoal making and (g) coconut fibre products.

8.13.1 Coconut Water Vinegar

Coconut water vinegar is a natural product resulting from the alcoholic and acetous fermentation of sugar-enriched coconut water. It contains 3–4 % acetic acid and is used as an indispensable commodity in any household. Vinegar derived from fermenting coconut water can be produced either on a commercial scale or as a village cottage industry. As a non-synthetic food product, coco water vinegar is widely preferred as table seasoning or as an ingredient in food processing. Coco water vinegar is processed by allowing filtered coconut water, mixed with other substances, to undergo fermentation and acetification at ambient temperature (28–32 °C). The first step of the process is done by straining the coconut water through filter cloth. The sugar content of coco water is then adjusted to 15°°Bx (162 g/l) by adding refined sugar into it. The mixture is pasteurised by heating to boiling point. The boiled mixture is then cooled and inoculated with the active dry yeast at 0.5 g/l. The mixture is then allowed to undergo alcoholic fermentation for 5–7 days. After the fermentation process, alcoholic coconut water is then transferred to another container with a faucet at its bottom. Mother vinegar or a starter culture is then added to about one-fourth its volume. The container is only filled up to three-fourths its capacity to provide headspace for effective acetic acid fermentation. The mixture is then stirred thoroughly, covered with clean cloth and allowed to undergo acetification for 7 days. The coco water vinegar is harvested by opening the faucet or by siphoning. The amount of vinegar harvested is equivalent to the amount of alcoholic coconut water added. The remaining vinegar will then serve as the starter for the next batch of alcoholic coconut water acidification. Since the process involves fermentation, care must be taken to ensure that all fermentation containers are either made of plastic or stainless steel. The process produces a natural product, which is highly acceptable, based on flavour, aroma and general acceptability. It contains 4.0 % acetic acid, which conforms to the Food and Drug Administration requirements.

The technology is simple, economic and an accelerated method of coconut water vinegar production. It can be easily adopted in the rural areas since no sophisticated equipment is needed and very little capital investment is required. The utilisation of coconut water which is considered a waste material in copra making or in desiccated coconut factories will certainly give an added income to the rural families in the coconut farming communities. It will provide productive use of the time and employment to the women in the coconut countryside.

8.13.2 Toddy Vinegar

When the coconut inflorescence is tapped, a very sweet juice or sap exudes from it. This is called coconut toddy in Malaysia, Sri Lanka, India and other countries. The coconut toddy contains as high as 16 % sucrose and can be had throughout the year. A characteristic of coconut sap is its spontaneous and rapid fermentation. No yeast is needed since there is a ready source of very active 'wild' yeast in the environment. The coconut sap starts alcoholic fermentation right away and becomes completely fermented within a day. Fresh coconut toddy can be used as a beverage. However, it becomes unpalatable if allowed to ferment for more than 24 h. After this period, acetic fermentation converts the alcohol into vinegar. In the traditional method of vinegar production, toddy is allowed to ferment in large acetifying vats for 10–14 weeks. When the vinegar reaches the maximum strength of about 4–7 % acetic acid, the clear supernatant liquid or vinegar is then transferred to closed casks for ageing up to 6 months. The aged vinegar is then bottled for household or commercial purposes.

8.13.3 Alcoholic Beverages from Coconut Sap

Sweet toddy or fresh sap undergoes spontaneous fermentation producing a common alcoholic drink 'fermented toddy'. The toddy becomes stale when the fermentation exceeds 24 h. Normally, the toddy is consumed within 12 h after the sap is collected. The nutritional value of toddy for thiamine and riboflavin resides mainly in the yeast-free fluid portion. Toddy also contains small amounts of protein, fat and other nutrients. Fermented toddy on distillation yields a strong alcoholic drink known as arrack. The range of recovery is 15–18 % of the original toddy. Normally, sweet toddy is allowed to undergo fermentation in loosely covered wooden or plastic container for 3–5 days before it is distilled. Reports revealed that analysed samples of arrack collected from several locations had an average value of total soluble solids of (°Bx) 13.52, a pH of 3.92 and an alcohol content of 42.65 (vol. per cent).

8.13.4 Home-Made Moulded Coconut Sugar

Moulded coconut sugar is edible sugar made from fresh coconut sap. Produced by small-scale cottage industries, moulded coconut sugar is used for edible purposes essentially as a sweetening agent in many traditional food preparations and food products especially in Indonesia. The concentration of total sugars in moulded coconut sugar is 80 % total soluble solids. The process of producing moulded coconut sugar starts from tapping or collection of coconut sap. But before this is done, the collection vessels are first washed with clean water, followed with hot water and then dried. Alternatively, the clean vessels are smoked using firewood for 10–15 min. The treatments are used to reduce microbial loads of vessels. To prevent spoilage of sap during tapping, the collection vessels are added with tablespoon in the form of paste, a few pieces of mangosteen bark or other natural preservatives such as special varieties of leaves. Once treated, the collection vessels are then ready to be used for tapping the coconut sap. Collection of coconut sap from the palm is done twice a day at 6–7 in the morning and at 4–5 in the evening. Although it is not a common practice by home processors, it is desired that the collected sap be tested for acidity using a pH indicator paper. This is because the fermented or spoiled coconut palm sap is no longer suitable for brown sugar manufacture. The collected coconut sap is then filtered through a muslin cloth to remove insects, ants and other contamination. It is then transferred into a cooking vessel. The next step involves evaporation of water from the sap to increase the concentration of the sap. Thus, the filtered sap is boiled in a cooking vessel at a temperature of 100–110 °C for 3 h. The material will then turn into a thick liquid. During boiling, foam will be formed. This should be discarded from the vessel. A few drops of cooking oil or grated coconut are added to the mash to prevent excessive foam formation. The mash is heated for another 1 h with occasional stirring. To avoid caramelisation of sugars, heating should be done slowly. When the mash has become very thick and suitable for moulding, the cooking vessel is lifted from the stove and cooled to 60 °C. The cooled mash is then poured into clean halves of coconut shell or bamboo vessels for cooling and setting.

8.13.5 Home-Made Coconut Jam

Referred to as coconut caramel spread by some South Pacific countries, especially in Samoa, coco jam is actually coconut milk cooked in brown sugar and glucose. Coco jam is the 'butter' in many coconut-producing countries, and it is commonly taken as a spread, biscuit sandwich, pancake syrup, sponge cake filling, doughnut spread, ice cream topping, fruit dessert topping and marinade syrup for meat. The concentration of total sugars in coconut jam is 75–76 % total soluble solids. To produce good quality coconut caramel spread, one starts with choosing 100 fully matured coconuts which are devoid of cracks or any damage. The selected nuts are dehusked, cut into halves and grated immediately. Freshly grated coconut meat has the characteristic coconut smell and must not have any off odour. It is also important that all containers used are thoroughly clean and the working place completely sanitary. Once the grated coconut is ready, coconut milk (33.3 kg) may then be extracted by adding water in the proportion of 1 part grated coconut to 0.5 part water (if pressing is to be done manually). However, no water is needed if pressing is done mechanically. Strain or filter the coconut milk through cheesecloth to remove any solid particle or foreign matter from the milk. Weigh the coconut milk (12 kg) and determine the amount of brown sugar (2.13 kg)

and glucose (1.06 kg) needed. Mix the sugar and the glucose with one-half of the total volume of the coconut milk, and boil slowly to dissolve the brown sugar and the glucose. Stir continuously for about 10 min, and maintain the cooking temperature at about 78–80 °C. Add the remaining half of the coconut milk extract when almost thick and boil for another 35 min until the temperature reaches to 100–102 °C. Stir the mixture frequently or almost continuously to prevent burning and continue to boil until done. The end point is reached when a drop of the mixture forms a soft ball in cold water. Strain the cooked mixture through a clean wire mesh and pack while hot. Packing is best done using clean and sterilised bottles for longer shelf life. Cool the bottles, label and seal.

8.13.6 Nata de Coco Production

Nata de coco is prepared from matured coconut water, a waste product in copra and desiccated coconut manufacturing units. It is a gelatinous delicacy formed by the action of a microorganism, *Acetobacter xylinum*, in a culture medium of coconut water. The culture solution is prepared by mixing coconut water with sugar and acetic acid at a stipulated proportion which is inoculated with *Acetobacter xylinum*, through a culture liquid. It is filled in glass jars, covered with thin cloth and kept for 2–3 weeks without any disturbance. During this period, a white- or cream-coloured jelly-like substance forms and floats on the top of the culture medium. It is harvested, cut into pieces and washed in pure water several times to remove all the acid. It is then immersed in flavoured sugar syrup for 12 h and packed in glass bottles. CDB has developed the know-how for the preparation of nata de coco and the same would be transferred after pilot-scale testing of the process. Nata de coco is also prepared from coconut skimmed milk. More details on this can be obtained from the technology section of the CDB at Kochi. Nata de coco is a white, gelatinous food product obtained from the action of microorganism *Acetobacter xylinum* on coconut water or coconut milk mixed with water, sugar and acetic acid. Quality nata is smooth, clear and chewy. It is sweetened by boiling it in sugar-water solution. Nata de coco is popular primarily for its food uses. It can be sweetened as desserts or candies. It is an excellent ingredient for sweet fruit salads, pickles, fruit cocktails, drinks, ice cream, sherbets and other recipes. Nata de coco also has some industrial uses. The process flow in the production of nata de coco is as follows:

1. Preparation of ingredients: Measure all ingredients in the formulation properly from the 28 l of tap water; get approximately 3 l for dissolving sugar and 2 l for extracting the coconut milk from the freshly grated coconut meat.
2. Milk extraction: Place the coconut meat in the basin and add half of the water set aside for extraction. Mix thoroughly and squeeze grated meat in water. Filter through a piece of cheesecloth. Repeat extraction using the remaining water and filter. Add second extract to the first.
3. Filtration of dissolved sugar: Filter dissolved sugar to remove impurities that might have entered accidentally into the sugar in stock.
4. Mixing: To the remaining 23 l of water, add the extracted coconut milk, dissolved sugar, glacial acetic acid and mother liquor. Stir thoroughly with a wooden ladle to get a homogenous mixture. Set aside a small portion as mother liquor for the next mixing.
5. Filling: Distribute the rest of the mixture into nata moulders at a level of approximately 3 cm high.
6. Fermentation: Arrange the nata moulders in the nata fermentation room. Cover with newspapers or similar materials. To maximise space, nata moulders can be placed one on top of the other to obtain several layers. Fermentation is completed after 8–10 days, depending on environmental conditions. Optimum temperature for nata production is between 23 and 32 °C.
7. Harvesting: Harvest by separating nata from the spent liquor.
8. Scraping: Clean nata by scraping the cream and the thin, white layer at the bottom part.
9. Using a blunt piece of plastic or bamboo.
10. Soaking: Place clean nata in a plastic container and keep immersed in water.
11. Syrup of nata de coco: To cook nata de coco in syrup, cut clean nata into cubes approximately 1 cm^3 or according to the customer specifications. Soak the nata for 1 or more days in several changes of water to remove the sour taste and smell. Drain the nata and boil in water for 5–10 min. Check if acid is totally removed. Add sugar equal to the weight of drained nata. Mix thoroughly and set aside overnight. The next day, stir the mixture to disperse any undissolved sugar. Add a small amount of water. Heat the mixture to boiling point while stirring occasionally. Add flavouring, if desired. Set aside overnight, and repeat the heating process until the nata is fully penetrated with sugar as evidenced by the clear and crystalline appearance of the sweetened nata. Pack the sweetened nata two-thirds full into sterilised preserving jars. Add syrup leaving a 0.25 in. airspace. Cover jar immediately with PVC lined caps. Sterilise bottled nata by immersing in boiling water for 30 min. Remove bottled nata and tighten the caps. Cool the jars in inverted position to further sterilise the caps and check for leakage. Wash cooled jars and wipe them thoroughly, place plastic seal and label. Store in a cool dry place.

8.14 Soap Making

Mixing oil with a solution of caustic soda in water makes soap. When the caustic soda is mixed with oil, a chemical reaction occurs and all the component fatty acids of the oil are changed into sodium salts, known familiarly as 'soap'. The oil, caustic soda and water used to make the soap have to be mixed together in correct proportions to ensure that the finished soap contains no excess alkaline which would cause a burning reaction on the skin. The oils used for soap making fall into two categories. In the first category are oils that are obtained from the kernels of different types of palms. The most commonly known oils in this category are coconut oil and palm kernel oil. They are known as 'lauric oils' because lauric acid is the major fatty acid that they contain. These fats make hard soap which produces fast-foaming lather. To make soap by the cold process, follow the following steps. Weigh 1 kg of NaOH and 2 kg of water, and pour the NaOH flakes in water and stir constantly until dissolved. Avoid inhaling the vapour over the solution as the mixture will become very hot. Set aside for cooling to a temperature of about 96 °F. Prepare 6 kg of coconut oil and slowly pour this into the caustic soda solution while constantly stirring the mixture in one direction. The mixture is kept stirred until it thickens to a desired consistency, approximately after 40 min to 1.75 h. Add any desired colour or essence. Stir and immediately pour into moulders and leave for about 24 h for saponification. Cut into desired sizes and dry or age for at least 1 week to complete the chemical reaction. Wrap if required and use the soap only after 1 week of ageing.

8.15 Coconut Wood-Based Products

The coconut wood because of its distinct grain characteristics is ideal for making wall panels, furniture, doors and windows, show pieces, etc. There are several small-scale units manufacturing a variety of articles from coconut wood. Coconut wood is a potential source for diverse purposes. It could be used for the manufacture of furniture, flooring tiles, wall panels, curios, building purposes, etc. Coconut being a monocot has marked structural differences from other dicot timbers. Effective utilisation of coconut wood is by and large depended on the basic understanding of characteristics that are unique with coconut wood and adoption of appropriate technologies for processing and end-use applications.

8.16 Coconut Leaves

Coconut leaves are plaited and used for thatching houses and sheds in rural areas. It is also used for thatching 'honeymoon huts' and such huts in town and cities. The Regional Research Laboratory at Thiruvananthapuram has standardised a simple process to extend the life of coconut leaf thatch up to 4 years. The process involves dipping plaits of coconut leaves for 5 min in copper sulphate (1 kg copper sulphate in 100 l water for 100 plaited leaves), draining out excess solution and stacking them overnight, followed by spraying with cashew nut shell liquid (3 kg CNSL + 0.5 kg kerosene) and then subjecting it to sun drying for 3–4 days. The treatment costs more than about 50 % of the initial cost for thatching, but it is economical because of durability and no recurring expenses for the next 4 years. Plaited coconut leaves are also used for making baskets and headgears and for erection of temporary fences. Plaiting of coconut leaves is a cottage industry in traditional coconut-growing states. Midribs of leaves are used to make brooms of different types which are used for cleaning rough grounds and floors. Brooms of midribs of coconut leaves are manufactured on a commercial scale in Tamil Nadu and Karnataka.

8.17 Coconut Shell-Based Products

Shell charcoal, shell-based activated carbon, shell powder, shell handicrafts, shell ice cream cups and bear glasses, ladles, forks, show pieces, shell buttons, etc. are the shell-based products available in the country.

8.17.1 Coconut Shell Charcoal Making

Charcoal making is based on the principle that coconut shell, wood and other carbonaceous materials can be converted into charcoal by incomplete burning. Limiting the amount of air used during the burning process produces incomplete burning. Thus, the quantity and quality of charcoal depend largely on how well the amount of air is regulated in the charcoal chamber. Charcoal making started with simple methods such as those employed in the backyard to make charcoal for household use. As the demand increased, more sophisticated methods were developed to produce charcoal in commercial scale. Today, small backyard and commercial kilns are being used. The two types of kiln used in charcoal making are the primitive or modified pit and the drum:

Primitive or modified pit – A simple pit is dug in the ground just enough to accommodate the desired number of coconut shells to be made into charcoal. The process mainly involves simple drying of coconut shells arranged in the hole and burned. Some farmers cover the pit from time to time while the shells burn. To control the fire, sprinkle the flame with enough water so as to put the fire totally out. The charcoal produced out of the modified pit method is suitable only for household use due to its poor quality.

Drum – A 55 gal drum open on one end and punched with four holes at the bottom is used in this method. This is then raised from the ground by two pipes to allow air entry through the holes. As a starter, remove the cover from the drum and place and burn a shovel full of shells on it. When already burning strongly, throw the shells into the drum. Throw in just enough fresh shells to put the flames but not the fire. Feed shells continuously at the top to assure that they will not burn fiercely. Slow burning gives the highest charcoal yield and the least ash. When burning reaches the top pile of shells, cover the drum with banana stalks or wet sack plastered with sand or mud. Never allow sand or mud to get inside the drum. The charcoal-filled drum should be left to cool overnight. It takes about four hours of shell burning to fill one drum. When properly attended to, one drum can yield 75–90 kg charcoal. Generally, it takes 3 t of shells to make 1 t of charcoal.

8.17.2 Uses of Coconut Shell Charcoal Activated Carbon

Activated carbon from coconut shell charcoal is a manufactured carbonaceous material having a porous structure and a large internal surface area. It can absorb a wide variety of substances. Activated carbon particles are capable of attracting molecules to their internal surface area and are therefore called adsorbents. The main characteristic of activated carbon is the extent of their internal pores. Among other products with internal pores used as adsorbents on a commercial scale are silica gel, zeolites, alumina and molecular sieves. The main difference between these adsorbents and activated carbon is the ability of activated carbon to adsorb an extremely wide spectrum of adsorbents. This is because activated carbon has different types and/or sizes of pores. Generally, coconut shell charcoal-based activated carbon is microporous, and depending on the size of the pores predominant, it will exhibit affinities for molecules of different sizes. Activated carbon is used for a wide array of purposes which include water, air and food purification, solvent recovery, pharmaceutical industry and catalyst support. The examples of use of coconut shell charcoal activated carbon in air purification are their uses in gas masks, cooker hoods and other filters.

8.17.3 Other Uses of Coconut Shells

Coconut shell is exploited in small-scale industries, e.g. manufacture of novelties. It has been in demand when ground into 'flour' for mosquito coils (insect repellent) and as filler for articles made of synthetic resin-plastics. With the present shortage of fuels, the shell itself and the charcoal made from it are gradually taking the place of liquefied petroleum gas (LPG) for home cooking.

8.18 Coir Industries and Production

Coconut husk is the basic raw material for coir industry. At present only 35 % of total husk is used, while there is scope for utilising 50 % of the husk (Viswanathan 1998). India is the premier coir-producing countries in the world earning foreign exchange of Rs. 239 crores (1997–1998).

8.18.1 Coconut Fibre Products

Coconut husk is the raw material for the coir industry. The coir fibre is extracted either by natural retting (microbiological process) or mechanical means. Roughly 10 % of the global annual production of coconut husks is used to extract coir fibre resulting in 480,000 mt of coir approximately. An average of 100,000 t of this total production (21 %) enters into the world trade. Majority of exports take the form of fibre that is then processed in consuming countries. Coir product exports take the form mainly of mats, matting, brushes and a very small quantity of needled felt and rubberised coir.

8.18.2 Extraction Process of Coconut Fibre

There are two distinct varieties of coir fibre, white fibre and brown fibre. The fibre extracted from green coconut husk by the natural retting process is known as 'white fibre', whereas fibre extracted mechanically from dry coconut husk is 'brown fibre'. Retting of coconut husk for the production of white fibre is a biological process which softens the husks paving way for easy extraction of fibre manually by beating with wooden mallets. It is normally done in the saline back waters that are bestowed with a gentle natural tidal action. There are three process of retting: net retting, pit retting and stake retting. In the areas where very good quality fibre is produced, net retting is practised. In this process, the husks are filled inside a net made out of coir yarn and toyed to the retting field. The bundles are then weighed down using mud. Stake retting is practised where there is a heavy current and fear of husks being washed off. In this method, husks are filled in the enclosures made out of bamboo stakes and covered with mud. In pit retting, the bottom of the pits are covered with mud and sides with plated coconut leaves or coir yarn nets. The entire pit or nets are filled with fully matured green husk. The bundles are piled appropriately in such a manner, and they are not disrupted and left for a period varying from 4 to 12 months.

After about 4 weeks of soaking, the water gets warm up and becomes cloudy and yellowish white, and the covering is formed on the surface. Exhale of gas bubbles is observed with the smell of hydrogen sulphide which subsides after a period of approximately 4 months. The retted husks are taken out from the soaking pits and beaten manually by wooden mallets to separate the fibre from the embedded pith. The extracted fibre is cleaned properly and dried under shade for further processing. The fibre is graded in accordance with its colour, length of fibre and other factors. Women of coconut farming families are greatly involved in these activities.

Flow Chart of Processing in the Production of Coir
Coconut Palm > Plucking of Coconut > Dehusking > Green Husks Dry Husk > Natural Retting > Crushing > Manual Extraction Soaking > of White Fibre > Mechanical Extraction > Spinning of Different Types of Yarn/Rope Decoration Defibering > Decorticated Bristle > Mattress > Weaving Coir Mats, > Fibre > Mattings > Carpets > Ropes > Curled Coir Curled Coir > Rubberised Coir

8.18.3 Coir Pith

Coir pith and coir ply, waste products obtained during the extraction of coir fibre from husk, are very light, highly compressible and highly hygroscopic. It is used as a soil conditioner, surface mulch/rooting medium and desiccant. Composted coir pith is an excellent organic manure for indoor plants as well as for horticulture crops. Several firms are manufacturing composted coir pith in the country. Compressed coir pith in the form of briquettes for easy transportation is also manufactured in the country. After 30 days of decomposition, it turns black manure (Table 8.12).

The total coir production during 1993–1994 was 239,100 t. The coir industry is basically export oriented; 60 % of the product is exported. In the year 1997–1998, India earned Rs. 239 crores from it (Singh 1998).

Table 8.12 Number of coir industry in different states of India

State	No. of units	State	No. of units
Kerala	5,124	Orissa	100
Karnataka	217	West Bengal	75
Tamil Nadu	262	Maharashtra	5
Andhra Pradesh	265	Pondicherry	6
		Goa	3
Total 6,531			

Source: Indian Coconut Journal 29 (4): 4–27

8.18.4 Coir Geotextiles

Coir geo fabrics are an inexpensive, ready-to-use, effective item for a variety of application including control of soil erosion, control of landslide, etc.

In Kerala the state government's decision to allow mechanisation of coir industry helped boost the production and export of coir and coir products.

8.19 Available Technologies for Product Diversification

The technology development centre functioning under CDB has been successful in developing technologies feasible for both at small-scale level and large-scale level (Podual et al. 1998.)

8.19.1 Feasible Technologies at Small-Scale Level

Apart from traditional coconut products like copra and coconut oil, technologies catering to small-scale sector for the manufacture of the following products are available today in the country:

- *Desiccated Coconut*
  Manufacture of DC 5,000 nuts/day adopting latest fluidised bed dryer technology involves an investment of Rs. 30–32 lakhs. The investment analysis has realised an internal rate of return of 37.5 % with a payback period of 3 years.

- *Tender Coconut Water*
  The CDB in collaboration with Defence Food Research Laboratory (DFRL) has developed the technology for packing tender coconut water in pouches/aluminium canes. The processing and packing of 10,000 tender nuts per day require investment of 35–40 lakhs. The profitability is about 20 %. The technology is available at a transfer fee of Rs. 3 lakhs.

- *Coconut Vinegar*
  The manufacture of coconut vinegar with a capacity of 500 l/day involves capital investment of about Rs. 7 lakhs and profitability about 20 %. The know-how fee for licensing the technology is Rs. 10,000.

- *Nata de Coco*
  This technology is available with CDB. It involves Rs. 2–2.5 lakhs for a unit to produce 50 kg of nata per day. The product has great export potentialities. The profit margin works out to be Rs. 12 per kg nata.

- *Coconut Shell Powder*
  12,000 shells yield about one ton shell powder capital investment about 2.2 lakhs. The product finds extensive demand in plywood, laminated board, mosquito oil and agarbatti industry.

- *Shell Charcoal*
  Now there are modern methods for production of shell charcoal, e.g. the waste heat unit technology. A unit of capacity of 30,000 nuts/day would involve an investment of Rs. 30 lakhs.

8.19.2 Feasible Technology at Large-Scale Level

Technologies suitable for large-scale set-up are today available in the country for manufacture of coconut milk powder, coconut oil, skimmed milk, shell-based activated carbon, etc.

- *Coconut Milk/Cream*
  For a raw material capacity of 10,000 ripe nuts, the yield is around 2,500 kg of cream. The capital investment is Rs. 1.49 crores; profit margin is about 30 %. Technology would be available from the CDB with a transfer fee of Rs. 3 lakhs.

- *Spray-Dried Coconut Milk Powder*
  The CDB in collaboration with CFTRI has developed this technology. The capital investment is about Rs. 2.36 crores. On an average 1,000 coconut can yield 99.6 kg of milk powder. This product has tremendous market potential.

Nandanasabapathy et al. (1999) reported from DFRL some recent R and D effort to produce some new products like coconut honey, coco sauce, coconut lemonade, coconut chutney, etc.

8.20 Coconut Oil and Coronary Heart Disease

Coronary heart disease (sym: ischaemic heart disease) has been defined as impairment of heart function due to inadequate blood flow to the heart compared to its needs, caused by obstructive changes in the coronary circulation to the heart. It is the cause of 25–30 % of death in most industrialised countries. The WHO has drawn attention to the fact that CHD is our modern epidemic. Depending on the rate of development of the ischaemia and its ultimate severity, CHD may manifest itself in many presentations: (a) angina pectoris (AP), (b) myocardial infarction (MI), (c) irregularities of the heart, (d) cardiac failure and (e) sudden death.

Coronary artery disease (CAD) is a complex degenerative disease that causes reduced or absent blood flow in one or more of the arteries that encircle and supply the heart. The principal cause of coronary artery disease is coronary artery atherosclerosis which is a disease of the arteries characterised by endothelial dysfunction, vascular inflammation and the build-up of lipids, cholesterol, calcium and cellular debris within the intima of the vessel wall. CAD refers to the presence of atherosclerotic changes with the walls of the coronary arteries, which causes impairment or obstruction of normal blood flow with resultant myocardial ischaemia. Ischaemia refers to a lack of oxygen due to inadequate perfusion, which results from an imbalance between oxygen supply and demand. Ischaemic heart disease (IHD) is the most common, serious, chronic, life-threatening disease in the USA.

Although any artery may be affected by atherosclerosis, the major targets are aorta, coronary and cerebral arteries. Atherothrombotic disease of the cerebral vessels is the major cause of brain infarcts, the so-called strokes. Fatty streaks develop as circulating monocytes migrate into intima, take up oxidised LDL from plasma become foam cell. When it dies, smooth muscle cells then migrate into and proliferate within the plaque. It encroaches into the lumen of the vessel. Plaque rupture may lead to rapid occlusion of vessel and cause most acute syndrome. The pattern of CHD in India has been reported to be as follows: (a) the peak period is attained between 51 and 60 years, (b) males are affected more than females, (c) hypertension and diabetes account for about 40 % of all cases and (d) heavy smoking is responsible in a good number of cases. The aetiology of CHD is multifactorial. The greater the number of risk factors present, the more likely one is to develop CHD. The principal risk factors are as below (Table 8.13):

Smoking has been identified as a major CHD risk factor with several mechanisms – carbon monoxide-induced atherogenesis and nicotine stimulation raising both blood pressure and myocardial oxygen demand. *Hypertension* accelerates the atherosclerotic process. In the Framingham study, men aged 45–62 with BP 160/95 mmHg had a more than fivefold greater risk of IHD. There is a triangular relationship between habitual diet, blood cholesterol-lipoprotein levels and CHD.

Serum cholesterol concentration is an important risk factor for CHD at levels 220 mg/dl or more (Keys 1980). When we look at the various type of lipoprotein, it is the level of low-

Table 8.13 Risk factors for CHD

Modifiable	Non-modifiable
Age	Cigarette smoking
Sex	High blood pressure, dietary factors
Family history	Elevated serum cholesterol
Genetic factor	Diabetes, haemostatic variables
Personality	Obesity, sedentary habits, stress, etc.

density lipoprotein (LDL) cholesterol that is most directly associated with CHD (Gordon et al. 1977), while very low-density lipoprotein (VLDL) has also been shown to be associated with premature atherosclerosis. High-density lipoprotein (HDL) cholesterol is protective against CHD (Gordon et al. 1977). HDL should be more than 30 mg/dl. Cholesterol/HDL ratio less than 3.5 has been recommended as a clinical goal for CHD prevention (Superko et al. 1985). *Regular physical exercise* increases HDL concentration and decreases the body weight and blood pressure, reduces blood clotting and promotes collateral vessels which are beneficial to cardiovascular health. The risk of CHD is two to three times higher in *diabetics. High alcohol* intake is a risk factor for CHD, hypertension and all cardiovascular diseases.

Genetic factors are determinants of an individual's TC and LDL levels determines the CHD risk. Type A behaviour is associated with competitive restlessness, hostility and impatience – are more prone to CHD than the calmer type B individuals.

Haemostatic factors – High levels of fibrinogen and factor VII are associated with increased risk of MI (coronary thrombosis), obesity, physical inactivity, etc. and are also risk factors.

Other dietary factors – Low levels of vit. C, vit. E and other antioxidants may enhance the production of oxidised LDL which favours CHD. By dietary modification and controlling the risk factors where applicable reduced the risk of CHD.

8.21   Some Facts About Fats

What people eat is not calories but food, and consideration of fads, flavour and variations of appetite can make nonsense of the dietician's theories. Good nutrition means 'maintaining a nutritional status that enables us to grow well and enjoy good health'. The food according to chemical composition is divided into protein, fats, carbohydrates, vitamins and minerals. In the Indian dietary they contribute to the total energy intake as follows: protein 7–15 %, fats 10–30 % and carbohydrates 65–80 %. In developed countries fats provide 30–40 % of the total energy intake. The WHO expert committee on prevention of CHD has recommended only 20–30 % of total dietary energy to be provided by fats.

Fats are classified as simple lipids, e.g. triglycerides; *compound lipids*, e.g. phospholipids; and *derived lipids*, e.g. cholesterol. Fats yield *fatty acids* and *glycerol* on hydrolysis. Fatty acids are divided into *unsaturated fatty acids* (e.g. lauric, palmitic, stearic acids, etc.) and *unsaturated fatty acids* which are further divided into *monounsaturated* (e.g. oleic acid) and *polyunsaturated fatty acids* (e.g. linoleic acid).

Saturated fatty acids contain no double bond, while unsaturated fatty acids contain one or more double bonds. Monounsaturated fatty acid contains one double bond. Polyunsaturated FA contains two or more double bond. *Essential fatty acids* (EFA) are those that cannot be synthesised by humans. They can be derived from food; linoleic acid is the most important EFA which serves a basis for production of other EFA (Table 8.14).

Table 8.14 Fatty acid content of different fats (%)

Fats	Saturated FA	Monounsaturated FA	Polyunsaturated FA
Coconut oil	92	6	2
Palm oil	46	44	10
Cotton seed oil	25	25	50
Ground nut oil	19	50	31
Safflower oil	10	15	75
Sunflower oil	8	27	65
Corn oil	8	27	65
Soya bean oil	14	24	62
Butter	60	37	3
Margarine	25	25	50

Rajan (2006)

8.22   Coconut Oil and Coronary Heart Disease

Coconut oil constitutes the most important source of dietary fat in many countries. But, of late, the consumption of coconut oil has been linked with the incidence of coronary heart disease, and sustained campaign against its use is going on. Being rich in saturated FA, coconut oil is often bracketed with animal fats such as lard, butter and egg products as having cholesterogenic properties (Rajan 2006) (Table 8.15).

Chemically all the saturated fatty acids are not alike in their dietary properties. The difference is basically in the length of the carbon chain or the number of carbons present in the fatty acid. The dominant FA present in coconut oil are medium-chain fatty acids (MCFA). These MCFA are absorbed directly through the portal vein into the liver and do not require the carnitine transport for their entry into the cells and subsequent metabolism. They are immediately made available to the body unlike long-chain FA which require emulsification in the intestine for absorption, and they reach the systemic circulation via the lymphatic system. Because of their distribution to all parts of the body before reaching to the liver, they are more prone to be deposited in the different tissues. In simple word, coconut fats are easily digestible and not circulate in the bloodstream and not stored. About 50 % of fatty acid in coconut oil is lauric acid (MCFA). MCFA are not

deposited in the adipose tissue but are readily oxidised in the body to provide quick energy (Kaunitz 2001) favourable for use in sports nutrition and in slimming diet products. MCFA are beneficial for people suffering from fat malabsorption and used in infant formulations; the lauric acid in coconut oil is used by the body as disease-fighting derivative monolaurin that babies make from the lauric acid they get from their mothers' milk.

Coconut oil and palm oil are called lauric oils which make them unique from other vegetable oils. On hydrolysis coconut oil yields 85 % FA and 15 % glycerol, whereas the corresponding proportion in FA production is more when other oils are consumed. The limitation in the use of coconut oils is that it contains very low amount of EFA (Ghosh and Ghosh 2008). Some fish oil, soya bean oil, sunflower oil, peanut oil, etc. are rich sources of EFA. This deficiency can be met up through the use of a variety of food, particularly fish.

In the developed countries, the unsaturated vegetable oils are generally consumed after hydrogenation. On hydrogenation, part of the unsaturated fats gets converted into saturated LCFA trans fatty acids that cause elevation in cholesterol (Enig 1990). Contrary to this coconut oil both in its natural state and hydrogenated form will have majority of its FA as MCFA which on consumption will not elevate cholesterol levels. Coconut oil is also a preferred oil in the household culinary preparations because it could be reheated in subsequent uses. But other vegetable oils are not amenable to such uses as they produce toxin to health in the process of heating, cooling and reheating.

As it is already mentioned earlier, an excess of cholesterol in the blood and of blood fats in general is caused not only by the fat component of a normal diet but by other factors as well. They are related to heredity; diseases of the thyroid gland, pancreas, liver, kidney, etc.; obesity; sedentary habits; prolonged mental stress; and intake of cholesterol-rich foods, e.g. dairy product, eggs, red meat, etc. The vegetable oil including coconut oil does not contain cholesterol as in animal food.

Table 8.15 Fatty acid composition of coconut oil (%) wt. of total fatty acids

Medium-chain saturated	%	Monounsaturated	%
Capric acid (C:10)	6	Oleic acid (C18:1:9)	6
Lauric acid (C:12)	47	Polyunsaturated	
Myristic acid (C:14)	18	Linoleic acid (omega 6C18:2)	2
Palmitic acid (C:16)	9	Omega 3 fatty acid	Nil
Stearic acid (C:18)	3		

The reported figures are:

Vegetable oil	Cholesterol	Animal fats	Cholesterol
Palm oil	18 ppm	Egg	5,000 ppm
Soybean oil	28 ppm	Butter	3,150 ppm
Corn oil	50 ppm	Cheese	1,100 ppm
Coconut oil	0–14 ppm	Lard	3,500 ppm
		Milk	150 ppm

It is a wide accepted fact that an excessive intake of cholesterol-rich food will lead to a gradual increase in the serum cholesterol level and cause blood lipid abnormalities. High serum cholesterol is one of the major risk factors of coronary heart disease. But the consumption of coconut oil has not been proved to have any relationship either with the incidence of the disease (CHD) or with hypercholesterolaemia or hyperlipidaemia. Coconut kernel contains 7–8 % dietary fibre which beneficially influences serum cholesterol (Sindhu and Rajmohan 2006).

Studies conducted in many countries by different workers on the effects of dietary FA on human health have revealed beneficial results from the use of coconut oil. In most of the studies, coconut oil has proved to be neutral. Many animal experiments show beneficial or harmless effects of coconut oil consumption. Coconut oil feeding produced significantly higher alpha lipoproteins (HDL) relative to sunflower oil feeding in rats (Hostmark et al. 1980). Total tissue cholesterol accumulation for animals on the safflower diet was six times greater than on coconut oil diet and twice that of soybean oil diet (Award 1981). CHD is unknown among Polynesian population whose staple diet is coconut. In an epidemiological and experimental observation, Kaunitz and Dayrit (1992) reviewed that dietary coconut oil does not lead to high serum cholesterol or high CHD. Prior et al. (1981) had shown that when a population lowered their intake of coconut oil, their total cholesterol and LDL increased and HDL (good cholesterol) decreased. Ng et al. (1991) reported that by supplying 75 % fat from coconut oil, total cholesterol increased 1.1 % but HDL cholesterol increased 21.4 % resulting LDL/HDL ratio decreased by 3.6 %.

From the experiments, it can be observed that the effect of coconut oil on serum cholesterol is that there may be a rise in total cholesterol, LDL and HDL which have lower serum cholesterol, but there is lowering of total cholesterol and LDL (bad cholesterol) in hypercholestrolemics. Clinical studies done at the New England Deaconess Hospital, a Harvard Medical School affiliate, show that coconut oil is neutral in its effects on blood lipids and will not cause an increase in cholesterol or cause cardiovascular disease. Coconut oil even increased the HDL (good cholesterol) reducing the risk for CHD.

Saturated fats may be considered as a major culprit for CHD because there is some association between serum cholesterol and CHD and between saturated fat and serum cholesterol. However, a fear complex has been created among the general public that consumption of coconut oil results in elevated cholesterol level. This myth was primarily due to equating coconut oil with saturated fat without knowing everything about saturated fat. The saturated fat in coconut oil is of short-chain and medium-chain FA. All saturated fats are not harmful. The real problematic fat is the hydrogenated vegetable fats or animal fats. Blackburn et al. (1989) concluded that when coconut oil is fed with other fats or supplemented with linoleic acid, coconut oil is a neutral fat in terms of atherogenicity. Kurup and Rajmohan (1994) also found no statistically significant alteration in the serum total cholesterol, HDL, LDL, HDL/total cholesterol ratio and LDL/HDL ratio of triglycerides from the baseline values.

The observation and other evidences suggest that coconut oil is more beneficial to human body as a dietary fat than the natural or hydrogenated forms of unsaturated vegetable oil. The incidence of coronary heart disease could be better linked with the consumption of cholesterol-rich food of animal origin and the sedentary habits associated with a tension ridden mechanical society.

8.23 Postharvest Losses and Disorders in Coconut

Lack of awareness and actual skills on coconut postharvest technologies has caused significant losses starting from the harvesting of the nuts, seasoning, drying and storage. While wastage and losses occur at different stages, the copra drying stage or the efficiency of the drying process at the farm level is the most critical stage as this affects subsequent losses in terms of product quality and reduced prices. Harvesting of immature nuts causes the production of rubbery copra with high moisture content. If one allows the nuts to fall naturally, without harvesting or picking the nuts from the tree, the losses due to overripe nuts or germinated nuts are likely to occur. This could be as high as 10 % of the total harvest especially with varieties that are early germinating. As the growing embryo utilises the stored food in the endosperm, the copra produced from germinated nuts would be thinner, lighter and with lower oil content. Losses due to pilferage and losses due to nuts that are hidden or covered by thick weeds or shrubs could also range from 5 % to 10 % of the total harvest if one does not regularly harvest his or her coconuts. To avoid these losses, it is recommended that the 45-day cycle of coconut harvesting be adopted. Seasoning of unripe nuts for 2–4 weeks should also be practised. Farm sanitation, e.g. weeding of thick shrubs and grasses in the spaces between coconut palms, is highly recommended to prevent losses due to uncollected nuts. As mentioned earlier, major postharvest losses are caused by improper drying of copra as a result of a lack of know-how on the proper drying technology and the lack of incentives to adopt the recommended copra dryers and the appropriate copra drying methods. Improperly dried copra or copra with high moisture content are prone to aflatoxin contamination. Coconut researchers have also identified beetles, cockroach, a moth and an earwig to be associated with deteriorating copra and copra cake. Studies reveal that after 1 year of storage, copra weight loss due to pests will be as high as 5–10 %. Spraying of suitable insecticides may be done, but this is not practised due to its prohibitive cost. Sanitary practices in the copra warehouse are the best recommended alternative to control these pests. Generally, these pests are considered a minor problem when compared to the attack of aflatoxin-related moulds or fungi. Other factors cited to contribute to copra/copra cake deterioration are presence of wet or improperly dried copra, rubbery copra, delays in transport, long storage period and unsanitary conditions in the farms and warehouse. Long storage time also favours the breeding of copra pests or the proliferation of aflatoxin-related moulds.

8.23.1 Postharvest Disorders

8.23.1.1 Fruit Cracking

Dehusked coconuts are susceptible to stress cracking in which transverse fissures develop, mostly on the bottom half of the nut. Cracks may vary in width from a fine fracture up to 1 cm (0.4 in.), which also splits the inner shell and results in leakage of the internal water. Stress cracks provide entry for fungi and bacteria which cause the water to turn sour and the meat to rot, rendering the fruit unsaleable. Younger dehusked coconuts have a lower rupture force than mature coconuts. Dehusked coconuts are also susceptible to cracking if they are exposed to more than an 8 °C (46 °F) temperature change within a few minutes or to extreme heat or cold. In addition to cracking, mechanical damage to immature coconuts will cause the white husk to turn brown.

8.23.1.2 Mould

Mould growth on the husk surface, caused by various species of fungi, is largely cosmetic and does not penetrate into the coconut meat. Four different moulds cause deterioration of copra. *Rhizopus* sp., under favourable conditions, destroys significant quantity of oil in the meat. *Aspergillus niger* has a low moisture requirement; 18–20 % is the optimum and 12 % is the minimum. It causes considerable damage to copra, and the loss of oil may be as much as 40 % of the total oil content. *Aspergillus flavus* flourishes in a moisture range of 8–12 %, and the oil loss may be more than 40 % of the total oil content of the copra. *Penicillium glaucum* grows well even on copra with a low percentage of moisture. Its growth is entirely superficial and causes practically no loss of oil. Fungi commonly associated with surface mould are various species of *Penicillium*. Mould is caused by moisture condensation on the coconut surface or storing the fruit at a RH above 90 %.

8.24 Coconut Industry: Strengths, Constraints and Possibilities

West Bengal ranks 5th in coconut production after the four major South Indian coconut-growing states (Hore 1999). According to Agarwal (1999), the production of coconut oil through crude process in undivided Bengal dates back to several centuries. Kolkata may be the principal distributor of coconut product in the eastern region, so there is a huge scope of developing this industry in WB. WB requires about 60,000 t of milling copra for production of coconut oil. There are more than 25 copra crushers in West Bengal. They depend on South Indian supply.

In W.B. more than 80 % of the total produce is consumed in green stage (Guha 1998). Podwal et al. (1998) reported that Philippines Ordinary, Kerasree (MYD × WCT) and Kerasankara (WCT × COD) have been identified as suitable cultivars for tender nut which have appreciable volume of water with sugar and mineral during the seventh month after fruit set. WB has a tremendous scope for developing coconut water processing industry. However, area under coconut in West Bengal should have to increase to meet up the demand for raw materials. There is at present 75 coir units in West Bengal for producing coir products. This industry has also good prospect. West Bengal Awadhoot Agro (P) Ltd processing plant has been inaugurated in Hasnabad, 24 Parganas (N), on 3 February 2006 with a processing capacity of 20,000 nuts/day. They market tender coconut water, coconut jam and coconut lemoled (Ghosh et al. 2000). Rural Bengal is traditionally an area of different handicrafts. If the people are properly trained, there is an enormous scope for developing coconut-based handicraft industry in West Bengal.

8.24.1 Strengths of Indian Coconut Industry

1. One of the leading producers of coconuts in the world producing 13 billion nuts per annum

2. Coconut area distributed in 18 states and 3 union territories under different agro-climatic conditions
3. 3,000 years' tradition in coconut cultivation
4. Premier coir manufacturing country in the world
5. Producer of best grade milling copra in the world yielding high-grade coconut oil known for its aroma and flavour
6. Large number of farmer's cooperative societies in primary processing and marketing
7. Government agencies such as Kerafed, State Trading Corporation, Kerala State Marketing Federation and Karnataka State Marketing Federation in manufacturing and marketing of branded coconut oil in small packs
8. Hundreds of reputed and established private firms in manufacturing and marketing of various coconut products including branded coconut oil in small packs
9. A wide range of coconut products both edible and nonedible available for export
10. Technical know-how and trained manpower for the manufacture of various coconut-based products
11. Availability of research support by reputed research organisations such as CSIR, ICAR and DRDO
12. Good number of cultivars/varieties having specific nut characteristics

8.24.2 Constraints to Coconut Industry

1. According to Thampan (1998) the major constraints which hinder the development of coconut industry are the inadequate support extended to technological development in the existing research institute and that technological research on coconut has received only a low priority than the food crop.
2. Though coir and coir products constitute an export commodity, the average export earnings from other products are not large enough to attract major investment in this sector.
3. The high price variation of coconut products in the international as well as domestic market essentially necessitates that the cost of

product be reduced. Therefore, low-cost product technology needs to be evolved.

4. Quality standard should be maintained by establishing quality control laboratories because export market gives tremendous stress on quality of the product.

5. Marketing system is not well organised in India. Market promotional activities are less. An awareness campaign on coconut products is essential.

6. By-products of coconut is not fully utilised. Markose (1999) and many others stressed on product diversification and by-product utilisation.

7. Participation of growers in the copra manufacturing is negligible. So a large number of middlemen hike the price of raw materials.

8. Development of a nontraditional sector like coconut-based industry, at its initial stages, should get economic support from the government.

8.24.3 Possibilities to Improve the Status of Coconut Industry

With a view to encourage the speedy development of coconut processing industry in India, the following suggestions are made (Thampan 1998):

1. A Coconut Technology Development Centre (CTDC) has to be established for extending techno-economic support to the coconut processing sector for its modernisation and integrated development.

2. The CTDC shall have the mandate to promote, sponsor, support and undertake technological and market research.

3. The coconut farmers in the major producing centres have to be organised under cooperatives to get uniform supply of raw materials at right price.

4. The coconut oil has its re-entry in areas where it has suffered partial substitution.

5. To organise a network of display cum sales outlets for the various processed products in all the states in the country.

6. Publicity campaigns for promoting the various coconut products are the prime need.

7. Partial mechanisation in the coir industry is essential to reduce the cost of production.

8. The desiccated coconut industry has to be reorganised and modernised.

9. The development strategy for coconut needs a new orientation with a thrust on product diversification, product development, marketing and export promotion.

10. Coconut processing sector can be strengthened by attention on nontraditional products.

11. There is an urgent need to undertake pilot-scale trials for commercialising some of the products. Refinement is needed in packaging and quality (Podual et al. 1998).

Coconut is a versatile crop which yields innumerable products right from the roots to the tip of the palms. This crop had a glorious past and tremendous future, sustaining millions of smallholders, processors and traders. More than ten million families directly or indirectly depend upon coconut for their livelihood (Nampoothiri et al. 1998). There is a growing consumer demand for desiccated coconut, coconut milk, canned coconut water and different other coconut-based products. Coconut is a subsistent crop which has provided the basic needs of a number of countries in the tropics for centuries. With the use of coconut oil in the production of soap and margarine in Europe in the nineteenth century, it was converted into a commercial crop. In the beginning of the twentieth century, copra was the king among the oil seeds. In East Indies it was known as green gold. However, the period after the Second World War saw the substitution of vegetable oils and oleochemicals for coconut oil in international trade. The increase in the output of coconut was marginal. Price of coconut oil fluctuated heavily due to frequent short supply situations. A campaign against coconut oil alleging that it causes cardiovascular diseases aggravated the situation. With the depressed price of coconut oil, coconut-producing countries have now moved from traditional products to the processing of value-added products. Consequently, recent years have seen coconut oil being further processed to produce coco chemicals. Export of coconut shell charcoal and activated carbon is on the increase though in small quantities; products like coconut cream, nata de coco, fibre dust, coconut powder, coconut water and geotextiles are finding their way into the international market. Another interesting feature that is becoming evident increasingly is the shift of the foreign markets from the traditional base to new areas. The newly industrialised countries in the east as Taiwan and South Korea are fast emerging as key importers of coconut products. The medical and other evidence that came to light in the last few years in defence of coconut oil has cleared the misconception and misinformation about it. Coconut products are also drawing attention as environmentally friendly. Research carried out has proved the adaptability of coconut oil as biodiesel. Coir is an excellent natural fibre which is strong, durable and biodegradable. Coir geotextiles are now becoming popular and are being used increasingly for erosion control particularly where land, bank reinforcements is required as well as for landscaping. Coconut shell which is a major by-product of coconut industry finds important uses in daily life in place of non-biodegradable plastic containers. Activated carbon produced out of coco shell charcoal is used for water purification, air purification and food purification. Fibre dust briquettes have found a place as a soil reconditioner and a suitable nutrient for landscaping and an ideal ready-made potting mixture. Coco peat, a 100 % renewable resource, is now replacing bog peat, the depletion of which environmentalist feel would destroy land forms, habitat of some unique fauna and flora in the UK. Coconut water is a safe drink in the world unadulterated and untouched by human hands. Coco wood is a renewable resource and an answer to depleting forests reserves. Coco shell, husk, trunk, coir dust and fronds are energy sources. Coconut is a smallholder crop and millions of rural people depend on it for survival. Its development particularly in postharvest activities could be the base for rural development in the coconut-producing countries.

Availability of manpower is the strength for opportunities of coconut industries in India.

There are many problems confronting coconut industries which require to be addressed in the right perspective to make the coconut industry more dynamic and competitive. Product diversification, adoption of stringent quality standards for coconut products and increased productivity are some of the ways to make this industry's fate bright. Apart from research on agro-techniques, research on processing technology needs to be undertaken. In conclusion it is pertinent to mention that there is bright future for coconut in India provided we prepare ourselves to meet the challenges and tap well the large domestic and export market.

References

Agarwal RK (1999) Copra crushing industry in West Bengal – an overview. Indian Coconut J 30(5):46–47

Alex G (1999) Export market of coconut product. Indian Coconut J 30(5):48–49

APCC (1996a) Coconut statistical yearbook. Asian and Pacific Coconut Community (APCC), Jakarta

APCC (1996b) In: Arancon RN Jr (ed) Coconut harvesting and copra making – coconut processing technology information document. Asian and Pacific Coconut Community (APCC), Jakarta

APCC (1996c) In: Arancon RN Jr (ed) Coconut food process – coconut processing technology information document. Asian and Pacific Coconut Community (APCC), Jakarta

Award AB (1981) Effect of dietary lipids on composition and glucose utilization by rat adipose tissue. J Nutr 111:34–39

Banzon JA, Velasco Jr (1982) Coconut: production and utilisation. Philippine Coconut Research and Development Foundation, Pasig

Bhat SG (1999) Coconut oil in cosmetics. Indian Coconut J 30(5):34–35

Blackburn GL, Kater G, Mascioli EA, Kowlchuk M, Babayan VK, Bistrian BR (1989) A reevaluation of coconut oil's effect on serum cholesterol and atherogenesis. J Philipp Med Assoc 65:144–152

Bosco SJD (1998) Coconut shell made product harvest and post harvest technology of plantation crops. ICAR-Central Plantation Crops Research Institute, Kasaragod, Kerala, India, pp 54–59

Dippon K (1996) Copra dryers and copra drying technologies. Proceedings of the XXXIII COCOTECH meeting. Asian and Pacific Coconut Community (APCC). Jakarta

Enig MG (1990) Fats and oils: understanding the functions and properties of partially hydrogenated fats and oils and their relationship to unhydrogenated fats and oils. PJCS XV(1):27–31

Ghosh DK (2008) Feasibility of processing and product diversification of coconut in India with special reference to West Bengal. Indian Coconut J 51(7):2–7

Ghosh DK, Bandopadhyay A (2010) A comparative study on the performance of some indigenous and exotic coconut germplasm in alluvial plains of West Bengal AbstractTS2 – PO2 pp32. International conference on coconut biodiversity for prosperity held at CPCRI, Kasaragod, Kerala, 27–31 October 2010

Ghosh DK, Ghosh A (2008) A simpler approach to understand coronary heart disease, fat facts and its correlation with consumption of coconut oil. Indian Coconut J 51(6):15–19

Ghosh DK, Hore JK, Sharangi AB (2000) Prospect of development of coconut based industry in West Bengal Abstract of papers, VIIth W B State Science & Technology Congress, Bot-Agri 11

Gordon T et al (1977) Am J Med 62:707

Guha S (1998) Tender coconut market in Kolkata. Indian Coconut J 28(12):9–13

Harrison's principles of internal medicines, 15th edn. McGraw-Hill, pp 1377–1140

Hore (1999) Coconut research in west Bengal. Indian Coconut J 30(4):1–6

Hostmark AT, Spydevold O, Eilertseen E (1980) Plasma lipid concentration and liver output of lipoprotein in rats fed coconut fat or sunflower oil. Artery 7:367–383

Kaunitz H (2001) The ABC of MCT'S, Coconut Today. Special Issue. UCAP

Kaunitz H, Dayrit CS (1992) Coconut oil consumption and coronary heart disease. Philipp J Intern Med 30:165–171

Keys A (1980) Seven countries, a multivariate analysis of death and CHD. Harvard University Press, Cambridge, MA

Kumar V, Cotran RS, Robbuis SL (1992) Basic pathology, 5th edn. Prism Books Pvt Ltd, India

Kurup PA, Rajmohan T (1994) Consumption of coconut oil and kernel and the incidence of atherosclerosis. Proceedings of the symposium on coconut and coconut oil in human nutrition, 27th March Kochi. India, pp 35–59

Markose VT (1999) Coconut industry on revival path. Hindu Surv Indian Agric 1999:91–92

Murray RK, Ganner KD, Mayes AP, Rodwell WV (1994) Harper's biochemistry, 21 eds

Nampoothiri KUK, Biddappa CC, Upadhyay AK (1998) Milestones in coconut research. Indian Coconut J 29(4):28–37

Nanda Kumar TB (1998) Recent development in coconut industry in India with special reference to Kerala. Indian Coconut J 28(11):6–13

Nandanasabapathy S, Sri Vatsa AN, Arya SS (1999) Development of coconut products of commercial value. Indian Coconut J 30(5):32–33

Ng TKW, Hanan K, Lim JB, Lye MS, Ishak R (1991) Non hypercholesterolemic effects of a palm oil diet in Malaysian volunteers. Am J Clin Nutr 53:101

PCA (1993) Proceedings of workshop on: village, small & medium scale processing of fresh coconuts. June 1993

PCA-CETC (1996) Insect pests of copra and copra meal. Cocoscope. Philippine Coconut Authority (PCA). Coconut Extension Training Centre, Davao City

PCARRD (1993) Philippines recommendations for coconut. In: Magat SS (ed) Philippine Council for Agriculture Forestry and Natural Resources Research and Development (PCAARD). Department of Science and Technology, Los Banos

Philippine Coconut Authority (PCA) Coconut Extension Training Centre, Davao City, Philippines. PCA-CETC (1994) Copra quality improvement. How can it be done? Cocoscope. Philippine Coconut Authority (PCA), Coconut Extension Training Centre, Davao City

Philippine Coconut Authority (PCA) Industrial Technology Development Institute (ITDI). Appropriate Technology International (ATI). Koninklijk Instituut voor de Tropen (KIT). PCA-CETC (1997) Coconut processing technology manual of procedures. April 1997. Guide prepared for the coconut processing technology skills development training for South Pacific Commission

Podual SK, Pillai AM, Nair B (1998) Technologies for product diversification and by product utilization in coconut. In: Harvest and post harvest technology of plantation crops. ICAR-Central Plantation Crops Research Institute, Kasaragod, Kerala, India, pp 70–73

Podwal M, Abu H Md, Chattopadhyay PK (1998) Evaluation of coconut cultivars for tender nut water for West Bengal. Indian Coconut J 29(1):3–6

Prior IA, Davidson F, Satmond CE, Ezochans K (1981) Cholesterol, coconuts and diet on Polynesian atolls: a natural experiment: the Pukapuka and Tokelau Island studies. Am J Clin Nutr 34:1552–1561

Punchihewa PG (1997) Status of the coconut industry. Proceedings of the seminar-workshop on coconut biotechnology, Merida, Mexico

Rajagopal V (1998) Produce and products of plantation crops. In: Harvest and post-harvest technology of plantation crops. ICAR-Central Plantation Crops Research Institute, Kasaragod, Kerala, India, pp 9–12

Rajan R (2006) Coconut oil – the healthiest oil on earth. Indian Coconut J 37:2

RP-UK (1992) Reduction in aflatoxin contamination of copra in the Philippines. Technical cooperation project report. Republic of the Philippines and United Kingdom Funded Project. Philippine Coconut Authority (PCA). Quezon City, Philippines

Setiawan YY, Breag G (1996) The projects of hot oil immersion drying technology in Indonesia. Proceedings of the XXXIII COCOTECH meeting. Asian and Pacific Coconut Community (APCC). Jakarta

Sindhu JA, Rajmohan T (2006) Coconut fiber – a natural hypolipidemic and hypoglycemic agent. Indian Coconut J 37(3):21–23

Singh HP (1998) Coconut industries in India challenges and opportunities. Indian Coconut J 29(4):4–27

Sundaram K, Hayes KC, Siru OH (1994) Dietary palmitic acid results in lower cholesterol than does a lauric–myristic acid combination in normolipemic humans. Am J Clim Nutr 59:841–846

Superko HR et al (1985) Am J Med 78:826

Swetman T (1996) Use of the Ram press to extract oil from dried coconut. Proceedings of the XXXIII COCOTECH meeting. Asian and Pacific Coconut Community (APCC). Jakarta

Thampan PK (1993a) Hand book on coconut palm, 3rd edn. Oxford & IBH Publishing co. Pvt. Ltd., New Delhi

Thampan PK (1993b) Processing of coconut products in India. Asian and Pacific Coconut Community (APCC), Jakarta

Thampan PK (1996a) Product diversification a must. Hindu Surv Indian Agric 1996:99–100

Thampan PK (1996b) Coconut for prosperity. Peekay Tree Crops Development Foundation, Kerala

Thampan PK (1998) Prospect for development of coconut industry in India. Indian Coconut J 30(5):22–24

Tillekaratne HA (1995) Processing of coconut products in Sri Lanka. Asian and Pacific Coconut Community (APCC), Jakarta

Vaz Antonel PC (1996) Coconut fibre processing and marketing. Proceedings of the XXXIII COCOTECH meeting. Asian and Pacific Coconut Community (APCC). Jakarta

Viswanathan R (1998) Coconut husk made product in harvest and post- harvest technology of plantation crops. ICAR-Central Plantation Crops Research Institute, Kasaragod, Kerala, India, pp 39–52

Product Diversification of Areca Nut, Cashew Nut and Oil Palm

9

J.K. Hore and M. Anitha

Abstract

India is endowed by Mother Earth with a wide range of plantation crops, which serve as the source for diversified products of commercial importance. Areca nut (*Areca catechu* L.), cashew nut (*Anacardium occidentale* L.) and oil palm (*Elaeis guineensis* Jacq.) have great potential at international market as foreign exchange earner. India is the largest producer of areca nut in the world. The products of areca nut, viz. fat, alkaloids, tannins and areca husk, have alternate uses with the potential for establishing small-scale industries. Cashew nut is an export-oriented crop grown for its nut. The by-products of cashew, viz. cashew nut shell liquid (CNSL), shell cake, testa and cashew apple, are commercially exploitable. CNSL has always been credited with words such as 'versatile raw material'. Cashew apple is a rich source of vitamin C, organic acids, antioxidants and minerals. The oil palm is recognised as the cheapest edible oil yielding crop. Main applications of palm oil are in edible food industry field, mainly as solid fat for margarine, shortening and cooking oil. Nonedible applications including soaps, oleochemical production and automobile energy sources are substantial and enlarging day by day. The relationship between value addition technologies, farm and agro-food processors may provide leeway to improve profitability.

9.1 Areca Nut

India is the largest producer of areca nut in the world. Areca nut is traditionally used as a masticator in the countries of India-Malayan Peninsula. The ethno-religious importance of areca nut in India is unique. The products of areca nut have alternate uses with the potential for establishing small-scale industries.

J.K. Hore (✉) • M. Anitha
Department of Spices and Plantation Crops,
Bidhan Chandra Krishi Viswavidyalaya,
PO. Mohanpur, Nadia 741252, West Bengal, India
e-mail: jkhore31@rediffmail.com;
anithamajji01@gmail.com

A.B. Sharangi and S. Datta (eds.), *Value Addition of Horticultural Crops: Recent Trends and Future Directions*, DOI 10.1007/978-81-322-2262-0_9, © Springer India 2015

The nut contains fat, polyphenols (tannins), polysaccharides, fibre and protein. Polyphenols decrease with maturity, whilst polysaccharides, fat and fibre increase. Among alkaloids, ripe areca nuts have a higher arecoline content of 0.2–0.3 % compared to tender nuts with 0.05–0.1 %.

9.1.1 Processed Nuts

Chali or Kottapak

Ripe nuts are dried in the sun for 35–40 days, dehusked and marketed as whole nuts. This is known as chali. The different grades of decreasing order of chali are Moti, Srivardhan, Jamnagar and Jini. The main producing areas of chali are Kerala, Karnataka, Assam and Maharashtra. Sometimes the fruits are cut longitudinally into two halves and sundried for 10 days (Chempakam 1998).

Kalipak

Nuts of 6–7 month old maturity are dehusked, cut into pieces, boiled with water or a diluted extract from previous boiling, coated with kali and dried. This is prevalent in Kerala and Karnataka. Kali coating can be repeated three to four times to get a good glossy appearance (Kali is the extract obtained after three to four batches of boiling and then concentrated tenfold; tannins are its major components). The kalipak is known by different names depending on the number, shape and size of the cuts: Api (without any cuts), Batlu (transverse cut into two halves), Choor (several longitudinal cuts), Podi (both longitudinal and transverse cuts) and Erazel (transverse thin slices).

Scented Supari

Here, dried areca nuts are broken into bits, blended with flavour mixture and packed in butter paper. In South India, kalipak is used for this purpose, by adding spices and synthetic flavours. Instead of raw spices, essential oils are used for easy blending. Rose essence as well as menthol is very common. Saccharin is occasionally used for sweetening, with additives like colour and flavour.

9.1.2 Use of Individual Constituents

1. *Tannins*

Polyphenols or tannins form the major constituents of the nut. Tender nut contains about 38–41 % whilst ripe nuts have only 16–22 %. Raghavan (1957) found that tannic acid or gallic acid from nuts when mixed with ferrous sulphate in warm distilled water gave black writing ink of acceptable quality. These are widely used for dyeing clothes, ropes, etc. Another possible use of tannins is as a food colour. In the present scenario of increasing ban on synthetic food colours, this natural product assumes greater significance. Tannins extracted from defatted areca nut were of better quality (Chempakam 1998).

2. *Areca nut fat*

The nut contains about 8–12 % fat, which is extractable with organic solvents like hexane or chloroform. The fact contains both saturated and unsaturated fatty acids and is highly rich in myristic acid. Hence, it can be a good indigenous source for preparing myristic acid and its derivatives. Refining of the fat can be done with alkali, which makes it as an edible fat, which is harder than cocoa butter. Sweets, savouries and biscuits prepared from this fat are comparable with those from vanaspati. The blended fat with cocoa butter in the ratio 1:1 can be substituted for cocoa butter or hydrogenated coconut oil.

3. *Alkaloid*

Areca nut has 1.5 % alkaloid, viz. nine closely related alkaloids (George and Robert 2006) including arecoline, arecaidine, arecaine, arecolidine, guvacine, isoguvacine, guvacoline, coniine and norarecoline. Among the alkaloids, arecoline (0.12–0.24 % in ripe nut) has been reported to be the main alkaloid having cholinergic muscarinic agonistic activity (Jayalakshmi and Mathew 1982). Lime facilitates its release from salt (Boucher and Mannan 2002). These phenolic compounds are extracted using organic solvents and are to be of greater pharmaceutical importance. They are anthelmintic. It is also used as a

CNS depressant drug. It inhibits the growth of *E. coli* and *S. aureus*.

Polyphenols are the large numbers of natural phenolic compound found abundantly in plants, possessing antioxidant activity. They are diverse in their chemical structure, nature and biological activity, capable of providing protection against oxidative stress and thus may play a significant role in the prevention or improvement of several clinical conditions like cancer, osteoporosis, neurodegenerative, cardiovascular diseases, diabetes mellitus, etc. (Kumaraswamy and Satish 2008). The blood sugar lowering effect of arecoline is mentioned in Ayurveda.

9.1.3 Areca Nut Husk

The husk of areca nut constitutes about 60–80 % of total weight of nut. The husk fibres are predominantly made of cellulose, with lower amounts of hemicelluloses, lignin, pectin, etc. Hard fibres are those adjoining the inner layer, whilst soft fibres are from the middle layer. Fibre extraction is done by soaking the husk for 3 weeks and beating with a mallet. Several processes have been developed for utilisation of areca husk for making hard boards, plastic and brown wrapping paper (Narayanamurthi and Singh 1964). CNSL can be used for tempering the boards which gives good water resistance, higher strength and less microbial growth (Chempakam 1998). Areca husk is used as a substrate for mushroom cultivation. Husk can be processed into insulating wool and felt in admixture with jute and candles (Raghavan and Baruah 1957). The wool is comparable in thermal conductivity, packing density, etc. with standard products like palco wool, defabricated teak bark, etc. Its utility in thermal installation, packing, etc. is promising. It is used as a cheap fuel and mulch. In Indonesia and the Philippines, areca nut husk is used for tooth brushes. It can also be used for the manufacture of thick boards' fluffy cushions and nonwoven fabrics. Activated charcoal and xylose can also be produced from areca husk (Rajagopal 1998).

9.1.4 Leaf Sheath

Leaf sheath with its high cellulose content and crude fibre makes a flexible and pliable material for heat moulding. From leaf sheath, domestic items like cups, plates and head caps are prepared. It is also used in manufacturing plyboards, veneer boards, picture mounts, decorative panels, teapoys, file boards, bags, etc.

9.1.5 Stem and Leaf

Due to its hardness, the areca stem is used as a building material, poles, rulers, shelves, nails, etc. Hollow stem can be used as drainage and irrigation pipes. Areca leaves are a good source of organic manure. They are mostly used for thatching purposes.

9.1.6 Areca Nut Wastes

Areca nut bunch waste and leaf stalk are pasteurised by soaking in a solution of 500 ppm formalin + 25 ppm bavistin (Chandramohan and Murthy 1991). The areca nut waste is found to be an ideal substrate for mushroom production. Mushroom cultivation is an economically feasible and eco-friendly process for bioconversion of these wastes into high-quality protein food.

9.2 Cashew Nut

Cashew (*Anacardium occidentale* L.) is popularly known as 'gold mine of wasteland'. Cashew is a fast growing, hardy and drought-resistant multipurpose tree species cultivated in many tropical countries. India is the largest producer, importer (raw nuts), processor, consumer and second largest exporter of cashew kernels in the world. India has been earning sizeable foreign exchange by exporting cashew kernels and had a pre-eminent position in the global cashew trade (Adiga and Kalaivanan 2013). Cashew kernel is known for its delicious, pleasant taste and nutritive value. It is an ideal appetising snack and a complete food in harmony with a modern diet. Roasting of kernel increases characteristics of aroma and improves its texture. Cashew proteins are complete with all essential amino acids. The nut fats are complete, very active and easily digestible. Hence, the nuts can be used by both adult and infant alike. Every part of cashew is useful to man. Cashew nut shell liquid is an important raw material in paint, minerals and wood industry. Cashew apple is highly nutritious and could be utilised for the production of alcoholic beverages, juice, syrup, jam, etc. Besides these cashew tree is a valuable source of raw materials for pulp industry and is a major source of fuel in Kerala. The by-products derived from cashew tree are discussed.

9.2.1 Cashew Apple

It is a highly potential nutritional source. The fruit is very juicy, and the expressed juice has a brix of 12–14° containing 10.15–12.5 % sugar (mostly reducing) and vitamin C – 261.5 mg/100 g (Nair et al. 1979). Nutritious cashew apple is still being wasted in India, even whilst the country faces umpteen nutritional problems. Kerala Agricultural University (KAU) has developed several technologies for effective utilisation of cashew apple and are getting good response to the products and technologies from cashew farmers, food processors and general public (Mini et al. 2008). Extraction of juice can be done by using screw press, basket press, cashew juice expeller or hydraulic press for maximum juice (Augustin 2001). KAU has developed techniques for preparing cashew apple products, viz. soft drinks, syrup, jam, candy, canned cashew apple, chutney, pickle vinegar and even fermented products like wine and alcohol (KAU 1996). Rotting of cashew apples and spoilage is a major problem in its utilisation. Presence of astringent and acrid principle that produces a tingling sensation on tongue forms another problem. Technologies are available to remove this astringent principle and for the effective utilisation of cashew apple.

1. *Cashew apple juice*
The cashew apple is highly perishable and requires prompt handling. Only good and sound fruits should be used for juice extraction. Juice is boiled with sugar and cooled. 0.8 % sodium benzoate is added as a preservative. The juice extracted from the cooked fruits is clarified by the addition of gelatine (430 mg/l) and stirring for 15 min to remove the astringent and acrid principles (Nagaraja 1998). After filtration, the brix of the juice in raised to 15° and acidity to 0.4 % by addition of sugar and citric acid. The juice is then boiled for a minute and preserved by overflow pasteurisation at 85–90 °C for 30 min (Nair et al. 1979).

In Brazil, concentrated juice from cashew apple ranks first in sales among tropical juices. It has contributed to human nutrition, particularly the poor people, by supplying low cost vitamin C. Cajuda, cajuuina, cajuvita and cajuaperativo are different cashew drinks in South Brazil. Cashew apple juice can also be blended with other fruit juices like lime, pineapple, orange, grape, watermelon and apple juices to produce mixed or composite beverages with improved consumer acceptance (Mathew et al. 2011).

2. *Cashew syrup*
Extraction of juice and removal of astringent principle are done in the same way in the pretreatment of juice. Sugar is added at the rate of 1.0–1.25 kg for every litre of juice. Twenty to twenty two grammes citric acid per litre and 0.08 % sodium benzoate are added to the juice. All the ingredients are mixed thoroughly and kept as such for 3–5 h so that clear syrup forms a separate layer which can be easily shiphoned. Bottling can be done as described for juice (Suganya and Dharshini 2011). Cashew apple soda was prepared by using syrup and carbonated chilled water. Syrup has a storage life of 1 year. Cashew apple drink is an RTS (ready-to-serve) beverage. Drink is marketed both in glass bottles and in attractive food grade pouches. Pasteurised drink in glass bottles

has a storage life of 3 months under ambient storage condition (Mathew et al. 2011).

3. *Cashew apple jam*

   Cashew apple is thoroughly cleaned by washing with water. The apple is immersed in 3 % salt solution for 3 days to reduce the tannin content, after which the fruits are steamed for 15–20 min at 0.7–1.05 kg steam pressure. Then the apples are crushed and mixed with sugar and boiled. A pinch of citric acid is added towards the end of the cooling process to improve the taste. Finally it is stored well in sterilised jam bottles. Mixed fruit jams by mixing the cashew apple pulp with equal quantity of banana/pineapple/mango/apple pulp can also be prepared (Jain et al. 1954; Mini and Jose 2007) for increased acceptability.

4. *Cashew apple candy*

   It is a sweet product, and quality apples with good shape are selected for candy preparation. As in jam preparation, tannin is removed from apples, cooked, pierced using fork and dipped in sugar solution. Steaming the cashew apple was found to be the most efficient way in removing the astringent and acrid principles. Pressure of steam and time of exposure vary from 2 to 6 kg and 5–15 min, respectively, according to the quality of apple and the product which is to be made. The astringent principle can also be removed by boiling in common salt solution (2 %) for 4–5 min. Concentration of sugar solution is gradually increased so as to reach 70° brix. After 2 weeks of soaking, sugar solution is drained out and candy is dried in shade. It takes about 2–3 weeks for making the final product. About 745 g candy can be obtained from 1 kg of cashew apple. Vitamin C content of the product is 28.4 mg/100 g.

5. *Canning of cashew apple*

   Firm fruit suitable for canning is first peeled by treatment in 0.5 % boiling sodium hydroxide solution for 3–4 min, followed by rinsing in water and a subsequent treatment for about 4 min in boiling 0.2 N solution of H_2SO_4. The fruit after washing is steamed for about 4 min at 2 kg steam pressure followed by cooling under a spray of water. After draining, the fruit is cut into values (lengthwise) and after trimming off the undesirable portions, canned in 40° brix syrup using an exhaust time of 4–5 min in boiling water or steam and a process time of half an hour in steam at 4 kg pressure (Nair et al. 1979). A good-quality salad pack is said to be obtained by mixing pineapple ring segments with cashew apples in equal proportions. Canned curried vegetables from raw green fruit of cashew in combination with potatoes (1:1) or potatoes and tomatoes (2:1) with or without tamarind are also reported.

6. *Cashew apple chutney and pickles*

   Cashew apple chutney can be prepared from fruits treated in 2 % brine as for jam followed by washing and steaming at 4.0 kg steam pressure for 5–7 min. Raw green fruit is steamed, washed and kept in 10 % brine for a week. After removing from salt water, it is again washed and pickle is prepared using oil, chilli powder, fenugreek powder, turmeric powder, ginger and garlic paste.

7. *Cashew apple vinegar*

   Cashew apple vinegar was prepared from cashew apple juice by adding sago, sugar and yeast along with mother vinegar. The brix of cashew apple juice was raised to 12°; the juice was pasteurised, cooled and inoculated with a pure strain of yeast, *Saccharomyces cerevisiae* or brewer's yeast for alcoholic fermentation. Addition of 0.05 % ammonium phosphate was found to help fermentation (Satyavati et al. 1963).

8. *Cashew apple chocolate*

   Cashew apple chocolate was prepared from cashew apple powder by adding milk powder, sugar and butter.

9. *Alcoholic beverages*

   In Goa, a sort of brandy called 'fenni' is made by fermenting apple juice. Goa is the only place in India where cashew liquor has been distilled, for the last three centuries or more. More recently, apples are used in Kerala for making cashew brandy. Good and sound cashew apples preferably by plucking from the trees should be used. The cashew apples should be washed, and tannin content is reduced. Eight litres of cashew apple juice is required to obtain one litre of fenni. In traditional method, the juice of cashew apples is collected in vats which when allowed to stay is acted upon by the bacteria present in the apple causing fermentation. The fermented juice is distilled in pot still to give 'arrack' which on further distillation produces 'fenni'. Time required for fermentation is 65–70 h. In fenni, alcohol content ranges from 40 to 45 % (Suganya and Dharshini 2011). Cashew wine is made in many countries throughout Asia and Latin America. It is a light yellow alcoholic drink, with an alcohol content of 6–12 %. Cashew apple wine can be mixed with fresh juices of orange, pineapple, tomato, grape and cashew apple as well as tender coconut water to produce wine coolers to serve as good health drink as they contain both wine with its medicinal properties and fruit juices with high amount of nutrients and minerals.

10. *Confectionery products*

Cashew apple can also be utilised for the preparation of tutti-frutti. One kilogramme of cashew apple on processing gives 715 g of tutti-frutti. The whole fruit can also be processed into nutritious toffee, a feasible dessert item with extended shelf life. Toffee could provide 7.5 g of protein and 442 K calories per 100 g. Cashew apple juice can be used for preparing frozen desserts and dairy confectionery items by optimisation of juice concentration and spray-drying. The pulp or the residue of apple can also be dried, powdered and sieved for use as cashew apple powder without juice. Ten to thirty percent dehydrated cashew apple powder can be used in various value-added products like wheat laddu, masala biscuits, sweet and masala doughnuts, sponge cake, steamed kabadu, tomato cashew apple powder soup, powder koftas, chocolates, sweet and hot bread products. Nutri-cashew, a ready mix, has been prepared using cashew apple powder for the elderly as high fibre fruit (drink)

food mix for instant use. A ready-to-serve beverage mix, fruit milk/lassi mix has been prepared from clarified juice by homogenisation, spray-drying and mixing with milk/lassi powder. Ten to fifteen percent clear and cool cashew juice mixed with skim milk powder can be spray-dried for the production of cashew milk powder and can be utilised for the preparation of products like milk shakes, ice cream and ice candy (Mathew et al. 2011).

11. *Nutraceutical properties*

Ascorbic acid, carotenoid pigments, minerals and host of other chemicals, which are of significance to human health, are contained in cashew apple. Cashew apple powder lipids are rich in unsaturated fatty acids, major ones detected being palmitoleic and oleic acid. A valuable by-product that can be obtained from cashew apple waste is pectin. Pectin is used in manufacturing jams, jellies, marmalades, preserves, etc. It is useful in thickening, texturising and emulsifying agent and finds numerous applications in pharmaceutical preparations, cosmetics, etc. The cashew apple pomace or the fruit waste has been identified as the ideal medium for pectinase enzyme production for *Aspergillus foetidus* (Mathew et al. 2011).

12. *Agricultural uses*

Considerable amount of cashew apple residue is obtained as waste when bulk quantities of cashew apple are utilised for the manufacture of soft drinks or fermented beverages on a commercial scale.

(a) *Vermicompost*

The cashew apple waste, which is highly perishable and seasonal, can be converted to value-added products with good manurial value without creating problems for disposal. Apple residue could be effectively utilised for the production of vermicompost of 1.69 % N, 0.44 % P and 0.58 % K using *Eudrilus eugeniae*. The pH of the compost from cashew apple is 8.9 and hence could be used as a good ameliorant for acidic soils.

(b) *Animal feed*

The ripened cashew apple or its residue could be utilised for the preparation of cattle feed, pig feed and poultry feed. Cashew apple is a promising feed source for dairy cows in Vietnam. Cashew apple or its residue could be preserved for long-term use as a cow feed by anaerobic ensiling. Cashew peel (7.6 % protein, 12.3 % fat and 59.2 % carbohydrate) is a good poultry feed. Apples are also dried and preserved as cattle feed for rainy season. Daily feeding of 3.4 kg fresh apple along with normal feed to cow is found useful (Mathew et al. 2011).

(c) *Pest management*

It is observed that cashew apple extraction is an effective insecticide against red palm weevil in coconut. Cashew apple and gum extract, in combined form or alone, acts as an effective repellent against leaf feeding pests of vegetables.

13. *Industrial use*

(a) *Biofuel*

It is estimated that cashew apple can yield 8–10 % of ethanol. This indicates that there is a huge potential of generating ethanol from cashew apple. The residue, after extracting juice for fenni preparation, is used as fuel in liquor industry in Goa.

(b) *Biogas*

Ripened fruit can be used as raw material for biogas plant.

(c) *Tannin extraction*

Cashew peel can be used for extraction of tannin (24–26 %) which in useful in leather industry (Mathew et al. 2011).

9.2.2 Cashew Kernel

Cashew kernel is the most popular nut used by the confectionary industry. At present, cashew kernels are mainly used as snack food in the roasted and salted form. The broken cashew kernels are mainly used in confectionary, bakery and chocolate industry. Many cashew recipes have been developed and are gaining popularity among housewives all over the world (Parthasarathy et al. 2006). Some value-added products of cashew kernel are discussed.

1. *Cashew kernel oil*

This is a light yellow edible oil that can be extracted from lower grade kernel by cold pressure extraction. It is a high-quality edible oil that has been favourably compared to olive oil. The kernels contain 35–40 % oil (Van Eijnatten 1991).

2. *Cashew kernel butter*

Kernel residues after extraction of kernel oil are used to produce cashew kernel butter, which is similar to peanut butter. The cake remaining after extraction of oil serves as animal feed (Van Eijnatten 1991). Central Food Technological Research Institute (CFTRI) at Mysore perfected technology for the extraction of cashew butter.

3. *Cashew kernel milk*

The sweetened and flavoured milk could be prepared from cashew kernel baby bits. The sweetened and vanillin flavoured cashew spread is also possible and most preferred than salted spread (Bhaskara Rao and Swamy 2002).

4. *Cashew kernel flour*

Lower grade kernels are processed into cashew flour which has high protein content and is easily digestible (Kurian and Peter 2007).

5. *Coated cashew kernel*

Results obtained so far at NRCC on value addition have indicated the possibility of developing sugar, honey and salt-coated baby bits, which are organoleptically acceptable. Baby bits are the lowest grade kernels marketed commercially (Bhaskara Rao and Swamy 2002). Baby bits coated with honey, cardamom essential oil and apple green colour were most preferred by tasters.

9.2.3 Cashew Shell

Cashew nut consists of the epicarp, mesocarp, endocarp, testa and kernel. The epicarp forms the epidermis; mesocarp is the thickest of the three layers. In the mesocarp, which has a honeycomb structure, there are ducts filled with a sticky resinous corrosive oil, the cashew nut shell liquid (CNSL). The endocarp is hard and is formed of a compact mass of sclerenchymatous cells. The CNSL and cashew nut shell cake are the by-products of cashew shell. Cashew nut shell contain about 35–47 % of tannin (Lokeswari et al. 2010)

9.2.3.1 Cashew Shell Oil

The CNSL is a naturally occurring phenol. Fresh CNSL contains 90 % anacardic acid, which is converted into cardanol on heating. The rest 10 % is cardol which is mainly responsible for vesicant property. It comes as a by-product from shell obtained during processing of nut. About 7 kg crude oil can be extracted from 100 kg shell. It is often considered as the better and cheaper material for unsaturated phenols. The CNSL is a versatile industrial raw material. It is used widely in formulation of resins, detergents, insecticides, bactericides, fungicides, herbicides, disinfectants, emulsifiers, dyestuffs, antioxidants, plasticisers, stabilisers, lubricants, polyurethane-based polymers, etc. (Kurian and Peter 2007). Major products for industrial application are cashew lacquer, insulating varnishes, electrical winding and electrical conductors impregnated with CNSL and cashew cement (CNSL reacted with formaldehyde). Its anti-termite and antimicrobial properties are well known since ancient times as it protects the bottom of the boat.

CNSL is indirectly used in the manufacture of typewriter rolls and industrial flooring titles. Lightweight plastics, composite panels suitable for partitions, flush doors, etc. have been developed using CNSL resins. (Salem and Peter 2010). Brake lining is one of the important products in which large quantity of CNSL is employed. Brake linings made from CNSL have excellent frictional and other desirable properties (Shivadasani 1972). The CNSL distillation residue and its blends with alkyd were used in the preparation of air drying varnishes, cycle enamels and primers (Nair et al. 1979). CNSL was used as a source for polyurethane industry. Polyurethane is an extremely versatile polymer. They find applica-

tion in construction and automobiles and are used as coatings, sealants, adhesives and elastomers (Howard 2002). Rice husk boards prepared from CNSL residue-based resin as binder are useful for false roofing, as insulating panels and for acoustic purpose.

9.2.3.2 Cashew Shell Cake

Shells from the steam-roasted and drum-roasted nuts are valuable by-products of cashew industry. Cashew nut shell cake has high burning properties and is popularly used as a substitute for wood. It can be used as an eco-friendly fuel, replacing wood for heating boilers and dryers in refineries and other industries. Cashew shell, after extraction of CNSL, can be used for the manufacture of particle boards in packing industry (Salem and Peter 2010).

9.2.4 Products from Bark and Stem

Bark contains an acrid sap of thick brown resin which becomes black on exposure to air. This is used as an indelible ink in marking and printing lines and cottons. It is used as a varnish, as preservative for fish nets and as flux to solder metals (Woodroof 1967). It contains plenty of tannins (4.9 %). Stem yields amber-coloured gum which is partly soluble in water. This is used in book binding as cardol, acts as a vesicant and has insect repellent properties.

9.3   Oil Palm

Oil palm has been recognised as the highest edible oil yielding crop. It is being observed as a crop of promise from the stand point of view of wiping out deficit oil pool of the country. The plant produces two distinct types of oil, i.e., palm oil from mesocarp and kernel oil from kernel. The two oils are different in composition, physical properties and usability (Hartley 1988). Both types are used for culinary as well as industrial purposes. Crude palm oil (CPO) being the richest source of carotenoid can contribute substantially to the nutritional requirement of masses.

The palm produces the highest yield of oil per unit area.

Palm oil is a natural source of antioxidant, vitamin E, constituents of tocopherols and tocotrienols. These natural antioxidants act as a scavenger of the damaging oxygen free radicals and are hypothesised to play a protective role in cellular ageing, atherosclerosis and cancer. Main applications of palm oil are in edible food industry field, mainly as solid fat for margarine, shortening and cooking oil. Nonedible applications including soaps, cosmetics, antibiotics, gums, candles, bakery products, oleochemical production and automobile energy sources are substantial and enlarging day by day (Ngo 1991). CPO exhibits two fractions when stored at room temperature (25 °C): top part is dark red colour liquid called olein and bottom part is lighter colour, semisolid in nature, called stearin. CPO is fractionated to olein and stearin and they are the two important ingredients for several value-added food product. Some diversified products are:

1. *Palmolein*

   Olein is refined to produce RBD (refined, bleached and deodorised) palmolein. Commercially available palmolein is actually RBD palmolein. This product is widely used as cooking oil all over the world. Unsaturated fatty acids content is higher in this than that of CPO, though some of the minor components, especially carotenoids, get degraded during the bleaching process. However, this is very stable at higher temperature, especially in the case of Indian style of cooking, and forms lesser smoke, foam and unhealthy polymers. It is widely used for deep-frying of commercial products like chips, fries and doughnuts and preparation of fast foods. During deep-frying, unsaturated liquid oils are not used due to their oxidation at very high temperature involved in the process. By virtue of its fatty acids composition and presence of vitamin E, palm oil render better oxidative stability and at the same time it is more economical. Palm oil is also used in bread industry, where around 2 % palm oil is mixed with the dough to soften the loaf (Mandal and Kochu Babu 2005).

2. *Red palmolein*

   During the process of refining and bleaching, palm oil loses one of its valuable components, carotenoids. To retain the carotenoids, CPO is refined in special conditions involving mild treatments. Hence, this is a high-value, nutrient-rich oil, which is not recommended for deep-frying or extreme heat treatment during cooking. Best use of it is for salad dressing (as salad oil), blending with margarines for natural colour, and mild cooking where the carotenoids would come into the food and also would impart attractive colour.

3. *Shortening*

   Bakery shortening is used for making different types of bakery products, which gives a characteristic of smooth and spongy texture of the products. It prevents to form any hard mass during baking. Presently, palmolein palm stearin and sometimes the palm kernel oil are the sources of bakery shortening. Palm-oil-based shortenings are proved to be used for high-quality bakery product and are widely used now. Similar to bakery shortening, frying shortening is required for commercial level frying of chips, French fries, etc., which gives a characteristic of crispness and texture.

4. *Vanaspati*

   This is produced by hydrogenating and saturating the liquid vegetable oils. It is reported that trans-fatty acids forms when the liquid oils are hydrogenated to make solid fats. Recent studies have raised concerns about the possible negative health effect of high dietary intake of trans-fatty acids. Hence, general reduction of trans-fatty acid component of food fat is advised. Interestingly, palm oil does not produce trans-fatty acids whilst using it for vanaspati making. It can be used solely or blending in different proportions with other vegetable oils. In many cases, refined palmolein can be used as such, or the texture can be obtained with mild hydrogenation. Use of palm oil also reduces the hydrogenation cost considerably during vanaspati making.

5. *Margarine*

   Recently, low-calorie margarine is becoming popular. Table margarine should have butter-like consistency at 25–30 °C. Different fractions of palm oil and palm kernel oils are highly suitable for the production of table as well as industrial margarines. Due to semi-solid nature of palm oil, little or no hydrogenation is required to produce margarine. Some of the fatty acid composition of palm kernel oil is similar to butter, which gave the butter-like feelings whilst eating the palm kernel-based margarine. All the palm oil-based margarines are also trans-free (Mandal and Kochu Babu 2005).

6. *Confectionary fat*

   Unique taste of chocolate is due to the coco butter, which is one of the main ingredients of chocolate making. More than 75 % of cocoa butter is made up of two-oleo-disaturated glycerides. Palm oil mid fraction (called PMF) and different blends of it with stearin is used as cocoa butter equivalent fats. This can be used for plain chocolate as well as for milk chocolates manufacturing.

7. *Dairy products*

   In many cases, vegetable oil substitutes the milk fat. Skimmed milk powder or sodium caseinate is reconstituted to milk using such substitute fat. These products have better keeping properties than those having butter fat and that is a particular advantage obtained with palm oil, when it is used as milk fat substitute. Any vegetable oil cannot be used to produce stable foam in ice cream, unless that has a butter fat-like properties. Palm kernel oil is used for this purpose because it is a satisfactory alternative to butter fat.

8. *Palm oil cake*

   The press cake, after extraction of oil from kernels, is an important livestock feed. It contains 18–19 % protein, rich in amino acid like organine and glutamic acid. According to Okeudo et al. (2005), the by-product from the mechanical expeller procedure is referred to as palm kernel cake, whilst that from the solvent extraction technique is called palm kernel meal. Palm kernel is processed by either one of two processes, viz. expeller press which yields the 'factory-type' palm kernel cake and an indigenous local technique that produces the 'cottage-type' palm kernel cake. With a crude protein content of 14–21 %, palm kernel cake is precluded as a protein source as in most of the cases, the crude protein level is less than 20 %. However, the crude protein content is enough to meet the requirement of most ruminants despite other negative attributes reported by other researchers (Hassan and Yeong 1999).

9. *Palm wine*

   It is produced from sap obtained by tapping male in inflorescence. Fresh sap is a sweet containing 4.3 g sucrose and 2.4 g glucose/100 ml and ferments quickly by action bacteria and natural yeast to produce a drink which has a milky flocculent appearance with slight sulphurous odour.

10. *Biopolymer*

    Empty fruit bunches (EFBs) can be a source of biopolymers such as polyhydroxyalkanoates (PHAs) and polylactate (PLA). Bioplastics have similar characteristics as petroleum derivatives used for packaging materials. In bioplastic production, sugar is obtained from the EFB which is used as a cheap carbon source in bacterial fermentation. Starch and cellulose are first converted into organic acids like lactic acids which are then polymerised to form bioplastic.

11. *Biofuel*

    EFB also has the potential as a source of lignin for the production of ethanol. The countries like Malaysia and Indonesia may have to meet the 1–2 % utilisation of gasoline (Shinichi et al. 2009). Increased utilisation of renewable energy resources, in particular oil palm wastes, is strategically viable as it can contribute to sustainable energy supply whilst minimising the negative impacts of energy generation on the environment. It will also solve the agriculture disposal problem in an environmentally friendly manner whilst recovering energy and higher-value chemicals for commercial applications like biofuel, helping the government to achieve its obligation to prolong the fossil fuel reserves.

12. *Oleo chemicals*

Palm oil is well known to contain high amount of carotenoids. Typical crude palm oil contains 500–700 ppm carotenes, which contribute to palm oil stability and nutritional value. Conventional processing of palm oil leaves about 3–7 % residual oil in the fruit fibre (Nik Norulaini et al. 2008) and a high content of carotenoids in the pressed palm fibres. Residual fibres from palm oil production contain between 4,000 and 6,000 ppm of carotenoids, about six times higher than that found in crushed palm oil (France and Meireless 1997, 2000).

Many palm oil processing by-products can be used as the basis of the oleochemical industry, one of which is β-carotene that are needed in the food, pharmaceutical and cosmetics industries (Ahmed et al. 2009). Palm oil waste contains β-carotene of 0.19 %, which can be isolated using the enzymatic reaction. Separation of carotenoids by enzymatic reaction method is better than the separation of carotenoids by column chromatography method. Enzymes like α-Amylase, β-amylase, and cellulose may help in release of carotenoids from the matrix of fibres and other carbohydrates in the CPO and POME waste (Hudiyono and Septian 2012).

13. *Miscellaneous use*

The central shoot or cabbage is edible. Leaves, petioles and rachis are used for fencing. Refuse after stripping bunches is used for mulching and manuring. Trunk is much softer and less easily seasoned than that of coconut palm and is more suitable for furniture, fibreboard and pulping (Kurian and Peter 2007). Mesocarp waste of oil palm is found to be an ideal substrate for organic matter production. The availability of mesocarp waste is about 3 tonnes per ha in oil palm plantations under irrigated condition. These organic wastes can also be converted to good-quality organic manure by composting

methods and by vermicomposting (Thomas 1998). Every 5 tonnes of EFB produces 1 tonne of pulp. EFB has been used as inorganic fertiliser by incineration methods and as organic manure directly thrown back and mulched in the field (Yosuf 2006).

These diversified products are in high demand throughout the world. Gradually, this would be an export oriented crop mainly due to these value-added products. The benefit of palm oil is multifaceted. When it gets popularised, our people got nutrition, palm oil industry gets a boost, farmers get the remuneration, it increases employment generation and ultimately economy develops.

References

Abdul SM, Peter KV (2010) Cashew – a monograph. Studium Press (India) Pvt. Ltd., New Delhi, pp 198–199

Adiga JD, Kalaivanan D (2013) Cashew. In: Dhillon WS (ed) Fruit production in India. Narendra Publishing House, Delhi, p 145

Ahmad AL, Chan CY, Sunarti AR (2009) Isolation of carotenes from palm oil mill effluent and its use as a source of carotenes. Presented at EuroMed 2008, Desalination for Clean Water and Energy Cooperation among Mediterranean Countries of Europe and the MENA Region, 9–13 November, Dead Sea, Jordan

Augustin A (2001) Utilization of cashew apple. In: Souvenir of world cashew congress – 2001, India. The Cashew Export Promotion Council of India, Cochin, p 57

Bhaskara Rao EVV, Swamy KRM (2002) Indian cashew industry. In: Singh HP, Balasubramanium PP, Venkatesh NH (eds) Directorate of Cashew Nut and Cocoa Development, Kochi, pp 30–41

Boucher BJ, Mannan N (2002) Metabolic effect of the consumption of *Areca catechu*. Addict Biol 7:103–110

Chandramohan R, Murthy VK (1991) Utilization of areca leaf as a substrate for cultivation of *Pleurotus sajor caju*. In: Nair MC (ed) India mushrooms. KAU, Vellani kkara, pp 140–142

Chempakam B (1998) Post harvest technology and product utilization in arecanut. In: Bosco SJD, Sairam CV, Muralidharan K, Amarnath CH (eds) Harvest and post-harvest technology of plantation crops. CPCRI, Kenaragod, pp 74–78

Franca LF, Meireles MAA (2000) Modeling the extraction of carotene and lipids from pressed palm oil (*Elaes guineensis*) fibers using supercritical CO_2. J Supercrit Fluids 18:35–47

Franca LR, Meireless MAA (1997) Extraction of oil from pressed palm oil (*Elaes guineensis*) fibers using supercritical CO_2. Ciência e Technologia de Alimentos 17:384

George WS, Robert F (2006) Bevacqua: *Areca catechu* L. (betel nut palm) species profiles for Pacific Island Agro forestry. August ver. 1.3

Hartley CWS (1988) The oil palm (*Elaeis guineensis* Jacq.), 3rd edn. Longman Group, New York

Hassan OA, Yeong SW (1999) By-products as animal feedstuffs. In: Gurmit S, Lim KH, Teo L, Lee DK (eds) Oil palm and the environment: a Malaysian perspective. Malaysian Oil Palm Growers' Council, Kuala Lumpur, Malaysia, pp 225–239

Howard GT (2002) Biodegradation of Polyurethane- A review. Int Biodeterior Biodegrad 49:245–252

Hudiyono S, Septian A (2012) Optimization of carotenoids isolation of the waste CPO using α-amylase, ß-amylase and cellulose. IOSR J Appl Chem 2(2):7–12

Jain NL, Das DP, Lal G (1954) Proc. Symp. Fruits and vegetable preservation industry. CFTRI, Mysore, pp 75–80

Jayalakshmi A, Mathew AG (1982) Chemical composition and processing. In: Bavappa KVA, Nair MK, Kumar TP (eds) The areca nut palm. Central Plantation Corps Research Institute, Kerala, pp 225–244

KAU (1996) Package of practices recommendation – crops. Kerala Agricultural University, Directorate of Extension, Mannuthy – 685601, Trichur, pp 65–67

Kumaraswamy MV, Satish S (2008) Antioxidant and anti-lipoxygenase activity of *Thespesia lampas* dalz & Gibs. Adv Biol Res 2(3–4):56–59

Kurian A, Peter KV (2007) Commercial crops technology. New India Publishing Agency, New Delhi, pp 63–98

Lokeswari N, Sriramireddy D, Pola S, Varaprasad B (2010) Production of 3, 4, 5- trihydroxybenzoic acid by solid-state fermentation. J Pharm Res 3:906

Mandal PK, Kochu Babu M (2005) Palm oil a high value edible oil and raw material for value added food products. Plant Hort Tech 6(1):39–41

Mathew Jose, Sobhana A, Mini C (2011) Technological advancements in cashew apple processing and potentials for commercial application. In: Souvenir and abstracts of 1st international symposium on cashewnut, December 9–12, 2011, Madurai, India, pp 61–69

Mini C, Jose M (2007) Multi uses of cashew apple. In: Proceedings of 6th national seminar on Indian cashew in the next decade-challenges and opportunities, May 18–19, 2007, Raipur, India, pp 45–52

Mini C, Jose M, Indira V (2008) Changes in chemical and microbial quality of mixed cashew apple jams during storage. J Plant Crops 36(3):496–499

Nagaraja KV (1998) Quality and processing aspect of cashew. In: Bosco SJD, Sairam CV, Mnralidharan K, Amarnath CH (eds) Harvest and post-harvest technology of plantation crops. CPCRI, Kasaragod, pp 111–113

Nair MK, Bhaskara Rao EVV, Nambiar KKN, Nambiar MC (1979) Cashew. ICAR, CPCRI, Kasargod, p 92

Narayanamurthi D, Singh J (1964) Final report of utilisation of areca husk. Forest Research Institute, Dehra Dun, p 33

Ngo V (1991) Palm oil marketing issues and developments in the 1990s. In: Proceedings of the PORIM international palm oil conference. Module IV Promotion and Marketing, Kualalampur, Malaysia, pp 102–106

Nik Norulaini Ahmad AO, Fatehah MO, Banana AAS, Zaidul ISM, Mohd O (2008) Sterilization and extraction of palm oil from screw pressed palm fruit fiber using supercritical carbon dioxide. Sep Purif Technol 60:272–277

Okeudo NJ, Eboh KV, Izugboekwe NV (2005) Growth rate, carcass characteristics and organoleptic quality of broilers fed graded levels of palm kernel cake. Int J Poult Sci 4(5):330–333

Parthasarathy VA, Chattopadhyay PK, Bose TK (2006) Plantation crops, vol 2. Naya Udyog, Kolkata

Raghavan V (1957) On certain aspects of the biology of areca nut (*Areca catechu* Linn.) and utilization of its by products in industry D. Phil. thesis. Gauhati University, India, p 193

Raghavan V, Baruah HK (1957) On areca nut and its scope. Arecanut J 8:5–9

Rajagopal V (1998) Produce and products of plantation crops. In: Bosco SJD, Sairam CV, Muralidharan K, Amarnath CH (eds) Harvest and post-harvest technology of plantation crops. CPCRI, Kenaragod, pp 9–13

Satyavati VK, Mookerji KK, Bandopadhyay GG (1963) Annual report, CFTRI, Mysore, Karnataka, India, 1962, p 433

Shinichi Y, Katsuji M, Shigeki S, Kenjil M, Shinya Y (2009) Ethanol production from oil palm empty fruit bunches in South Asian countries considering xylose utilization. J Jpn Inst Energy 88(10):923–926

Shivadasani HB (1972) Cashewnut shell liquid in break linings. Indian Cashew J 8(3):7–9

Suganya P, Dharshini R (2011) Value added products from cashew apple on alternate nutritional source. Int J Curr Res 3(7):177–180

Thomas GV (1998) Utilization of waste of plantation industry. In: Bosco SJD, Sairam CV, Muralidharan K, Amarnath CH (eds) Harvest and post-harvest technology of plantation crops. CPCRI, Kenaragod, p 142

Van Eijnatten CLM (1991) Edible fruits and nuts. In: Verheij EWM, Coronel RE (eds) Plant resources of South East Asia No. 2. Pndoc-DLO, Wageningen, p 446

Woodroof JG (1967) Tree nuts: production, processing, products, vol 1. The Agri. Publ. Co. Inc., West Port, p 356

Yusoff S (2006) Renewable energy from palm oil: innovation on effective utilization of waste. J Clean Prod 14(1):87–93

Value Addition of Fruits and Vegetables Through Drying and Dehydration

10

S. Datta, A. Das, S. Basfore, and T. Seth

Abstract

Fruits and vegetables are highly perishable in nature; postharvest losses vary from 40 to 50 % in developed countries like India, which badly affects the availability of fruits and vegetables to the consumers. Drying and dehydration are the methods to preserve the perishable raw commodity against deterioration and to reduce the cost of packaging, handling, storing and transporting the material by converting it into a dry solid, thus reducing its weight and also, usually but not necessarily, its volume. Advances in vegetable and fruit drying and dehydration techniques are helpful in the development of novel value-added products which meet the quality standard, stability and functional requirement coupled with the economy of the country. Dried and dehydrated fruits and vegetables can be successfully used for different food preparations and for the preparation of instant soups, baking, dairy and confectionery products. In this chapter, different aspects of drying and dehydration of fruits and vegetables like predrying treatments, drying methodology and postdrying treatments have been discussed.

10.1 Introduction

Horticulture accounts for 30 % of India's agricultural GDP (gross domestic product) from 8.5 % of cropped area (Narayana 2014). India is the second largest producer of fruits and vegetables with first rank in the production of ginger and okra and second in bananas, papayas, mangoes, etc. (Anonymous 2013). The horticulture export industry offers an important source of employment for developing countries. Cultivation of fruits and vegetables is substantially more labour-intensive than growing cereal crops and offers

S. Datta (✉) • S. Basfore • T. Seth
Uttar Banga Krishi Viswavidyalaya,
Pundibari, Cooch Behar, West Bengal, India
e-mail: suchanddatta@rediffmail.com

A. Das
Bidhan Chandra Krishi Viswavidyalaya,
Mohanpur, Nadia, West Bengal, India

more postharvest opportunity and development of value-added products (Joshi et al. 2004). Fruits and vegetables play an important role in human diet, nutrition and health. They are important sources of essential dietary nutrients, vitamins, minerals, crude fibres, etc. India produces 162.12 million metric tons of vegetables and 81.29 million metric tons of fruits during the years 2012–2013 (Anonymous 2014). Due to perishable nature of fruits and vegetables, postharvest losses noticed were close to 50 % of the total fruits and vegetables produced in India, which badly affects the availability of fruits and vegetables to the consumers. In monetary terms, the fruits and vegetables of Rs. 13,300 crores are spoiled every year in India (Singh et al. 2014). Vegetables and fruits are available during specific seasons, and they are perishable. Hence, majority of them are not available during off-season. The abundance of indigenous vegetables during the rainy seasons and poor transport and storage systems lead to their high postharvest losses but scarcity in the dry seasons. Effective preservation and storage of these vegetables would not only reduce their postharvest losses but make them available throughout the year with higher prices (Seidu et al. 2012). To overcome this problem, dehydration technique has been developed by which vegetables in dehydrated form are preserved for a longer period and are made available during off-season. Drying or dehydration is a simple, low-cost way to preserve food that might otherwise be spoiled. Drying is the removal of moisture from food to a certain level at which microorganisms cannot grow. Dehydration is the removal of moisture under controlled temperature and humidity condition without impairing the inherent quality of the products. In ancient times, fruits and vegetables were exposed to direct sunlight for drying (Ghoshal 2012). The calorific content of the dried and dehydrated food is more in per unit area as compared to fresh food, because nutrient becomes concentrated as water is removed. Dried and dehydrated fruits and vegetables can be used successfully for different food preparations and for preparation of instant soups, baking, dairy and confectionery products (Savo et al. 2012). India is also a prominent exporter of dried and preserved

Table 10.1 Dried and dehydrated products prepared from fruit and vegetable crops

Sl. No.	Name of the crop	Dried and dehydrated products
1.	Mango	Mango powder, mango lather
2.	Banana	Banana chips, dried banana
3.	Grape	Raisin
4.	Date palm	Dried date palm
5.	Apple	Dried apple
6.	Pear	Dried pear
7.	Peach	Dried peach
8.	Anola	Dried shred of anola
9.	Pomegranate	Anardana (dried pomegranate)
10.	Fig	Dried fig
11.	Potato	Potato chips, potato flour, papad
12.	Tomato	Dried tomato, tomato powder
13.	Chilli	Dried chilli, chilli powder
14.	Cauliflower	Dried cauliflower
15.	Cabbage	Dried cabbage
16.	Garden pea	Dried pea
17.	Turnip	Dried turnip
18.	Onion	Dehydrated onion, onion powder
19.	Garlic	Dehydrated garlic, garlic powder
20.	Pumpkin	Pumpkin lather

vegetables to the world. The country has exported 56,158.40 MT of dried and preserved vegetables to the world for the worth of Rs. 742.74 crores during the years 2013–2014. Similarly, India also earns a sustainable amount of foreign money by exporting dried and dehydrated fruit products. Dried and dehydrated fruits and vegetables are mainly exported in Germany, Russia, the United Kingdom, the United States, France and Brazil. A number of different dried and dehydrated products prepared from fruits and vegetables have been presented in Table 10.1.

10.2 Advantages of Drying and Dehydration

1. It is the most nutritious way to preserve food which maintains its quality.

2. It is the easiest way to preserve the food materials.
3. It makes the seasonal fruits and vegetables available throughout the year in hygienic condition.
4. Dried or dehydrated fruits and vegetables fetch higher price as compared to fresh fruits and vegetables.
5. Glass, plastic or metal airtight containers can be used to store the dehydrated fruits and vegetables.
6. It can be stored for longer time.
7. Dried and dehydrated foods can be stored without refrigeration or sealing.

10.3 Preparation of Fruits and Vegetables for Drying

Fruits and vegetables are dried by either the sun or by providing heat artificially. The fruits and vegetables should be washed properly under cold running water, drained and cut away bruised and fibrous portions, and all inedible portions like seeds, stems, etc. should be removed properly. Dried or dehydrated fruits and vegetables can be produced by a variety of processes. These processes differ primarily by the type of drying method used, which depends on the type of food and the characteristics of the final product. In general, dried or dehydrated fruits and vegetables undergo the following process steps: predrying treatments, such as size selection, peeling, colour preservation, drying or dehydration and using natural or artificial methods, and postdehydration treatments, such as conditioning, inspection and packaging.

10.3.1 Predrying Treatments

The objective of the predrying treatment is to check the enzymatic reaction of the living tissue of the fruits and vegetables. These changes manifest themselves by affecting adversely the colour, flavour, etc. of the product. Predrying treatments prepare the raw product for drying or dehydration and include raw product preparation, enzyme

inactivation and colour preservation (Somogyi and Luh 1988). Pretreating vegetables by blanching in boiling water or citric acid solution is recommended to enhance the quality and safety of the dried vegetables. Blanching helps slow or stop the enzyme activity that can cause undesirable changes in flavour and texture during storage. Blanching also relaxes tissues so that pieces dry faster, helps to protect the vitamins and colour, and reduces the time needed to refresh vegetables before cooking. In addition, research studies have shown that pretreating vegetables by blanching in water or citric acid solution enhances the destruction of potentially harmful bacteria during drying, including *Escherichia coli*, *Salmonella* species and *Listeria monocytogenes*.

10.3.1.1 Selecting and Pretreating Fruits and Vegetables

Raw product preparation includes selection and sorting, washing, peeling (some fruits and vegetables), cutting into the appropriate form and blanching (for some fruits and most vegetables). Use only ripe and good-quality fruits and vegetables. Select the fruits and vegetables individually. Discard damaged, rotted and diseased and insect-infested fruits and vegetables (Somogyi and Luh 1986, 1988). Fruits and vegetables are selected; sorted according to size, maturity and soundness; and then washed to remove dust, dirt, insect matter, mould spores, plant parts and other materials that might contaminate or affect the colour, aroma or flavour of the fruits or vegetables. Prunes and grapes are dipped in an alkali solution to remove the natural waxy surface coating which enhances the drying process. Next, the product is cut into the appropriate shape or form (i.e. halves, wedges, slices, cubes, nuggets, etc.), although some items, such as cherries and corn, may bypass this operation. Inactivation by heat treatment (blanching) is usually done by immersing the fruits and vegetables in hot water at 95–100 °C for few minutes or exposing it to steam.

10.3.1.1.1 Blanching

Blanching is mostly practised in the case of the vegetables only. Blanching is one of the most commonly used pretreatment to inactivate the enzyme responsible for quality deterioration of processed vegetables, and it can be carried out in high temperature (Shivhare et al. 2009). Blanching is briefly precooking food in boiling water or steam before processing or drying the food to destroy most of the enzymes responsible for the breakdown of food. Blanching also shortens the drying time and kills most of the spoiling organisms. There are two basic methods of blanching, viz., steam blanching and boiling water blanching generally used for vegetables and syrup blanching for syrup. In steam blanching, a large stainless steel pot and a steamer rack are all that is needed to properly steam blanch the produce. The steaming time for the produce is reduced to approximately 15–20 s to a few minutes per loaded steamer rack. The produce is not submerged in the boiling water, keeping the texture and vitamin content more intact, whereas in boiling water blanching, the produce is dipped into the boiling water for 30 s or more and removed with a slotted spoon. Drain the wet produce in a colander before laying out to dry, placing into freezer bags to be vacuum sealed, or placing in sterilized jars for hot water canning. This method of blanching is good at destroying the enzymes to reduce spoilage, but many water-soluble vitamins (like vitamin C) and minerals will be lost if the blanching time is too long. Water blanching is mostly used for vegetable blanching. Syrup blanching is mostly used for blanching of the fruits only. It is done to check the browning of fruits like apple, banana, pear, peach, etc.

Vegetables contain catalase, peroxidase, polyphenol oxidase (PPO) and many other enzymes; at the time of cutting and peeling, enzymatic reactions are accelerated and discolouration will often occur. Enzymes can be inactivated by heating or acidification to low pH values.

10.3.1.1.2 Colour Preservation or Sulphuring

The final step of the predehydration treatment is the colour preservation also known as sulphuring. Vegetables undergo browning due to enzymatic or nonenzymatic reactions that occur during processing and storage. The majority of fruits are treated with sulphur dioxide (SO_2) for its antioxidant and preservative effects. The presence of SO_2 is very effective in retarding the browning of fruits, which are not inactivated by the sufficiently high heat normally used in drying. In addition to preventing browning, SO_2 treatment reduces the destruction of carotene and ascorbic acid, which are the important nutrients from fruits. Sulphuring of dried fruits must be closely controlled so that enough sulphur is present to maintain the physical and nutritional properties of the product throughout its expected shelf life, but not so large that it adversely affects flavour. Although dried fruits commonly use SO_2 gas to prevent browning, this treatment is not practical for vegetables. Instead, most vegetables (potatoes, cabbage and carrots) are treated with sulphite solutions to retard enzymatic browning (DiPersio et al. 2005, 2007). In addition to colour preservation, the presence of a small amount of sulphite in blanched, cut vegetables improves storage stability and makes it possible to increase the drying temperature during dehydration, thus decreasing drying time and increasing the drying capacity without exceeding the tolerance for heat damage.

10.3.1.2 Selecting and Pretreating of Fruits

Select fresh and fully ripened fruits because immature produce lacks flavour and colour, and overmature produce can be tough and fibrous or soft and mushy. Thoroughly wash and clean fruits to remove dirt or spray. Sort and discard any fruit that shows decay, bruises or mould affected. Such defects can affect all foods being dried. Pretreating fruits prior to drying is highly recommended. Pretreating helps to keep light-coloured fruits from darkening during drying and storage, and it speeds the drying of fruits with tough skins, such as grapes and cherries. Fruits such as grapes, prunes, small dark plums, cherries, figs and firm berries have tough skins with a wax-like coating. To allow inside moisture to evaporate, crack or 'check' skins before drying whole fruits. Lye peeling with alkali solution also enhances the speed and quality of the drying process. To crack skins, dip fruit in briskly boiling water for 30–60 s, and then dip in very cold water. Drain on absorbent towels before placing on drying trays.

Research studies have shown that pretreating with an acidic solution or sodium metabisulphite dip also enhances the destruction of potentially harmful bacteria during drying, including *Escherichia coli* O157:H7, *Salmonella* species and *Listeria monocytogenes*. Several methods of pretreatments of fruits are discussed below.

10.3.1.2.1 Ascorbic Acid Pretreatment

Ascorbic acid (vitamin C) is an important antioxidant that prevents fruit from darkening and enhances destruction of bacteria during drying. Pure crystals usually are available at supermarkets and drug stores. Stir 2.5 tablespoons (34 g) of pure ascorbic acid crystals into 1,000 ml of cold water. Vitamin C tablets can be crushed and used (six 500 mg tablets is equal to 1 teaspoon of ascorbic acid). One thousand millilitres of solution treats about 10 quarts of cut fruit. Cut peeled fruit directly in ascorbic acid solution. Soak for 10 min, remove with a slotted spoon, drain well and dehydrate. Commercial antioxidant mixtures are not as effective as ascorbic acid but are more readily available in grocery stores. Follow directions on the container for fresh cut fruits.

10.3.1.2.2 Citric Acid or Lemon Juice Pretreatment

Citric acid or lemon juice may also be used as antidarkening and antimicrobial pretreatments. Prepare the citric acid solution by stirring 5 g of citric acid into 1,000 ml of cold water. For the lemon juice solution, mix equal parts of lemon juice and cold water (i.e. 1 cup of lemon juice and 1 cup of water). Cut the peeled fruit directly into the citric acid or lemon juice solution. Allow the fruits to soak for 10 min, then remove with a slotted spoon, drain well and dehydrate.

10.3.1.2.3 Sodium Metabisulphite Pretreatment

Sulphur and sulphite compounds have been used for centuries to prevent discolouration and reduce spoilage during the preparation, dehydration, storage and distribution of many foods.

10.4   Drying or Dehydration

Drying or dehydration is the removal of the majority of water contained in the fruit or vegetable and is the primary stage in the production of dehydrated fruits and vegetables. Dehydration of fruit and vegetables is one of the oldest forms of food preservation techniques known to man and consists primarily of establishment engaged in sun drying or artificially dehydrating fruits and vegetables. Dehydration of fruits and vegetables also lowers the cost of packaging, storing and transportation by reducing both the weight and volume of the final product. Given the improvement in the quality of dehydrated foods, along with the increased focus on instant and convenient foods, the potential of dehydrated fruits and vegetables is greater than ever.

During the time of drying, the following aspects should be taken cared of:

1. Select the drying method and equipment that is right for drying. Foods can be dried in a conventional oven, a commercial dehydrator or in the sun. Drying times vary with the method and foods chosen. Be sure to read the instructions with your dehydrator.

2. Maintain 60–65 °C temperature with circulating air during the time of drying. Remove enough moisture as quickly as possible to prevent spoilage. A drying temperature of 80–85 °C allows moisture to be removed quickly without adversely affecting food's texture, colour, flavour and nutritive value. If the initial temperature is lower, or air circulation is insufficient, foods may undergo undesirable microbiological changes before drying. Optimum temperature and humidity should be maintained. If the temperature is higher or humidity is too low, nutrients can be lost or moisture may be removed too quickly from the product's outer surface. This causes the outer surface to harden and prevents moisture in the inner tissues from escaping.

3. Knowledge about the optimum drying stage is important because some foods are more pliable when cool than warm. Foods should be pliable and leathery or hard and brittle when

sufficiently dried. Some vegetables actually shatter if hit with a hammer. At this stage, they should contain about 10 % moisture. Because they are so dry, vegetables do not need conditioning like fruits.

10.4.1 Methods of Drying

Several drying methods are commercially available, and the selection of the optimal method is determined by quality requirements, raw material characteristics and economic factors. There are three types of drying processes: (1) sun and solar drying, (2) atmospheric dehydration including stationary or batch processes (kiln, tower and cabinet driers) and continuous processes (tunnel, continuous belt, belt-trough, fluidized-bed, explosion puffing, foam-mat, spray, drum and microwave-heated driers) and (3) subatmospheric dehydration (vacuum shelf, vacuum belt, vacuum drum and freeze driers). The following types of drying methods are followed for drying of fruits and vegetables.

10.4.1.1 Sun Drying

Sun drying is the evaporation of water from products by sun or solar heat, assisted by the movement of surrounding air. To be successful, it demands a rainless season of bright sunshine and temperatures above 98 °F coinciding with the period of product maturity. Sun drying requires considerable care. Products must be protected from insects and must be sheltered during the night. In the case of vegetable drying, pretreated vegetables are spread in thin layer on trays. Expose the trays to the sun, but only for 1 or 2 days. Direct sun on vegetables can cause sunburn or scorching. Drying can be completed in the shade. Fruits are placed in direct sun and turned occasionally. A light covering of cheesecloth or screen suspended above the food will keep it away from insects. Several days in direct sun are sufficient to make fruit about two-thirds dry. At this stage, stack the trays in the shade where there is good air circulation and continue drying until leathery. This method is relatively slow, because the sun does not cause rapid evaporation of moisture. Reduced drying times may be achieved by using a solar dryer.

10.4.1.2 Air Drying

Air drying is an alternative to sun drying for such products as herbs and chilli peppers. The material is tied into bunches or strung on a string and suspended out of the sun until dry. This can be in a shady porch, shed or corner of the kitchen. Enclosing produce in a paper bag protects it from dust and other pollutants. Some herbs can be dried simply by spreading on a dish towel or tray and leaving on the counter for 2 or 3 days.

10.4.1.3 Dehydrators

Dehydrators with thermostatic controlled heat and forced air circulation are available from a number of commercial sources. They can also be constructed from a variety of materials available to the home carpenter. Dehydrators require (1) an enclosed cabinet, (2) a controlled source of heat and (3) forced air to carry away the moisture. Venting to allow intake and exhaust of air is necessary. Price is not a full-proof method of measuring the quality of a commercial dehydrator. In the commercial dehydrator, thermometer should measure temperatures from 130 to 180 °F. Place the thermometer on one of the shelves inside a working dehydrator. Desirable dryer temperatures are 140–160 °F. Controls to adjust temperature should be accurate. Uniformity of temperature inside the dehydrator is during the drying procedure. Temperature uniformity can be measured by checking the temperature front and back and top and bottom of the dehydrator. The airflow through the dehydrator is also important. Designs of dehydrators vary but all will have an air intake and exhaust. The intake for air is frequently on the bottom or back and the exhaust on the top or front of the dehydrator. With the dehydrator turned on, light a match or a candle, and holding it in the outflow of air, slowly move it towards the dehydrator. The airflow should blow it out at 2–4 ft from the exit port.

10.4.1.4 Oven Drying

Oven drying is harder to control than drying with a dehydrator; however, some products can be quite successfully dried in the oven. It typically takes two to three times longer to dry food in an oven. Thus, the oven is not as efficient and uses more energy. Set the oven at the lowest setting, preferably around 150 °F and leave the door open 2–3 in. A small fan positioned to the side of the oven door blowing inward will help to remove moist air.

10.4.1.5   Freeze Drying

Freeze drying, a relatively new process of drying in vacuum at very low temperatures, ensures the preservation of all thermolabile compounds in the initial raw material, and final low content of moisture provides microbiological stability and permanent preservation of products. All preservation technologies have reduced the amount of biologically active compounds, such as vitamin C and phenolic compounds. Related to the chemical composition of fresh fruits, the decrease of tested parameters was affected by freezing for 15 %, freeze drying for 28–32 % and conventional drying for 45–48 %. The process of freeze drying is one of the methods for preservation of raw materials of plant origin (fruits, vegetables, spices and herbs). Final product has a high nutritional value (as fresh fruits and vegetables). Freeze-dried products do require special storage conditions, e.g. absence of light, packaging materials with low gas permeability, inert atmosphere, etc. (Savo et al. 2012). Nacheva et al. (2012) reported that the freeze-dried vegetables contain moisture from 2 to 5 % and taste and aroma complex preserved to the highest degree. The carried out gamma sterilization ensures a high microbial purity of the vegetables and guarantees for their long-term preservation (up to 5 years) in polymer packing, under usual conditions.

10.4.1.6   Solar Energy Dryer

Solar drying is an affordable and cost-effective alternative for preservation of food and agricultural crops. The external heating media is not required thus able to overcome global energy crises. It not only saves energy but also saves time, occupies less area, improves quality of product, makes the process more efficient and protects the environment (Deshmukh et al.

2011). Recent developments of solar dryers such as solar tunnel dryer, improved version of solar tunnel dryer, roof-integrated solar dryer and greenhouse-type solar dryer are used for drying of fruits, vegetables, spices, medicinal plants and fish (Janjai and Bala 2012). Isiaka et al. (2012) studied on the effect of selected factors on drying process of tomato in a forced-convection solar energy dryer. The results further revealed that drying rate increases with a decrease in slice thickness and increase in airflow rate. Drying of 15, 20 and 25 mm slice thickness of tomato was achieved in time range of 21–24, 27–29 and 30–50 h, respectively. Kiremire et al. (2010) studied on the effects of vegetable drying techniques on nutrient content, a case study of southwestern Uganda. The results showed that some nutrients were lost during the drying process but in general, the nutrient content remained high. With respect to retention of the nutrient content in vegetables, solar drying was found to be the best of the other methods, namely, sun drying and oven drying. Based on this study, solar drying is recommended as a method for vegetable preservation.

10.4.1.7   Osmotic Dehydration

Osmotic treatment describes a preparation step to further processing of foods involving simultaneous transient moisture loss and solids gain when immersing in osmotic solutions, resulting in partial drying and improving the overall quality of food products. As the process is carried out at mild temperatures and the moisture is removed by a liquid diffusion process, phase change that would be present in the other drying processes will be avoided, resulting in high-quality products and also lead to substantial energy savings. Several techniques such as microwave heating, vacuum, high pressure, pulsed electric field, etc. may be employed during or after osmotic treatment to enhance performance of the osmotic dehydration (Maftoonazad 2010). Levent and Ferit (2012) reported that in the case of osmotic dehydration, water activity, colour, solid content, water loss and solid gain of samples varied with the variation of the concentration of solution, temperature and time.

10.4.1.8 Microwave-Related Drying

Microwave-related drying (like microwave hot air, microwave vacuum, microwave freeze and microwave osmotic) is a rapid dehydration technique that can be applied to specific foods, particularly to fruits and vegetables. Increasing concerns about product quality and production costs have motivated the research to investigate and the industry to adopt combination drying technologies. The advantages of microwave-related combination drying include shorter drying time, improved product quality and flexibility in producing a wide variety of dried products (Karaaslan 2012).

Microwave-assisted freeze drying (MFD) is a rapid dehydration technique that can be applied to certain foods, particularly to seafoods, solid soup and fruits and vegetables. MFD involves much less drying time and energy consumption than conventional freeze-drying methods. Currently, this technology has been successfully used to dry many food materials and has a potential in the food industry. Increasing concerns over product quality, energy savings and production costs have motivated researchers and the industry to adopt MFD technologies. The advantages of MFD include shorter drying time, energy savings, improved product quality and flexibility in producing a wide variety of dried products. However, current applications are limited to small categories of foods due to high start-up costs and relatively complex technology compared to conventional freeze drying (Duan et al. 2010).

10.5 Postdrying Treatment and Storage

10.5.1 Conditioning

When the food is taken from the dehydrator, the remaining moisture may not be distributed equally among the pieces because of their size or their location in the dehydrator. Conditioning is the process used to equalize the moisture. It reduces the risk of mould growth. After food is dried, it should be conditioned for a week or 2 to assure that pieces are uniformly dried. To do this, loosely pack the dried food to about two-thirds full in a large nonmetal container. Shake the jars daily to separate the pieces and check the moisture condensation. Place the container in a warm, dry, well-ventilated place away from animals, insects and dust. Cover with a clean cloth and stir once or twice daily. The food can be left in this way for 1–2 weeks. After conditioning, package and store the fruits and vegetables.

10.5.2 Pasteurization

At the end of the conditioning or curing time, foods dried in the sun should be pasteurized to prevent possible infestation from insects or insect eggs. This is done by reheating dried foods on trays in an oven at 160 °F for 30 min or 175 °F for 15 min. Normally, heat brittle, dried vegetables are pasteurized for 10 min and fruits for 15 min. Allow pasteurized food to cool to room temperature before packaging for storage. Insects or insect eggs also may be killed by placing the packaged dried vegetables in the home freezer for 48 h.

10.5.3 Packaging

Packaging is common to most all dehydrated products and has a great deal of influence on the shelf life of the dried product. Packaging of dehydrated fruits and vegetables must protect the product against moisture, light, air, dust, microflora, foreign odour, insects and rodents; provide strength and stability to maintain original product size, shape and appearance throughout storage, handling and marketing; and consist of materials that are approved for contact with food. Cost is also an important factor in packaging. Package types include cans, plastic bags, drums, bins and cartons and depend on the end use of the product. Dried foods are susceptible to insect contamination and reabsorption of moisture and must be properly packaged and stored immediately. Heat and light have an adverse effect on the quality of dried foods. Dried foods must be protected from moisture absorption and from insect infestation. Pack dried foods in small amounts in dry, sealed glass jars (preferable dark coloured) or in moisture vapour-proof freezer cartons or plastic bags. Bags may be heat sealed or closed with twist tapes. Containers should be filled as full as possible without crushing.

For packaging the dried foods, the following aspects should be taken cared of:
1. First, cool completely.
2. Warm food causes sweating which could provide enough moisture for mould to grow.
3. Package dehydrated foods in tightly sealed containers, such as moisture-proof freezer containers or Ziploc-type bags, or dark-scalded (sterilized) glass jars.
4. Choose the right containers like glass jars, metal cans or boxes with tight-fitted lids or moisture vapour-resistant freezer cartons that make good containers for storing dried foods.
5. Heavy-duty plastic bags are acceptable, but keep in mind that they are insect and rodent proof. Plastic bags with a 3/8 in. seal are best to keep out moisture.
6. Fruit that has been sulphured should not touch metal. Place the fruit in a plastic bag before storing it in a metal can. Sulphur fumes will react with the metal and cause colour changes in the fruit.
7. Pack as tightly as possible without crushing.
8. Pack food in amounts that will be used in a recipe. Every time a package is reopened, the food is exposed to air and moisture that will lower the quality of the food.

Foods that seem 'bone dry' when packed can be spoiled by reabsorption of moisture during storage. Any moisture that collects in the inside of glass jars or plastic containers can be noticed. If this happens, the food can be rescued by heating to 150 °F for 15 min and resealing the container.

10.5.4 Storing the Dried Foods

Dried foods should be stored in a dark, dry, cool place. Food quality is affected by heat. Low temperatures extend the shelf life of the dried product. Most dried fruits can be stored for 1 year at 60 °F, 6 months at 80 °F. Vegetables have about half the shelf life of fruits. Use foods within 6–12 months for best quality. Check dried foods frequently during storage to see if they are still dry.

10.5.4.1 Dehydration Ratio (DR)
Dehydration ratio is the amount of water present in fresh product to the amount of water present in dehydrated product or simply, it is the ratio of the weight of the fresh sample (B) used for drying to the weight of the dehydrated sample (A). DR = B/A = B:A

10.5.4.2 Rehydration Ratio (RR)
Rehydration is achieved by keeping the dehydrated product in water for 30 min duration so as to restore the fresh-like appearance of the dehydrated product. It is the ratio of the weight of dehydrated onion to the weight of rehydrated onion. If the weight of the dehydrated sample (A) and drained weight of the rehydrated sample (B) are equal, then the rehydration ratio (RR) can be written as: RR = A/B = A: B

10.5.4.3 Reconstitution of Dehydrated Fruits and Vegetables
The following steps should be followed for reconstitution of dehydrated fruits and vegetables:
1. Boiling – To reconstitute vegetables, pour one cup of vegetables into 1 cup of boiling water, and let it soak for 5–20 min. To reconstitute fruit, mix 1 cup of water with 1 cup of dried fruit in a pan and simmer until the fruit is tender. Cover dried vegetables with cold water and let them soak until they are nearly restored to their original texture (0.5–2 h). If they are soaked longer than 2 h, the vegetables should be refrigerated.
2. Cooking – For vegetable dishes, mix 1 cup of liquid for every cup of dried food. Fruit dishes work best if you simmer the fruit in water before cooking. Add more water during cooking if needed. Salt can be added to vegetables any time – during soaking, while cooking or before serving.

3. Soaking – Put your food in a pan and cover with water. Allow 1–2 h for the rehydration process to occur. Powdered onions, herbs, greens, cabbage and tomatoes do not need soaking.
4. Steaming – Place your dehydrated food in the steamer. The dry food will absorb the moisture from the steam and plump right up.

10.6 Fruit and Vegetable Drying Guide

Step-by-step directions of some fruits and vegetables drying are discussed below.

10.6.1 Fruits

Most of the fruits are dried at a temperature range of 60–80 °C.

Apples: Wash, peel, trim out blemishes, cut into slices, soak in water with lemon juice or ascorbic acid or steam blanch for 3–5 min or syrup blanch for 10 min; drying time is 6–12 h.

Banana (*green*): Peel, blanch in boiling water after peeling and slice to 12 mm thick. During slicing, keep in 0.1 % ICMS solution to avoid discolouration, sulphur for 2 h and dry for 8–10 h.

Mango (*green*): Peel and cut into pieces of 12 mm thick, sulphur for 2 h in fumes and then dry at temperature of 45–50 °C for a period of 30 h.

Grapes: Select seedless grape, wash properly, remove stem, blanch for 15–30 s in hot water and dry for 6–12 h.

Pear: Wash, peel, slice, steam blanch or syrup blanch, sulphur and dry for 30–36 h.

Peaches: Select fully ripe fruits, wash, blanch in boiling water for 30 s, sulphur and dry for 36–48 h.

10.6.2 Vegetables

All vegetables except onions and peppers and mushrooms should be washed, sliced and blanched. For drying, vegetables should be spread in single layers on trays. Depending on drying conditions, drying times may take longer. For drying vegetables, drying temperature should be maintained at 60–80 °C.

Beans: Wash and cut into 1 in. pieces. Steam blanch for 3 min and dry for 6–12 h until brittle.

Peas: Dry for 5–14 h until brittle.

Cauliflower: Remove outer leaves, cores and stalks and cover leaves and stems. Break flowers into small pieces. Water blanch for 5 min and dry for 6–14 h.

Cabbage: Remove outer leaves and cores, shred longitudinally into 0.5 cm thick, water blanch for 2–3 min with 1 % sodium bicarbonate and dry for 6–12 h.

Broccoli: Cut and dry for 4–10 h.

Carrots: Peel, slice or shred. Dry for 6–12 h until almost brittle.

Beets: Cook and peel beets, cut into 1/4 in. pieces and dry for 3–10 h until leathery.

Corn: Cut corn off cob after blanching and dry for 6–12 h until brittle.

Mushrooms: Brush off, don't wash and dry at 90° for 3 h and then 125° for the remaining drying time. Dry for 4–10 h until brittle.

Onions: Peel, slice into 3–5 mm thick, pretreat and dry for 6–12 h until crisp.

Tomatoes: Dip in boiling water to loosen skins, peel, slice or quarter. Dry for 6–12 h until crisp.

Potatoes: Peel and cut into 0.25 cm thick for fry and 0.5–1 cm for curry, water blanch for 3 min and dry for 6–12 h until crisp.

References

Anonymous (2013) Press Information Bureau Government of India, Ministry of Agriculture, MP: SS: CP: Vegetables and Fruits, 17th December, 17:20 IST

Anonymous (2014) Indian Horticulture Database, National Horticulture Board, Ministry of Agriculture, Government of India, Gurgaon, India

Deshmukh AW, Wasewar KL, Verma MN (2011) Solar drying of food materials as an alternative for energy crisis and environmental protection. Int J Chem Sci 9(3):1175–1182

DiPersio P, Kendall P, Yoon Y, Sofos J (2005) Influence of blanching treatments on Salmonella during home-type dehydration and storage of potato slices. J Food Protect 68:2587–2593

DiPersio PA, Kendall PA, Yoon Y, Sofos JN (2007) Influence of modified blanching treatments on inactivation of Salmonella during drying and storage of carrot slices. Food Microbiol 24:500–507

Duan X, Zhang M, Mujumdar AS, Wang R (2010) Trends in microwave-assisted freeze drying of foods. Dry Technol 28(4):444–453

Ghoshal MK (2012) Fabrication and performance study of a photovoltaic integrated solar dryer. Int J Agric Eng 5(1):73–76

Isiaka M, El Okene AMI, Muhammed US (2012) Effect of selected factors on drying process of tomato in forced convection solar energy dryer. Res J Appl Sci Eng Technol 4(19):3637–3640

Janjai S, Bala BK (2012) Solar drying technology. Food Eng Rev 4(1):16–54

Joshi PK, Gulati A, Birthal PS, Tewari L (2004) Agriculture diversification in South Asia: patterns, determinants and policy implications. Econ Polit Wkly 39(24):2457–2467

Karaaslan S (2012) Microwave related drying of fruits and vegetables. Ziraat Fakultesi Dergisi Suleyman Demirel Universitesi 7(2):123–129

Kiremire BT, Musinguzi E, Kikafunda JK, Lukwago FB (2010) Effects of vegetable drying techniques on nutrient content: a case study of South-Western Uganda. Afr J Food Agric Nutr Dev 10(5):2587–2600

Levent IA, Ferit A (2012) Partial removal of water from red pepper by immersion in an osmotic solution before drying. Afr J Biotechnol 11(6):1449–1459

Maftoonazad N (2010) Use of osmotic dehydration to improve fruits and vegetables quality during processing. Recent Pat Food Nutr Agric 2(3):233–242

Nacheva I, Miteva D, Todorov Y, Loginovska K, Tsvetkov T (2012) Modern high technology solutions for quality and long-term vegetable preservation. Bulg J Agric Sci 18(2):161–165

Narayana CK (2014) A step towards prevention of food losses. Curr Sci 106(1):10

Savo I, Dragan M, Miladin B (2012) Specificities of fruit freeze drying and product prices. Econ Agric 59(3):461–471

Seidu JM, Bobobee EYH, Kwenin WKJ, Frimpong R, Kubge SD, Tevor WJ, Mahama AA (2012) Preservation of indigenous vegetables by solar drying. J Agric Biol Sci 7(6):407–415

Shivhare US, Gupta M, Basu S, Raghavan GSV (2009) Optimization of blanching process for carrot. J Food Process Eng 32:587–605

Singh V, Hedayetullah M, Zaman P, Meher J (2014) Postharvest technology of fruits and vegetables: an overview. J Postharvest Technol 2(02):124–135

Somogyi LP, Luh BS (1986) Dehydration of fruits. In: Woodroof JG, Luh BS (eds) Commercial vegetable processing, 2nd edn. Published by AVI Publishing Company, Westport Connecticut, USA

Somogyi LP, Luh BS (1988) Vegetable dehydration. In: Luh BS, Woodroof JG (eds) Commercial vegetable processing, 2nd edn. Published by Van Nostrand Reinhold, Westport Connecticut, USA

Value Addition of Fruits and Vegetables Through Packaging

11

G. Mandal

Abstract

Packing of fruits and vegetables is an important step in the long and complicated journey from grower to consumer. Bags, crates, baskets, cartons, bulk bins, and palletized containers are convenient containers for handling, transporting, and marketing of fresh produce. To optimize the shelf life of fruits and vegetables, regardless of the packaging system, maintenance of appropriate temperatures throughout distribution and storage is another key factor as every 10 °C increase in temperature reduces shelf life by two to three times. The active packaging allowing a controlled interaction between the produce, package, and internal gaseous environment, thus, extends shelf life, improves fruit and vegetable safety or provides superior sensory quality, and inhibits pathogens. Fresh or minimally processed fruits and vegetables respire at rates specific to the fruit or vegetable species, variety, growth, harvest and storage history, and conditions of the surrounding environment. Various films provide breathable package that can keep gas levels at an optimum, which significantly improves shelf life. The labels on packaging materials assist buyers to know who, what, and how much as well as quality assurance code that help to find the source of the problem. More than 1,500 different types of packages are used for different produce, and the number continues to increase as the industry introduces new packaging materials and concepts.

11.1 Introduction

India, the second largest producer of fruits and vegetables in the world, aims to double its horticulture production to 500 million tonnes by 2020. It contributes nearly 28 % of GDP and 54 % of export share in agriculture from the cultivated

G. Mandal (✉)
Palli Siksha Bhavana, Visva-Bharati, Sriniketan, Birbhum, West Bengal, India
e-mail: gmciphet@gmail.com

A.B. Sharangi and S. Datta (eds.), *Value Addition of Horticultural Crops: Recent Trends and Future Directions*, DOI 10.1007/978-81-322-2262-0_11, © Springer India 2015

area share of 8.5 % only. Postharvest operations are assuming greater importance due to higher yields and increased cropping intensity. Due to introduction of modern technology, yield levels have substantially increased resulting in a marketable surplus, which has to be stored till prices are favorable for sale. But India loses about 35–45 % of the harvested fruits and vegetables during handling, storage, transportation, etc., a loss estimated at Rs 40,000 crores per year. India wastes fruits and vegetables every year equivalent to the annual consumption of the United Kingdom.

Packaging fresh fruits and vegetables is one of the more important steps in the long and complicated journey from grower to consumer. Bags, crates, hampers, baskets, cartons, bulk bins, and palletized containers are convenient containers for handling, transporting, and marketing fresh produce. More than 1,500 different types of packages are used for produce, and the number continues to increase as the industry introduces new packaging materials and concepts. The trend in recent years has moved toward a wider range of package sizes to accommodate the diverse needs of wholesalers, consumers, food service buyers, and processing operations.

11.2 Function of Packaging

There is an old adage that "If you package it right, you can sell just about anything." It's no different for packaging fruits and vegetables— they must be packaged so customers will buy them. Proper packaging is especially important when a grower is selling to a wholesale buyer.

11.2.1 Containment

The container must enclose the produce in convenient units for handling and distribution. The produce should fit well inside the container, with little wasted space.

11.2.2 Protection

The package must protect the produce from mechanical damage and poor environmental conditions during handling and distribution. Because almost all produce packages are palletized, produce containers should have sufficient stacking strength to resist crushing in a low temperature, high humidity environment. Produce destined for export markets requires that containers be extra sturdy. Airfreighted produce may require special packing, package sizes, and insulation. Marketers who export fresh produce should consult with freight companies about any special packaging requirements. Each fresh fruit and vegetable commodity has its own requirements for temperature, humidity, and environmental gas composition. Produce containers should be produce friendly—helping to maintain an optimum environment for the longest shelf life. This may include special materials to slow the loss of water from the produce, insulation materials to keep out the heat, or engineered plastic liners that maintain a favorable mix of oxygen and carbon dioxide.

11.2.3 Identification

The package must identify and provide useful information about the produce. It is customary (and may be required in some cases) to provide information such as the produce name, brand, size, grade, variety, net weight, count, grower, shipper, and country of origin. It is also becoming more common to find included on the package nutritional information, recipes, and other useful information directed specifically at the consumer. In consumer marketing, package appearance has also become an important part of point of sale displays. Universal Product Codes (UPC or bar codes) may be included as part of the labeling.

11.3 Types of Packaging Materials

11.3.1 Wood Pallets

Wood pallets literally form the base on which most fresh produce is delivered to the consumer. About 40 % of total are single-use pallets. Because many are of a nonstandard size, the pallets are built as inexpensively as possible and discarded after a single use. Although standardization efforts have been slowly underway for many years, the efforts have been accelerated by pressure from environmental groups, in addition to the rising cost of pallets and landfill tipping fees.

Depending on the size of produce package, a single pallet may carry from 20 to over 100 individual packages. Because these packages are often loosely stacked to allow for air circulation or are bulging and difficult to stack evenly, they must be secured (unitized) to prevent shifting during handling and transit. Plastic stretch film is also widely used to secure produce packages. A good film must stretch, retain its elasticity, and cling to the packages. Plastic film may conform easily to various size loads. It helps protect the packages from loss of moisture, makes the pallet more secure against pilferage, and can be applied using partial automation.

11.3.2 Pallet Bins

Substantial wooden pallet bins of milled lumber or plywood are primarily used to move produce from the field or orchard to the packing house. Most pallet bins are locally made; therefore, it is very important that they be consistent from lot to lot in materials, construction, and especially size. For example, small differences in overall dimensions pallet bin can add up to big problems when several hundreds are stacked together for cooling, ventilation, or storage. It is also important that stress points be adequately reinforced. The average life of a hardwood pallet bin that is stored outside is approximately 5 years. When properly protected from the weather, pallet bins may have a useful life of 10 years or more.

11.3.3 Wirebound Crates

Although alternatives are available, wooden wirebound crates are used extensively for snap beans, sweet corn, and several other commodities that require hydrocooling. Wirebound crates are sturdy and rigid and have very high stacking strength that is essentially unaffected by water. Wirebound crates come in many different sizes and have a great deal of open space to facilitate cooling and ventilation. Wirebound crates are not

generally acceptable for consumer packaging because of the difficulty in affixing suitable labels.

11.3.4 Wooden Crates and Lugs

Wooden crates, once extensively used for apples, stone fruit, and potatoes, have been almost totally replaced by other types of containers. The relative expenses of the container, a greater concern for tare weight, and advances in material handling have reduced their use to a few specialty items, such as expensive tropical fruit. The 5, 10, and 20 kg wooden lugs still used for bunch grapes and some specialty crops are being gradually replaced with less costly alternatives.

11.3.5 Wooden Baskets and Hampers

Wire-reinforced wood veneer baskets and hampers of different sizes were once used for a wide variety of crops from strawberries to sweet potatoes. They are durable and may be nested for efficient transport when empty. However, cost, disposal problems, and difficulty in efficient palletization have severely limited their use to mostly local grower markets where they may be reused many times.

11.3.6 Corrugated Fiberboard

Corrugated fiberboard (often mistakenly called cardboard or pasteboard) is manufactured in many different styles and weights. Because of its relativity low cost and versatility, it is the dominant produce container material and will probably remain so in the near future. The strength and serviceability of corrugated fiberboard have been improving in recent years. Most corrugated fiberboards are made from three or more layers of paperboard manufactured by the kraft process. To be considered paperboard, the paper must be thicker than 0.008 in. The grades of paperboard are differentiated by their weight (in pounds per 1,000 sq ft) and their thickness. Kraft paper made from unbleached pulp has a characteristic brown color and is exceptionally strong. In addition to virgin wood fibers, kraft paper may have some portion of synthetic fibers for additional strength, sizing (starch), and other materials to give it wet strength and printability. Most fiberboards contain some recycled fibers. Law may specify minimum amounts of recycled materials and the percentage is expected to increase in the future. Tests have shown that cartons of fully recycled pulp have about 75 % of the stacking strength of virgin fiber containers. The use of recycled fibers will inevitably lead to the use of thicker walled containers.

11.3.7 Double-Faced Corrugated Fiberboard

Double-faced corrugated fiberboard is the predominant form used for produce containers. It is produced by sandwiching a layer of corrugated paperboard between an inner liner and outer liner (facing) of paperboard. The inner and outer liners may be identical or the outer layer may be preprinted or coated to better accept printing. The inner layer may be given a special coating to resist moisture. Heavy-duty shipping containers, such as corrugated bulk bins that are required to have high stacking strength, may have double- or even triple-wall construction.

Both cold temperatures and high humidity reduce the strength of fiberboard containers. Unless the container is specially treated, moisture absorbed from the surrounding air and the contents can reduce the strength of the container by as much as 75 %. New anti-moisture coatings (both wax and plastic) are now available to substantially reduce the effects of moisture.

Waxed fiberboard cartons (the wax is about 20 % of fiber weight) are used for many produce items that must be either hydrocooled or iced. The main objection to wax cartons is disposal after use—wax cartons cannot be recycled and are increasingly being refused at landfills. Industry sources suggest that wax cartons will eventually be replaced by plastic or, more likely, highly controlled forced-air cooling and rigid temperature, and humidity maintenance on many commodities will replace the use of ice and hydrocooling.

There are numerous styles of corrugated fiberboard containers. The two most used in the produce industry are the one-piece, regular slotted container (RSC) and the two-piece, full telescoping container (FTC). The RSC is the most popular because it is simple and economical. However, the RSC has relatively low stacking strength and therefore must be used with produce, such as potatoes, that can carry some of the stacking load. The FTC, actually one container inside another, is used when greater stacking strength and resistance to bulging are required. A third type of container is the Bliss box, which is constructed from three separate pieces of corrugated fiberboard. The Bliss box was developed to be used when maximum stacking strength is required. Glue, staples, or interlocking slots may close the bottoms and tops of all three types of containers.

11.3.8 Pulp Containers

Containers made from recycled paper pulp and a starch binder are mainly used for small consumer packages of fresh produce. Pulp containers are available in a large variety of shapes and sizes and are relatively inexpensive in standard sizes. Pulp containers can absorb surface moisture from the product, which is a benefit for small fruit and berries that are easily harmed by water. Pulp containers are also biodegradable, made from recycled materials, and recyclable.

11.3.9 Paper and Mesh Bags

Consumer packs of potatoes and onions are about the only produce items now packed in paper bags. The more sturdy mesh bag has much wider use. In addition to potatoes and onions, cabbage, turnips, citrus, and some specialty items are packed in mesh bags. Sweet corn may still be packaged in mesh bags in some markets. In addition to its low cost, mesh has the advantage of uninhibited air flow. Good ventilation is particularly beneficial to onions. Supermarket produce managers like small mesh bags because they make attractive displays that stimulate purchases.

However, bags of any type have several serious disadvantages. Large bags do not palletize well and small bags do not efficiently fill the space inside corrugated fiberboard containers. Bags do not offer protection from rough handling. Mesh bags provide little protection from light or contaminants. In addition, produce packed in bags is correctly perceived by the consumer to be less than the best grade. Few consumers are willing to pay premium price for bagged produce.

11.3.10 Plastic Bags

Plastic bags (polyethylene film) are the predominant material for fruit and vegetable consumer packaging. Besides the very low material costs, automated bagging machines further reduce packing costs. Film bags are clear, allowing for easy inspection of the contents, and readily accept high-quality graphics. Plastic films are available in a wide range of thicknesses and grades and may be engineered to control the environmental gases inside the bag. The film material "breathes" at a rate necessary to maintain the correct mix of oxygen, carbon dioxide, and water vapor inside the bag. Since each produce item has its own unique requirement for environmental gases, modified atmosphere packaging material must be specially engineered for each item. Research has shown that this packaging extends the shelf life of fresh produce considerably. The explosive growth of precut produce is due in part to the availability of modified atmosphere packaging.

In addition to engineered plastic films, various patches and valves have been developed that affix to low-cost ordinary plastic film bags. These devices respond to temperature and control the mix of environmental gases.

11.3.11 Shrink Wrap

One of the newest trends in produce packaging is the shrink wrapping of individual produce items. Shrink wrapping has been used successfully to package potatoes, sweet potatoes, apples, onions, sweet corn, cucumbers, and a variety of tropical fruit. Shrink wrapping with an engineered plastic wrap can reduce shrinkage, protect the produce from disease, reduce mechanical damage, and provide a good surface for stick-on labels.

11.3.12 Rigid Plastic Packages

Packages with a top and bottom that are heat formed from one or two pieces of plastic are known as clamshells. Clamshells are gaining popularity because they are inexpensive, versatile, provide excellent protection to the produce, and present a very pleasing consumer package. Clamshells are most often used with consumer packs of high-value produce items like small fruit, berries, mushrooms, etc. or items that are easily damaged by crushing. Clamshells are used extensively with precut produce and prepared salads.

11.3.13 Master (Flat)

Master (flat) is a type of carton designed to contain smaller, usually consumer-sized, units of produce. Masters designed to contain 8 quarts or 12 pints or half pints are commonly used with strawberries and other small fruit.

11.3.14 Lug

Lug is a sturdy container, often wholly or partially of wood, designed to have high stacking strength. Lugs are often used for soft fruits such as grapes, berries, or tomatoes that are easily damaged by crushing.

11.4   Cooling of Fruits and Vegetables

A key element to optimizing shelf life, regardless of the packaging system used, is the maintenance of appropriate temperatures throughout distribution and storage. A 10 °C rise in temperature can reduce shelf life by two to three times.

11.4.1 Types of Cooling

Precooling

Precooling is the practice of cooling bulk produce prior to grading and packaging although the term has been used to describe cooling at any time before transport. The practice of true precooling is gradually fading since precooled product has an opportunity to warm during subsequent operations.

Room Cooling

Room cooling is the practice of storing bulk or packaged produce in a refrigerated room for an indefinite period. The cooling rate of room cooling may be sufficient for many, not very perishable produce items such as potatoes, cabbage, or root crops; however, it has proven unsatisfactory for many very perishable items such as strawberries, blueberries, and snap beans.

Hydrocooling

Hydrocooling is cooling freshly harvested produce by flooding, immersing, or spraying with large quantities of cold water. Compared to other cooling methods, hydrocooling is fast and generally thorough. Hydrocooling is limited to those produce items that will tolerate liquid water.

Icing

Icing is cooling fresh produce by the addition of crushed ice or slurry of crushed ice and water over the top of a load or to each individual package. This method is limited to those produce items that are not harmed by contact with ice.

Forced-Air Cooling

Forced-air cooling is a commonly used cooling method that utilizes specially constructed portable fans or rooms to draw chilled air horizontally through pallets or stacks of packaged produce. A properly designed forced-air cooling system is fast, energy efficient, and relatively inexpensive. It may be utilized on most types of produce.

Vacuum Cooling

Vacuum cooling is fast and may be utilized very effectively on packaged produce. It is generally more effective on those items with a large surface area to weight ratio such as lettuce and various greens.

11.5   Standardization of Packaging

Produce package standardization is interpreted differently by different groups. The wide variety of package sizes and material combinations is a result of the market responding to demands from many different segments of the produce industry. Selecting the right container for fresh produce is seldom a matter of personal choice for the packer. For each commodity, the market has unofficial but nevertheless rigid standards for packaging; therefore, it is very risky to use a nonstandard package.

11.5.1 Types of Packaging

11.5.1.1   Count Packing

A packing method in which a certain specified number of sized and graded items are placed in the carton.

11.5.1.2   Face Packing

A packing method where most of the container is loose or volume-filled except for top layer. Items in the top layer are arranged in an orderly pattern for appearance.

11.5.1.3   Mixed Load

A single truck load of fresh produce consisting of two or more products shipped together. The use of mixed loads reduces transportation costs, but care must be exercised to prevent ethylene or odor contamination.

11.5.1.4 Weight Packing
A packing method where a specified minimum weight, but not necessarily number of produce items, is packed in a container.

11.5.1.5 Vibration Fill Packing
A packing method designed to reduce bruising and scuffing of the produce. After filling, the package is gently vibrated to maximize product density and stability.

11.5.1.6 Volume-Fill Packing
A packing method where a specified volume of produce is packed in the container. Both the number of individual produce items and the weight may vary in volume-fill packing.

11.5.1.7 Layer Packing
A packing method where the entire package of produce is packed in orderly layers.

11.5.1.8 Field Packing
A packing method in which all harvesting, grading, and packing functions are performed at the same time in the field or orchard.

11.5.1.9 Hundredweight
A unit of 100 lb. abbreviated cwt.

11.5.1.10 Modified Atmosphere Packaging
A method of packaging in which the produce is packed in a sealed container into which a specific mix of gases is introduced. The container only prevents the gases from escaping and does not regulate the mix.

11.5.1.11 Controlled Atmosphere Package
An engineered package where the interaction between the produce and the packing material actively regulates a beneficial mix of environmental gases.

11.5.1.12 Active Packaging
Active packaging contains active component, allowing a controlled interaction between the food, package, and internal gaseous environment, thus extends shelf life, improves fruit and vegetable safety, or provides superior sensory quality.

11.5.1.13 Antimicrobial Packaging
These packaging systems contain a component that inhibits molds and bacteria. Some examples of antimicrobial agents that have been successfully incorporated into films and have shown efficacy include carbon dioxide, sulfur dioxide, grapefruit seed extract, nisin, lysozyme (an enzyme found in eggs), and allyl isothiocyanate, which is a component found in horseradish oil. Challenges exist with respect to optimizing the release of the antimicrobial agent and ensuring minimal flavor carryover while maintaining the desired gas permeation rates. However, developments are currently underway and a few antimicrobial films are now commercially available.

11.5.1.14 Equilibrium Modified Atmosphere Packaging (MAP)
Fresh or minimally processed (ready-to-eat) fruits and vegetables respire at rates specific to the fruit or vegetable species, variety, growth, harvest and storage history, and conditions of the surrounding environment. If an inappropriate package is used, CO_2 can build up, creating conditions that lead to anaerobic respiration and subsequent quality reductions (e.g., off-flavors caused by fermentation products). In addition, an anaerobic atmosphere (less than 2–3 % oxygen) creates conditions that increase the risk for growth of anaerobic pathogens such as *Clostridium botulinum*, known for its production of deadly toxin. By determining the rate of carbon dioxide production and oxygen uptake for a specific product, one can select packaging film that will allow maintenance of an acceptable

amount of oxygen, CO2, and nitrogen. Various film manufacturers can provide a "breathable" package that can keep gas levels at an optimum, which significantly improves product shelf life.

11.5.1.15 Temperature Controlled Packaging

An interesting approach taken by one film manufacturer (Landec) has been to physically "program" their patented films so that gas permeability changes with temperature—at abusive temperatures, gas exchange through the polymer is able to increase accordingly, providing a longer shelf life for produce.

11.6 Ethylene Removal

Ethylene gas, which acts as a plant hormone, is produced by fruits and vegetables during ripening and is also found in the environment.

It plays a role in normal ripening, but excessive exposure can radically reduce the shelf life of produce, in some cases inducing undesirable reactions such as development of bitter flavors and loss of chlorophyll (yellowing of greens). For this reason, produce in the table below should be separated (e.g., don't store apples next to lettuce). Ethylene absorbers that will reduce these negative effects are commercially available in the form of both sachets and films. While sachets have some drawbacks due to limited consumer acceptance, ethylene-removing films are growing in popularity. These films are based on the incorporation of powdered materials such as minerals (e.g., Peakfresh ethylene-scavenging films, approved for use in Canada and the USA). The mechanism by which these films work is not well understood; despite this, the films have reportedly allowed increases in the shelf life of certain products such as strawberries, lettuce, and broccoli (the latter can be shipped iceless in non-waxed cartons).

11.6.1 MAP and Ethylene Scavenging in Synergy

Attainment of equilibrium modified atmosphere and scavenging of ethylene have both been shown to improve the shelf life of produce. What about combining these two techniques to allow true optimization of conditions for shelf life extension? Combining an optimized gas mix with the use of ethylene scavengers inside packages of high-value fruit and vegetable products is expected to provide a synergistic extension of produce shelf life. This is a promising area for research investments for innovative companies.

11.7 Labeling of Packed Commodities

Produce containers should be labeled at each end and on the top with the following information:
1. Commodity (potatoes, apples, etc.)
2. Size, count, and/or net weight (50 count, 32 kg, etc.)
3. Grade (US Fancy, US #1, etc.)
4. Packer/grower/shipper (name and address)
5. Packing date (23.01.91, etc.)
6. Quality assurance or control code (harvest date information, bin number, packing time, packer number, inspector number, gassing room number for tomatoes or honeydew melons, etc.).

It is most important to label the end of the container; this part of the box will be most visible when containers are stacked. The first four items let the buyer know who, what, and how much. The last two items help with quality assurance or control of the produce. If there are problems, a quality assurance code system will help find the source of the problem.

11.8 Grades and Sizes

Besides having a specific size container, each crop is packed to a uniform grade and size. The difference between No. 1 and lower grades is the percentage of fruits or vegetables that do not pass the grade standards. Sizing is an important part of packing a crop "right." The buyer wants a container with uniformly sized fruits and vegetables and also wants them labeled appropriately. Size may be designated as the number or count in the containers, but some crops use actual produce size in length or diameter. Number or count is the number of fruits or vegetables that will fit into the container.

precooled produce must be stored at optimum storage temperatures until it is delivered to the buyer.

11.9   Conclusions

Washing, drying, sorting, grading, and packing crops into the right container is a good start, but it will not be good enough for most wholesale buyers and brokers. The packed produce must be at a proper storage temperature when it is delivered before most buyers will accept it. Ideally, the produce should be precooled rapidly to storage temperature before or after packing. Precooling will ensure the longest produce shelf life. Packed and

References

Mitchell FG, Guillou R, Parsons RA (1972) Commercial cooling of fruits and vegetables, manual 43. California Agricultural Experiment Station, Davis

Paine FA (ed) (1987) Modern processing, packaging and distribution systems for food. Van Nostrand Reinhold Company, New York

Parsons RA, Mitchell FG, Mayer G (1972) Forced-air cooling of palletized fresh fruit. Trans ASAE 15(4):729–773

Pierce LC (1987) Vegetables: characteristics, production and marketing. Wiley, New York

Breeding Methods for Quality Improvement in Horticultural Crops

12

S. Debnath and S. Guha

Abstract

Quality improvement of crop plants is a significant component in nutritional and food security. It is always a challenge both in field and horticultural crops. Several conventional as well as modern breeding methods are practised to triumph over the challenge. Recent advancements in the field of agricultural biotechnology have created a new domain to complement the methods of plant breeding. So, plant breeding, an interdisciplinary science, is moving towards a new horizon. Here these methods of improving quality in horticultural crops have been discussed with several achievements obtained in different fields. It would help to deduce a comprehensive and comparative idea amongst the methods of plant breeding available for designing a proper route to meet the demand. Different varieties with such genetic improvement have already been released and successfully cultivated which have also been focused here. Biotechnological methods employed for quality improvement through genetic modification to breed new crop types have been briefly discussed.

S. Debnath (✉)
Department of Genetics and Plant Breeding,
Faculty of Agriculture, Uttar Banga Krishi
Viswavidyalaya, PO. Pundibari, Cooch Behar
736165, West Bengal, India
e-mail: sandip22gene@yahoo.co.in

S. Guha
Department of Spices and Plantation Crops,
Faculty of Horticulture, Bidhan Chandra Krishi
Viswavidyalaya, PO. Krishi Viswavidyalaya,
Mohanpur, Nadia 741252, West Bengal, India

12.1 Introduction

Plant breeding can be defined as an art, science and technology for changing the genetic makeup of plants for the betterment of their economic use to serve the mankind. Plant breeding intends to improve the characteristics of crop plants so that they become agronomically and economically more desirable. The specific objectives vary immensely depending on the crop under consideration.

The journey of human civilisation also involved taking steps to improve plant breeding with a goal of helping the society. The major goal is to meet the hunger of the increasing population worldwide. Aside from that, quality improvement is one of the major challenges faced by people dealing with plant breeding, as it is a demand for today. The number of varieties has been evolved in different crop plants with the application of classical genetics both in pre- and 'post-Green Revolution' era. Progresses that were made in the field of molecular biology have sharpened the basic breeding tools and intensified the prospects of confidence to serve the civilisation. Successful breeding of new plant varieties undergoes various basic procedures which are very important to note.

The following are the major steps of plant breeding:
1. Creation of variety
2. Selection of elite genotypes
3. Evaluation
4. Release
5. Multiplication
6. Distribution of the new variety

Horticultural crop breeders more often are concerned with quality traits rather than yield which is solely the biomass produced in the field. Quality includes flavour, colour, shape, size, degree of damage, nutrient levels and traits that permit greater perceived food safety or environmental sustainability.

Various quality traits may be grouped into distinct categories:

Type of traits	Description	Examples
Morphological	Appearance of the produce	Fruit size, shape, colour
Organoleptic	Palatability of the produce	Taste, aroma, juiciness, softness
Nutritional	Value of the produce for human or animal nutrition	Protein content, oil content, vitamin content, mineral content
Consumer preference	Increases the marketability of the produce	Keeping quality, storability

12.2 Plant Breeding Methods

Suitable breeding procedure to be adapted to specific condition is always determined from the reproductive behaviour of the particular crop. Based on the mode of pollination and reproduction, crop plants are divided into three groups, viz.:
1. Self-pollinated
2. Cross-pollinated
3. Vegetatively (asexually) propagated

12.2.1 Self-Pollinated Species

Self-fertilising species are known as self-pollinated or autogamous species. In these plants, the development of seed takes place by self-pollination (autogamy). Hence, self-pollinated species are also known as inbreeders. Various plant characters such as homogamy, cleistogamy, chasmogamy, bisexuality, etc. promote self-fertilisation.

Some important features of autogamous species are:
1. Regular self-pollination occurs in them.
2. They are homozygous, i.e. they are true breeding type.
3. Generally do not have recessive deleterious genes, because deleterious genes are eliminated due to inbreeding by way of gene fixation.
4. In autogamous species, new gene combinations are not possible due to self-pollination.

12.2.1.1 Breeding Methods Performed in Autogamous Species
1. Plant introduction
2. Pure line selection
3. Mass selection
4. Pedigree method
5. Bulk method
6. Single seed descent method
7. Backcross method
8. Heterosis breeding
9. Mutation breeding

10. Polyploidy breeding
11. Distant hybridisation
12. Transgenic breeding
   Recurrent selection, disruptive selection, diallel selective mating and biparental mating are used for population improvement in self-pollinated crops.

12.2.2 Cross-Pollinated Species

Cross-fertilising species produce seed by cross-pollination (allogamy) hence, also known as allogamous species or out breeders. Different flower characters that promote cross-pollination include dichogamy, monoecy, dioecy, hetero-styly, herkogamy, self-incompatibility and male sterility.

Some important features of out breeders are:
1. They are random mating type. In such population (panmictic), each genotype has equal chance of mating with all the other genotypes.
2. Individuals are heterozygous in their genetic makeup.
3. Individuals have deleterious recessive gene which is hidden by masking effect of dominant genes.
4. Out breeders exhibit high degree of inbreeding depression on selfing.
5. Cross-pollination allows new gene combinations.

12.2.2.1 Breeding Methods Performed in Allogamous Species

1. Plant introduction
2. Mass and progeny selection
3. Backcross method
4. Heterosis breeding
5. Synthetic breeding
6. Composite breeding
7. Polyploidy breeding
8. Distant hybridisation
9. Transgenic breeding
   Recurrent selection, disruptive mating and biparental mating are used for population improvement.

12.2.3 Asexually Propagated Species

Asexually propagated plants are multiplied by means of vegetative plant parts like stem or root cuttings or by others. These species are found in both self- and cross-pollinated groups. Generally asexually propagated species are highly heterozygous and have broad genetic base with wide adaptability.

12.2.3.1 Methods of Breeding Asexually Propagated Species

1. Plant introduction
2. Clonal selection
3. Heterosis breeding
4. Mutation breeding
5. Polyploidy breeding
6. Distant hybridisation
7. Transgenic breeding

12.2.4 Major Breeding Methods Practised for Quality Improvement in Horticultural Crops

The elemental discoveries of Darwin and Mendel established the scientific basis for genetics and plant breeding. The recent introduction of advances in biotechnology, genomic research and molecular marker applications along with conventional plant breeding practices has strengthened the foundation of modern plant breeding, an interdisciplinary science.

Quality improvement in horticultural crops can have several contributions to the society. It enables breeders to incorporate desired genes into elite cultivars, thereby improving their value considerably. It offers unique opportunities for improving nutritional quality and bringing other health benefits. Many crops have been bred to improve traits such as higher nutritional status or better flavour and to reduce bitterness or anti-nutritional factors. Besides, transgenic horticulture can be also used for vaccine delivery. In this section different breeding methods have been described in relation to specific achievements gained in different important crops.

12.2.4.1 Plant Introduction

Introduction is the earliest of the methods practised in plant breeding. It is the transposition of established plants (crops) from one area taken to the newer area where plants were never grown earlier for human consumption.

12.2.4.1.1 Types of Plant Introduction

Plant introduction can be classified on the basis of criteria like adaptation and utilisation.

Primary: Introduced variety is adapted to the new environment, and it is directly released commercially in the new area without any genotypic alteration.

Secondary: The introduced variety is subjected to genetic alteration to select the elite one.

Several crop plants were introduced in our country, dating back to the movement of people across the borders. For example, apple, pear and walnut were introduced from Central Asia to several parts of India. Cherries and grapes were introduced from Afghanistan in 1300 A.D. by the Muslim invaders. Similarly, potato, chili, pineapple and papaya came here through the Portuguese. Two popular vegetables, viz. cabbage and cauliflower, were brought from the Mediterranean region. Today, plant introduction is carried out through organisational mode by the supervision of the Govt. of India. The principal institute to execute the role is NBPGR, New Delhi. This organisation deals both the introduction of agricultural and horticultural crops in our country. The International Plant Genetic Resources Institute (IPGRI), Rome, has similar duty globally.

12.2.4.1.2 Important Achievements in Quality Improvement Aspects

12.2.4.1.2.1 Direct Release as Variety

Bonneville (pea): It is a variety introduced from the USA about 120 cm high. The seeds are wrinkled, and the green pods are ready for picking in about 90 days. Yield is approximately 100 Q of green pods/ha.

Early Badger (pea): It is a variety introduced from the USA where the seeds are wrinkled, having both early and dwarf types. Green-coloured pods are about 7 cm long which are ready for picking at 60–65 days after sowing. The average yield is 80–85 Q/ha.

Marglobe (tomato): This is a late-maturing variety, and plants are of indeterminate type. Fruits are large, round, smooth and juicy.

Sioux (tomato): It is a high-yielding American variety with dwarf, spreading and determinate types of growth habit. The fruits that are medium to large in size appear to be whitish green in colour on ripening. It is suitable for growing in hilly regions of India. It yields about 250 Q/ha.

12.2.4.1.2.2 Selection from Introduced Material

Japanese white (radish): The skin is snow white, and the flesh is crisp, solid and mildly flavoured. It matures in 45 days. It is suitable for growing between October and December on plains and during July to September on hills.

12.2.4.1.2.3 Varieties Through Hybridisation

Pusa Ruby (tomato): This variety evolved from a cross between Meeruti and Sioux. The plants are medium and determinate. The fruits are medium sized, lobed and uniformly red when ripe, and yield is about 300 Q/ha.

Pusa Early Dwarf (tomato): This variety evolved from a cross between Meeruti and Red Cloud. Red Cloud is an early dwarf variety which was introduced from the USA. Fruits are flattened, medium large, uniformly red and ribbed which mature in 60–70 days after transplanting. The yield is about 300–400 Q/ha.

Pusa Kesar (carrot): Pusa Kesar is a selection from a cross between Local Red and Nantes Half Long. Pusa Kesar contains high amount of carotene (about 38 mg/100 g). A spectacular feature of this variety is that the roots persist about a month longer in the field than Local Red without showing any sign of bolting. It can tolerate higher temperature than Nantes.

12.2.4.2 Pure Line Selection

A pure line is defined as the progeny of a single, homozygous self-pollinated plant. Pure line selection is a method of plant breeding in which new variety is developed by selection of single best (elite most) plant progeny amongst the traditional varieties or land races.

Pure line varieties are homozygous and homogeneous as they are genetically similar and true breeding type. Such varieties hold narrow genetic base, so they are more susceptible to diseases and have poor adaptability. A pure line breeding method is generally applied for self-pollinated crops. But it also has application to develop inbred lines that are used to make hybrids in self- or cross-pollinated crops.

12.2.4.2.1 Steps in Pure Line Selection

1. Selection of individual plants from a local/traditional variety, land races or some other mixed population.
2. Individual plant progenies are grown in rows.
3. Selected progenies are harvested individually.
4. In subsequent years, replicated yield trials are conducted.
5. After 4–6 rounds, superior plant is offered as a new cultivar (Table 12.1).

12.2.4.2.2 Important Achievements in Quality Improvement Aspects

Arka Saurabh (tomato): This variety has been developed at IIHR, Bangalore, through selection. Plants are semi-determinate type in growth. The fruits are firm round, medium large, deep red and nipple tipped. Fruits are suitable both for consumption as fresh in the markets as well as for processing purposes.

Pusa Purple Long (brinjal): It is a selection from the mixed Batia cultivar grown in Punjab, Delhi and western Uttar Pradesh, India. It is an early-maturing cultivar where the mature fruit is ready for harvesting in 100–110 days. Fruits are long (20–25 cm), purple, glossy and tender. The average yield is 281 Q/ha.

Pant T-3 (tomato): This variety has been developed at Pantnagar Agricultural University by pure line selection. The plants are semi-

Table 12.1 Detailed procedure of pure line selection

Years of practice	Description
First year	A number of individual plants (about 200–3,000) are selected from a local variety or some other mixed population based on their phenotype. Seeds are always harvested separately
Second year	Seeds from each single plant are grown separately with proper spacing in distinct rows. Here the lines are evaluated for the characters under consideration. Though in this first season, it is not always possible to judge the value of each line separately, but certainly visually poor, weak and defective progenies are rejected. Numbers of progenies are drastically reduced in this step; fewer are carried to the next step
Third year	Seeds harvested at second year are planted for a preliminary yield trial. Standard plots are kept for comparison; undesirable progenies are rejected
Fourth to seventh years	Replicated yield trials are conducted at multi-locations. Standard plots are kept for comparison; undesirable progenies are rejected
Eighth year	Best progeny is released as a new pure line variety. Seed is multiplied for distribution

determinate type with large foliage. It is suitable for sowing in winter season. Fruits of Pant T-3 are suitable for processing. The average yield is about 200 Q/ha.

12.2.4.3 Pedigree Selection

It is one of the most popular breeding methods worldwide. This method is termed so that its variants require a lot of documentation, viz. detailed records of the origin of the selected individuals or lines are maintained. Such records enable us to know from which F_2 plant the selected progeny originated. In this method, essentially a plant to row system is also adapted to develop near pure lines followed by their performance testing (Table 12.2).

Table 12.2 Detailed procedure of pedigree selection

Steps involved	Description
Hybridisation	Hybridisation between superior parents
F_1 generation	Seeds obtained through hybridisation (F_1 seeds) are planted, maintaining proper sowing distance. Seeds of about 20–40 plants are harvested in bulk and advanced to F_2 generation
F_2 generation	Selection is the principal tool carried out in this step. About 10,000–20,000 plants are grown from F_1 seeds (F_2 plants). About 500 plants are selected and harvested separately
F_3 generation	About 30–40 progenies are raised from each of the selected plant of F_2 generation. About 100–400 superior plants (the number may vary, preferably less than those selected in F_2 generation) are selected
F_4 generation	Seeds from F_3 generation are space planted. Plants with desirable characters are selected in number much less than those selected in F_3 generation
F_5 generation	Individual plant progenies are planted in multi-row (3 or more) plots so that superior plants (about 50–100) can be selected by comparison
F_6 generation	Individual plant progenies are planted in multi-row (3 or more) plots. Plants are selected based on visual evaluation; progenies showing segregation should be eliminated
F_7 generation	Preliminary yield trials with minimum 3 replications and a check. Quality tests are conducted
F_8 to F_{12} generation	Multi-location yield trials with replications are conducted. Tests for quality and disease resistance are conducted
F_{10} or F_{13} generation	Seed multiplication for distribution

12.2.4.3.1 Important Achievements in Quality Improvement Aspects

Pusa Kranti (brinjal): It is a less-seeded variety having the dwarf and spreading types of growth habit. Fruits are oblong, 15–20 cm long and dark purple with shinning green calyx. The variety is good for both autumn and spring plantings under North Indian conditions.

Pusa Early Dwarf (tomato): This variety was developed by pedigree selection from a cross between Meeruti and Red Cloud. This variety is dwarf in nature having 50–55 cm in height. Fruits are flattened, medium large, uniformly red and ribbed. Fruits are ready to harvest at 60–70 days after transplanting. The yield is about 300–400 Q/ha.

12.2.4.4 Heterosis Breeding

Heterosis is defined as superiority of F_1 in one or more characters over its better-parental, mid-parental and standard varietal values. Heterosis breeding is the development of hybrid varieties for genetic betterment of character under improvement. Heterosis is fully exploited in hybrids and partially in synthetics and composites.

Commercial hybrid seed production requires several crossing techniques which are specific to crop's floral biology and flowering behaviour (Table 12.3). A few crossing techniques are adopted for commercial hybrid seed production.

12.2.4.4.1 Hand Emasculation and Pollination

Hybrid seeds are produced manually. Modification of the plant flower structure by removal of the male organ from the female plant before the onset of anthesis is a regular practice. But this method is strictly applicable when the male and female parts of a single flower or plants are separate. This method is a practice in bisexual perfect flowers. The sterility of female line is created by removal of anther (male part), and the line is used as female line to be dusted with the pollen of desired male parent source.

12.2.4.4.2 Self-Incompatibility (SI)

Self-incompatibility passes up self-fertilisation through recognition of self-pollen in or on stigma on the female pistil. Successful pollination and fertilisation occur when pollen comes from other plants carried by different physical factors like wind, pollinators, etc. So, self-incompatibility prevents self-pollination (inbreeding) and promotes cross-pollination (out breeding). It is genetically controlled by one or more multi-allelic loci

Table 12.3 F_1 seed production method adopted in important horticultural crops

Name of the crop	Method of hybrid development
Bitter gourd (*Momordica charantia*)	Hand-pollinated hybrids
Broccoli (*Brassica oleracea*)	Self-incompatibility system hybrids
Cabbage (*Brassica oleracea*)	Self-incompatibility system hybrids
Carrot (*Daucus carota*)	Cytoplasmic male-sterile system hybrids
Cauliflower (*Brassica oleracea*)	Self-incompatibility system hybrids
Celery (*Apium graveolens*)	Genetic male-sterile system hybrids
Chinese cabbage (*Brassica rapa*)	Self-incompatibility system hybrids
Mustard (*Brassica juncea*)	Self-incompatibility system hybrids
Cucumber (*Cucumis sativus*)	Hand-pollinated hybrids
Brinjal (*Solanum melongena*)	Hand-pollinated hybrids
Gourd (*Benincasa hispida*)	Hand-pollinated hybrids
Leek (*Allium porrum*)	Cytoplasmic male-sterile system hybrids
Luffa (*Luffa angulata* and *L. cylindrica*)	Hand-pollinated hybrids
Melons (*Cucumis melo*)	Hand-pollinated hybrids
Okra (*Abelmoschus esculentus*)	Hand-pollinated hybrids
Onion (*Allium cepa*)	Cytoplasmic male-sterile system hybrids
Pakchoi and pe-tsai (*Brassica rapa*)	Self-incompatibility system hybrids
Peppers (*Capsicum annuum*)	Hand-pollinated hybrids
Pumpkin (*Cucurbita moschata*)	Hand-pollinated hybrids
Radish (*Raphanus sativus*)	Self-incompatibility system hybrids
Sweet corn (*Zea mays*)	Hand-pollinated hybrids and cytoplasmic male-sterile system hybrids
Tomato (*Lycopersicum esculentum*)	Hand-pollinated hybrid
Turnip (*Brassica rapa*)	Self-incompatibility system hybrids
Watermelon (*Citrullus lanatus*)	Hand-pollinated hybrids
Zucchini (*Cucurbita pepo*)	Hand-pollinated hybrids

Source: Modified from Tay (2002)

and executed by a series of complex cellular biochemical interactions between the self-incompatible pollen and pistil.

12.2.4.4.3 Dicliny

Dicliny occurs in unisexual flowers and has two types. In monoecious plants, male and female flowers are borne on the same plant, e.g. cucurbits, maize, castor and coconut. In dioecious plants, male flowers are borne on different plants, e.g. papaya and cannabis.

12.2.4.4.4 Male Sterility

A female plant in which no viable male gametes are borne is the basic requirement of hybrid breeding. The objective of emasculation is to make a plant devoid of pollen grains (male reproduction unit) so that it is converted into a pure female. But, for large-scale commercial F_1 seed production, hand emasculation is not too easy to perform. The substitute is to identify or generate a line that is unable to produce viable pollen. This is a reliable way to inhibit self-pollination in female plants, and seed formation is biologically dependent upon foreign pollen source. Male sterility can be defined as the failure of plants to produce functional anthers, pollen or male gametes. Male sterility can be either genetic or cytoplasmic or cytoplasmic genetic. This prevents selfing and promotes cross-pollination. Sterility is controlled by either nuclear or cytoplasmic gene or both.

In commercial hybrid seed production, female plants are male-sterile line (A) crossed with male fertility restorer line (R) to obtain F_1 hybrid.

12.2.4.4.5 Important Achievements in Quality Improvement Aspects

Pusa Anmol (brinjal): It is a hybrid cultivar developed from a cross between Pusa Purple Long and Hyderpur from IARI, New Delhi.

Sweet-72 (tomato): It is a cross between 'Pusa Red Plum' and 'Sioux'. Fruits are with green stem end, round, uniformly red and less acidic. The variety is semi-determinate and suitable for growing in Central India.

PG-13 (pea): This is a hybrid variety of pea suitable for growing in Punjab. It is dwarf, its flowers are white and seeds are creamy white and slightly wrinkled. It matures in 125 days and yields 25 Q of dry grains/ha.

H-41 (sweet potato): This is a hybrid developed by the Central Tuber Crops Research Institute (CTCRI), Trivandrum. This fibre-free variety obtained a cross between a Japanese variety Norin and a local type. The flesh tastes sweet and boils easily. The crop may be harvested in 120 days and about 370 Q/ha of tuber yield may be obtained.

H.42 (sweet potato): This is also a hybrid developed at the CTCRI, Trivandrum. It is also a fibre-free cultivar. The variety is a cross between an indigenous cultivar Vella Damph and an American variety Triumph. This hybrid has the property of resistance to the weevil attack. It yields about 350 Q of tubers/ha.

Kohir Safeda (guava): It is a hybrid with few soft seeds and white flesh which is obtained from a cross between of Kohir and Allahabad Safeda. The vigorous tree-type growth of the variety bears larger fruits.

Hisar Safeda (guava): The heterotic combination is a cross between 'Allahabad Safeda' and 'Seedless', which has upright growth with a compact crown. Its fruits are round, weighing about 92 g each; pulp is creamy white with less seeds, which are soft, 13.4 % TSS and 185 mg/100 g of ascorbic acid.

Hisar Surkha (guava): It is a hybrid of 'Apple Colour' x 'Banarasi Surkha'. Tree is medium in height having broad to compact crown. The fruits are round, each weighing about 86 g. The pulp is pink in colour having 13.6 % TSS, 0.48 % acidity and 169 mg/l00 g of ascorbic acid. Yield is about 94 kg/tree/year.

12.2.4.5   Mutation Breeding

Genetic variation is the primary requirement of plant breeding. Sometimes the desired variation is not obtained for the trait under improvement. Mutagenic agents (radiation or chemicals) can be applied successfully to induce mutations and to create genetic variations from the existing population in which desired mutants may be selected. Induced mutation has been proven as a technique of generating variation within a crop variety (Table 12.5).

12.2.4.5.1 Physical Mutagens

Physical mutagens include the use of various types of radiation, viz. X-rays, gamma rays, alpha particles, beta particles, fast and thermal (slow) neutrons and ultraviolet rays (Table 12.4).

Table 12.4 Physical mutagens (radiations) and their properties

Type of radiation	Properties
X–rays	SI, penetrating and non-particulate
Gamma rays	SI, very penetrating and non-particulate
Alpha particles	DI, particulate, less penetrating and positively charged
Beta rays particles	SI, particulate, more penetrating than alpha particles and negatively charged
Fast and thermal neutrons	DI, particulate, neutral particles, highly penetrating
Ultraviolet rays	Nonionising, low penetrating

DI Densely ionising, SI Sparsely ionising

Table 12.5 Detailed procedure of mutation breeding

Year	Generation	Characterisation
1st		Seeds, pollen, vegetative parts or tissue cultures are treated by physical (radiation) or chemical mutagens
2nd	$M_1(M_1V_1)$	Plants grown from treated seeds (M_1) or vegetative propagula (M_1V_1)
3rd	$M_2(M_1V_2)$	Population of plants grown from seeds (M_2) or vegetative parts (M_1V_2) harvested from M_1 or M_1V_1 respectively. Selection of desired mutants may start in this generation or later
4th–9th	M_3–M_8 (M_1V_3–M_1V_8)	Continuing selection, genetic confirmation, multiplication and stabilisation of field performance of mutant lines
10th–13th	Next 2–3 generations	Comparative analyses of mutant lines during different years and in different locations
14th onwards	Next 2–3 generations	Official testing before release as a new variety

Source: Novak and Brunner (1992)

12.2.4.5.1.1 X-Rays

X-rays was discovered by Rontgen in 1895. They are sparsely ionising and highly penetrating. They are generated by X-ray machines. X-rays are able to break chromosomes and produce all types of mutations at nucleotide level, viz. addition, deletion, inversion, transposition, transitions and transversions. X-rays were first used by Muller in 1927 for induction of mutations in *Drosophila*. In plants, Stadler in 1928 used X-rays for the first time for induction of mutations in barley.

12.2.4.5.1.2 Gamma Rays

Gamma rays have shorter wavelength than X-rays. It is more penetrating than X-rays. They are generated from radioactive decay of some elements like 14C, 60Co, radium, etc. Of these, cobalt 60 is commonly used for the production of gamma rays. Gamma rays also cause mutations both at chromosomal and gene levels.

12.2.4.5.2 Chemical Mutagens

The chemical mutagens can be divided into four groups, viz. (1) alkylating agents, (2) base analogues, (3) acridine dyes and (4) others.

12.2.4.5.3 Important Achievements in Quality Improvement Aspects

NBRI, Lucknow, is the pioneering institution to work on somatic mutation in chrysanthemum by using a Cobalt 60 (60Co) radioactive gamma irradiation source to breed a number of lines by changing one or few characters of an outstanding cultivar without altering the remaining and/or unique part of the existing genotype.

Mutation breeding has also been successfully used in roses for the development of new varieties.

12.2.4.6 Clonal Selection

Plants obtained from a single plant by means of vegetative propagation are called as clones. Clonal selection is an art and science of choosing the desired clones (superior) from a group of clones for cultivation. The examples of clones are cuttings in roses, tubers in potato, bulbs in onion and suckers in banana.

12.2.4.6.1 Salient Feature of Clones and Clonal Selection

1. Clones are phenotypically similar and genotypically identical.
2. Genotypically clones are heterozygous so that hybrid vigour can be exploited for a number of generations.
3. Clones remain genetically stable for any number of generations.

12.2.4.6.2 Method

- Healthy plant parts of healthy plants selected on the basis of phenotypical characters of interest after performing field tests. Selection within a clone is less effective unless there are desirable mutations. Selection is always made amongst the clones.
- The selected superior clones are tested for 3 years by multi-location trials, and the best performing clones are released as new varieties.

12.2.4.6.3 Important Achievements in Quality Improvement Aspects

Mango: Extensive survey of Dashehari orchards around Maliabad in Uttar Pradesh has resulted in the isolation of best clone, viz. Dashehari-51, with higher yield and regular bearing potential.

12.2.4.7 Biotechnological Methods

Plant biotechnology refers to a spectrum of technologies playing altogether, which arose through a journey from conventional plant selection and moving through a phase when humans intervene in cellular environment and its biological processes for its genetic alteration and improvement. A number of biotechnological methods known as recombinant DNA technology, genetic engineering (GE), transgenic modification or genetic modification (GM) are practised to improve quality in horticultural crops. In a nutshell, genes from one species are transferred into other species across barriers for the trait under consideration. There are several ways of transgenic development like chemical procedures, electroporation, particle (microprojectile) bombardment, *Agrobacterium*-mediated transformation, etc.

12.2.4.7.1 Chemical Procedures

Plant protoplasts are treated with polyethylene glycol and protoplasts readily take up DNA from their surrounding medium. After successful taking up, the foreign DNA segments can stably be integrated into the plant's chromosomal DNA. Protoplasts are then cultured under ambient growth conditions that allowed them to grow cell walls, start dividing to form a callus, develop shoots and roots and regenerate into whole plants.

12.2.4.7.2 Electroporation

As plant cell walls restrict the movement of macromolecules across it, protoplast is used in the electroporation method for introduction of foreign DNA. Suspension of protoplasts is applied with electrical pulses where the DNA of interest is placed between electrodes in an electroporation cuvette. Electrical pulses induce the development of temporary micropores in cell membranes allowing DNA to enter the cell and then to the genome.

12.2.4.7.3 Particle Bombardment

Particle bombardment is a technique used to introduce foreign DNA into cells. Gold or tungsten particles (1–2 μm) coated with the DNA of interest are used for the transformation. In this case, the coated particles are loaded into a particle gun and accelerated to a high speed (released by the electrostatic energy from a droplet of water exposed to high voltage or by using pressurised helium gas) to the target like plant cell suspensions, callus cultures or tissues. The projectiles enter through the plant cell walls and membranes. After the successful entry of microprojectiles into the cells, the DNA of interest gets released from the surface of the particle for subsequent incorporation into the plant's genome.

12.2.4.7.4 *Agrobacterium*-Mediated Plant Transformation

Agrobacterium tumefaciens is a plant pathogenic bacterium found in soil. It causes 'crown gall disease' in a variety of dicotyledonous plants. Plants infected with this bacterium develop large tumour-like swellings (galls). After infection, the bacterium transfers part of its DNA to the plant cells. This bacterial DNA successfully integrates into the plant's genome, resulting in the appearance of tumours and associated alterations in plant metabolism. So, it is the natural genetic engineer. The transformation principle of the bacterium is used in the transgenic development in plants, and *Agrobacterium*-mediated transformation is the most widely used method for plant genetic engineering.

12.2.4.7.5 Important Achievements in Quality Improvement Aspects

Application of genetic engineering to control fruit ripening: The researchers have developed a genetically modified tomato variety that slows down the natural softening (ripening) process. Pectin occurs in many fruits and chemically responsible to fruit firmness. The pectin in ripening tomatoes is degraded by the action of an enzyme called polygalacturonase. As the pectin gets destroyed, the cell walls of tomato fruit dissociate and ultimately soften. The scientists became able to reduce the amount of this enzyme in tomatoes, which slowed the rate of cell wall breakdown and produced a firmer fruit for a longer period of time. Therefore, the Flavr Savr tomato spends more days on the vine. As a result, sugars are transported to the fruit, increasing the concentration of sugar and more flavour. Thus, the Flavr Savr tomato remains rigid enough and have a longer shelf life than regular tomatoes (Table 12.6).

Table 12.6 Lists of some remarkable researches regarding quality improvement through genetic modification

Target traits	Crop	Achievement	Method applied
Vitamins and carotenoids	Potato	Significant increases in carotenoid levels, with up to 14-fold more β-carotene	Silencing the first step in the beta-epsilon branch of carotenoid biosynthesis
	Sweet potato	Identification of quantitative trait loci (QTL) for dry matter, starch content and β-carotene content	
	Lettuce	Increased enzyme activity and conversion of γ-tocopherol to the more potent α form. (Conversion of γ-tocopherol to α-tocopherol in vegetable crops could increase their value and importance in human health because vitamin E reduces the risk of several serious disorders)	Expressing a cDNA encoding γ-tocopherol methyltransferase from *Arabidopsis thaliana* to improve tocopherol composition
Folates	Tomato	Fruits contained approximately 25-fold more folate	Fruit-specific overexpression of GTP cyclohydrolase I that catalyses the first step of pteridine synthesis, and aminodeoxychorismate synthase that catalyses the first step of p-aminobenzoate (PABA) synthesis
Flavonoids	Tomato	Production of novel flavonoids in tomato fruit	Introduction of heterologous, flavonoid pathway genes stilbene synthase, chalcone synthase, chalcone reductase, chalcone isomerase and flavone synthase
Sweet taste	Tomato	Taste of tomato fruit sweetened	Transformation of tomato with the thaumatin gene from the African plant katemfe (*Thaumatococcus daniellii*)
	Potato	Potato plants that produce insulin	Production of insulin by the expression of the 1-*SST* (sucrose:sucrose 1-fructosyltransferase) and 1-*FFT* (fructan:fructan 1-fructosyltransferase) genes from globe artichoke
Tearless onion	Onion	Tearless onion	By silencing the lachrymatory factor synthase gene using RNAi technology. The conversion of valuable sulphur compounds to the tearing agent was inhibited

Source: João Silva and Rodomiro (2012)

References

João Silva D, Rodomiro O (2012) Transgenic vegetable breeding for nutritional quality and health benefits. Food Nutr Sci. doi:10.4236/fns.2012.39159

Novak FJ, Brunner H (1992) Plant breeding: Induced mutation technology for crop improvement. IAEA Bulletin 4:25–33

Tay D (2002) Vegetable hybrid seed production. In: McDonald M, Contreras S (eds) Proceedings of the international seed seminar: trade, production and technology. Pontificia Universidad Católica de Chile, Facultad de Agronomía e Ingeniería Forestal, Departamento deCiencias Vegetales, Santiago, 15–16 October 2002, pp 128–139

Suggested Readings

Singh BD (2005) Plant breeding principles and methods. Kalyani Publishers, Ludhiana

Singh P (2005) Essentials of plant breeding. Kalyani Publishers, Ludhiana

Value Addition of Non-timber Forest Products: Prospects, Constraints, and Mitigation

13

Sumit Chakravarty, Anju Puri, Mohit Subba, Tanusri Dey, Prakash Rai, Gopal Shukla, and Nazir A. Pala

Abstract

NTFPs are primary or supplemental source of livelihood mainly in the poor and developing nations. NTFPs contribute significantly to forest economy of developed and industrialized nations also. NTFPs can be processed or value added into consumer-oriented products. They have commercial importance and can contribute to the economic development of a region or a nation. Commercialization or value addition of NTFPs is now promoted as an approach to rural development especially in tropical forest areas. Unfortunately, commercialization of NTFPs has exploited and deprived the collectors. The chapter describes the prospects, constraints, and strategies to overcome these constraints of NTFP development and its value addition. The successful NTFP commercialization will be transparent, equitable, and sustainable which will have a positive impact on poverty reduction, gender equality, resource access, tenure, and management which demands aggressive policy interventions. The effective policy interventions can improve the stake of all stakeholders involved in NTFP from collection to value addition and ultimately improve collection, value addition, income, and livelihood without sacrificing the principles of sustainable forest conservation.

13.1 Introduction

Forest resources are natural capital for livelihood of rural poor that sustain them with capabilities, assets, and means of living (de Beer and McDermott 1989; Falconer and Arnold 1989; Falconer 1990; Ruiz-Perez and Arnold 1996; Plotkin and Famolare 1992; Neumann and Hirsch 2000). Some of these forest resources are non-timber forest products (NTFP) or non-wood forest products (NWFP) which include all biological and natural materials except timber extracted for

S. Chakravarty (✉) • M. Subba • T. Dey • P. Rai • G. Shukla • N.A. Pala
Department of Forestry, Faculty of Horticulture, Uttar Banga Krishi Viswavidyalaya, Pundibari, Cooch Behar, West Bengal, India
e-mail: c_drsumit@yahoo.com

A. Puri
Baring Union Christian College, Batala, Punjab, India

A.B. Sharangi and S. Datta (eds.), *Value Addition of Horticultural Crops: Recent Trends and Future Directions*, DOI 10.1007/978-81-322-2262-0_13, © Springer India 2015

human use (Tewari 1989; Lynch 1992; Subedi 1997; Shiva 1998; Byron and Arnold 1999; Warner 2000; Belcher 2003; Dubois 2003; Surayya et al. 2005; Pandey 2009). Some have used the term to encompass service functions rendered by forestlands. NTFPs are components of forest ecosystem that exist in nature and are generally not cultivated. They are non-timber but can be made of wood. NTFPs are plants or plant parts and animals (whole, parts or artifacts) that have perceived economic or consumption value sufficient to encourage their collection and removal from the forest. NTFPs which are harvested from within and on the edges of natural and disturbed forests may be all or part of living or dead plant, lichens, fungi, or other forest organisms. It represents a diversity of potential products sought after by a wide variety of people on a continuum of scales and intensities (Adepoju and Salau 2007). NTFPs have economic and social values even higher than timber products in some areas (Peters et al. 1989; Padoch 1992).

The products include fuelwood; charcoal; lac; broad range of edibles (fruits, honey, spices, fungi/mushroom, juices); medicinal (roots, leaves, bark, fruits, flower), decorative (leaves and twigs), and handicraft goods (rattan, vines, bamboo, grasses); and wood carved or woven into pieces of art or utilitarian objects harvested from woodlands/forests and wildlife products such as bones and skins for subsistence, social, cultural, ritual or religious, spiritual, ornamental, leisure, science, and revenue purposes (Anonymous 1990a; Hammett and Chamberlain 1997; Richards 1997; Chamberlain et al. 1998; Morse 2003; Narendran et al. 2001; Thandani 2001; Cooks and Wiersum 2003; Adepoju and Salau 2007; McLain 2008). Like timber, NTFPs can be further processed or value added into consumer-oriented products. Tropical rain forests are particularly abundant with plants yielding edible items, essential oils, gums, medicines, resins, tannins, colorants, and many others. The indigenous knowledge of the communities, the commercial viability of the extracted products, the abundance of NTFPs, and the proximity to forested areas decide the diversity of products extracted from the forests (Narendran et al. 2001). Service functions include grazing, watershed protection, provision and management of wildlife habitats, and tourism (Narendran et al. 2001; Ambus et al. 2007; Stainsby 2009).

The gathering of NTFP is as old as the human species itself. Wildcrafters are the people who harvest wild-grown NTFPs, and they are as diverse and widespread as the products they gather. The gatherers were reported to have a passion and satisfaction while harvesting, using, and selling the NTFPs and its products (Schlosser et al. 1991; Emery et al. 2002, 2006; Jahnige 2002; Wilsey and Nelson 2008). These have been used for subsistence by communities living near forests from time immemorial. The vast majority of NTFPs are consumed directly by the collectors/wildcrafters and their families. Some are important mainstays in the household economies. A smaller but considerable number of NTFPs are produced for sale or barter which include fruits, nuts, and vegetables primarily traded in local and regional markets, while large quantities of "bush meats" are traded in urban markets (Brown and Williams 2003). Other products find demand in more distant markets. Mushrooms, for instance, collected in remote China are sold the next day in supermarkets at Tokyo, and herbal medicines and essential oils are sold in the growing Western health and beauty markets. Professional wildcrafters collect and sell NTPFs for profit (Vaughan et al. 2013).

NTFPs are primary or supplemental source of livelihood mainly in the poor and developing nations (de Beer and McDermott 1989; Nepstad and Schwartzman 1992; Arnold and Ruiz Perez 1998; Warner 2000; Pandey 2009; Vaughan et al. 2013) while for others they are a means for recreation (Vaughan et al. 2013). Even in developed and industrialized nations like the USA, NTFPs contribute significantly to their forest economy earning billions of dollars (Duryea 1988; Schlosser et al. 1991; Dix et al. 1997; McLain and Jones 1997; Alexander 1999; Chamberlain et al. 2002; Jones and Lynch 2002; Taylor and Foster 2003; Lynch 2004; McLain and Jones 2005; Muir et al. 2006; Butler 2008). Most of the countries have their forestry sector significantly contributed by NTFPs but unfortunately has informal nature of market, lack of comprehensive product tracking and inventory making its total monetary value estimation difficult and hence undervalued (Tewari and Campbell 1995; Chamberlain et al. 2009; Anonymous 2011). In India, during the 1990s, the Ministry of Environment and Forests estimated that US$ 10 billion worth of 220 million tons of fuelwood, 250 million tons of grass and green fodder, and 12 million m³ of timber are removed from forests annually (Mukherjee 1994).

13.1.1 NTFP and Livelihood

The number of products available from NTFP is considered to be staggering. The United Nations and FAO had claimed at least 150 NTFPs in the international market (Adepoju and Salau 2007). In India, over 3,000 plant species produce economically significant products (Tewari 1994). In West Bengal, 72 plant species and three animal species were used extensively for fuel, fodder, medicine, household articles, rituals, ornaments, and recreation by the local peoples (Malhotra 1992). About 165 plant species are in trade from Nepal, of which 20 constitute about 80 % of the value (Shrestha-Acharya and Heinen 2006). In Sri Lanka, villagers collect 47 plant species for food, medicine, spices, and construction materials from Knuckles National Wilderness Area (Gunatilake et al. 1993). It was reported that about 1,200 species in Indonesia, 400–500 species in South Africa, and 1,500 species of wild plants in temperate forest were wild harvested and sold in local and international markets have medicinal value (Anonymous 2001). In Michigan, Upper Peninsula, 138 products from 80 species were identified (Emery 1998a). Classifying these products into like categories is an important first step of the NTFP enterprise/industry (Adepoju and Salau 2007).

NTFPs are often common property resources in traditional systems or as de facto open-access resources in state forest lands (Arnold 1995). These are likely to be available for direct consumption or sale when crops fail due to drought or disease or when shocks hit the household such as unemployment, death, or disease (Cavendish 1989). They can be broadly classified into edibles and nonedibles. The major categories of NTFPs are as follows:

13.1.2 Edible Products

Edible products include mushrooms, wild fruits and berries, nuts, ramps (wild onions), fiddleheads, herbs, ferns, syrup, spices, and honey. Most of these are sold as specialty foods, since most of these products are not traded widely and are usually collected and consumed by the collectors/harvesters themselves. Hence, it is difficult to assess their economic magnitudes (Adepoju and Salau 2007).

13.1.3 Forest Botanicals (Medicinal and Dietary Supplements)

These includes wild plants used for medicinal or pharmaceutical purposes, aromatic plants, and plants for dying which are often sold to herbalists or for craft purposes. These are traded as botanical products (Foster 1995).

13.1.4 Decorative Products

Evergreens, vines, berries, foliage, pine boughs, cones, grapevines, ferns, flowers, moss, mistletoe, and holly are sold as decorative products and may appear in floral arrangements, dried flower decorations, and ornaments (Hammett and Chamberlain 1998).

13.1.5 Specialty Wood Products

These are woods used for special purposes like woodcraft, handicrafts, carvings and turnings, musical instrument containers (baskets), special furniture pieces, utensils, barrels, baseball and cricket bats, canes and gun, and tool handles. These are considered nontraditional if these are produced directly from trees (i.e., tree need not be cut down to produce these items) and not from lumber or timber purchased from mills (Adepoju and Salau 2007).

13.1.6 Native Wild Plants

Seeds, cones, and native plants are collected for nursery sale.

13.1.7 Wood Byproducts

Chips, sawdust, shavings, bark, charcoal, and fuelwood are byproducts of wood sold for a variety of uses like landscaping, animal bedding, or as a source of domestic energy.

13.1.8 Tourism, Aesthetics, and Recreation

Rural and forest land or wilderness areas attract people for its aesthetic beauty or have potential for tourism and recreation (camping, fishing, picnic, sightseeing, trekking, animal and bird watching, boating).

13.1.9 Livelihood

Due to restrictions on felling of trees in forest and other environmental pressures, the attention on NTFP has increased (Surayya et al. 2005). Many NTFPs do not have scope for commercial development but are extremely important in millions of households. Unfortunately, this has not been adequately recognized as is evidenced from strategies suggested in poverty reduction strategy papers (Oksanen and Mersmann 2002). There is strong evidence that the poorest of the rural people or forest dwellers are most dependent on NTFPs (Neumann and Hirsch 2000; Surayya et al. 2005; Panda 2013) that poor frequently use NTFPs as an "employment of last resort" (Angelsen and Wunder 2003) and that NTFPs can serve an important safety net function

(McSweeney 2004). This can be explained in terms of the economic characteristics of forest-dependent people and of the product themselves (Cavendish 1989). A 2000 World Bank report estimated that one fourth of world's poor satisfy/fulfill their livelihood requirements/needs directly or indirectly from forests (Pandey 2009). In India, over 50 % of forest revenues and 70 % of export income from forest come from NTFP (Mishra et al. 2002).

NTFPs have commercial importance and can contribute to the economic development, livelihoods, and benefits directly flow directly and consistently to local communities (household economy) though a temporary way to make ends meet (Yadav and Roy 1991; Chowdhuri et al. 1992; Malhotra et al. 1992; Campbell and Tewari 1995; Everett 1996a; Greene et al. 2000; Emery 2001; Emery et al. 2002; Roy 2003; Murphy et al. 2005; Pierce and Emery 2005; Ghosal 2011; Panda 2013). Daily collection of NTFPs like food, fuel, and craft materials is done by women, while men hunt, collect timber, building and fencing poles, honey and products found deep in forest (Neumann and Hirsch 2000). Common raw and processed NTFPs are collected by women but may include products traditionally collected by men (Shillington 2002; Saxena 2003; Shackleton and Shackleton 2004; Kalu and Rachael 2006) as it is flexible and has low or no cost market entry (Emery 2002). The processing of honey into the traditional honey beer (*mbote*) and its selling is exclusively a female-based local enterprise in Zambia (Mulenga and Chizhuka 2003). It is this feature of NTFPs that has led to the widespread promotion of these products particularly by agencies interested in sustainable development as tools for enhancing gender equity and empowering and benefiting women (Neumann and Hirsch 2000; Shillington 2002). NTFPs are especially important to women in developing countries from Latin America to Asia and Africa (Gbadebo and Gloria 1999). A societal perspective needs to be developed for livelihood analysis of NTFPs which requires a broader approach usually for some type of social cost-benefit analysis and is necessarily a normative exercise because it involves judgment about what outcomes are socially preferable (Costanza and Folke 1997).

Between 50 and 75 % of the households in rural area gather a diversity of products. Landless and indigenous communities were reported to have a larger involvement in the extraction of NTFPs as compared to land-owning farmer and wage earners from the organized sectors (Narendran et al. 2001). Income from agriculture and other vocations has led to a lower dependence on NTFPs by the small farmers and other categories (Hegde et al. 1996). In recent years, harvesting for supplemental or full income has increased among minorities and migrant workers (Emery 1998b; Emery and O'halek 2001; Emery et al. 2002; McLain et al. 2005; Arora 2008). A substantial number of NTFPs are harvested for subsistence and household income from tropical and temperate forests around the world (Vaughan et al. 2013). In developing countries, the subsistence sustainable livelihood needs of poor rural households are met from plant and animal products of forests which are a source of a variety of food items that supplement and complement agricultural crops especially during adverse seasons, fuels for cooking, and a wide range of traditional medicines and other hygiene products that provide required inputs for income and safeguard from vulnerability (Falconer and Arnold 1989; Warner 2000; Dubois 2003; Adepoju and Salau 2007). In addition to subsistence and income-generating potential, NTFPs also provide food security to large low-income populations, their cattle, and other domestic animals particularly during droughts or famines (Anonymous 1989). Collectors or wildcrafters also harvest edible items or material for crafts without sales in mind but to use personally, to be gifted or donated (Love and Jones 1997; Love et al. 1998; Jahnige 2002; Emery et al. 2006; Robbins et al. 2008).

Southern Asia has a long history of human use of forest products (Bawa and Godoy 1993). NTFPs provide about 40 % of the total official forest revenues and 55 % of forest-based employment, sustaining nearly 500 million people living in and around forests (Anonymous 1990b). In Orissa, more than 45,000 tons of *kendu* (*Diospyros melanoxylon*) leaves are gathered annually by 1.8 million women (Mallik 2001). The total women labor engaged in the collection of forest produce in Orissa was about 300 million women days (Saxena 2003). NTFP collection in Madhya Pradesh generated employment equivalent to 233.8 million man days, while it was estimated to be up to 1,062.7 million man days for the whole of India (Khare and Rao 1993). Many village communities in West Bengal, India, derive about 17 % of their annual incomes from these collections (Malhotra et al. 1991). There are few dominant NTFPs like *Emblica officinalis*, *Sapindus emarginatus*, tree moss, *Tamarindus indica*, and *Acacia sinuata* that are extracted by a large number of households along with others in many parts of India contributing almost 75 % of the total NTFP revenues generated (Ganesan 1993) or earning US$45–74 as cash income annually (Malhotra 1992; Uma Shankar et al. 1996; Narendran et al. 2001). In India, collector's share in consumer's purchase was around 45–81 % in case of NTFPs at tribal areas of India (Naidu et al. 2003; Varadha et al. 2003; Pal et al. 2009).

An estimated 50 million Indians living in and around the forests depend on NTFP products (Anonymous 1987), and their NTFP collections contribute 12–22 % of their household income (Malhotra 1992; Narendran et al. 2001). Extraction value of fuelwood, fodder, and honey from Palni Hills, an offshoot of the Western Ghats, was about US$ 70/ha (Appasamy 1993), while annual flow of NTFPs and services per hectare from tropical deciduous forests in India ranged from US$220 to 357 (Chopra 1993). Tribal women in Madhya Pradesh collect NTFPs worth more than ☐ 21 billion (INR) annually (Tewari and Campbell 1995). In the year 2010–2011, export earnings from lac and lac products was around ☐ 211.13 cores INR in India (Pal et al. 2012). Annual income per hectare from NTFP collection by ten Forest Protection Committees under the Joint Forest Management programs ranged between US$8 and 186 (Malhotra et al. 1991). Small-scale forest-based enterprises based on NTFPs provide up to 50 % of income for 20–30 % of the rural labor force in India (Tewari and Campbell 1995). About 35 % of the income of tribal households in India was earned from the collection of unprocessed NTFPs (Tewari and Campbell 1995).

Sale of timber, firewood, and NTFPs together contributes nearly 10 % of Nepal's annual national GDP (Sharma et al. 2004). Economic crisis in Cameroon and South Africa forced women to depend on NTFPs for cash (Brown and Lapuyade 2001; Shackleton et al. 2008), while NTFPs are the only source of cash income for women in isolated rural areas (Schreckenberg and Marshall 2006). During the dry season of Somalia, pastoralists earn from selling of gums and resins which contributes up to a third of their cash income (Lemenih et al. 2003; Lemenih and Kassa 2010a, b). In Ghana, 49–87 % of villagers earn their household income from NTFP sale (Chege 1994). African shea butter and honey are often supplementary and complementary to other income and activities (Mickels-Kokwe 2006; Elias and Carney 2007). In South Africa, 85 % of households used products such as wild spinaches, fuelwood, wooden utensils, and edible fruits (Charlie and Sheona 2004). In Nigeria, food security of rural dwellers is improved by growing trees in the home gardens and on farms, while leaves, rattan, honey, sap, and gums are important source of income (Okafor et al. 1994). The annual combined value of consumption and sale of forest goods from a Central American rain forest ranged from US$18 to 24/ha (Godoy et al. 2000). Forest plants are also harvested for nonmarket reasons like floral, greenery, and decorations (Nelson and Williamson 1970; Thomas and Schumann 1993; Emery 2001; Greenfield and Davis 2003). In Georgia, the total estimated farm gate value of pine straw for 2001 through 2005 was over $132 million with an average annual value of $26.5 million (Harper et al. 2009). Likewise, the leaves of salal and beargrass from Pacific Northwest forests provide tens of millions of dollars to the region's economy every year (Chamberlain 2002). Exports of moss and lichen from the USA amount to $14 million (Vance 1995). The worth of USA herbal market was $970 million while globally it was $60 billion (Goldberg 1996), while sale of medicinal herbs increased from $1.5 billion in 1995 to $5 billion in 2000 (Mater 1993).

13.2 NTFP Value Addition: New Dimension for Sustainable Development

Early interest in NTFPs was encouraged by the belief that NTFP commercialization that added sufficient value to forest products could contribute to forest conservation (Nepstad and Schwartzman 1992). Sustainable forestry consists of conservation, sustainable management, and sustainable utilization of forest resources. Compatible management is the practice of managing forests for both timber and non-timber values including NTFPs (Titus et al. 2004). The value of the forests is based on its harvesting, processing, and marketing of wood and non-wood products and services. NTFPs have been widely promoted as a contribution to the sustainable development of tropical forest resources (de Beer and McDermott 1989; Wickens 1991; Nepstad and Schwartzman 1992; Arnetz 1993; Arnold and Ruiz Perez 1998). This interest is based on earlier perceptions that forest exploitation for NTFPs can be more benign than for timber (Myers 1988), together with a growing recognition of the subsistence and income generation contribution made by many NTFPs to rural livelihoods (Ruiz Pérez et al. 2004).

Within the context of new international commitments to address rural poverty, such as the United Nations Millennium Development Goals, NTFP commercialization is recognized as having the potential to achieve dual conservation and development goals by increasing the value of forest resources to local communities (Wollenberg and Ingles 1998; Neumann and Hirsch 2000). At the ecosystem level, the hypothesis that increasing NTFP value could provide incentives for forest conservation has not been confirmed. It is to be realized that the people who benefit from NTFP production are major agents of deforestation or that they have influence over those agents. Thus, low-intensity NTFP production is economically the most rewarding use of forest. In practice, this linkage is often missing. The beneficiaries of NTFP development activities are not the main agents of deforestation and have no control or can influence over decisions to log or convert forest. Increased value does not automatically translate into effective incentives for conservation (Salafsky and Wollenberg 2000).

The collection of NTFPs is now becoming commercialized due to opportunities of a much wider market (May 1991; Richards 1992; Narendran et al. 2001) and valuation of NTFP products (de Beer and McDermott 1989; Padoch and de Jong 1989; Peters et al. 1989; Schwartzman 1989; Malhotra et al. 1991; Tewari and Campbell 1995). The extension of markets to more remote areas has created both the demand and opportunity for increased incomes by the collectors. Globalization and growing interests in various kinds of natural products such as herbal medicines, wild foods, handcrafted utensils, and decorative items have increased demand and trade in these products. Moreover, developmental projects have ensured to increase income opportunities through production, processing, and trade of NTFPs. Few NTFPs have large and reliable markets, and these tend to be supplied by specialized producers/collectors using more intensive production systems (Belcher et al. 2003). Unfortunately, still, the majority of traded NTFPs are sold in relatively small quantities and for relatively low prices by the collectors.

The sustainability of NTFP extraction/collection for long-term ecological integrity of forests is debatable. This sustainability principally contributes to the economic well-being of the forest people and involves them in conservation of biodiversity (Uma Shankar et al. 1996). Concern about NTFP sustainability is increasing among collectors/wildcrafters (Love et al. 1998; Lynch and McLain 2003). Many collectors are adopting conservation practices like replanting seeds, gathering after plants have reproduced, collecting only a portion of the plant population, selecting old or unhealthy plants, removing marketable portions with minimal damage to the residual plant, and rotating harvest locations (Emery 2001; Love and Jones 2001; Emery et al. 2002; Jones and Lynch 2002). Sustainable exploitation of these resources can improve livelihood of people living in and around forests through income and employment. Concerns about sustainability have also led to

increased interest in forest farming of NTFPs (Workman et al. 2003; Kays 2004; Strong and Jacobson 2006; Butler 2008; Chamberlain et al. 2009). But these people are unaware of the potential of these resources for income generation as they lack access to information on processing possibilities. The United Nations Conference on Environment and Development (UNCED) in 1992 was instrumental in focusing international attention to these following factors as Agenda 21, Chap. 11 for implementation-combating deforestation includes the promotion and development of NWFPs through value addition, domestic processing, and promotion of small-scale forest-based enterprises for rural income and employment:

- The growth of green consumerism in developed countries
- More open international markets
- Increasing awareness of biodiversity conservation and sustainable and protective use of forest resources
- Realization, based on growing number of reports of the potential of multiuse forest management to generate rural income and in many cases NWFP yield more income over time than timber species from the same resource
- Growing recognition of the need to involve people living near forests for sustainable management of forest resources

As a result of this, many countries have started to implement Agenda 21 and incorporate NWFP programs. Decisions on whether or not to permit NTFP extraction would depend on a variety of considerations including its importance to the local economy, possibility of alternative sources of income to the people, ecological impacts of NTFP extractions, and legal status of the forests (Narendran et al. 2001). Revenues from NTFPs have been growing faster than revenues from timber in the past, as in India it was 40 % higher than those for timber, while its export earnings account for 60–70 % of total export earnings from forest products, and this proportion is rising. There is considerable scope for increasing exports further by exploiting untapped resources as the current production of most NTFPs is estimated to be about 60 % of the potential production (Tewari and Campbell 1995). For instance, the production of nonedible fibers and flowers is only 7 and 12 %, respectively, of the potential production (Gupta et al. 1982). NTFPs involve a large variety of seasonal products; returns are frequent and relatively continuous. Moreover, local processing or value addition of NTFPs can increase off-farm rural employment opportunities (Anonymous 1991a, b; Tewari and Campbell 1995). Management and development of NTFP resources is essential for the following reasons as reported by Tewari and Campbell (1995):

– Forest management focused on the production of NTFPs may be ecologically and economically sustainable provided that extraction rates do not exceed the maximum sustainable yield.
– NTFPs are a vital source of livelihood for a large proportion of poor living in or close to the forest in most tropical countries.
– In addition to subsistence and income-generating potential, NTFPs also provide food security to large low-income populations, their cattle, and other domestic animals particularly during droughts or famines.

13.2.1 Challenges for NTFP Development and Value Addition

NTFP collectors/harvesters/wildcrafters were not able to contribute in formal NTFP management as they are marginalized section in the society due to their sociocultural norms and decentralization of industries (Savage 1995; Antypas et al. 2002; Lynch et al. 2004; Arora 2008). The factors that make NTFPs important in the livelihoods of the poor also limit the scope for NTFPs to lift people out of poverty (Angelsen and Wunder 2003) or were termed as "poverty trap," i.e., forced to be in a weak bargaining position relative to traders who provide them with transport, market connections, and even credit (Neumann and Hirsch 2000). Markets of many of these products are small and are inferior. They tend to disperse with annual or seasonal fluctuation in quality and quantity of production. They are also subjected to overexploitation of the naturally

regenerating resource base over time (being an open-access resource), production on plantations outside forests, and increased competition from synthetic substitutes (Homma 1992; Belcher et al. 2003). The problem further becomes acute for remote locations for wild NTFPs in terms of its market inaccessibility. These collectors, because their penury relies on NTFPs with no better alternatives, are unable to use these resources to break themselves free out of their poverty. However, NTFPs have high values which tend to be appropriated by people with more power, more assets, and better connections which happen either through coercion and physical control of trade or are mostly achieved through domestication when market forces lead to intensified and specialized production (Dove 1993).

NTFP commercialization and value addition has not been successful neither in fulfilling expectations of local income generation nor leading to improved conservation of resources (Godoy and Bawa 1993; Peters 1996; Chamberlain 2002; Sheil and Wunder 2002). Several studies reported that certain NTFP species or groups of species are being overused and degraded (Edwards 1994, 1996a, b; Hertog 1995; Malla et al. 1995; Karki 1996; Sharma 1996). The reasons for this overuse and degradation are complex but include the lack of control over these resources, rural poverty, and social and cultural traditions (Subedi 1997). Forest managers interested in developing NTFP programs or consultancies often are ill-equipped because they lack time, money, personnel, and/or technical information (Gautam and Watannabe 2002; Lynch et al. 2004).

Promotion of trade in traditional NTFPs may not be beneficial always (Schreckenberg and Marshall 2006; Hasalkar and Jadhav 2004; Neumann and Hirsch 2000) as in Karnataka, India, participation of women decreased significantly with mechanization of extraction and processing when factory-type units were established (Hasalkar and Jadhav 2004). Promoting NTFP trade without gender consideration can create competition between men and women as was experienced for marula fruit collection in South

Africa and Namibia for commercial production of a liqueur "Amarula cream" (Wynberg et al. 2003; Shackleton and Shackleton 2005, 2010). There are some reports of NTFP regulation and management to track collection and harvest for its inventorying (Love et al. 1998; Alexander et al. 2001; Antypas et al. 2002; Jones and Lynch 2002; Lynch and McLain 2003) but with limited success (Lynch and McLain 2003; Muir et al. 2006) because such regulations, permit embargo, and management efforts reduced the autonomy of collectors forcing them to be exploited by highly capitalized companies (McLain et al. 2008); their needs, norms, and ability to afford permits to collect from national forests were not fully understood and overlooked (McLain 2000; Jones and Lynch 2002). As a result, the NTFP enterprise generally remains small, flexible, tax free, regulation free, and low cost (Alexander et al. 2002) which complicates management strategies (Charnley et al. 2007). Further, it was also believed that the tradition and culture of collectors may prevent forest managers from learning their practices, preferences, and needs (Bailey 1999). The challenges/problems related to increasing NTFP exploitation for commercialization and development are discussed below.

13.2.2 Limited Information and Knowledge

A major challenge related to development of NTFPs is the limited availability of documentation related to sustainable harvesting levels (Tewari and Campbell 1995) and market information or knowledge (Newman and Hammett 1994; Von Hagen et al. 1996; Chamberlain 2002; Vance 2002). Marketing information on NTFPs was ignored in research and management due to their inherent geographic fragmentation and lower returns (Mater 1993). There is deficiency of efficient inventories and scientific information on social and economic benefits to be derived from appropriate industrial utilization of NTFPs, economics of NTFP management, trade and marketing in different forest types, biological production functions for most NTFP species,

traditional harvesting and utilization patterns, and impacts of commercialization and changing use patterns on the state of NTFPs and related activities (McLain and Jones 2005; Bih 2006; Shrestha-Acharya and Heinen 2006). Baseline ecological data on many NTFP species is limited, and other areas critical to sustainable management remain poorly understood including harvesting techniques, sustainable yields, and monitoring (Shanley et al. 2002). Information on standing volume and rates of harvest are limited as they are very less monitored, and thus planning appropriate management strategies is unlikely (Alexander et al. 2001; Burkhart and Jacobson 2009; Chamberlain et al. 2002; Jones and Lynch 2002; Kerns et al. 2002; Lynch et al. 2004; McLain and Jones 2005). Monitoring is a problem because collectors reluctantly cooperate in this activity/exercise fearing exposure of their harvest locations and distrust forest managers (Lynch et al. 2004), while the managers are hesitant to perform this exercise believing these collectors as hidden or nonexistent (Emery 2001; Love and Jones 2001; Emery et al. 2002; Jones and Lynch 2002; Emery and Pierce 2005; Pierce and Emery 2005).

13.2.3 Loss of Resource Base

Deforestation is looming large on the future of NTFPs. There is every possible chance of unsustainable exploitation with increasing development of NTFPs (Homma 1992). Collectors may disregard traditional harvesting practices with increasing demand of NTFPs. Many tribals in Madhya Pradesh are now prematurely harvesting *chironji* (*Buchanania lanzan, B. latifolia*) fruits and overexploiting them to such an extent that is disturbing its natural regeneration in the wake of tremendous increase in demands of *chironji* seeds or Cuddapah almond which is a substitute for almond in various delicacies (Tewari and Campbell 1995). Similarly, faulty practice of collecting *mahua* flowers in West Bengal (breaking the apical twigs which affects flowering in the following year) and repeatedly burning mahua forests in Central India to simplify the collection of yellow flowers from the forest floors are damaging the natural stock and regeneration, threatening extinction of the species by AD 2020 (Anonymous 1992; Tewari and Campbell 1995). Unsustainable exploitation and felling of trees for indiscriminate collection of raw materials for incense stick like *Machilus macrantha* (*gulmavu*), *Ailanthus malabarica* (*halmaddi*), *Boswellia serrata*, and *Garcinia cambogia* (*uppage*) in Karnataka has extensively damaged the natural stock of these trees (Parameswarappa 1992).

13.2.4 Profitability Issues

Communities encounter many constraints that reduce profitability of NTFP-based activities (Kabra 1983; Peters et al. 1989; Padoch 1992; Everett 1996b; Kant et al. 1996; Taylor et al. 1996; Robbins 1998; Teel and Buck 1998; Neumann and Hirsch 2000; Alexander et al. 2001; Emery 2001, 2002; Love and Jones 2001; Emery et al. 2002; Spero and Fleming 2002; Escobal and Aldana 2003; Prakash 2003; Morsello 2004; McLain et al. 2005, 2008). These are:

- NTFPs are collected in small quantity by traditional methods and marketed in raw form without grading and standardization.
- NTFP markets are informal and imperfect in nature and therefore operate blindly on market conditions with no formal regulations to protect or record the production, distribution, and consumptions of the commodities.
- These markets are rarely reported to the government, taxed, or regulated for safety standards and working conditions.
- Harvesting profits are influenced by changes in transportation costs and volatile labor shifts.
- Long and difficult trade pathways due to involvement of intermediaries who buy products from the community for a low price and sell to the consumers/consumer center at higher price.
- Inefficient and inequitable marketing system as it is often controlled by urban-based and

capital-intensive industrial units or manufacturer and exporter.

- Lack of competitive pricing on harvested NTFPs so returns are low or no value adding to collected products.
- Relationships of dependency that harvesting of NTFPs could bring between the community and the middlemen that they are dealing with.
- Dependency is usually verified in the exploitation of only few products rather than the greater variety of NTFPs available from the forest that results in only seasonal income from harvesting.
- General lack of information about market conditions or lack of price intelligence.
- No notification for sale in unregulated markets.
- Lack of basic infrastructure like transport, storage, and processing.
- Market inaccessibility.
- Lack of skills, knowledge, formal education, and credit necessary for market entry.

13.2.5 Policy and Institutional Challenges

Typically, NTFPs have been ignored by policy. NTFPs have been historically been neglected by national governments (Adepoju and Salau 2007). They are often covered by forest regulations designed for timber management or are not considered at all. Institutional and organizational processes need to be better understood to help communities manage NTFPs as part of a larger livelihood strategy while maintaining an equitable distribution of responsibilities and benefits (Tewari and Campbell 1995). A very good example of a policy and institutional response that proved inappropriate is governmental intervention in the NTFP industry in India through setting up the Forest Development Corporations (FDCs). Unfortunately, the functioning of these cooperatives has often been detrimental to the interest of tribal people, and such organizations have not been cost effective, resulting in tribal people receiving only 10–40 % of the sale price in the nearest NTFP market (Chambers et al. 1990). Industrial development of NTFP has been lacking financial support and incentives to the entrepreneurs as a result of low priority that governments and banks have placed on forest product processing (De Silva and Atal 1995).

13.2.6 Tenure and Ownership Issues

Unless access and usufruct rights are given to users, there is little incentive to manage NTFPs sustainably (Tewari and Campbell 1995). Some of the Indian states nationalized many NTFPs, but unfortunately production levels of some NTFPs declined sharply following nationalization by reducing and delaying the remuneration to the collectors (Chambers et al. 1990). This situation sometimes encourages black markets and further higher margins required to cover the costs of illegal activities.

13.2.7 Size of Enterprise Development

Another challenge associated with the increased exploitation of NTFPs is a shift from small-scale to large-scale activities. If not carefully planned and managed, this shift can produce undesirable results particularly in terms of benefits to local people (Tewari and Campbell 1995). Small-scale enterprises have limited access to institutional finance, are bereft of tax incentives, are exposed to highly risky market environments, and face income-sharing problems. Further, with the expansion of NTFP market, capacity to increase local processing for realizing value-added benefits disrupts traditional patterns of management, income distribution, and the division of labor (Anonymous 1991a, b).

13.2.8 Synthetic Substitutes

Cheaper synthetic substitutes are always a threat to markets of NTFPs.

13.2.9 Non-realization of Financial Benefits

Forest managers or owners are unable to capture multiple non-timber financial benefits due to absence of markets, inaccessible or inability to create viable market, and stakeholders inadequately compensated in the absence of markets for the non-timber benefits they provide to the society (Mitchell 2009).

13.2.10 Marketability

Many non-timber goods and services do not meet the criteria of scarcity and excludability in economic terms, that is, there must not be such an abundance of supply relative to demand that no one is willing to pay for the goods or service and it should be possible to exclude people from enjoying its benefit for free (Merlo et al. 2000; Mitchell 2009).

13.2.11 Problems in Value Addition

Problems associated with processing or value addition of NTFPs in developing countries (De Silva and Atal 1995) are:

- Uncertain supply due to natural disasters and wide fluctuations in market demand
- Poor harvesting (indiscriminate) and postharvest treatment practices
- Inefficient processing techniques leading to low yields and poor quality products
- Poor quality control procedures
- Lack of R&D on product and process development
- Difficulties in marketing
- Lack of local market for primary processed products
- Lack of downstream processing facilities
- Lack of trained personnel and equipment
- Lack of facilities to fabricate equipment locally
- Lack of access to latest technological and market information

Fortunately, many of these obstacles have been successfully managed, and industries based on essential oils, tannins, and medicinal plants have thrived particularly in China and India and are competing globally.

13.3 Benefits and Opportunities of NTFP Development

13.3.1 Sustaining Livelihood and Resource Conservation

The important benefit of NTFPs in sustaining people's livelihoods as a source of income is widely accepted (Falconer 1990; Scoones et al. 1992; Kant et al. 1996) and is one of the two driving forces behind donor support to NTFP value addition/commercialization initiatives, the other being sustainable resource conservation (Schreckenberg et al. 2006). The demand and market values of NTFPs have been steadily increasing with declining revenues from timber which have encouraged the stakeholders to seriously consider its values and potential (Savage 1995) through sustainable exploitation which is an immediate and profitable way of integrating the use and conservation of forest efficiently (Peters et al. 1989; Browder 1992). The increasing focus of development policy on poverty reduction has, however, brought with it a need for unequivocal evidence about whether and how much NTFP development (value addition and commercialization) can contribute to poverty reduction (Wunder 2001; Arnold 2002; Shackleton et al. 2008). NTFP activities are often considered to be attractive to resource-poor people particularly women because they generally have low technical entry requirements and can provide instant cash in times of need and the resources is often freely accessible (Neumann and Hirsch 2000). Opportunities for home-based processing can open new doors for both urban and rural women (Kalu and Rachael 2006; Carr 2008). Paradoxically, it has been suggested that these poor attractive features of NTFP activities are a "poverty trap," i.e., also responsible for making the activities economically inferior (Angelsen and Wunder 2003). Ruiz Pérez et al. (2004) and Schreckenberg et al. (2006) reported that the CEPFOR (Commercialization of NTFP in Mexico and Bolivia: Factors Influencing Success) project had no such "poverty traps" but identified three types of NTFP activities that reduce poverty:

- NTFP activities provide "safety nets" by ensuring collectors against greater poverty and reducing their vulnerability to risk particularly during crisis and unusual need when subsistence agriculture or cash crops fail or when illness hits the family.
- Provides "gap-filling" NTFP activities that ensures supplementary 7–95 % cash household income in addition to more important farm and off-farm activities which are performed regularly often in the nonagricultural season. However, this income may not be significant, but the timing of this income is relatively important and significant due to income spreading and making poverty more bearable through improved nutrition and ensuring the source of earning.
- The activities are "stepping stone" to make people less poor only in areas well integrated into the cash economy that some collectors are able to pursue a "specialized" strategy earning more than 50 % of the total household income, and these collectors are better off than their peers.

13.3.2 Community-Based and Participatory

Local communities for their livelihood are actively involved with NTFPs and are also knowledgeable about it. They regularly monitor the condition of forest and thus are custodian of the forest. Many forest users are now looking forward to gain formal control of their resources and initiate activities to gain financially from harvesting and processing the NTFPs (Maharjan 1994; Hertog 1995; Edwards 1996a; Karki 1996).

13.3.3 Variety of Marketable Products and Marketing

NTFP development provides local communities with a variety of marketable products which include bamboo and rattan crafts, weaving products and paper. In the Philippines, handicraft enterprises have been jointly established by community-based organizations and NGOs. They have craft centers where harvesters can bring the raw and semi-processed materials. The centers act as a meeting point for harvesters and weavers and are involved in product development, design, assembly, finishing, and marketing. They also provide training, marketing, and strategy development services for producer groups. This has led to an array of handicrafts from indigenous and rural communities being marketed throughout the country, even reaching markets in Europe and the USA and receiving press attention alongside Gucci, Prada, and Hermes (see www. cmcrafts.org).

13.3.4 Opportunities for NTFPs and Tourism Products

The suggestions enlisted below which are adapted from Mitchell (2009) can stimulate recognition of NTFPs from a commodity basis to a specialized goods or tourism service. Each of these suggestions requires either that a product be moved closer to the consumer (by adding value through processing or packaging), that the product be differentiated from other similar products (wild, fresh from the forest), or that the product be coupled with an experience that consumer wants. The addition of the experience is what creates the "tourism" value (Mitchell 2009).

Examples of NTFP-related tourism opportunities
Product based
For gift shops (e.g., wild foods, crafts, cultural items such as drums)
For restaurants (e.g., specialized sauces, condiments, preserves, and other wild food products)
For local stores, bed and breakfast operations, tour companies (e.g., foodstuffs, soaps, decorative items)
For spas (e.g., salves, essential oils)

Educational
Craft based
Using forest botanicals for salves
Making honey and beeswax candles
Building bird houses
Making wreath or swags
Creating moss art (e.g., picture frames, table decorations)
Producing dried floral arrangements
Culinary
Cooking workshops focusing on morels, chanterelles, wild berries, etc.
Walks (e.g., "wild foods" or "wild grocery walk") and collection/preparation
Preserving methods
Interpretive: guided or self-guided
Native plant tours
Nature walks (e.g., forest, shore, alpine, desert)
Forest ecology tours
Mushrooms (collection/identification)
Wildlife tours
Cultural
First nations historical use of an area
Ethnobotany (the study of human uses of plants)
Culturally modified trees
Canoe and drum making
Harvesting/resource management
Cooking practices
Other
Food festivals
Craft shows (e.g., Christmas, floral, carving)

13.4 Value Addition and Commercialization of NTFPs

Commercialization of NTFPs has been widely promoted as an approach to rural development in tropical forest areas (Schreckenberg et al. 2006) through combinations of technical and capacity-building interventions to improve raw material production, processing, trade, and marketing and through improved policy and institutional frameworks. Resource tenure is a key factor, and considerable effort has been invested to help communities gain recognized rights and responsibilities to manage and use forest resources. Only a few developing countries have capabilities of large-scale commercial NWFPs processing.

Mostly, NWFPs have been exported to industrialized countries for processing or value addition to final products (Tewari and Campbell 1995). Normally, the following products have the processing potential of commercially important NTFPs:

– Essential oils and oleoresins
– Medicinal products
– Vegetable oils (small-scale production)
– Tannins
– Dyes/colorants
– Sweetening agents
– Gums and balsams
– Waxes
– Fiber boards

Most of these can be produced on a large scale and only a few on a small scale due to requirements for capital investment, trained personnel, and infrastructure. A value chain describes the range of activities required to bring a product from the producer to the consumer, emphasizing the value that is realized and how it is communicated (Schreckenberg et al. 2006). The NTFP value chain may include a number of different activities from harvesting of the wild resource to cultivation of the resource, various degrees of processing or value addition, storage and accumulation of the product at different points in the chain, transport, marketing, and sale (Kaplinsky and Morris 2000; Kassa et al. 2011; Shackleton et al. 2011). In many NTFP value chains, both men and women may either be involved independently at different stages or together for certain functions (Schreckenberg and Marshall 2006), but in many cases, women may be subordinate to men restricted out of cultural traditions and traditional roles in their homes to carry out activities that have limited visibility, i.e., lack of control over key functions (Chabala 2004; Gausset et al. 2005; Shackleton et al. 2007; Carr 2008; Elmhirst and Resurreccion 2008; Adedayo et al. 2010; Shackleton et al. 2011). Unfortunately, very little about the success of NTFP commercialization has been reported (Marshall et al. 2006). According to Schreckenberg et al. (2006), value chains can be successful in terms of:

– Volumes or values traded via different routes and incomes generated – both overall as a contribution to local and national economies and for the individual actors concerned
– Governance of the chain – the rules governing the relationships (preferably transparent) between different actors and the sharing of benefits (preferably equitable) between them
– Sustainability of the chain or the ability to deliver a consistent supply to meet demand over the long term – incorporating not just social and economic but also environmental sustainability
– Achievement of a range of locally defined objectives

13.4.1 Requirements for Value Addition Enterprise

Small-scale processing by rural communities needs access to centralized downstream factories and a regular market for the primary products. The link between secondary processing and consumer industries must be strengthened to ensure healthy local chain of operations from farmer to consumer as a viable alternative to export. The main requirements for establishing a NTFP-processing industry are financial resources, available raw materials, and a ready market for the finished products (De Silva and Atal 1995). The following requirements briefed below (Anonymous 1991c, d, 1994; Wickens 1991; Anand 1993; Tangley 1993; Schreckenberg et al. 2006; Carr 2008) can effectively manage the challenges to NTFP development and problems of its value addition:

13.4.2 Harvesting and Postharvest Treatment

Norms of the collectors for sustainability should be the main criteria for its harvesting which include harvesting only what is required without damaging the forest, selective gathering without disturbing future stock, protecting sites from over-harvest, and promoting growth by harvesting techniques and propagation (Fearnside 1989; Padoch 1992; Emery 1996). Creating a regular but limited demand with processing units supporting a supply from sustainable harvesting can control indiscriminate harvesting of NTFPs. Impacts on biodiversity due to regular harvesting of NTFPs should be assessed prior to initiating such processing. Domestication of the species to be processed should be encouraged for ensuring steady and environmentally sustainable supply. Raw materials which are to be kept after harvesting have to be dried and stored properly to prevent any deterioration and infestation. The following factors which can differ from raw materials to raw materials can influence the yield:

- Stage of harvesting (maturation, flowering stage)
- Time of harvesting (early morning, evening, etc.)
- Rate of drying (avoid decomposition)
- Temperature of drying (avoid decomposition)
- Moisture content after drying (avoid molds growth)
- Storage conditions (prevent hydrolysis, oxidation, infestation)
- Storage time before processing (loss of oil)

13.4.3 Raw Material Supply

Nursery approach on raw materials should be initiated in order to safeguard the natural flora. Afforestation programs associated with local processing (e.g., through agroforestry) should consider species with multiple uses. This ensures a product mix and can also provide farmers with various primary products for subsistence or sale.

13.4.4 Market Strategies

Among the three important channels (producer/collector-consumer, producer-retailer, and producer-wholesaler) identified for marketing of NTFPs, most of the NTFPs (more than 85 %) were disposed through second channel (Beohar 2003). NTFP for economic growth and forest conservation can be effectively developed

through proper understanding and knowledge of its market (Fox 1994) and marketing chains (Padoch 1992; Edwards 1996c). Marketability decides the failure or success of industries. Providing a weigh scale and information on commodity prices and quality requirements of wholesale buyers in a trading center can help remote producers gain a better bargaining position. Intermediaries from the NTFP market can be abolished through sale of products directly to end users (Neumann and Hirsch 2000; Letchworth 2001) which will reduce risk through improving harvesting consistency and developing resistant to price competition (Gold et al. 2004; Shackleton et al. 2007). Effective communication between all the stakeholders of NTFP can enhance market flexibility and sustainability (Vaughan et al. 2013). Inventory and tracking data can be combined with NTFP networks which will help to tailor gathering or improve market strategies (Vaughan et al. 2013). Systematically tracking NTFPs will require flexibility in using primary and secondary data or with mixed method analysis (Greenfield and Davis 2003; McLain 2008).

There is possibility to increase profitability of NTFPs by group marketing, cooperative marketing, establishment of processing units, and provision of infrastructural facility for marketing. Improved market extension network will improve the dissemination of current market prices of NTFPs. Quality conscious collectors and other stakeholders in NTFPs will have better bargaining power and can be able to enter the export market. Investment on and promotion of industrial processing of NTFPs should carefully consider marketability and the use of these as products for import substitution. Collective investment in a building for storage or in a drying machine gives producers of perishable commodities more flexibility in their marketing.

Improvements in processing and marketing to improve product quality and reach more valuable markets add value, create more income downstream in the market chain, and increase demand and earnings for raw material producers. Alternatively, its local utilization downstream processing and development of new products can be encouraged. The planning of products for industrial production should consider a situation where regular supply of NTWP to the world market reduces supply due to its increased local utilization. Governments must seriously promote trade activities to advertise specific products and also negotiate marketing agreements. Production scale should be decided on the basis of local and national demands along with the possibilities for secondary processing or used in the manufacture of other consumer products such as soaps, cosmetics, and pharmaceuticals. In terms of improving market outlets, the following activities are recommended:

– Conduct feasibility studies for new ventures in terms of investments, marketability, sustainability, and economic viability and potential for joint ventures.
– Increase entrepreneur's awareness about the potential of NTFP industrial processing through workshops, symposia, and exhibitions.
– Promote trade through participation in trade fairs, etc.

13.4.5 Increasing Competitiveness of Products

All attempts should be made to minimize production costs and improve the quality of the products in order to be in a better position to compete in the world markets. Any threat from synthetic products should be taken as a challenge for vigorous R&D work to improve the economic competitiveness of the product.

13.4.6 Research and Development

The R&D work should be encouraged to tap the full potential of NTFP. There should be wide research objectives like developing superior propagation materials, developing higher-yielding and disease-resistant varieties through genetic improvement, improving agrotechnology, developing new products, and marketing of finished products. Creating infrastructure and facilities will encourage R&D. Governments and

donor agencies should give priority to the following activities:

– R&D on process and product development. Appropriate technology adaptable at forest or rural locations
– Development of downstream processing to produce value-added products
– Strengthening of R&D institutes including provision of pilot plant processing facilities for testing viability and for training personnel
– Setting up facilities for design and fabrication of process equipment for appropriate technology as well as pilot plants for scale-up operations
– Develop indigenous scientific and technological capabilities in industrial utilization of NTFPs by training overseas.

13.4.7 Recognizing Rights

Developed countries strengthen their industrial research to develop synthetic substitutes as soon as a new useful natural product is discovered. It is thus vital for developing countries to safeguard the property rights of the original resource with international conventions.

13.4.8 Information on Resource Base (Inventory) and Monitoring

Inventorying and sustainable development for NTFP needs effective monitoring system which is only possible through voluntary participation and amicable relationship among the stakeholders (Lynch 2004; Lynch et al. 2004; Emery and Pierce 2005; McLain and Jones 2005) which can be developed by declaring targeted incentives and maintaining a harmonious balance between large-scale commercial and subsistence-focused forest management (Charnley et al. 2007; Rist et al. 2011). Information on NTFP marketing and development of government and NGO capacity for its marketing is critical to realize optimal profit and balance within the limits of forest production system (Everett 1996a).

Complete and systematic information on resource base and marketing systems is critical for a country to plan development of NTFP industries so it is necessary to:

- Involve local user/collectors and representatives of trade and industry so that all the important NTFP resources used/required by local communities and by trade and industry or for export are identified/documented.
- Compile the collected information based on its use and prioritize items for its social benefits and/or large volume internal or export trade as raw material for industry, including information on use and value of exported species.

13.4.9 Policy Issues

Several countries like Sweden, Finland, Lithuania, Latvia, Estonia, Russia, and Poland have protocols to inventory and value NTFP stocks, while the FAO and European Union provide comparative information for such countries (Von Hagen et al. 1996). Following these countries and international organizations, the national governments should enact the following policies to develop the NTFP enterprise (Pierce 1999; Saxena 2003; Burkhart and Jacobson 2009; Vaughan et al. 2013):

- Facilitate free market economic policies by removing bureaucratic controls and other restrictions.
- Reform regulatory and legislative controls on processed NTFP. For instance, the NTFP Policy of Orissa state has given many responsibilities to Gram Panchayats in terms of monitoring and regulating the NTFP trade which was a newfound role of this grassroot institution.
- Include NTFP production into afforestation and reforestation projects and other agroforestry programs.
- Increasing the role of forest certification will improve market and can create opportunities to better monitor standing and harvested volumes.

13.4.10 Financial and Infrastructural Support

NTFP-processing units in rural areas need financial resources and other infrastructural support through fund support from governments, aid organizations, and international development agencies. Funds for such projects should be disbursed under proper supervision to ensure maximum benefits. Proper implementation of these projects for higher returns from NTFPs through cooperative organizations can be ensured through regular participation of farmers, government officials, NGOs, and other agencies (Kant 1997). Faster accessibility to the sites and providing easy access to credit and tax incentives will attract private entrepreneurs to invest NTFP industries in rural areas. Moreover, institutional development, R&D needs of agroforestry, and process technology for efficiently achieving potential benefits of NTFP will need additional fund allocation. The national governments should be proactive to provide legislative, financial, and administrative services to support NTFP industries and to enhance transfer of technology and human resource development. They should lobby with international organizations for technical assistance and support for the management, conservation, and development of forests. An area requiring more attention by both national governments and nongovernment stakeholders is access to micro-credit particularly to informal women traders.

13.4.11 Polyvalent Pilot Plant

The gap between laboratory and industry can be bridged by pilot-scale processing which allows chemical engineers to translate bench-scale findings to industrial-scale outputs. UNIDO's polyvalent pilot plant design includes all engineering drawings, specifications, and bills of quantities so that it can be fabricated in all countries with facilities for stainless-steel welding. The plant allows for simplicity of design, installation, operation, maintenance, and repair.

13.4.12 Quality Criteria

Processed products must comply with national and international specifications. Quality has to start with the use of good quality raw materials and postharvest treatment that avoids contamination. Machinery and processes also require ISO 9000 validation. ISO 14000 series will have environmental dimension. The awareness of quality criteria is increasing in the developing countries, and new regulations governing safety of products, quality specifications, and good manufacturing procedures are being enacted.

13.4.13 Stages of Value Addition

According to De Silva and Atal (1995), steps in the process include:

- Selection of NTFPs for processing based on facilities available and marketability.
- Fabrication or procurement of equipment and provision of required services (water, energy, chemicals).
- Adaptation or development of agronomical practices, harvesting, and postharvest treatment.
- Training in processing methods and quality control.
- Actual processing often with assistance from experts, NGOs, or international agencies.
- Packaging and storage of finished products.
- Marketing outlets (local or export). In some cases, primary processed product can be used as a raw material for downstream processing of fragrances, isolates, and flavors.

The Fig. 13.1 shows a scheme for establishing processing industries based on NTFPs. Proper coordination of the multidisciplinary activities is vital for the success of industries into this field. De Silva and Atal (1995) enlisted the following stages progressively for processing as an example of medicinal plants:

Stage I

1. Harvesting of authentic material
2. Good postharvest treatment

Stage II

3. Stage I followed by comminution of raw materials
4. Packaging of powders as uniform doses
5. Formulation of pills from powder
6. Production of medicinal wines under controlled conditions which will need a sugar source

Stage III

7. 1, 2, and 3 followed by aqueous extraction
8. Preparation of standardized extracts (liquid and solid)
9. Formulation into dosage forms, capsules, and sachets

Stage IV

10. Stage III followed by conversion into other dosage forms such as tablets, syrups, and ointments

Stage V

11. 1, 2, and 3 followed by preparation of extracts with other solvents
12. Fractionation of extracts
13. Activity screening of extracts
14. Formulation into dosage forms

Stage VI

15. Isolation of pure phytopharmaceuticals from 8, 11, or 12
16. Conversion into semisynthetic drugs
17. Formulation of 15 and 16 into dosage forms including injections

Stage VII

18. New drug development

The first stage requiring no energy sources, sophisticated packaging material, or background technical education can take place in remote villages with products transported to nearby towns for further processing. However, rural producers will need training only in harvesting methods, postharvest treatments, and simple packaging. The other stages will depend on the pace of local development. In most rural areas, Stage II is desirable. When all the needed facilities are available, this can to be upgraded to Stage III. Stage VI is possible in developing countries that can afford to invest large sums of money or establish joint ventures with companies from industrialized countries. Apart from a few like India and

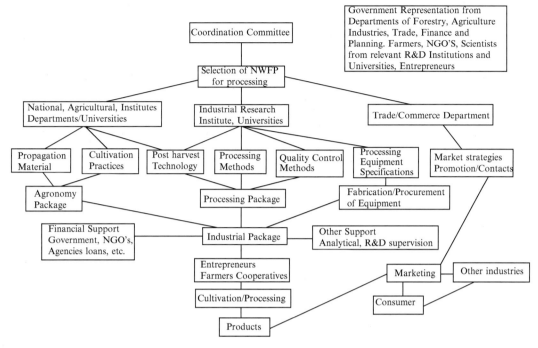

Fig. 13.1 Scheme for the development of the NTFP processing (Adapted from De Silva and Atal, 1995)

China, most developing countries have neither facilities nor money for the initiation of Stage VII. Increasing sophistication in processing can follow the same pattern for production of essential oils, again depending on the facilities available. Primary processing in that case will require an energy source and a supply of water. Mobile distillation units can be introduced if roads exist for easy access.

13.5   Value-Added NTFPS

Value addition or processing potential of some commercially important NTFPs reported by several workers (De Silva and Atal 1995; Martin 1921; Mantell 1949; Hill 1951; Anonymous 1954, 1975, 1986, 1994; Humphrey 1964; George 1966; Godin and Spensley 1971; Ikram 1978; Balick 1979; Rembert 1979; Duke 1981; Atal and Kapoor 1982a, b; De Silva 1982, 1993; Tcheknavorian and Wijesekera 1982; Princes 1983; Trease and

Evans 1983; Negi et al. 1984; Sharma 1984; Grainge and Ahmed 1986; Chawla 1987; Singh 1988; Okafor 1991; Soni 1991; Gupta and Guleria 1992; Mugedo and Waterman 1992; Wijesekera 1992; Bielenber 1993; Chaudhry 1993; Dwivedi 1993; Anand 1993) is discussed.

13.5.1 Essential Oils

Steam distillation is the widely used technique in the extraction of essential oils. Production of some expensive essential oils from flowers such as jasmine and rose takes place through a process called enfleurage. Some parts of plants that contain very minute amounts of essential oils such as flowers are extracted using low boiling solvents such as hexane, pentane, or petroleum ether. Certain finished perfumery products like colognes use the alcohol-soluble fraction of concretes called *absolutes*. The extraction of spices such as ginger, pepper, chili, and cardamom using solvents and the removal of these solvents as in the case of concretes yield oleoresins, which contain not only the essential oil compounds but other flavor principles of the spice such as pungency, waxes, and other solvent-soluble extractives. Essential oils could be further processed or rectified to add value. This is done by a process of fractionation. Pure isolates of added value such as citronellal and citral can be produced through fractionation. These isolates are further processed using chemical methods to produce high value aroma chemicals which have an export market. These aroma chemicals can be used in blending of perfumes and flavors for local industries.

13.5.2 Medicinal Plant Products

Medicines prepared using traditional methods are still used by the practicing healers in the developing countries. About 20 % of the drugs in modern pharmacopeias are also plant derived, either as pure phytopharmaceuticals extracted from plants or as synthetic derivatives of them. The medicines for internal use prepared in the traditional manner involve simple methods such as hot or cold water extraction; expression of juice after crushing; powdering of dried material; formulation of powder into pastes via such a vehicle as water, oil, or honey; and even fermentation after adding a sugar source. The value of medicinal plants as a source of foreign exchange for developing countries depends on the use of those plants as raw materials in the pharmaceutical industry.

13.5.3 Vegetable Oils

Many forest trees possess fatty oil containing seeds which could be processed to give vegetable oils. Edible oils are used as cooking oils and in the food industry. Bulk of the oil is used in soap making on both small and large scale. Some are used as components of other industrial products after secondary processing. Many of the oils are industrially produced on a commercial scale. The primary processing of seeds to yield the oil is dry expression avoiding contamination and taking

care not to subject to temperatures leading to decomposition. Sometimes boiling the crushed raw material in water will yield the solid fat on cooling. Apart from cooking oil and domestic lighting, vegetable oils have also been used as a fuel in small diesel engines, enabling a certain degree of mechanization in rural areas. Although the oils are primarily exported, they can be further processed to yield much more valuable products for a number of industries.

13.5.4 Tannins

Tannins are a group of non-crystallizable compounds widely distributed in plants, but usually localized in specific parts such as beans, nuts, fruits, barks, and stems. In addition to combining with animal skins to form strong and flexible leather, tannins also react with salts of iron to form dark-blue or greenish-black compounds, the basis of common inks. Tanning materials are often utilized in oil drilling to reduce the viscosity of the drill without reducing the specific gravity and in the production of pharmaceuticals. The main industrial uses of tannins are leather, dyes, inks, antioxidants, lubricants, and drugs.

13.5.5 Dyes and Colorants

Coloring for food, textiles, paper, and paints was originally obtained from plant and mineral resources. Many of the forest resources are rich in dyes and pigments and hence could be sustainably harvested for commercial use.

13.5.6 Sweetening Agents

Many countries have sources of plant-based sweeteners other than industrially produced sugar. These can be good substitutes in rural areas as people do not always have access to refined sugar. Some of the resources for getting the sweeteners are the sap of palm flowers and parts of plants such as leaves of stevia, arils of *Thaumatococcus daniellii*, and bark sap of the maple tree. The final products are syrups, powders, and solids. In addition, honey obtained as a NWFP is used as a sweetening agent. The primary processing of these raw materials is simple and can be carried out at rural level. Processing as a small-scale industry can be cost effective, and the product has a local market as well as a secondary use in confectioneries. The remaining liquor (molasses) could be fermented to yield alcoholic beverages and vinegar.

13.5.7 Gums

Gums are natural hydrocolloids mostly produced by plants as a protective after injury. They have diverse applications in pharmaceutical, cosmetic, food, and textile industries. As a food additive, a wide range of toxicological evaluation is needed to satisfy the international regulatory committees concerned with the safety of food and with specifications of their identity and purity. Gums and resins are used in industries for paper, textiles, adhesive, pharmaceutical, food, and perfumery, as well as in paints, coatings, printing, detergents, and cosmetics.

13.5.8 Balsams

Balsams are resinous mixtures containing large amounts of benzoic acid and cinnamic acids or esters of these acids. They are used in medicine and other consumer industries. These are mainly pathogenic products obtained as exudates from trees.

13.5.9 Waxes

Natural waxes are NWFPs of commercial value used as components of industrial products like candles, varnishes, pharmaceuticals, and cosmetics. Some of them are collected, melted, and formed into cakes or pieces. Some waxes such as candelilla can be obtained by solvent extraction. These too can be processed at rural level for income generation.

13.5.10 Fiber Boards

Some parts of plants or residues after extraction of the main product could be used to produce other products such as fiber board boxes and handmade paper. Pine needles are one such source of lignocelluloses which could be converted to fiber board for use as packing material. The process for the production of fiber boards from pine needle used in the sub-Himalayan region in India is simple and can be adapted by other countries. The board is converted into packing boxes using wooden battens which are stapled with a machine specially designed for the purpose. A few perforations or slots are given to the board for fruit breathing. Fiber angles could also be used in place of wooden battens for converting the board into a box.

13.6 Factors Promoting Success

13.6.1 Innovation

Risk and vulnerability can be managed through innovation which occurs as a response to resource scarcity and market competition (Kaplinsky and Morris 2001; Schreckenberg et al. 2006).

13.6.2 Resource Scarcity

Scarcity stimulates communities to innovate techniques of improved resource management and/or domestication(Schreckenberg et al. 2006).

13.6.3 Market Competition

Value chains can be upgraded or improved through market competition (Schreckenberg et al. 2006) by:
– Process upgrading like accumulation of larger quantities of the product, introduction of new technology, and reduction of transport costs increases production efficiency.
– Product upgrading or product quality improvement and introduction of new products pro-

vide access to a more specialized market niche and protect against substitution.

- Functional upgrading is changing the mix of activities within a value chain or by an individual actor, e.g., a cooperative of producers may take on new roles of credit provision, capacity building, and marketing.
- Chain upgrading, i.e., switching over to a new value chain. Most NTFP value chains are demand driven, and switching to a new chain solely on existing supply may not ensure success.

13.6.4 Collaboration and Cooperation

Social networking can effectively enhance trust, collaboration, and cooperation between the NTFP stakeholders (Banana 1996; McLain and Jones 1997, 2001; Touchette 1998; McLain 2000; Neumann and Hirsch 2000; Scott 2000; Letchworth 2001; Smith et al. 2003; Gold et al. 2004; Lynch et al. 2004; Kelly and Bliss 2009; Nybakk et al. 2009; Sirianni 2009) and also encourages research, regulatory direction, coordinated monitoring, and sharing of market and technical information, leverages competition, enforces industry norms, and increases profits (Banana 1996; Touchette 1998; Zaheer et al. 1998; Neumann and Hirsch 2000; Letchworth 2001; Gold et al. 2004; Emery and Pierce 2005; Vaughan 2011). With advancement of electronic communication online, social networks for NTFP are developing in countries like the USA which will reduce opportunity cost by allowing self-selected asynchronous participation (Finholt et al. 1990; Wittig and Schmitz 1996; Boyd and Ellison 2007).

Involving community-level organization was instrumental for success of NTFP commercialization (Schreckenberg et al. 2006). Traditional knowledge of a product and the resource determine a community's interest and capacity to successfully commercialize an NTFP. Many community-based producer organizations also provide additional benefits by contributing funds to community projects. Clear opportunities are

recognized for enhancing women's control and obtaining greater recognition for activities through greater recognition or cooperation (Carr 2008). Collective action to pool raw materials and efforts can improve women's ability to negotiate better prices and income (Shackleton et al. 2011). Organized women collectors have better access to resources, external support, venues for selling, training, and credit opportunities (Anonymous 2000). Collective action can also reduce the need to travel, i.e., raw materials can be bulked up to sufficient quantities to make it worthwhile for buyers to travel to source (Shackleton et al. 2011) and also can joint promotion and marketing activities reducing unwanted competition for each other (Pettenella et al. 2007).

13.6.5 Entrepreneurs

Entrepreneurs are important players in a value chain as they bridge information gaps (lack of information on price, quality, quantity, and market prevents poor people to enter the NTFP trade), identify new market niches, help gain physical access to markets, provide training and information to ensure product quality, help organize communities, and advance capital to ensure consistent product supply (Schreckenberg et al. 2006). However, concerns of exploitation by these intermediaries due to their position of power were also reported (Schwartzman 1992).

13.6.6 Conducive Legislative and Policy Environment

Countries have either very little policy or legislation specific to NTFPs (as in Bolivia) or over-regulation with NTFPs (Mexico) that creates complications on its application and implementation (Schreckenberg et al. 2006). NGO support can be important to overcome such constraints, particularly if it is coordinated with relevant government programs such as grants or subsidies for specific permit or certification requirements.

13.6.7 Capacity Building

Training provides an ideal opportunity to become more active in NTFP commercialization (Carr 2008). However, training often excludes women because of their perceived minor roles or the timing of sessions to conflict with women's domestic work (Shackleton et al. 2011). While case-specific opportunities to enhance women's roles exist, there is also a need to look beyond just the product trade to understand women's role in family life and economic life in general (Esplen and Brody 2007; Carr 2008). Policies need to support women to become more economically active which may include further education and training, access to credit, and recognition of inadvertent gender-biased regulations (Shackleton et al. 2011). Generally, trainings are focused on production at one end of the chain and the export trade at the other. There has been little business skills training for people involved in activities between these two extremes. Such trainings including women, however, sometimes fail to address their basic concerns such as poor literacy, awareness of cultural and gender biases in mixed groups, or focus on building skills that will boost their confidence (Kalu and Rachael 2006; Shackleton et al. 2011). More efforts need to be made to specifically address women's concerns when designing training packages.

13.7 Conclusion

Commercialization or value addition of NTFPs has been widely promoted as an approach to rural development in tropical forest areas. Unfortunately, commercialization of NTFPs has exploited and deprived the collectors who are mainly illiterate, poverty stricken, ignorant, impoverished, and unorganized. Moreover, donor investments in the development of NTFP resources have often failed to deliver the expected benefits in terms of poverty alleviation and improved conservation of natural resources (Schreckenberg et al. 2006). Marketing of NTFPs should be developed through private sector like more liberal and open agricultural marketing.

Markets need to be developed for simple home-made processed items which will encourage the collectors to value add their collection and improve income. NTFP commercialization will be successful only when it is a transparent, equitable, and sustainable activity that has a positive impact on poverty reduction, gender equality and resource access, tenure, and management. It is more likely if producers and/or processors collaborate with one another and with socially minded entrepreneurs, if there is a realization of the need for continuous innovation, and if there is external support to enable producers and traders overcome various barriers to entry including legislative constraints, inconsistent quality and quantity of products, and lack of market information. This demands aggressive policy interventions that will strengthen the socioeconomic situation of these poor and marginalized people. These policy interventions can be improving market access through Internet-based advertisement, order processing, shipping of products, enlarging the scale of operations (through NTFP-based forest management or cultivation of NTFP species in agroforestry systems), developing local markets for raw or value-added NTFPs, and improving the bargaining power of the collectors/local marketing agents by establishing local NTFP agency (Greene et al. 2000). The effective policy interventions will improve the stake of all stakeholders involved in NTFP from collection to value addition and ultimately improve income and value addition without sacrificing the principles of sustainable forest conservation.

References

Adedayo AG, Oyun MB, Kadeba O (2010) Access of rural women to forest resources and its impact on rural household welfare in North Central Nigeria. For Policy Econ 12:439–450

Adepoju AA, Salau AS (2007) Economic valuation of non-timber forest products. MPRA paper no. 2689. MPRA. Available in http://mpra.ub.uni-muenchen.de/2689/

Alexander SJ (1999) Who, what, and why: the products, their use, and issues about management of non-timber forest products in the United States. In: Davidson-Hunt I, Duchesne LC, Zasada JC (eds) Forest communities in the third millennium: linking research, businesses, and policy toward a sustainable non-timber forest products sector. USDA Forest Service North Central Research Station, North Carolina, USA-217, pp 18–22

Alexander SJ, McLain RJ, Blatner KA (2001) Socio-economic research on non-timber forest products in the Pacific Northwest. J Sustain For 13:95–103

Alexander SJ, Weigand J, Blatner KA (2002) Non-timber forest product commerce. In: McLain RJ, Jones ET, Weigand J (eds) Non-timber forest products in the United States. University Press of Kansas, Lawrence, pp 115–150

Ambus L, Davis-Case D, Mitchell D, Tyler S (2007) Strength in diversity: market opportunities and benefits from small forest tenures. Br Columbia J Ecosyst Manage 8:88–99

Anand N (1993) Factors having a bearing on the industrial utilization of medicinal plants for the production of plant based medicines. UNIDO, Geneva

Angelsen A, Wunder S (2003) Exploring the Forest Poverty Link: key concepts, issues and research implications, CIFOR Occasional Paper No. 40. CIFOR, Bogor

Anonymous (1954) Natural plant hydrocolloids, Advances in chemistry series 11. American Chemical Society, Washington, DC

Anonymous (1975) Herbal pharmacology in Peoples Republic of China. American Chemical Society, Washington, DC

Anonymous (1986) Status report on prominent tree-borne oilseeds. Regional Research Lab, Jammu

Anonymous (1987) Documentation on forest and rights, vol I. National Center for Human Settlements and Environment, New Delhi

Anonymous (1989) Household food security and forestry: an analysis of socio-economic issues. FAO Community Forestry Note No. 1, Rome

Anonymous (1990a) The major significance of minor forest product. The local use and value of forest in the West African humid forest zone. Community Forestry Note 6. FAO, Rome

Anonymous (1990b) The World Bank in the forest sector: a global policy paper. World Resources Institute. Wasteland News 8: 6–12

Anonymous (1991a) Case studies in forest based small scale enterprises in Asia: rattan matchmaking and handicrafts. FAO Community Forestry Case Study No. 4, Bangkok

Anonymous (1991b) Woman's role in dynamic forest based small scale enterprises: case studies on uppage and lacquerware from India. FAO Community Forestry Case Study No. 3. Rome

Anonymous (1991c) Report of 2nd UNIDO workshop on essential oil industry, Manila, Philippines

Anonymous (1991d) Design options for a polyvalent pilot plant for the distillation and extraction of medicinal and aromatic plants. UNIDO, Geneva

Anonymous (1992) Role of NTFP in forest fringe dweller's economy and current status of forest flora: a case study

at Raigarh Forest Protection Committee under Bankura South Division. Ford Foundation Project Report. Rama Krishna Mission Lokashiksha Parishad, Delhi

Anonymous (1994) Development programmes on the industrial utilization of medicinal and aromatic plants. UNIDO, Geneva

Anonymous (2000) The gum collectors: struggling to survive in the dry areas of Banaskantha. Self Employed Women's Association (SEWA), Ahmadabad

Anonymous (2001) Sustainable management of non-timber forest resources. CBD Technical Series No. 6, Secretariat of the convention on biological Diversity, Montreal, Canada, 30p

Anonymous (2011) National report on sustainable forests-2010. FS-979. USDA Forest Service, Washington Office, Washington, DC, 214p

Antypas A, McLain RJ, Gilden J et al (2002) Federal non-timber forest products policy and management. In: McLain RJ, Jones ET, Weigand J (eds) Non-timber forest products in the United States. University Press of Kansas, Lawrence, pp 347–374

Appasamy P (1993) Role of non-timber forest products in a subsistence economy: the case of a joint forestry project in India. Econ Bot 47:258–267

Arnetz F (1993) Non-timber products: their sustainable development of tropical forests. Trop For Manag Update 3(2):3–7

Arnold JEM (1995) Poverty and conservation. Society and non-timber forest products in tropical Asia, Occasional Paper No. 19. East West Center, Honolulu

Arnold JEM (2002) Clarifying the links between forests and poverty reduction. Int For Rev 4:231–234

Arnold JEM, Ruiz Perez M (1998) The role of non-timber forest products in conservation and development. In: Wollenberg E, Ingles A (eds) Incomes from the forest: methods for the development and conservation of forest products for local communities. CIFOR/IUCN, Bogor, pp 17–42

Arora D (2008) California porcini: the new taxa, observations on their harvest, and the tragedy of no commons. Econ Bot 62:356–375

Atal CK, Kapoor BM (1982a) Cultivation and utilization of medicinal plants, Part I. Regional Research Laboratory, Jammu

Atal CK, Kapoor BM (1982b) Cultivation and utilization of aromatic plants, Part II. Regional Research Laboratory, Jammu

Bailey B (1999) Social and economic impacts of wild-harvested products. PhD dissertation, West Virginia University, Morgantown, WV, 103p

Balick MJ (1979) Amazonian oil palms of promise: a survey. Econ Bot 33:142–128

Banana AY (1996) Non-timber forest products marketing: field testing of the marketing information system methodology. In: Leakey RRB, Temu AB, Melnyk M, Vantomme P (eds) Domestication and commercialization of non-timber forest products in agroforestry systems. FAO, Rome, pp 218–225

Bawa KS, Godoy R (1993) Introduction to case studies from South Asia. Econ Bot 47:248–250

Belcher BM (2003) What isn't an NTFP? Int For Rev 5:161–168

Belcher BM, Ruiz-Perez M, Achdiawan R (2003) Global patterns and trends in NTFP development. Paper presented to the international conference on rural livelihoods, forests and biodiversity, Bonn, Germany, 19–23 May

Beohar B (2003) Marketing of non-wood forest products (NWFPs) in Mandla tribal district of Madhya Pradesh. Indian J Agric Mark 17:66

Bielenber C (1993) Feasibility of Jatropha oil as a substitute for diesel fuel in male. Appropriate Technology International, Washington, DC

Bih F (2006) Assessment methods for non-timber forest products in off-reserve forests: case study of Goaso district, Ghana. PhD dissertation, Faculty of Forestry and Environmental Sciences, University of Freiburg, Germany

Boyd D, Ellison N (2007) Social network sites: definition, history and scholarship. J Comput Med Commun 13:11

Browder JO (1992) Social and economic constraints on the development of market-oriented extractive reserves in Amazon rain forests. In: Nepstad DC, Schwartzman S (eds) Non-timber forest products from tropical forests. The New York Botanical Gardens, New York

Brown K, Lapuyade S (2001) A livelihood from the forest: gendered visions of social, economic and environmental change in Southern Cameroon. J Int Dev 13:1131–1149

Brown D, Williams A (2003) The case for bushmeat as a component of development policy: issues and challenges. Int For Rev 5:148–155

Burkhart EP, Jacobson MG (2009) Transitioning from wild collection to forest cultivation of indigenous medicinal forest plants in eastern North America is constrained by lack of profitability. Agrofor Syst 76:437–453

Butler BJ (2008) Family forest owners of the United States, 2006. USDA Forest Service General Technical Report NRS-27, North Research Station, Newton Square, PA, 72p

Byron N, Arnold JEM (1999) What futures for the people of the tropical forests? World Dev 27:789–805

Campbell JY, Tewari DD (1995) Increased development of non-timber forest products in India: some issues and concerns. J Sustain For 3:53–79

Carr M (2008) Gender and non-timber forest products: promoting food security and economic development. International Fund for Agricultural Development, Rome

Cavendish W (1989) Rural livelihoods and non-timber forest products. Paper presented at CIFOR workshop on the contribution of non-timber forest products to socio-economic development, Zimbabwe, October

Chabala C (2004) Forest management in Zambia: a focus on women's access to forest resources management in Chief Chiwala's area, Masaiti district. Master's thesis, Southern African Regional Center for Women's Law, University of Zimbabwe, Harare

Chamberlain JL (2002) The management of national forests of Eastern United States for non-timber forest products. PhD dissertation, Virginia Polytechnic Institute and State University, Blacksburg, VA, 250p

Chamberlain JL, Bush RJ, Hammett AL (1998) Non-timber forest products: the other forest products. For Prod J 48:2–12

Chamberlain JL, Bush RJ, Hammett AL et al (2002) Eastern national forests: managing for non-timber products. J For 100:8–14

Chamberlain JL, Mitchell D, Brigham T et al (2009) Forest farming practices. In: Garrett HE (ed) North American agroforestry: an integrated science and practice, 2nd edn. American Society of Agronomy, Madison, pp 219–255

Chambers R, Saxena NC, Tushar S (1990) To the hands of the poor: water and trees. Oxford & IBH Publishing, New Delhi, 273p

Charlie S, Sheona S (2004) The importance of non-timber forest products in rural livelihood security and as safety nets: a review of evidence from South Africa. S Afr J Sci 100:588–664

Charnley S, Fischer AP, Jones ET (2007) Integrating traditional and local ecological knowledge into forest biodiversity conservation in the Pacific Northwest. For Ecol Manage 246:14–28

Chaudhry DC (1993) Agarwood from Aquilaria malaccensis. Minor Forest Products News 3(4), CMFP, ICFRE, Dehra Dun, India

Chawla JS (1987) Utilization of forest and agro-industrial residues. Holzforschung Holzverwerchung 39:121–125

Chege N (1994) Africa's non-timber forest economy. World Watch, Washington, DC, pp 19–23

Chopra K (1993) The value of non-timber forest products: an estimation for tropical deciduous forests in India. Econ Bot 47:251–257

Chowdhuri MK, Chakraborty SM, Mukherjee AK (1992) Utilization of forest produces- a study among forest dwellers. Bull Cult Res Inst 18:44–47

Cooks ML, Wiersum KF (2003) The significance of plant diversity to rural household in the Eastern Cape Province of South Africa. For Trees Livelihood 13:39–58

Costanza R, Folke C (1997) Valuing ecosystem services with efficiency, fairness and sustainability as goals. Island Press, Washington, DC, pp 49–68

de Beer JH, McDermott MJ (1989) Economic value of non-timber forest products in Southeast Asia with emphasis on Indonesia. Malaysia and Thailand. Netherland Committee for the IUCN, Amsterdam

De Silva KTD (1982) Development of drugs from plants. Proc Sri Lanka Assoc Adv Sci 38:55–71

De Silva KTD (1993) UNIDO development programmes on industrial utilization of medicinal and aromatic plants. Acta Horticult 333:47–54

De Silva T, Atal CK (1995) Processing, refinement and value addition of non-wood forest products. Available in http://www.fao.org/docrep/v7540e/V7540e19.htm

Dix ME, Hill DB, Buck LE et al (1997) Forest farming: an agroforestry practice. National Agroforestry Center, USDA Forest Service, Rocky Mountain Station, Vancouver, 7p

Dove M (1993) A revisionist view of tropical deforestation and development. Environ Conserv 20:17–24

Dubois O (2003) Forest based poverty reduction: a brief review of facts, figures challenges and possible way forward. In: Oksanen T, Pajari B, Tuomasjukka T (eds) Forests in poverty reduction strategy: capturing the potential. EFI proceeding no. 47, Finland, pp 65–85

Duke JA (1981) Handbook of the legumes of world economic importance. Plenum Press, New York

Duryea ML (1988) Alternative enterprises for your forest land: forest grazing, Christmas trees, hunting leases, pine straw, fee fishing, and fire wood. Circular 810, Florida Cooperative Extension Service, IFAS, University of Florida, Gainesville, FL, 30p

Dwivedi AP (1993) Forests: the non-wood resources. International Book Distributors, Dehra Dun

Edwards DM (1994) Non-timber forest products and community forestry. Project report G/NUKCFP/12, 36p

Edwards DM (1996a) Non-timber forest products from Nepal: aspects of the trade in medicinal and aromatic plants. FORESC monograph 1/96, 141p

Edwards DM (1996b) Non-timber forest products and community forestry: are they compatible? Banko Jankari 6:3–8

Edwards DM (1996c) The trade in non-timber forest products in from Nepal. Mt Res Dev 16:383–394

Elias M, Carney J (2007) African Shea butter: a feminized subsidy from nature. Africa 77:37–62

Elmhirst R, Resurreccion BP (2008) Gender, environment and natural resource management: new dimensions, new debates. In: Resurreccion BP, Elmhirst R (eds) Gender and natural resource management: livelihoods, mobility and interventions. Earthscan, Sterling

Emery M (1996) Livelihood diversity: non-timber forest products in Michigan's Upper Peninsula. Northeastern Forest Experiment Station, Michigan

Emery MR (1998a) Seeing, gathering, managing special forest products and public land management. In: Proceedings of special forest product: working together in a changing world. Western Forestry & Conservation Association, Portland, Oregon, 31p

Emery MR (1998b) Invisible livelihoods: non-timber forest products in Michigan's Upper Peninsula. PhD dissertation, Rutgers, the State University of New Jersey, 291p

Emery MR (2001) Gatherers, practices and livelihood roles of non-timber forest products. Non-timber forest products, Fact sheet no. 6, 5p

Emery MR (2002) Space outside the market: implications of NTFP certification for substance use (US). In: Shanley P, Pierce AR, Laird SA, Guillen A (eds) Tapping the green market: certification and management of non-timber forest products. Earthscan Publishing Ltd, Sterling, pp 302–312

Emery M, O'halek SL (2001) Brief overview of historical non-timber forest product use in the U.S. Pacific Northwest and Upper Midwest. J Sustain For 1:25–30

Emery MR, Pierce AR (2005) Interrupting the Telos: locating subsistence in contemporary US forests. Environ Plan 37:981–993

Emery MR, Ginger C, Newman S et al (2002) Special forest products in context: gatherers and gathering in the Eastern United States, GTR-NE-306. USDA Forest Service Northern Research Station, Newtown Square, 59p

Emery MR, Martin S, Dyke A (2006) Wild harvests from Scottish woodlands: social, cultural and economic values of contemporary non-timber forest products. Forestry Commission, Edinburgh, 40p

Escobal J, Aldana U (2003) Are non-timber forest products the antidote to rainforest degradation? Brazil nut extraction in Madre De Dios, Peru. World Dev 31:1873–1887

Esplen E, Brody A (2007) Putting gender back in the picture: rethinking women's economic empowerment. BRIDGE (development-gender). University of Sussex, Brighton

Everett Y (1996a) Building capacity for non-timber forest products in the Trinity Bioregion: lessons drawn from international models. Rural Development Forestry Network Paper 20a, 13p

Everett Y (1996b) Research and activities focused on non-timber forest products in the Hayfork AMA. Watershed Research and Training Center, USDA Forest Service Pacific Southwest Research Station, California

Falconer J (1990) The major significance of "minor" forest products: examples from West Africa. Appropr Technol 17:13–16

Falconer J, Arnold JEM (1989) Household food security and forestry: an analysis of socio-economic issues, Community Forestry note 1. FAO, Rome

Fearnside PM (1989) Extractive reserves in Brazilian Amazonia: an opportunity to maintain tropical rain forest under sustainable use. Bioscience 39:387–393

Finholt T, Sproull L, Keisler S (1990) Communication and performance in ad hoc task groups. In: Galegher J, Kraut R, Egido C (eds) Intellectual team work: social and technological foundations of cooperative work. Lawrence Erlbaum Association, Hillsdale, pp 291–325

Foster S (1995) Forest pharmacy: medicinal plants in American forest. Forest History Society, Durham, 57p

Fox J (1994) Introduction: society and non-timber forest products in Asia. Soc Nat Resour 8:189–192

Ganesan B (1993) Extraction of non-timber forest products, including fodder and fuelwood in Mudumalai, India. Econ Bot 47:268–274

Gausset Q, Yago-Quattara EL, Belem B (2005) Gender and trees in Peni, South-Western Burkina Faso. Women's need, strategies and challenges. Dan J Geogr 105:67–76

Gautam KH, Watannabe T (2002) Silviculture for non-timber forest product management: challenges and opportunities for sustainable forest management. For Chronicles 78:828–830

Gbadebo JO, Gloria U (1999) The non-wood forest products of Nigeria. A report produced as output of the EC-FAO partnership program (1998–2000) – project gcp/int/679/ec

George W (1966) The commercial products of India. Today and Tomorrow Printers, New Delhi

Ghosal S (2011) Importance of non-timber forest products in native household economy. J Geogr Reg Plan 4:159–168

Godin VJ, Spensley PC (1971) Oils and oils seeds. Tropical Products Institute, London

Godoy RA, Bawa KS (1993) The economic value and sustainable harvest of plants and animals from the tropical forest: assumptions, hypotheses and methods. Econ Bot 47:215–219

Godoy RA, Wilkie D, Overman H et al (2000) Valuation of consumption and sale of forest goods from a Central American rain forest. Nature 406:62–63

Gold MA, Godsey LD, Josiah SJ (2004) Markets and marketing strategies for agroforestry specialty products in North America. In: Nair PK, Rao MR, Buck LE (eds) New vistas in agroforestry: a compendium for the 1st World Congress of Agroforestry. Kluwer Academic Publishers, Dordrecht, pp 371–384

Goldberg C (1996) From necessity, new forest industry rises. The New York Times: national report, March, 12p

Grainge M, Ahmed S (1986) Potential of neem tree for pest control and rural development. Econ Bot 40:2201–2209

Greene SH, Hammett AL, Kant S (2000) Non-timber products marketing systems and market players in Southwest Virginia: crafts, medicinal and herbal and specialty wood products. J Sustain For 11:19–39

Greenfield J, Davis J (2003) Collection to commerce: Western North Carolina non-timber forest products and their markets. A report to the USDA Forest Service, 104p

Gunatilake HM, Senaratne AH, Abeygunawardena P (1993) Role of non-timber forest products in the economy of peripheral communities of knuckles national wilderness area of Sri Lanka: a farming systems approach. Econ Bot 47:275–281

Gupta T, Guleria A (1992) Non-wood forest products in India. Oxford and IBH Publishing Co. Pvt. Ltd, New Delhi

Gupta R, Banerji P, Guleria A (1982) Tribal unrest and forestry management in Bihar, CMA monograph no. 98. Indian Institute of Management, Ahmadabad, 88p

Hammett AL, Chamberlain J (1997) Sustainable development of non-timber forest products: a case study in Southwest Virginia. In: Proceedings of Forest Products Society annual meeting, Vancouver, British Columbia

Hammett AL, Chamberlain JL (1998) Sustainable use of non-traditional forest product. Alternative forest based income opportunities. In: Jonathan SK (ed) Natural resources income opportunities on private land conference. Publication Division, University of Maryland, Hagerstown, pp 141–147

Harper RA, McClure ND, Johnson TG et al (2009) Georgia's forests, 2004. Resource Bulletin SRS-149,

USDA Forest Service, Southern Research Station, Asheville, 78p

Hasalkar S, Jadhav V (2004) Role of women in the use of non-timber forest produce: a review. J Soc Sci 8:203–206

Hegde R, Suryaprakash S, Achoth L et al (1996) Extraction of non-timber forest products in the forests of Biligiri Rangan Hills, India. 1. Contribution to rural income. Econ Bot 50:244–251

Hertog WD (1995) Trees and people in balance: forest utilization in Salyan district. SNV, Nepal

Hill AF (1951) Economic botany. McGraw Hill Book Co., Ltd., New York

Homma (AKO) 1992 The dynamics of extraction in Amazonia: a historical perspective. In: Nepstad DC, Schwartzman S (eds) Non-timber forest products from tropical forests: evaluation of a conservation and development strategy. Advances in economic botany, vol 9. New York Botanical Garden, New York, pp 23–32

Humphrey FR (1964) Occurrence and industrial production of Rutin in S. E. Australia. Econ Bot 18:195–253

Ikram H (1978) Compendium of medicinal plants. Pakistan Council of Scientific and Industrial Research, Karachi

Jahnige P (2002) The hidden bounty of the urban forest. In: McLain RJ, Jones ET, Weigand J (eds) Non-timber forest products in the United States. University of Kansan Press, Lawrence, pp 96–107

Jones ET, Lynch K (2002) The relevance of sociocultural variables to non-timber forest product research, policy, and management. In: McLain RJ, Jones ET, Weigand J (eds) Non-timber forest products in the United States. University Press of Kansas, Lawrence, pp 26–51

Kabra KN (1983) Dependence and dominance. Indian Institute of Public Administration, New Delhi

Kalu C, Rachael E (2006) Women in processing and marketing of non-timber forest products: case study of Benin City, Nigeria. J Agron 5:326–331

Kant S (1997) Integration of biodiversity conservation and economic development of local communities. J Sustain For 4:33–61

Kant S, Nautiyal JC, Berry RA (1996) Forests and economic welfare. J Econ Stud 23:31–43

Kaplinsky R, Morris M (2000) A handbook for value chain research. IDRC, Ottawa

Kaplinsky R, Morris M (2001) A handbook for value chain research. Report prepared for IDRC

Karki S (1996) Investigating non-timber forestry products opportunities in Nepal. Nepal Australia Community Forestry Project, 16p

Kassa H, Tafera B, Fitwi G (2011) Preliminary value chain analysis of gums and resins marketing in Ethiopia: issues for policy and research. Policy Brief. CIFOR, Bogor, Indonesia

Kays JS (2004) Alternative income opportunities: needs of county agents and foresters in the mid-Atlantic region. J Ext 42:2RIB6

Kelly EC, Bliss JC (2009) Healthy forests, healthy communities: an emerging paradigm for natural resource-dependent communities? Soc Nat Resour 22:519–537

Kerns BK, Liegel L, Pilz D et al (2002) Biological inventory and monitoring. In: McLain RJ, Jones ET, Weigand J (eds) Non-timber forest products in the United States. University Press of Kansas, Lawrence, pp 237–272

Khare A, Rao AVR (1993) Products of social forestry: issues, strategies and priorities. Wasteland News 6:7–17

Lemenih M, Kassa H (2010a) Opportunities and challenges for the production and marketing of gums and resins in Ethiopia. Working Paper, CIFOR, Bogor, 88p

Lemenih M, Kassa H (2010b) Socio-economic and environmental significance of dry land resources of Ethiopia and their development challenges. J Agric Dev 1:71–91

Lemenih M, Abebe T, Olsson M (2003) Gum-resins from some Acacia, Boswellia and Commiphora species and their economic contributions in Liban zone, Ethiopia. J Arid Environ 55:465–482

Letchworth B (2001) An industry of wildcrafting, gathering and harvesting NTFPs: an insider's perspective. In: Davidson-Hunt I, Duchesne LC, Zasada JC (eds) Forest communities in the third millennium: linking research, business and policy toward a sustainable non-timber forest products sector, GTR-NC-217. USDA Forest Service Northern Research Station, Newtown Square, pp 128–132

Love TJ, Jones ET (1997) Grounds for argument local understandings, science and global processes. In: Special forest products harvesting. USDA Forest Service General Technical Report, pp 163–185

Love TJ, Jones ET (2001) Why is non-timber forest product harvest an 'issue'? J Sustain For 13:105–121

Love TJ, Jones ET, Liegel L (1998) Valuing the temperate rainforest: wild mushrooming on the Olympic Peninsula Biosphere Reserve. Ambio Spec Rep 9:16–33

Lynch OJ (1992) Securing community based tenurial rights in the tropical forests of Asia. Issues in development, Report WRI, Washington

Lynch KA (2004) Workshop guide and proceedings: harvester participation in inventory and monitoring of non-timber forest products. National Commission on Science for Sustainable Forestry, Washington, DC, 151p

Lynch KA, McLain RJ (2003) Access, labor and wild floral greens management in Western Washington's forests. USDA Forest Service Pacific Northwest Research Station. GTR-PNW-585, 61p

Lynch KA, Jones ET, McLain RJ (2004) Non-timber forest product inventorying in the United States: rationale and recommendations for a participatory approach. Institute for Culture and Ecology, Portland, 50p

Maharjan MR (1994) Chiraito cultivation in community forestry. Project report B/NUKCFP/13

Malhotra KC (1992) Joint forestry in West Bengal. Paper presented at a seminar on forests for economic development and recreation. Maxmueller Bhavan, Madras

Malhotra KC, Deb D, Dutta M et al (1991) Role of non-timber forest products in village economy: a household survey in Jamboni Range, Midnapore, West Bengal. Indian Institute of Bio-social Research and Development, Kolkata

Malhotra KC, Deb D, Dutta M, Vasulu TS et al (1992) Role of non-timber forest produce in village economy. Indian Institute Bio-social Research and Development (IBRAD), Kolkata

Malla SB, Shakya PR, Rajbjandari KR et al (1995) Minor forest products of Nepal: general status and trade. Forest Resource Information System Project Paper no. 4. Forestry Sector Institutional Strengthening Program, Ministry of Forest and Soil Conservation Kathmandu, Nepal. HMGN/FINNIDA, 27p

Mallik RM (2001) Commercialization of NTFPs in Orissa: economic deprivation and benefits to primary collectors. Paper presented at South and East Asian Countries NTFP Network (SEANN) Workshop on NTFPs and biodiversity: SEANN Agenda for conservation and development in the 21st century, Manila, Philippines

Mantell CL (1949) The water soluble gums: their botany, sources and utilization. Econ Bot 3:3–31

Marshall E, Schreckenberg K, Newton AC (eds) (2006) Commercialization of non-timber forest products: factors influencing success. Lessons learned from Mexico and Bolivia and policy implications for decision-makers. UNEP-World Conservation Monitoring Centre, Cambridge

Martin G (1921) Perfumes, essential oils and fruit essences. Technical Press, Ludgate Hill

Mater C (1993) Minnesota special forest products. Minnesota Department of Natural resources, Forestry Division, Ironton

May PM (1991) Building institutions and markets for non-wood forest products from Brazilian Amazon. Unasylva 42:9–16

McLain RJ (2000) Controlling the forest understory: wild mushroom politics in central Oregon. PhD dissertation, University of Washington, Seattle, WA, 468p

McLain RJ (2008) Constructing a wild mushroom panopticon: the extension of nation-state control over the forest understory in Oregon, USA. Econ Bot 62:343–355

McLain RJ, Jones ET (1997) Challenging "community" definitions in sustainable natural resource management: the case of wild mushroom harvesting in the USA, Gatekeeper Series No. 68. International Institute for Environment and Development. Sustainable Agriculture Programme, Portland, 16p

McLain RJ, Jones ET (2001) Expanding non-timber forest product harvester/buyer participation IN Pacific Northwest policy. In: Emery MR, McLain RJ (eds) Non-timber forest products: medicinal herbs, fungi, edible fruits and nuts and other natural products from the forest. Food Products Press, New York, pp 141–161

Mclain RJ, Jones ET (2005) Non-timber forest products management on national forests in the United States, GTR-PNW-655. USDA Forest Service, Pacific Northwest Research Station, Portland, 85p

McLain RJ, McFarlane EM, Alexander SJ (2005) Commercial morel harvesters and buyers in Western Montana: an exploratory study of the 2001 harvesting season, GTR-PNW-643. USDA Forest Service, Pacific Northwest Research Station, Portland, 38p

McLain RJ, Alexander S, Jones ET (2008) Incorporating understanding of informal economic activity in natural resource and economic development policy, GTR-PNW-755. USDA Forest Service Pacific Northwest Research Station, Portland, 53p

McSweeney K (2004) Forest product sale as natural insurance: the effects of household characteristics and the nature of shock in eastern Honduras. Soc Nat Resour 17:39–56

Merlo M, Milocco E, Panting R et al (2000) Transformation of environmental recreational goods and services provided by forestry into recreational environment products. For Policy Econ 1:127–138

Mickels-Kokwe G (2006) Small-scale woodland-based enterprise with outstanding economic potential: the case of honey in Zambia. CIFOR, Bogor, 82p

Mishra M, Surayya T, Mishra R (2002) Sustainable harvesting, value addition and marketing of selected non-timber forest products: a case study of Koraput and Malkangiri districts, Orissa state. RCNAEB, Bhopal

Mitchell DA (2009) NTFP, tourism and small scale forestry: income opportunities and constraints. British Columbia Ministry for Range, Forestry Science Program and Center for Non-Timber Resources, Royal Roads University, Victoria. Land management handbook 63. www.for.gov.bc.ca/hfd/pubs/Docs/Lmh/Lmh63.htm

Morse R (2003) Preface. In: Proceedings of Hidden forest values: the first Alaska-side non-timber forest products conference and tour, Alaska Boreal Forest Council, comps. USDA Forest Service, Pacific Northwest Research Station General Technical Report. GTR-PNW-579, Portland, p ii, 150p

Morsello C (2004) Trade deals between corporations and Amazonian forest communities under common property regimes: opportunities, problems and challenges. In: Proceedings of 10th biennial conference of the International Association for the Study of Common Property (IASCP), the commons in an age of global transition: challenges, risks and opportunities, Universidad Nacional Autonoma de Mexico, Oaxaca, Mexico, 9–13 August 2004

Mugedo JZ, Waterman PG (1992) Sources of tannins. Econ Bot 46:55–63

Muir PS, Norman KN, Sikes KG (2006) Quantity and value of commercial moss harvest from forests of the Pacific Northwest and Appalachian regions of the US. Bryologist 109:197–214

Mukherjee AK (1994) India's forests: a status report: concepts, definitions, trends, controversies. Paper presented at the international workshop on India's forests management and ecological revival, New Delhi, 10–12 February

Mulenga AM, Chizhuka F (2003) Industry profile of honey in Zambia. Lusaka (Mimeograph)

Murphy IK, Bhat PR, Ravindranath NH (2005) Financial valuation of non-timber forest product flows in Uttara Kannada District, Western Ghats, Karnataka. Curr Sci 88:1573–1579

Myers N (1988) Tropical forests: much more than stocks of wood. J Trop Ecol 4:209–221

Naidu MR, Ravi Kumar KM, Murthy PSS (2003) Temporal variation in the marketing of minor forest in tribal areas of Andhra Pradesh- a case study. Indian J Agric Mark 17:39–49

Narendran K, Murthy IK, Suresh HS (2001) Non-timber forest product extraction, utilization and valuation: a case study from the Nilgiri Biosphere Reserve, Southern India. Econ Bot 55:528–538

Negi JS, Prabhakar DB, Chawla JS (1984) Whole tree utilization: pine twigs for fibre boards. Holzforschung Holz verwerschung 36:77–80

Nelson TC, Williamson MJ (1970) Decorative plants of Appalachia: a source of income. Agriculture Information Bulletin 342, USDA Forest Service, Washington, DC, 31p

Nepstad D, Schwartzman S (1992) Non-timber forest products from tropical forests. Evaluation a conservation and development strategy, Advances in economic botany. The New York Botanical Garden, New York, pp 9–12

Neumann RP, Hirsch E (2000) Commercialization of non-timber forest products: review and analysis of research. CIFOR, Bogor

Newman DH, Hammett AL III (1994) Non-timber forest products market development as a means of tropical forest protection. In: Bentley W, Gowen M (eds) Forest resources and wood-based biomass energy as rural development assets. Winrock International, Oxford and IBH Publishing CO, New Delhi, p 347

Nybakk E, Crespell PE, Lunnan A (2009) Antecedents to forest owner innovativeness: an investigation of the non-timber forest products and services sector. For Ecol Manage 257:608–618

Okafor JC (1991) Improving edible species of forest products. Unasylva 165 42:17, 165

Okafor JC, Omoraradion FI, Amaja (1994) Non-timber forest products (Nigeria). Consultancy paper prepared by the Tropical Forest Action Program (TFAP) Forest Management, Evaluation and Co-ordination Units (FORMECU) and Federal Department of Forestry (FDF) Abuja, Nigeria, 8p

Oksanen T, Mersmann C (2002) Forests in poverty reduction strategies. Draft report. PROFOR, USDA, USA

Padoch C (1992) Marketing of non-timber forest products in Western Amazonia: general observation and research priorities. In: Nepstad DC, Schwartzman S (eds) Non-timber forest products from tropical forests. Cronwell press, Trowbridge

Padoch C, de Jong W (1989) Production and profit in agroforestry: an example from the Peruvian Amazon. In: Browder JG (ed) Fragile lands of Latin America: strategies for sustainable development. Westview Press, Boulder, pp 102–113

Pal G, Jaiswal AK, Bhattacharya A (2009) An analysis of price spread in marketing of lac in Madhya Pradesh. Indian J For 32:581–584

Pal G, Jaiswal AK, Bhattacharya A (2012) Lac, plant resins and gums statistics at a glance- 2012. Indian Institute of Natural Resins and Gums, Ranchi (Communicated)

Panda SK (2013) Provision of livelihood opportunity in the fringe forest: some experiences of Tripura. Indian For 139:187–192

Pandey R (2009) Forest resource utilization by tribal community of Jaunsar, Uttarakhand. Indian For 135:655–662

Parameswarappa S (1992) Agarbathi industry in Karnataka: some thoughts on raw materials. My For 28:143–146

Peters CM (1996) The ecology and management of non-timber forest resources, World Bank technical paper no. 322. The World Bank, Washington, DC

Peters CM, Gentry AH, Mendelsohn RO (1989) Valuation of an Amazonian rainforest. Nature 339:655–656

Pettenella D, Secco L, Maso D (2007) NWPP&S marketing: lessons learned and new development paths from case studies in some European countries. Small Scale For 6:373–390

Pierce AR (1999) The challenges of certifying non-timber forest products. J For 97:34–37

Pierce AR, Emery MR (2005) The use of forests in times of crisis: ecological literacy as a safety net. For Trees Livelihoods 15:249–252

Plotkin M, Famolare L (eds) (1992) Sustainable harvest and marketing of rain forest products. Conservation International-Island Press, Washington, DC

Prakash B (2003) Marketing of minor forest produce in India: trends, constraints and prospects. Indian J Agric Mark 17:28–38

Princes LH (1983) A new oilseed crop. Econ Bot 37:478–492

Rembert DJ (1979) The indigo of commerce in colonial North America. Econ Bot 33:128

Richards DC (1992) Ecoprotection enterprise, hope hyperbole or hoax. Draft paper prepared for the Biodiversity Support Program

Richards RT (1997) What the natives know: wild mushrooms and forest health. J For 95:5–10

Rist L, Shanley P, Sunderland T, Sheil D et al (2011) The impacts of selective logging on non-timber forest products of livelihood importance. For Ecol Manage 268:57–69

Robbins CS (1998) American ginseng: the root of North America's medicinal herb trade. TRAFFIC North America, Washington, DC, 94p

Robbins P, Emery MR, Rice JL (2008) Gathering in Thoreau's backyard: non-timber forest product harvesting as a practice. Area 40:265–277

Roy SB (ed) (2003) Contemporary studies in natural resource management in India. Inter-India Publications, New Delhi

Ruiz Pérez M, Belcher B, Achdiawan R (2004) Markets drive the specialization strategies of forest peoples. Ecol Soc 9:4. Available in http://www.ecologyandsociety.org/Journal/vol9/iss2/art4

Ruiz-Perez M, Arnold JEM (1996) Current issues in non-timber forest products research. CIFOR, Indonesia

Salafsky N, Wollenberg E (2000) Linking livelihoods and conservation: a conceptual framework and scale for assessing the integration of human needs and biodiversity. World Dev 28:1421–1438

Savage M (1995) Pacific Northwest special forestry products: an industry in transition. J For 93:6–11

Saxena NC (2003) Livelihood diversification and non-timber forest products in Orissa: wider lessons on the scope for policy change? Working paper 223. Overseas Development Institute, London, 57p

Schlosser WW, Blatner KA, Chapman R (1991) Economic and marketing implications of special forest products harvest in the coastal Pacific Northwest. West J Appl For 6:67–72

Schreckenberg K, Marshall E (2006) Women and NTFPs: improving income and status? In: Marshall E, Schreckenberg K, Newton AC (eds) Commercialization of non-timber forest products: factors influencing success. UNEP-WCMC, Cambridge, 136p, Available in http://quin.unep-wcmc.org/forest/ntfp/gender.cfm?displang=eng

Schreckenberg K, Marshall E, Newton A et al (2006) Commercialization of non-timber forest products: what determines success? Forestry Briefing No. 10 ODI London

Schwartzman S (1989) Extractive reserves in the Amazon. In: Browder JG (ed) Fragile lands of Latin America: strategies for sustainable development. Westview Press, Boulder, pp 150–163

Schwartzman S (1992) Social movements and natural resource conservation in the Brazilian Amazon. In: Counsell S, Rice T (eds) The rainforest harvest: sustainable strategies for saving the tropical forests. Friends of the Earth, London, pp 207–212

Scoones I, Melnyk M, Pretty J (1992) The hidden harvest: wild foods and agricultural systems: a literature review and annotated bibliography. IIED, London

Scott J (2000) Social network analysis, 2nd edn. Sage Publications, London, 208p

Shackleton CM, Shackleton SE (2004) The importance of non-timber forest products in rural livelihood security and as safety nets: a review of evidence from South Africa. S Afr J Sci 100:658–664

Shackleton SE, Shackleton CM (2005) The contribution of marula (Sclerocarya birrea) fruit and fruit products to rural livelihoods in the Bushbuckridge district, South Africa: balancing domestic needs and commercialization. For Trees Livelihoods 15:3–24

Shackleton SE, Shackleton CM (2010) Exploring the role of wild natural resources in poverty alleviation with an emphasis on South Africa. In: Hebinck P, Shackleton CM (eds) Reforming land use and resource use in South Africa: impact on livelihoods. Taylor & Francis, Routledge

Shackleton SE, Shanley P, Ndoye O (2007) Visible but invisible: recognizing local markets for non-timber forest products. Int For Rev 9:697–712

Shackleton SE, Campbell B, Lotz-Sisitka H et al (2008) Links between the local trade in natural products, livelihoods and poverty alleviation in a semi-arid region of South Africa. World Dev 36:505–526

Shackleton S, Paumgarten F, Kassa H (2011) Opportunities for enhancing poor women's socio-economic empowerment in the value chains of three African non-timber forest products. Int For Rev 13:136–151

Shanley P, Pierce AR, Laird S et al (eds) (2002) Tapping the green market: certification and management of non-timber forest products, People and plants conservation series. Earthscan Publications Limited, London

Sharma OP (1984) Chemistry and technology of catechu and kutch. International Book Distributor, Dehra Dun

Sharma P (1996) Non-wood forest products and integrated mountain development: observations from Nepal. Business seminar on medicinal herbs, essential oils and other non-timber forest products. DEG/NGCCI, 11p

Sharma UR, Malla KJ, Uprety RK (2004) Conservation and management efforts of medicinal and aromatic plants in Nepal. Banko Jankari 14(2):3–11

Sheil D, Wunder S (2002) The value of tropical forests to local communities: complications, caveats and cautions. Conserv Ecol 6:9

Shillington LJ (2002) Non-timber forest products, gender and households in Nicaragua: a commodity chain analysis. MSc thesis, Virginia Polytechnic Institute and State University, Virginia, 103p

Shiva MP (1998) Inventory of forest resources for sustainable management and biodiversity conservation. Indus Publishing Company, New Delhi

Shrestha-Acharya R, Heinen J (2006) Emerging policy issues on non-timber forest products in Nepal. Himalaya 26:51–53

Singh B (1988) Effect of removal of pine needles from the forest floor. Indian For 114:761–769

Sirianni C (2009) The civic mission of a federal agency in the age of networked governance. Am Behav Sci 52:933–952

Smith R, Beckley T, Cameron S (2003) Building partnerships for the sustainable management of non-timber forest products. In: Proceedings of XII World Forestry Congress: Area B-Forests for the planet, paper no. 329. Quebec City, Canada, 307p

Soni PL (1991) Carbohydrates. Surya Publications, Dehra Dun

Spero V, Fleming C (2002) Case study: Rio Grande National Forest. In: McLain RJ, Jones ET, Weigand J (eds) Non-timber forest products in the United States. University of Kansas, Lawrence, pp 108–114

Stainsby M (2009) British Columbia maple syrup comes on-stream. Vancouver Sun, March 14

Strong N, Jacobson M (2006) A case for consumer-driven extension programming: agroforestry adoption potential in Pennsylvania. Agrofor Syst 68:43–52

Subedi BP (1997) Utilization of non-timber forest products: issues and strategies for environmental conservation and economic development. Workshop theme paper for the workshop on the utilization of NTFPs for environmental conservation and economic development in Nepal. Asia Network for small scale agricultural resources, Kathmandu, Nepal, 29 March 1997

Surayya T, Pethiya BP, Bhattacharya P et al (2005) Strategic role of non-wood forest products (NWFPs) and microfinance in reducing the poverty- case studies- India. Paper presented at 17th Commonwealth Forestry Conference, Colombo Sri Lanka

Tangley L (1993) Marketing biodiversity products: the Tagua initiative. Conservation International, Washington, DC

Taylor EL, Foster CD (2003) Managing your East Texas forest for the production of pine straw. Texas Coop Ext Publ 805–113, 11p

Taylor F, Mateke SM, Butterworth KJ (1996) A holistic approach to the domestication and commercialization of non-timber products. In: Leakey RRB, Temu AB, Melnyk M, Vantomme P (eds) Domestication and commercialization of non-timber forest products in agroforestry systems. FAO, Rome, pp 75–85

Tcheknavorian A, Wijesekera ROB (1982) Medicinal and aromatic plants for industrial development. UNIDO report IO 505, Geneva

Teel WS, Buck LE (1998) From wildcrafting to intentional cultivation: the potential for producing specialty forest products in agroforestry systems in temperate North America. In: Josiah SJ (ed) Proceedings of the North American conference on enterprise development through agroforestry: farming the forest for specialty products. Center for Integrated Natural Resource and Agricultural Management. Minneapolis, pp 7–24

Tewari DN (1989) Dependences of tribals on forests. Gujarat Vidyapith, Ahmadabad

Tewari DD (1994) Developing and sustaining non-timber forest products, policy issues and concerns with special reference to India. J World For Resour Manag 7:151–178

Tewari DD, Campbell JY (1995) Developing and sustaining non-timber forest products: some policy issues and concerns with special reference to India. J Sustain For 3:53–79

Thandani R (2001) International non-timber forest product issues. In: Emery MR, McLain RJ (eds) Non-timber forest products: medicinal herbs, fungi, edible fruits and nuts, and other natural products from the forest. Food Products Press, New York

Thomas MG, Schumann DR (1993) Income opportunities in special forest products: self help suggestions for rural entrepreneurs. USDA Forest Service, Agriculture Information Bulletin, 666, Washington, DC, 206p

Titus BD, Kerns BK, Cocksedge W (2004) Compatible (or co-) management of forests for timber and non-timber values. In: Proc. Can. Inst. Forestry/Institut Forestier du Canada and the Soc. Am. For. Joint 2004 Annu. Gen. Meet. Convention "One Forest under Two Flags- Une Foret Sous Deux Drapeaux", Edmonton, Alta., Canada, 2–6 October 2004. Available in http://bookstore.cfs.nrcan.gc.ca/detail_e.php?recid= 12584726

Touchette C (1998) The role and importance of regional and national trade associations in marketing. In: Josiah SJ (ed) Proceedings of the North American conference on enterprise development through agroforestry: farming the forest for specialty products, Center for

Integrated Natural Resource and Agricultural Management, University of Minnesota

Trease GE, Evans WC (1983) Pharmacognosy, 14th edn. Brown Publications, Oral Read, London

Uma Shankar K, Murali S, Uma Shaanker R et al (1996) Extraction of non-timber forest products in the forests of Biligiri Rangan Hills, India. 3. Productivity, extraction and prospects of sustainable harvest of Amla, *Phyllanthus emblica* (Euphorbiaceae). Econ Bot 50:270–279

Vance NC (1995) Medicinal plants rediscovered. J For 93:6–7

Vance NC (2002) Ecological considerations in sustainable use of wild plants. In: McLain RJ, Jones ET, Weigand J (eds) Non-timber forest products in the United States. University Press of Kansas, Lawrence, pp 151–162

Varadha RS, Swaminathan LP, Vanangamudi K (2003) Collection and marketing of non-timber forest products (NTFPs)- an economic analysis. Indian J Agric Mark 17:1–72

Vaughan RC (2011) Group analysis of collaborative conservation. MS thesis, Virginia Polytechnique Institute and State University, Blacksburg, VA, 145p

Vaughan RC, Munsell JF, Chamberlain JL (2013) Opportunities for enhancing non-timber Forest Products Management in the United States. J For 111:26–33

Von Hagen B, Weigan JF, McLain R (1996) Conservation and development of non-timber forest products in the Pacific Northwest: an annotated bibliography. General Technical Report PNW-GTR-375. USDA Forest Service Pacific Northwest Research Station, Portland, OR, 246p

Warner K (2000) Forestry and sustainable livelihoods: what part can forests and forestry play in reducing poverty? Unasylva 202:3–12

Wickens GE (1991) Management issues for development of non-timber forest products. Unasylva 42:3–8

Wijesekera ROB (1992) Practical manual on the essential oils industry. UNIDO, Geneva

Wilsey DS, Nelson KC (2008) Conceptualizing multiple non-timber forest product harvest and harvesting motivations among balsam bough pickers in northern Minnesota. Soc Nat Resour 9:812–827

Wittig MA, Schmitz J (1996) Electronic grassroots organizing. J Soc Issues 52:53–69

Wollenberg E, Ingles A (eds) (1998) Incomes from the forest. Methods for the development and conservation of forest products for local communities. CIFOR, IUCN, Bogor

Workman SW, Bannister ME, Nair PKR (2003) Agroforestry potential in the southeastern United States: perceptions of landowners and extension professionals. Agrofor Syst 59:73–83

Wunder S (2001) Poverty alleviation and tropical forests – what scope for synergies? World Dev 29:1817–1833

Wynberg R, Laird S, Shackleton S (2003) Marula commercialization for sustainable and equitable livelihoods. For Trees Livelihoods 13:203–215

Yadav G, Roy SB (1991) Significance of non-timber forest produces (NTFPs): availability and its utilization pattern in rural community of Midnapur, West Bengal. Technical paper, IBARD, Kolkata

Zaheer A, McEvily B, Perrone V (1998) Does trust matter? Exploring the effects of interorganizational and interpersonal trust on performance. Organize Sci 9:141–159

Sumit Chakravarty, Anju Puri, Nazir A. Pala,
and Gopal Shukla

Abstract

In the era of scientific advancement, technological upliftment, and
modernization, poor people of the country still depend upon wood for
their primary energy source. Up to the nineteenth century, wood was
irreplaceable as the most important fuel and raw material for construction,
agriculture, crafts, and shipbuilding. Assessments of the scope for greenhouse gases (GHG) mitigation through wood use at the different levels can
be done with a combination of approaches from different disciplines.
Treatment of carbon stored in harvested products varies among international, national, and voluntary project-based mitigation programs, and to
recognize carbon storage in wood products within international protocols
has been ongoing for the past several years. The role of carbon sinks in
harvested wood products (HWPs) is generally accepted, and one of the
themes discussed in the negotiations of a post-2012 agreement is the
possible inclusion of HWPs in accounting of CO_2 emission after 2012.
In the present paper, we are discussing about the various national and
international value-added wood products for storing or preserving carbon
for a long time and ultimately useful option for mitigating climate change.

14.1 Introduction

The earth's climate is changing, and there is a general agreement that it is largely due to the increasing emissions of greenhouse gases (GHG),
especially CO_2 from burning fossil fuels and loss
of tropical forests. Every year, 3.3 billion tonnes of
carbon are added to the atmosphere (Anonymous
2000). According to Intergovernmental Panel for
Climate Change (IPCC), many climate change

S. Chakravarty (✉) • N.A. Pala • G. Shukla
Department of Forestry, Uttar Banga
Krishi Viswavidyalaya, Pundibari 736 165,
West Bengal, India
e-mail: c_drsumit@yahoo.com

A. Puri
Barring Union Christian College,
Batala, Punjab, India

A.B. Sharangi and S. Datta (eds.), *Value Addition of Horticultural Crops: Recent Trends
and Future Directions*, DOI 10.1007/978-81-322-2262-0_14, © Springer India 2015

impacts can be reduced, delayed, or avoided through mitigation, and those efforts and investments in the next 20–30 years will have a large impact. If action is delayed, it increases the risk of more severe climate change impacts. Many forest-related activities can help to mitigate climate change. They include reducing deforestation globally and converting non-forested areas to forests, replacing fossil fuels with bioenergy, using more wood products instead of energy-intensive building materials, and managing forests so they absorb and store more carbon. The amount of CO_2 released through harvesting is small compared to what is typically experienced through forest fires and other natural disturbances. The carbon stored in wood products as an offset to emissions was shown to be significant (Lippke et al. 2010).

Once an area is harvested and then regenerated either naturally or by planting, the forest begins to store carbon again. This combination of harvest and regrowth along with the fact that most wood products have a lighter environmental footprint and store carbon for longer periods indicates that the sustainable management practices can lower GHG emissions. It is far better for the environment if the world's growing demand for harvested wood after processing or value adding such as in building, paper, and other wood products (or harvested wood product, i.e., HWP) relies on fibers from sustainably managed forests rather than turning to products that require more fossils or are from a less reliable source. Using wood products that store carbon instead of building materials that require more fossil fuel to manufacture can help slow climate change. At the end of their first life, HWPs can be easily reused, recycled, or used as a carbon neutral source of energy. As environmental awareness grows, building and other wood product professionals are realizing that wood is an excellent choice for green construction and other material designs which minimize the use of energy, water, and materials and reduce impacts on human health and the environment. Wood is light in weight, yet strong. It has excellent load-bearing and thermal properties, is easy to work with, and is well suited for large and small projects. Wood adds warmth and beauty to any building and products, enhancing

the well-being of occupants/users. Storage of carbon within wood products has far been ignored by policy analysts. The time is right, and strong opportunities exist for carbon protocols and markets for carbon credits to recognize the carbon storage benefits of HWPs (Bowyer et al. 2008).

Forest harvesting for wood products alters the natural carbon cycle between forest ecosystems and the atmosphere. On a country scale, harvesting may significantly change the net carbon sink-source balance related to a country's forest resources and wood utilization (Winjum et al. 1998). The effect of harvesting and utilization of wood products on forest carbon budget was recognized during the early 1990s (Subak et al. 1993; Anonymous 1995a). Several country-level carbon budgets for the forest sector including forest harvest for wood products for Canada (Kurz et al. 1992), Finland (Karjalainen et al. 1994, 1995; Karjalainen 1996), Germany (Burschel et al. 1993), and the United States (Heath et al. 1996; Row and Phelps 1996) have been estimated. In addition to this, a stand-level analysis incorporating carbon sequestration accounting in wood products from managed forests has also been developed (Nabuurs and Mohren 1993).

14.2 Harvested Wood Products (HWPs)

HWPs are wood-based materials harvested from forests which are used for products or to value add such as construction timber, furniture, plywood, paper, and paper-like products or for bioenergy. HWPs exclude logging residues that are left at harvest sites (Suadicani 2010). The flow chain of products based on harvested wood is illustrated in Fig. 14.1. The inflow to the stock of wood products in use is based on a domestic production chain from the forest to the consumer and import and export at different places in the chain (see figure). Import and export have become more important in the flow chain of products based on harvested wood. As time elapses, some of the wood products are taken out of use, termed here (figure) as outflow. Roughly, two-thirds of

Fig. 14.1 The flow chain of products based on harvested wood (Source: Suadicani 2010)

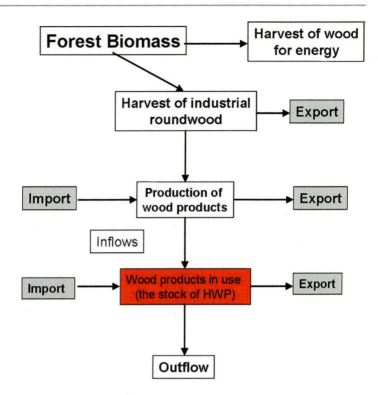

the inflow comes from solid wood products and one-third from paper and paper products, but because the lifetime for paper and paper products is so short, 94–97 % of the stock consists of solid wood products (Suadicani 2010).

Carbon leaves the forest as roundwood, fiber, and residues. Wood qualities determine the pathways and flows through sawmills and processing plants before reaching consumers as value added/ usable such as lumber, molding and millwork, specialty products like medium-density fiberboard, and paper. Many products become recycled after fulfilling their initial function which extends their utility (Mader 2007). Most of the roundwood carbon consumed in developing countries was for fuelwood and charcoal (79 %). However, the proportion varied, ranging from 69 % for Brazil to 89 % for India. In contrast, roundwood carbon consumption was predominant in industrial roundwood for developed countries. Worldwide, the portion of the total used as fuelwood and charcoal was 54 %. Of the worldwide 949 Tg carbon in net roundwood consumed, about 60 %

was within developing countries. Developing countries exported twice as much roundwood as they imported. Developed countries consumed about 40 % of the world total roundwood and imported 1.2 times as much roundwood as they exported (Winjum et al. 1998). The global production of wood waste from the conversion of industrial roundwood to commodities was estimated to be 88 Tg carbon with developed countries producing about twice as much as developing ones. Globally, the amount of wood waste accounted for 20 % of the industrial roundwood consumed with developing countries producing slightly more waste than developed countries, 24 and 19 %, respectively (Winjum et al. 1998).

14.2.1 Recognition of an HWP as Carbon Storage

Treatment of carbon stored in harvested products varies among international, national, and voluntary project-based mitigation programs (Ingerson 2011).

Efforts to recognize carbon storage in wood products within international protocols have been ongoing for the past several years (Bowyer et al. 2010). Despite these actions, there is still no agreement on language regarding HWPs, and this continued uncertainty is hindering the possibilities for wood to be a more significant part of carbon solution (Ingerson 2011). However, though carbon comprises half the mass of dry wood in long-lived wooden structures, furniture, finished goods, and a myriad of other durable wood products and also that carbon is stored for long periods within such products are recognized by global climate protocol negotiators. Further, it is being appreciated that value-added wood also consume substantially less energy than similar products from other raw materials which is translated to lower carbon emissions linked to such manufacturing.

Despite all these recognitions, wood products continue to be hampered in climate protocols by a Kyoto era default assumption that all of the carbon stored in harvested wood products manufactured in any given year is totally neglected by an equal volume of carbon released to the atmosphere through degradation of such products within the same time frame. In effect, the current policy assumption is that all of the carbon contained within trees is released at the moment of harvesting (Bowyer et al. 2010). Kyoto's clean development mechanism has so far approved only afforestation projects in the forest category with no accounting for harvested wood pools. To date, the European trading system accepts no forest offsets of any kind. The New South Wales Greenhouse Gas Abatement Scheme in Australia accepts afforestation credits but considers harvest to result in an immediate emission. In the United States, cap-and-trade bills introduced in both the house of representatives (2009) and senate (2009) explicitly listed wood products as integral to forest management offsets. Some voluntary carbon offset protocols also accept HWP carbon as a forest offset pool (Anonymous 2009a, b, 2010a). There is less unanimity regarding the potential to increase carbon sequestration by changing the management of existing forests or by increasing the storage of carbon off-site in HWPs or landfills.

Variations in treatment are not surprising as international and national climate policies are swayed by other private and public values (Ingerson 2011).

Unfortunately, carbon accounting till date could not be incorporated into climate protocols as the related issues were technical and also not based on sound research. However, still a few technical questions remain, but the primary obstacles linked with HWPs in climate negotiations are mired in international politics. In the case of internationally traded wood, what nation should get credit for carbon stored in harvested wood products, the nation in which trees were grown or the nation in which the wood is used? Will acknowledging carbon storage in wood lead to more forest harvesting? Will recognizing the longevity of stored carbon in discarded products within landfills encourage waste and discourage durability and recycling? Which nation, moreover, should be assigned to a burden or penalty when products begin to deteriorate and release carbon? These kinds of questions explain part of the reticence in dealing with the stored carbon issue (Bowyer et al. 2008). Japan, for instance, has expressed concern that it might be disadvantageous should the status quo relative to harvested wood products change (Mitchell 2003).

More fundamentally, there is no recognition at this point of the substitution effect and little likelihood that an issue like this could be effectively considered by international climate negotiators given the complicating reality of competing industries in addition to sometimes conflicting national agendas (Bowyer et al. 2008). These technical issues/questions as long as kept unresolved will jeopardize the international protocols to come into a sound resolution (Bowyer et al. 2010). Carbon benefits of wood products can be recognized by resolving the international debate to understand the magnitude of the opportunity for carbon storage in wood. Further on a state, the provincial or national level to create a stronger platform for recognition is simply an introduction of carbon tax, if not internationally. Uniform carbon tax if applied to all carbon emissions will systematically encourage all highly energy-efficient products (or at least those that are fossil fuel efficient) while discouraging products with lower energy efficiency (Sathre 2007).

14.2.2 International Carbon Protocols

Four approaches were identified to reduce forest-related CO_2 emissions (Anonymous 2001; Bowyer et al. 2008):

- Allow the growing forest to absorb CO_2 (through photosynthesis) and store it as wood in the forest.
- Harvest the forest before it burns or decomposes and store the carbon in less rapidly decomposing forest products.
- Use wood products as substitutes for aluminum, steel, concrete, brick, and other products that consume much greater quantities of fossil fuels (and release more carbon) in their manufacture.
- Establish new forests on non-forested sites.

Currently, forest-related strategies available for earning carbon credits toward compliance with the Kyoto Protocol are limited to the fourth option given above. The role of carbon sinks in HWPs is generally accepted, and one of the themes discussed in the negotiations of a post-2012 agreement is the possible inclusion of HWPs in accounting of CO_2 emission after 2012 (Hetsch 2008). Although there are many studies reporting carbon storage by HWPs (Dewar 1990; Koch 1991; Kurz et al. 1992; Burschel et al. 1993; Dixon et al. 1994; Karjalainen et al. 1994; Gjesdal et al. 1996; Harmon et al. 1996; Row and Phelps 1996; Schlamadinger and Marland 1996; Alexander 1997; Boyle and Lavkulich 1997; Micales and Skog 1997; Brown et al. 1998, 1999; Kohlmaier et al. 1998; Winjum et al. 1998; Apps et al. 1999; Skog and Nicholson 2000; Birdsey and Lewis 2003; Perez-Garcia et al. 2004; Gret-Remaney et al. 2008; Lippke et al. 2010; Stockmann et al. 2012), till now, the potential of storing carbon in wood products or of avoiding emission through product substitution (option 2

and 3 given above) is not recognized by the climate negotiators. It was earlier assumed that the fate of harvested wood once it is harvested, all of its carbon stock is released and the net stock of carbon in long-lived wood products remains unchanged (Sedjo and Amano 2006).

Fortunately, now recognizing this assumption as wrong, there are now strong initiatives on accounting for carbon storage within harvested wood products (Pingoud et al. 2003; Ruddell et al. 2007). In December 2007, the Chicago Climate Exchange (CCX) Committee on Forestry approved new protocols for carbon sequestration associated with long-lived wood products and managed forests. These efforts had helped inform international discussions about carbon storage in wood products (Bowyer et al. 2008). The carbon storage issue was taken up for discussion as a part of the United Nations Framework Convention on Climate Change (UNFCCC 2009). The carbon sink-source balance is an important component in country-level carbon budget analyses, a current activity within signatory nations to the UNFCCC (Dixon et al. 1996). Quantifying the HWP carbon pool size and understanding the significance of forests and the forest sector on forest carbon storage and GHG emission reduction are important especially among the signatory countries of UNFCCC and the Kyoto Protocol (Nabuurs and Sikkema 2001; Kurz et al. 1992; Birdsey and Lewis 2003; Heath and Skog 2004). Although there is a tentative discussion about the carbon storage under the UNFCCC, the outcome is far from certain.

Standards for forest carbon offset projects will define the carbon rights associated with wood, and that definition will in turn influence the relative emphasis on accumulating carbon in forests versus accumulating carbon in furniture, homes, and landfills or burning wood to generate energy. Since offsets under a cap-and-trade policy confer new carbon property rights on some parties, such programs should be carefully designed to meet public objectives including protecting the full suite of ecosystem services that forest provides (Ingerson 2011).

14.2.3 HWP as Carbon Storage

Wood products as wooden construction materials (single products or building components) are carbon stores, rather than carbon sinks, unlike their competitors are part of the carbon cycle giving climatic and other environmental benefits (Buchanan and Honey 1994; Karjalainen et al. 1994; Buchanan and Levine 1999; Pingoud and Perala 2000; Werner et al. 2005; Petersen and Solberg 2003, 2004; Perez-Garcia et al. 2005a; Gustavsson et al. 2006a; Lippke and Edmonds 2006; Taverna et al. 2007; Gret-Remaney et al. 2008; Upton et al. 2007; Valsta et al. 2008; Stockmann et al. 2012). Trees as they grow absorb CO_2 and store it as carbon and thus are an important carbon sink. Wood is a complex chemical compound that is about half the carbon by dry mass. This is referred to as carbon sequestration or biosequestration with specific reference to plant-based carbon storage (George 2008). This carbon remains in the wood, even after the tree is harvested and value added to manufactured products (Heath et al. 1996, 2011; Eckert 2007; Bratkovich et al. 2011). When a tree is cut down, 40–60 % of the carbon is left behind in the forest, and the rest is removed with logs which are then converted into forest/wood products (Anonymous 2007a). About half the dry weight of a wood product is carbon, and 1 tonne of carbon represents 3.67 tonnes of CO_2 (Anonymous 2006a). The amount of CO_2 stored per tonne dry wood is 1.28 tonnes for hardwood and 1.24 tonnes for softwood (George 2008). Each year, the amount of carbon stored in HWPs increases at a rate of between one-quarter and one-half of the total annual biological carbon sequestration by the world' forest (Kellomaki and Karjalainen 1996; Nabuurs 1996). The global pool of carbon in HWPs grows about 139–140 Mt/year (Brown et al. 1998; Winjum et al. 1998; Pingoud et al. 2003).

After tree harvest, carbon storage is influenced by the manufacturing, use, and ultimate disposition of forest products (Gustavsson et al. 2006b; Mader 2007). Roughly half of the biomass removed from the forests is processed into structural products for construction (lumber and plywood) and the remaining half as coproducts (wood chips, bark, sawdust, and hog fuel) used primarily in short-term applications or energy production (Bowyer et al. 2002; Manriquez 2002). Short-lived products like paper enters the waste stream quickly and decompose (at 10 % decomposition/year). These coproduct uses may involve additional processing, and energy use is assumed to carry their own share of environmental burden for those uses. Long-lived structural products for construction do not decompose if properly maintained, although there will be removals for social obsolescence (at 1 %/year). For instance, half of the 80-year-old US housing stock is still in use (Lippke et al. 2003). Thus, as more and more houses are built of wood, the carbon storage in these houses will last longer than the rotation age of that tree species, and consequently product carbon pool will accumulate from rotation to rotation. If the short-lived products are used as a biomass source for producing energy (cogeneration), net electrical energy is added to the electrical grid, displacing fossil fuels, thereby providing another source of reduced emissions. Since the substitutes for short-lived wood products are likely to consume more energy than their wood-based counterpart, the carbon stored in short-lived products would be greater than their conversion value (Bowyer et al. 2002; Manriquez 2002; Lippke et al. 2003).

Forest products such as structural lumber, furniture, and some paper products store carbon for decades. A typical 2,400-sq-ft wood-framed house contains 29 metric tonnes of carbon or the equivalent of offsetting the GHG emissions produced by driving a passenger car over 5 years, i.e., about 12,500 l of gasoline (Anonymous 2008a). The wood carbon continues to be stored through their life cycle, reused, recycled, and recovered by burning or decay. It is released only if the tree, wood, paper, or any wood product produced from a tree burns or decomposes. Recent reports estimated that the average life of wood products varies from 2 months for newspapers to 75 years for structural timber. The longer the life of these products, the greater the benefit is to the environment. Carbon accounts for about 49 % of the mass of a wood product. European wood product stock was estimated to be about

60 million tonnes (CEI-Bois Webpage 2014), and thus these global wood stocks can play a significant role in reducing climate change.

Carbon sequestration of US forests was 595 million tonnes annually (Anonymous 2008b; Skog 2008). The rate of carbon accumulation within wood products in use and in landfills was estimated at about 60–103 million metric tonnes annually which is about 15 % of that contained within standing trees and about 6 % of that in forest ecosystems. Annual additions to stocks of harvested wood products in use and within landfills were 12.4–17.0 % of annual additions to forests (Anonymous 2008b, 2009c; Heath and Smith 2004; Heath and Skog 2004). In the United States, 90 % of homes are wood-framed buildings, while wood furniture, cabinets, flooring, and trim are dominant in US homes, and consequently the carbon storage in the housing stock is increasing with the increasing number of housing units along with other uses (Koch 1991; McKeever and Phelps 1994; Skog and Nicholson 2000; Bowyer et al. 2010). In New Zealand, an increase in wood usage in building industry was reported (Baird and Chan 1983; Maugham and Clough 1986; Buchanan and Levine 1999; Honey and Buchanan 1992) and also in Finland (Pajakkala and Niemi 1989; Lehtinen 1990; Matilainen et al. 1991; Nippala and Skogberg 1991; Nippala and Jaakkonen 1993, 1996; Karjalainen et al. 1994; Haahtela and Kiiras 1995; Nippala et al. 1995; Pingoud et al. 1996; Pussinen et al. 1997; Perala and Nippala 1998; Vainio et al. 1998, 1999; Pingoud and Perala 2000; Pingoud et al. 2001). In Canada, also the use of HWPs is steadily increasing (Apps et al. 1999; Chen et al. 2008) with projection of a large long-term increase in carbon storage and atmospheric GHG reductions (Colombo et al. 2005, 2007). This scenario of wood storage in housing units in the USA and a few other countries is not a global case. Consequently, the proportion of carbon stored in wood to overall net carbon emissions is far smaller globally than in nations that produce and use large quantities of wood products. Globally, the volume of roundwood value added to finished forest products annually is about 8 % of the net additions to atmospheric carbons. Thus, the global quantity of carbon stored in long-lived forest products is roughly equivalent to 4 % of the overall net carbon emissions (Bowyer et al. 2010).

In wood processing, wood residues are created. Typically, 40–60 % of log biomass is lost to residues during processing to green rough sawn boards (Anonymous 2006a). Wood residues also contain stored carbon. The way or methods the wood residues are disposed influence the storage and duration of this carbon. Wood residues generally end up to other wood products (paper and panelboard), are burned for energy, and are burned to waste and to landfill, but only burning to waste results in no carbon saving. The release of carbon when wood is burnt to waste is treated as carbon neutral under carbon accounting rules. Carbon accounting considers that this released carbon after burnt to waste was taken up from the atmosphere initially as a normal part of the carbon cycle. This carbon is simply released after burning the previously stored carbon in that wood, rather contributing to the addition to the total amount of carbon in the carbon cycle. The majority of wood residues end up in the production of other products, storing carbon in the wood for the life of that product.

The amount of carbon stored in forest products declines gradually over time even with recycling and reusing, but much remains functional even after 100 years. Durable solid wood and particleboard products like saw wood, veneer, plywood, and structural and nonstructural panel cycle more slowly than paper products (Skog and Nicholson 2000; Penman et al. 2003). Wood residues in landfills were earlier believed to decompose and release its stored carbon rapidly, but a recent study reported otherwise that wood recovered from a landfill after 46 years decomposed very less (less than 4 %) indicating wood in landfill stores carbon for a very long time (Gardner et al. 2002). Similarly, various solid wood products also have longer carbon cycling period (Mader 2007). For greenhouse purposes, 100 years is considered as the threshold for permanency (Row and Phelps 1996; Miner 2006; Mader 2007; Ingerson 2011). Lumber, plywood, and other solid wood products can store carbon for many years ranging from 10 years for shipping pallets to 100 years or more for buildings (Skog and Nicholson 2000). Sawmill wastes are almost entirely used for paper or energy. Paper products have a relatively brief duration often releasing their carbon in less than a year, but paper is often recycled (in developed countries) reducing the carbon release as well as reducing the demand for wood from the forest. Overall, wood product harvests from natural tropical forests generally release more carbon than harvests from plantations and temperate and boreal forests do (Gorte and Ramseur 2008).

The treatment of wood-based products at their end of life has important CO_2 implications (Mader 2007). Wood-based products might be used for energy purposes, recovered for other material applications (recycled or reused), or landfilled as solid waste (Mader 2007). Wood products disposed to landfill, as influenced by their service life and storage, permanently store carbon (George 2008). The small amount of carbon stored in landfill relates to wood residues created during the production of the wood product. The remaining stored carbon is from use of wood residues for energy, displacing fossil fuel consumption. A small amount of the carbon stored in use is released over the product's life. But the majority remains stored until the end of the product's useful life. A part of wood is reused/recycled after the product's service life where carbon is stored without increasing forest consumption (Mader 2007), but the majority ends up to landfill where it is permanently stored (carbon stored for 100 years) with only minimal loss having carbon equivalent to 1.57 tonnes CO_2 per tonne of hardwood input at the start of the period (George 2008). Thus, wood products have a significant potential to reduce GHG emission. Further, this emission abatement is not constrained by land and water availability that apply to forest sequestration.

14.2.4 Benefit in Using HWPs

14.2.4.1 Lighter Footprint

Life cycle assessment studies show that wood building products and other wood products have

a lighter environmental footprint compared with alternative materials and offer environmental advantages at every stage. Wood products emit less GHG over their lifetime than building materials and other alternates of wood products such as steel, concrete, aluminum, or plastic indicating environmental benefits of wood (Anonymous 2008c). Substituting a cubic meter of wood for concrete blocks or bricks results in a significant saving of 0.75–1.0 tonne of CO_2 (Anonymous 2004). A recent life cycle assessment compared the environmental impacts of homes framed with wood, steel, and concrete and reported the production of steel- and concrete-framed homes generated 26 and 31 % more GHG emission, respectively, than their wood-framed counterparts. The production of the steel and concrete homes consumes 17 and 16 % more embodied energy and releases 14 and 23 % more air pollutants, respectively, than a wood-framed home (Bowyer et al. 2005).

14.2.4.2 Energy Efficient

Lesser energy is required to extract process and transport wood products as wood for buildings which require less energy to construct and operate over time. Imminently, less fossil fuel energy will be consumed resulting into less emission of GHG. The structure of wood which is cellular captures a lot of air in its tiny pockets that improves its natural thermal efficiency. Thus, wood can resist the flow of heat 400 and 10 times better than steel and concrete, respectively (Canadian Wood Council Webpage 2014). Steel and concrete structure needs more insulation to achieve the same thermal performance as wood framing. Wood energy in Canada has one-three thousandth the energy intensity of the cement or steel industries per unit of GDP (Office of Energy Efficiency, Canada Webpage 2002).

Paper manufacturing has long played an important role in Canada's emission reductions by using a fiber leftover from the sawmilling process that once was burned or sent to landfills. Today, the nations' pulp and paper industry has improved its environmental performance by shifting its energy needs away from fossil fuels to wood chips and residues and by using wood fiber more efficiently. Using these biomass as fuels instead of fossil fuels recycles carbon rather than introducing geologic carbon to atmosphere and thus helps to reduce the buildup of CO_2 (Miner 2007). As a result, 60 % of the industry's energy requirements are met by renewable resources. The industry emits 45 % fewer GHG now than it did 15 years ago (CEI-Bois Webpage 2014). Today, almost one-third of Canada's fiber supply for new paper comes from recovered paper, and about 85 % of the fiber used to make new paper and paperboard comes from a combination of sawmill residues and recycled paper (CEI-Bois Webpage 2014).

14.2.4.3 Substitution for Other Materials

While the carbon store effect of wood products helps keep CO_2 out of the atmosphere, an even greater carbon gain comes from the substitution effect of using wood in place of other more fossil fuel-intensive materials (Hall et al. 1991; Kohlmaier et al. 1998; Schlamadinger and Marland 1996; Marland and Schlamadinger 1995, 1997; Borjesson and Gustavsson 2000; Glover et al. 2002; Sharai-Rad and Welling 2002; Meil et al. 2004; Petersen and Solberg 2005; Gustavsson et al. 2006b; Nebel et al. 2006; Mader 2007; Seidl et al. 2007; Sathre and O'Connor 2008; Lippke et al. 2010; Pingoud et al. 2010; Ingerson 2011). Data differ according to material as well as to country (because of differences in energy sources); however, it is generally recognized that considerable CO_2 savings can be made by using wood (Anonymous 2006b) where it is appropriate instead of other materials like plastics, aluminum, steel, concrete, gypsum, and brick (Petersen et al. 2004; Gustavsson et al. 2006b; Lippke et al. 2010). The production of a cubic meter of wood releases around 1.1 tonnes less CO_2 emissions than the production of an equivalent amount of fossil fuel-intensive materials like steel, concrete, or plastics. This amount added to the 0.9 tonne of CO_2 stored in the wood means that every cubic meter of wood substituting for fossil fuel-intensive materials saves a total of 2 tonnes of CO_2 (CEI-Bois Webpage 2014). Life

cycle assessment showed that the energy-intensive alternatives to wood require as much as 250 % more energy to produce an equivalent amount of wood product and they are not renewable (Lippke et al. 2004). In order to encourage wood use to generate GHG benefits through substitution, it must replace alternative materials in the construction of final products and have lower GHG emission than those alternatives. The potential for wood to substitute for other materials will depend on the extent to which wood is already used. The GHG impact of wood product substitution also depends on the relative emissions associated with wood versus the materials it replaces (Ingerson 2011).

14.2.4.4 Durable and Adaptable

Wood products are durable and thus reduce environmental demands. Wood after decades or even centuries of use can be reused in new buildings/products that too are requiring little or no energy. Wood-framed buildings and other wood products can be easily adapted to meet new needs and extend their life. Wood residue from the production of lumber can be remanufactured into high-value composite products like medium-density fiberboard, finger-jointed lumber, and other wood-hybrid composite lumber. Wood residue is also chipped into mulch for landscaping and agricultural uses (CEI-Bois Webpage 2014).

14.2.4.5 Recycled Content

Almost one-third of the fiber supply for new paper can be derived from recovered paper, and more than four-fifths of the fiber used to make new paper and paperboard can be derived from sawmill residues and recycled paper as is done in Canada (CEI-Bois Webpage 2014). Used paper and paperboard products are prominent disposed materials. On decaying these materials release methane, an even more potent GHG than CO_2. For instance, in North American municipal landfills, used paper and paperboard products are the largest single disposed materials which are significantly used in paper recovery thus diverting these used papers from landfill lessening the environmental load.

14.3 Wood as Bioenergy

Wood was the world's main source of energy until the mid-1800s and continues to be a major source of energy in much of the developing world. But now, industrialized nations are again considering wood as a source of bioenergy. Bioenergy is clean renewable energy derived from biomass that can include forest harvesting and sawmill residues, agricultural residues, urban and industrial and organic waste, or dedicated energy crops. It is an environment-friendly and sustainable alternative to traditional energy. Bioenergy has no net GHG emission because the CO_2 produced is recycled by plants which absorb it for photosynthesis and cellular respiration.

Using biomass from wood and forest residues is a better choice for biomass than using agricultural crops for fuel. Cellulose fibers are an excellent choice for heat and electricity because they have higher efficiency than conventional agricultural feedstocks. The advantages of wood over other sources of biomass include a longer storage life and lower storage costs, higher bulk density (lower transport costs), less intensive use of water and fertilizers, and an established collective system (CEI-Bois Webpage 2014). After it has been used as a product, wood can be combusted to produce energy that can replace fossil fuels releasing the solar energy it had stored chemically without any effect on the global carbon balance. Wood as a source of energy is not only renewable but also is low in CO_2 emissions. Wood residues burnt for energy save greenhouse gas emission that otherwise would have been met with fossil fuels significantly saving emission equivalent to the greenhouse intensity of the displaced fuel (Anonymous 2005a). In addition, such savings are permanent (Miner and Lucier 2004).

14.4 Mitigate Climate Change and Use HWPs

When it comes to mitigating climate change, the world's forest is part of the solution. Wood products are an efficient way of extending the storage of the forest carbon sink. To increase their efficiency and have them play a greater role in the mitigation of climate change, we need to:

- Increase the market share of wood products through promotion and technical innovation.
- Extend the life of wood products.
- Recycle more wood products to extend their carbon storage still further.
- Recover the energy stored in wood by burning at the end of the use of the products.

Government and business leaders of other countries especially the developing countries can help by developing policies and procurement processes that encourage the use of more forest products from well-managed forests following the many jurisdictions that are taking action already in many European countries (CEI-Bois Webpage 2014):

- In many European countries, the legislation aimed at reducing GHG emissions often leads to the increased use of wood or consideration of wood as a preferred building material. Changes in national building regulations are encouraging multistory wood buildings; for example, a nine-story all-wood apartment building in England is the world's tallest wood residential structure.
- In France, the government requires that new public buildings must have at least 0.2 m^3 of wood for every 1 m^2 of floor area. This encourages designers to identify opportunities to use wood as a structural material as well as for floors, doors, molding, or other design features.
- As part of its promotion of a carbon neutral public service, the government of New Zealand is requiring that wood or wood-based products be considered as main structural materials for new government-funded buildings up to four stories.
- In Canada, the government of British Columbia and Quebec is moving to policies that will encourage the use of wood in public buildings.
- Members of the European Union have agreed on a building target to reach a 20 % share of renewable energy sources (i.e., biomass, biogas, wind, solar, hydro- and geothermal energy) in their total energy output by 2020.

A carbon accounting system has been constructed for forest management from the forest to end use markets. The system can be used by land managers to predict the impact of management treatments on carbon flows and stocks. Carbon in the forest and in products, the impact of cogeneration from biomass, and the substitution among competing product alternatives are all included in order to capture all changes in flows. The leakage of forest carbon into product streams is critical influencing carbon storage in renewable wood products. Forest intensively managed is higher in productivity, and hence, more extractable products will result in more carbon storage but only if all flows are recognized. It is the carbon stored in wood products that provides the opportunity to store more carbon and reduce emissions by using less fossil fuel-intensive products (Bowyer et al. 2002; Manriquez 2002; Lippke et al. 2003).

14.5 Carbon Trading on HWPs

Regulations have been enforced by the governments around the world to reduce GHG emissions (George 2008). National governments have supported several measures like:

– Research and development of low emission technologies
– Rebates or subsidies to encourage the adoption of low emission technologies
– Education and behavior change programs

Carbon pricing in the form of a carbon tax or carbon trading or some combination of the two that puts a price on carbon generally is being advocated and considered by many experts and stakeholders to achieve deep cut in GHG emission. Carbon taxes are administratively simple and provide certainty in terms of cost but are environmentally uncertain. It is difficult to estimate the required level of tax to achieve a desired level of emission reduction. Moreover, it accrues higher economic costs for the same level of emission reduction of carbon to carbon trading. In contrast, carbon trading accrues a lower overall cost allowing flexibility about where reductions

are achieved. This has resulted carbon trading to become a preferred policy response where compulsory reductions in emissions are considered. Key design issues relevant to the impact on wood products and their competitors (George 2008) are:

– Emission targets/carbon prices
– Emission coverage
– Permit allocation method and concessions
– Treatment of carbon stored in wood products

14.5.1 Emission Targets/Carbon Prices

Carbon prices are fixed by the interaction of supply and demand for permits where:

– Demand is influenced by the reduction target set, i.e., the larger the absolute reduction in emissions required, the higher is the carbon price.
– Supply is influenced by the cost of abatement in liable sectors and other sectors able to generate tradable offsets, i.e., the lower the abatement cost, the lower is the carbon price.

Reports indicate that carbon prices initially would be around $6–12 rising to around $28–34/$tCO_2$-e by 2030 (Anonymous 2006c).

14.5.2 Emission Coverage

Carbon trading proposals typically apply only to selected sectors of the economy or sources of emissions.

14.5.3 Permit Allocation Method and Concessions

A variety of permit allocations will involve some combination of:

– Auctioning, i.e., liable parties will pay a market price for tradable permits through auctioning
– Free allowances, i.e., permits will be allocated to affected parties at no cost

14.5.4 Treatment of Carbon Stored in HWPs

Carbon stored in wood products is no different to other forms of offsets. Unfortunately, there are no offset activities recognized conceptually under wood product storage. The need of a reliable estimate on the amount of carbon stored and workable approaches to scheme participation that avoid excessive transaction costs is the main barrier to this recognition. With improving the understanding on carbon stored in wood products, sufficiently accurate and robust approaches were developed which are gradually enabling the recognition of wood products in national accounts. TimberCAM (Timber Carbon Accounting Model) is now used to estimate carbon storage in wood products by a particular producer and to support participation in trading. However, it still requires significant information to be available to the producers about what products the wood will ultimately be used in (George 2008). Apart from equity and other broad social concerns, incorporating harvested wood carbon as a pool in forestry offset projects introduces technical complications that must be addressed to ensure accurate accounting and conservative crediting (Anonymous 2006d). Similar challenges arise for international projects under a post-Kyoto agreement, projects credited under national cap-and-trade climate policies, or projects offered through voluntary offset markets (Ingerson 2011). Issues related to the measurement challenges include:

14.5.4.1 Boundaries

A forest-based offset project is usually defined by geographic location (Ingerson 2011). Crediting carbon after it leaves that location requires the collaboration of multiple unrelated parties whose decisions affect the amount of carbon retained. Inclusion of this pool in forest offsets will require clarifying legal rights to carbon and determining when associated emissions that are not otherwise regulated under a systematic cap must be reported as part of a project.

14.5.4.2 Measurement

Protocols typically call for periodic field sampling and monitoring of wood flow for each project, but this task is much difficult than monitoring forest pools (Anonymous 2008d, 2009a, b, 2010a).

14.5.4.3 Impermanence

Emission saving should be permanent, but this requirement poses a challenge to wood products as stored carbon in it is likely to be released even if its storage life is very long. As mentioned earlier for greenhouse purposes, 100 years is considered as the threshold for permanency, so the forest owners must put systems in place to guarantee that carbon will remain stored for this period. A viable option to manage this impermanence is to manage credits in a pool. Storage in wood products is added to the pool every year and is removed from the pool after it has reached the end of its life. Credits are claimed only for the net increase in carbon storage (George 2008). Few wood products have useful lives of 100 years. Unfortunately, this long-term obligation of 100 years may discourage participation by individual producers. However, if carbon storage after the disposal of the product is taken into account, this 100-year storage for permanency period will be met for much of the wood products. This will also help the independent estimation of carbon storage duration without information on product service life and make carbon trading cheaper by lowering the transaction cost as detailed product destination information is not required. Long-term storage depends upon the behavior of consumers which is likely to change considerably over the decades normally considered a proxy for true permanence (Ingerson 2011).

14.5.4.4 Eligibility

Eligibility for wood product sequestration credits is restricted to forests that meet the criteria set out under the Kyoto Protocol and Marrakech Accords known as "Kyoto-compliant forests" or forests that have been induced by human activity on land that was not under forest on 31 December 1989. There is no such restriction under the UNFCCC which adopts comprehensive carbon accounting. Offsets are only briefly addressed in each set of proposals for a national carbon trading scheme. Thus, it is not yet clear whether harvested wood products would be eligible for crediting under either protocols (George 2008). Wood product credits are likely to be available only for some wood uses, those that are long lived. It is unlikely that wood used in paper production would be eligible, and this is a large proportion of hardwood production (George 2008).

14.6   Carbon Accounting of HWPs

The "default" approach under carbon accounting rules adopted for accounting for commitments under Kyoto protocols is to treat all carbon stored in a growing tree as released at the time the tree is harvested (Anonymous 2003a). The default assumption ignores the reality that carbon within long-lived wood products can be stored for many decades or even centuries. In terms of being able to fully evaluate the potential for wood products to be a part of the carbon solution, this default assumption eliminates a significant proportion of the wood product life cycle (Bowyer et al. 2010). However, the IPCC has approved accounting approaches that recognize that carbon remains stored in wood products after harvest (George 2008). These can be applied where the wood product pool is increasing in size, and thus, the store of carbon in wood products is increasing.

The IPCC (Anonymous 1997, 2000, 2003b, 2006e; Brown et al. 1998) has evaluated four approaches for estimating and reporting HWP carbon stock size and greenhouse gas emissions. They are:

- Stock change approach in which changes in HWP carbon pools are estimated in HWP-consuming countries regardless of where the HWPs are produced
- Production approach in which HWP pool carbon changes are attributed to the HWP-producing countries regardless of where the HWPs are used

- Atmospheric flow approach in which net emissions or removals of carbon to or from the atmosphere, respectively, are estimated within each country's national boundary
- Simple decay approach in which net emissions or removals of carbon to or from the atmosphere, respectively, are estimated and reported when but not where they occur if HWPs are traded

The stock change and production approaches focus on carbon stock changes in carbon pools, whereas the atmospheric flow and simple decay approaches estimate and report gross carbon fluxes to or from the atmosphere (Cowie et al. 2006). The simple decay approach and the production approach lead to the same national HWP carbon balance because they both attribute HWP carbon changes to HWP-producing countries. The stock change and atmospheric flow approaches also provide similar estimates as they both estimate and report HWP carbon changes in HWP-consuming countries. All four approaches will result in similar global HWP carbon stock estimates so long as all data sources are complete, accurate, and consistent.

National GHG inventories that comply with the UNFCCC may report carbon stored in HWPs following the guidelines developed by IPCC (Anonymous 2003a, 2006e). The United States Environmental Protection Agency Inventory of GHG emissions and Sinks reported the pool of carbon stored in wood products and landfills from timbers harvested in the United States by approximately 88×10^6 t of CO_2 e in 2008 (Anonymous 2010b). National reporting under UNFCCC allows voluntary reporting of carbon stored in wood product pools in accordance with IPCC methods. For instance, the increase in carbon stored in harvested wood products reduced Australia's GHG emissions by 5 million tonnes in 2005. It was estimated that the Australia's pool of wood products in service during 2005 was storing 96.5 million tonnes of carbon which is equivalent to 354 million tonnes of CO_2 (Anonymous 2006f). For national carbon accounting purposes, wood products can be

Table 14.1 Wood product pools used in Australia's national carbon accounting

Pool	Life (years)	Products
Very short-term products	3	Paper and paper products
		Softwood—pallets and cases
		Plywood—form board
Short-term products	10	Hardwood—pallets and palings
		Particleboard and medium-density fiberboard (MDF)—shop fitting, miscellaneous
		Hardboard—packaging
Medium-term products	30	Plywood—others (noise barriers)
		Particleboard and MDF—kitchen and bathroom cabinets, furniture
		Preservative-treated pine—decking and palings
		Hardwood—sleepers and miscellaneous
Long-term products	50	Preservative-treated pine—poles and roundwood
		Softwood—furniture
		Hardwood—poles, piles, and girders
Very long-term products	39	Softwood—framing, dressed products (flooring, lining, moldings)
		Cypress—green framing, dressed products (flooring, lining)
		Hardwood—green and dried framing, flooring and boards, furniture timber
		Plywood—structural, flooring, bracing, lining
		Particleboard and MDF—flooring and lining
		Hardboard—Weathertex, lining, bracing, underlay
		Preservative-treated pine—sawn structural timber

assigned to pools according to service life as is done in Australia given below in Table 14.1 (Anonymous 2006g). It is assumed that carbon stored within each pool is released at an increasing rate over the product life (George 2008).

In addition to tracking the harvested wood carbon pools in national-level GHG inventories,

it is also recommended to encourage the use of wood products as a climate change mitigation strategy (Winjum et al. 1998; Perez-Garcia et al. 2005b; Malmsheimer et al. 2008; Skog 2008; Tonosaki 2009). Unfortunately, inventories are insufficient to identify policy actions that can successfully lower emissions as they do not identify the effect of changes in one element (like harvested wood carbon) on other elements like forest carbon stocks or emissions from energy and waste sectors (Ingerson 2011). Therefore, conducting/encouraging detailed life cycle assessments (LCA) for wood products will supply information about the net impact on atmospheric GHG concentrations.

14.7 LCA of HWPs

A comprehensive life cycle assessment (LCA) for a solid wood product begins with the harvesting of trees and ends up with the disposal and decomposition of wood products made from those trees (Glover et al. 2002; Sharai-Rad and Welling 2002; Meil et al. 2004; Johnson et al. 2005; Perez-Garcia et al. 2005b; Puettmann and Wilson 2005; Gower et al. 2006; Winistorfer et al. 2005; Eriksson et al. 2007; Upton et al. 2007; Hennigar et al. 2008; Lippke et al. 2010; Puettmann et al. 2010; Ingerson 2011). Wood products have different service lives depending on their use. After reaching the end of their service life, HWPs are recycled, burned, or discarded in landfills (Chen et al. 2008). A proportion of the wood is wasted at each link in the processing chain. Waste wood may be value added or incorporated into other products, burned, or let to decompose in the landfills. Some carbon is lost before or when HWPs are placed in their end uses. It was estimated that there is 5 % construction lumber loss in the first year resulting from fitting and shaping (Kurz et al. 1992). There is 8 % loss for a solid HWP and 5 % for paper and paperboard products as they are transferred to end uses (Skog and Nicholson 2000). Lumber remains in use for a long time, whereas most paper and paper products are disposed more quickly (Chen et al. 2008).

The fate of the harvested wood at each link in the processing chain was reported in detail (Smith et al. 2006; Ingerson 2011). Transportation of wood to mills, industrial processes that transform wood into a variety of products, and delivery to customers and eventually to landfills also release GHG. LCAs of primary products for GHG emission are abundantly reported (Liski et al. 2001; Kline 2005; Milota et al. 2005; Wilson and Sakimoto 2005; Rivela et al. 2006, 2007; Bergman and Bowe 2008), whereas secondary manufacturing and later product life stages are reported scarcely (Crumpler 1996; Anonymous 1998, 2003c; Sharai-Rad and Welling 2002; Nebel et al. 2006; Ingerson 2011). Carbon is released through decomposition or combustion of waste material during the successive steps in the wood product chain—harvest, primary processing, secondary processing and construction, product use and maintenance, and finally disposal. Paper, however, is generally acknowledged to be a net emitter of GHGs (Forster et al. 2007; Skog et al. 2008; Zhang et al. 2008), thus not considered for LCA (Ingerson 2011).

It was estimated that there are 30 % round-wood logging wastes in the whole United States (3–84 % for the state level) (Anonymous 2007b, 2008e). Roots, stumps, and small limbs left on-site which constitute 19 % of the total tree volume (Li et al. 2003) are generally not considered as logging residues. If these wastes are considered, the quantity of logging waste increases up to 40 % of standing tree volume (Ingerson 2011). Most logging wastes release their carbon within a few years (Zhang et al. 2008). The saw-log portion of the harvested wood gives long-term carbon storage benefits, whereas wood used as fuel or converted to pulp loses its carbon rapidly. Harvested wood used for fuel and pulp contributes 24 % of the original tree volume (Ingerson 2011). The remaining portion of the harvested wood is converted by milling at a sawmill or other primary processing facilities into long-lived products. Primary mill waste varies considerably depending on the product and equipment used. Processing residues from primary mills may be burned on-site for energy, used to make pelleted wood fuel, converted to

structural panels or paper, or dumped or land-filled (Ingerson 2011). Wood leaving after the primary mill as lumber or panels may be processed to secondary products like furniture, flooring, and windows. Raw lumber that goes directly to the construction site will generate smaller amount of scrap ranging between 1 and 5 % (Anonymous 1995b, 2005b; Wilson and Boehland 2005).

Normally, the same wood material doesn't produce both secondary processing and construction wastes. Structural wood products are generally primary processed only. Secondary products like windows, doors, cabinets, and furniture are installed at the construction site without further losses (McKeever 2002). Wood after processing or after value addition begins to lose carbon, i.e., some of them are disposed gradually. If these value-added or processed wood are to contribute in the mitigation of climate change, the best option is to extend their service period as longer as possible. It was reported that in the USA, about three-fifths of primary solid wood products are processed to long-lived products such as building products or furniture (Smith et al. 2006). Shorter-lived solid wood products are pallets, shipping containers, and miscellaneous manufacturing like matches, Popsicle sticks, and toothpicks (Ingerson 2011).

The service life of value-added or processed wood products is uncertain, and permanency for them is considered generally up to 100 years (Miner 2006). Half-life is the time the HWP takes one-half of the carbon in it to be removed from use (Skog and Nicholson 2000; Anonymous 2003b, 2006e; Woodbury et al. 2007). The half-life for major HWP end use categories ranges from 67 to 100 years for construction lumber and from 1 to 6 years for paper and paper products (Kurz et al. 1992; Winjum et al. 1998; Skog and Nicholson 2000; Anonymous 2003b, 2007c; Skog et al. 2004). By year 100, 1 % of carbon is predicted to remain in this HWP category because it has few long-lived products like books and construction papers (Chen et al. 2008). As HWP carbon leaves the in-use category, it is relocated to one of the landfill, energy, or emissions (Kurz et al. 1992). Recycling extends the lifetime of

HWPs especially for paper products (Pingoud et al. 2003).

Disposed wood products with wood waste from mills and construction sites continue to store carbon over a considerable period of time in the landfills (Kurz et al. 1992; Doorn and Barlaz 1995; Barlaz 1998; Micales and Skog 1997; Borjesson and Gustavsson 2000; Pingoud et al. 2003; Anonymous 2006e; Ximenes et al. 2008). Commonly, wood waste in some countries ends up in open-air disposal sites producing less methane but releases carbon relatively quicker (Eleazer et al. 1997; Barlaz 1998). However, wood waste that is burned for energy rather than that ends up in landfills releases its CO_2 immediately and thus doesn't contribute to long-term carbon storage. Bioenergy production from all types of HWPs has also increased (High and Skog 1990; Huber et al. 2005; Zerbe 2006). In addition to this, processing and transport of wood products require energy primarily through fossil fuels that emit GHG of CO_2e amounting to the total of about 3.70 considering all the links in the processing chain (Liski et al. 2001; Sharai-Rad and Welling 2002; Meil et al. 2004; Johnson et al. 2005; Kline 2005; Milota et al. 2005; Wilson and Sakimoto 2005; Winistorfer et al. 2005; Anonymous 2006h; Gower et al. 2006; Nebel et al. 2006; Rivela et al. 2007; Bergman and Bowe 2008; Ingerson 2011).

14.8 Conclusion

Storage of carbon in HWPs needs close scrutiny before it is universally accepted as a climate mitigation strategy. All HWPs may not help build long-lived carbon stores, while high-value addition processing and transport emissions may undermine any gains of emission reductions from HWPs. International efforts have led to widely supported strategies for accounting HWP carbon pool within national-level greenhouse gas inventories. Unfortunately, inventories are yet insufficient to identify policy actions that can lower total GHG emissions. Thus, detailed and systematic LCA exercises are to be supported and encouraged for generating this information. Although encouraging more HWP use and subsequently its

disposal to landfill for accumulating more carbon is compatible with increasing carbon in standing trees/forests, there is some trade-off between these two approaches of reduction mitigation strategy. The treatment of HWP carbon pool in offsets and other mitigation activities will help tip the balance between these two approaches.

Further, policy makers should consider the full range of public costs and benefits that flow from each approach (Ingerson 2011). Strong emphasis should be given on efficiency and product longevity rather than simply on increased volume. It should be clearly understood that encouraging the use of HWPs will need more processing and value addition that will rely on continued fossil fuel inputs and more space for housing and landfills. This will ultimately destroy the remaining carbon-fixing vegetation. National forest carbon policies must balance multiple interests and multiple forest values. Climate policy developers must proceed with caution, continuously evaluate results, accommodate emerging science, and modify policies to best satisfy the climate mitigation potential of the forest sector. Therefore, both positive and negative climate consequences must be considered while encouraging the universal use of HWPs as a climate mitigation strategy.

References

Alexander M (1997) Estimation of National carbon stocks and fluxes of wood based products. MSc dissertation, University of Surrey

Anonymous (1995a) IPCC guidelines for national greenhouse gas inventories, volumes 1, 2 and 3. IPCC Secretariat, Geneva, Switzerland

Anonymous (1995b) Residential construction waste management: demonstration and evaluation. Prepared for US Environmental Protection Agency Office of Solid Waste, Washington DC. National Association of Home Builders Research Center, Upper Marlboro, MD. Available in www.toolbase.org/PDF/CaseStudies/resi_constr_waste_manage_demo_eval.pdf

Anonymous (1997) Revised IPCC 1996 guidelines for national greenhouse gas inventories, vol. 3. Greenhouse gas inventory reference manual. Chapter 6. Waste. IPCC, Bracknell

Anonymous (1998) Estimating emissions from generation and combustion of 'waste' wood. Wood Waste and Furniture Emissions Task Force. Division of Air Quality,

North Carolina Department of Environment and Natural Resources

Anonymous (2000) IPCC good practice guidance and uncertainty management in national greenhouse gas inventories. UN Intergovernmental Panel on Climate Change and Institute for Global Environmental Strategies, Hayama, Japan

Anonymous (2001) Forest products conservation and recycling review, vol 13, no. 2/3. USDA Forest Service. http://www.fpl.fs.fed.us/tmu/resources/documents/nltr/nltr0301.htm

Anonymous (2003a) Good practice guidance for land use, land-use change and forestry. Institute for Global Environmental Strategies (IGES) for the IPCC, Japan. Available in http://www.ipcc-nggip.iges.or.jp/public/gpglulucf/gpglulucf_contents.htm

Anonymous (2003b) Wood waste recycling in furniture manufacturing- a good practice guide. Waste and Resources Action Program. BFM Ltd. Banbury, Oxon, UK. Available in http://www.bfmenvironment.co.uk/images/wood%20waste%20recycling.pdf

Anonymous (2003c) Estimation, reporting and accounting of harvested wood products. Tech. Pap. Distrib. Gen. FCCC/TP/2003/7. UNFCCC, UN office at Geneva, Geneva, Switzerland. Available in unfccc.int/resource/docs/tp/tp0307.pdf

Anonymous (2004) Using wood products to mitigate climate change. International Institute for Environment and Development

Anonymous (2005a) Full fuel cycle emission factors for electricity end use from Australian Greenhouse office. AGO factors and methods workbook, pp 43–45. Available in http://www.greenhouse.gov.au/workbook/pubs/workbook-2005.pdf

Anonymous (2005b) Residential C&D waste study. Houston-Galveston Area Council and Texas Commission on Environmental Quality. Houston Advanced Research Center. Available in http://www.recyclecddebris.com/rCDd/Resources/WasteStudy

Anonymous (2006a) Forests, wood and Australia's carbon balance. Forest and Wood Products Research and Development Corporation and the Cooperative Research Center for Greenhouse Accounting. Available in www.fwprdc.org.au

Anonymous (2006b) Energy and greenhouse gas impacts of substituting wood products for non-wood alternatives in residential construction in the United States. Technical Bulletin No. 0925. National Council for Air and Stream Improvement, Inc. Research Triangle Park, NC

Anonymous (2006c) Possible design for a National GHG emissions trading scheme. National Emissions Trading Taskforce, Australia. Available in www.emissionstrading.nsw.gov.au

Anonymous (2006d) International Standard ISO 14064-2, First edition. Greenhouse gases Part 2. Specification with guidance at the project level for quantification, monitoring and reporting of greenhouse gas emission reductions or removal enhancements. International Standards Organization

Anonymous (2006e) IPCC guidelines for national green house gas inventory. Institute for Global Environmental Strategies (IGES) for the IPCC, Japan

Anonymous (2006f) National inventory report 2005, vol 2. Australian Greenhouse Office, 125p. Available in http://www.greenhouse.gov.au/inventory/2005/national-report.html

Anonymous (2006g) Australian methodology for the estimation of GHG emissions and sinks 2005—land use, land use change and forestry. Australian Greenhouse Office, pp 127–128. Available in http://www.greenhouse.gov.au/inventory/methodology/index.html

Anonymous (2006h) Solid waste management and green house gases: a life-cycle assessment of emissions and sinks, 3rd edn. EPA 530-R-02-006. US EPA, Washington, DC. Available in http://epa.gov/climatechange/wycd/waste/SWMGHGreport.html

Anonymous (2007a) Does harvesting in Canada's forests contribute to climate change? Canadian Forest Service. Available in http://cfs.nrcan.gc.ca/news/473

Anonymous (2007b) Timber product output. USDA Forest Service, Washington DC. Available in http://ncrs2.fs.fed.us/4801/fiadb/rpa_tpo/wc_rpa_tpo.ASP

Anonymous (2007c) Inventory of US greenhouse gas emissions and sinks: 1990–2005. Publ. EPA 430-R-07-002. US EPA, Washington, DC

Anonymous (2008a) Wood and climate change. FP Innovations, Canterbury

Anonymous (2008b) Carbon stocks and stock changes in US forests. In: US agriculture and forestry greenhouse gas inventory: 1995–2005. Technical Bulletin 1921. United States Department of Agriculture. http://www.usda.gov/oce/climate_change/AFGG_Inventory/4_Forest.pdf

Anonymous (2008c) A synthesis of research on wood products and GHG impacts. FP Innovations, Canterbury

Anonymous (2008d) Guidance for agriculture, forestry and other land use projects. Voluntary carbon standards. Available in http://www.v-c-s.org/docs/Guidance%20for%20AFOLU%20Projects.pdf

Anonymous (2008e) Forest resources of the United States, 2007. WO-xxx. USDA Forest Service, Washington, DC. Available in http://fia.fs.fed.us/programfeatures/rpa/

Anonymous (2009a) Chicago climate exchange offset project protocol: forestry carbon sequestration projects. Chicago Climate Exchange. Available in http://chicagoclimatex.com/docs/offsets/CCX_Forestry_Sequestration_Protocol_Final.pdf

Anonymous (2009b) Forest project protocol version 3.1. Climate Action Reserve. Available in http://www.climateactionreserve.org/how/protocols/adopted/forest/current

Anonymous (2009c) Inventory of US Greenhouse gas emissions and sinks: 1990–2007. EPA Publication 430-R-09-004. United States Environmental Protection Agency. http://www.epa.gov/climatechange/emissions/downloads09/GHG2007entire_report-508.pdfl

Anonymous (2010a) American carbon registry forest carbon project standard version 2.0. Winrock International. Available in http://www.americancarbonregistry.org/

carbon-accounting/ACR%20Forest%20Carbon%20 Project%20Standard%20v2.0%20-%20Public%20 Comment%20Draft%20021910.pdf

Anonymous (2010b) Inventory of US green house gas emissions and sinks, 1990–2008. US EPA, Washington, DC. Available in http://www.epa.gov/climatechange/ emissions/usinventoryreport.html

Apps MJ, Kurz WA, Beukema SJ et al (1999) Carbon budget of the Canadian forest product sector. Environ Sci Policy 2:25–41

Baird G, Chan SA (1983) Energy costs of houses and light construction buildings. Report No. 76. New Zealand Energy Research and Development Committee, New Zealand

Barlaz MA (1998) Carbon storage during biodegradation of municipal solid waste components in laboratory-scale landfills. Global Biogeochem Cycles 12:373–380

Bergman RD, Bowe SA (2008) Environmental impact of producing hardwood lumber using life-cycle inventory. Wood Fiber Sci 40:448–458

Birdsey R, Lewis G (2003) Carbon in US forests and wood products, 1987–1997: state by state estimates, General Technical Report GTR-NE-310. USDA-Forest Service. http://www.fs.fed.us/ne/newtown_ square/publications/technical_reports/pdfs/2003/ gtrne310.pdf

Borjesson P, Gustavsson L (2000) Green house gas balances in building construction: wood versus concrete from life-cycle and forest land-use perspectives. Energy Policy 28:575–588

Bowyer J, Briggs D, Lippke B (2002) Life cycle environmental performance of renewable industrial materials: CORRIM Phase I Interim Research Report. CORRIM Inc. Box 352100, Seattle, WA 98195, 400p

Bowyer J, Briggs D, Lippke B et al (2005) Life cycle environmental performance of renewable materials in the context of residential building construction. Consortium for Research on Renewable Industrial Materials (CORRIM), Seattle. www.corrim.org/ reports/phase1/, 60 pp þ15 chapter modules of approximately 1300 pp

Bowyer J, Bratkovich S, Lindburg A (2008) Wood products and carbon protocols carbon storage and low energy intensity should be considered. Dovetail Partners Inc. www.dovetailinc.org

Bowyer J, Bratkovich S, Howe J et al (2010) Recognition of carbon storage in harvested wood products: a post-Copenhagen update. Dovetail Partners Inc. www. dovetailinc.org

Boyle CA, Lavkulich (1997) Carbon pool dynamics in the lower Fraser basin from 1827 to 1990. Environ Sci Policy 2:25–41

Bratkovich S, Sherrill S, Howe J (2011) Carbon and carbon dioxide equivalent sequestration in urban forest products. Dovetail Partners Inc. www.dovetailinc.org

Brown S, Lim B, Schlamadinger B (eds) (1998) Evaluating approaches for estimating net emissions of carbon dioxide from forest harvesting and wood products. IPCC/OECD/IEA Program on National Greenhouse Gas Inventories. Meeting report, 5–7 May 1998, Dakar, Senegal. IPCC, Paris

Brown S, Lim B, Schlamadinger B (1999) Evaluating approaches for estimating net emission of carbon dioxide from forest harvesting and wood products. Meeting report, Dakar, Senegal 5–7 May 1998. IPCC/ OECD/IEA Program on National Greenhouse gas inventories. Organization for Economic Cooperation and Development, Paris, France, 55p

Buchanan AH, Honey BG (1994) Energy and carbon dioxide implications of building construction. Energy Build 20:205–217

Buchanan AH, Levine SB (1999) Wood-based building materials and atmospheric carbon emissions. Environ Sci Policy 2:427–437

Burschel P, Kursten E, Larson BC (1993) Present role of German forests and forestry in the national carbon budget and options to its increase. Water Air Soil Pollut 70:325–340

Canadian Wood Council Webpage (2014) Embodied energy of wood products. Canadian Wood Council. http://www.cwc.ca/NR/rdonlyres/FD8693D4-C735-44CA-959C-178D43FE092A/0/Quickfacts_ Sustainable_Building_Series_05.pdf

CEI-Bois Webpage (2014) Tackle climate change, use wood. www.cei-bois.org

Chen J, Colombo SJ, Ter-Mikaelian TT (2008) Future carbon storage in harvested wood products from Ontario's Crown forests. Can J For Res 38:1947–1958

Colombo SJ, Parker WC, Luckai N et al (2005) The effects of forest management on carbon storage in Ontario's forests. Climate Change Research Report CCRR-03. Ontario Ministry of Natural Resources, Applied Research and Development Branch, Sault Ste. Marie, ON

Colombo SJ, Chen J, Ter-Mikaelian M (2007) Carbon storage in Ontario's forests 2000–2100. Climate Change Research Information Note CCRN-06.Ontario Ministry of Natural Resources, Applied Research and Development Branch, Sault Ste. Marie, ON

Cowie A, Pingoud K, Schlamadinger B (2006) Stock changes or fluxes? Resolving terminological confusion in the debate on land-use change and forestry. Clim Policy 6:161–179

Crumpler P (1996) Industrial wood waste. In the source. Georgia Department of Natural Resources, Pollution prevention assistance division. Available in http:// www.p2ad.org/documents/tips/industww.html

Dewar RC (1990) A model of carbon storage in forests and forest products. Tree Physiol 6:416–428

Dixon RK, Brown S, Houghton RA (1994) Carbon pools and flux of global forest ecosystems. Science 263:185–190

Dixon RK et al (1996) Green house gas mitigation strategies: preliminary results from the US country studies program. Ambio 25:26–32

Doorn MRJ, Barlaz MA (1995) Estimate of global methane emissions from landfills and open dumps. Publ. EPA-600/R-95-019. US EPA, Office of Research and Development, Washington, DC

Eckert PJ (2007) Carbon sequestration in Sierra Pacific Industries forests: a watershed example. Tetra Tech EC, Inc., Bothell

Eleazer WE, Odle WS, Wang YS (1997) Biodegradability of municipal solid waste components in laboratory-scale landfills. Environ Sci Technol 31:911–917

Eriksson E, Gillespie AR, Gustavsson L (2007) Integrated carbon analysis of forest management practices and wood substitution. Can J Forest Res 37:671–681

Forster P, Ramaswamy V, Artaxo P, Bernsten T, Betts R, Fahey DW, Haywood J, Lean J, Lowe DC, Myhre G, Nganga J, Prinn R, Raga G, Schulz M, Van Dorland R (2007) Changes in atmospheric constituents and in radiative forcing. In: Solomon S, Qin D, Manning M, Chen Z, Marquis M, Avery KB, Tignor M, Miller HL (eds) Climate change 2007: the physical science basis, Contribution of Working Group I to the 4th assessment report of the IPCC. Cambridge University Press, Cambridge, pp 129–234

Gardner W, Ximenes F, Cowie A (2002) Decomposition of wood products in the Lucas heights landfill facility. In: Proceedings of the third Australian conference on life cycle assessment. Broadbeach, Queensland. Available in www.env.gov.bc.ca/epd/codes/landfill…/inventory_ggg_landfills.pdf

George A (2008) Impact of carbon trading on wood products. Project no. PR07.1059. Forest and Wood Products Australia Ltd., Victoria 8005, Melbourne. Available in www.fwpa.com.au

Gjesdal SFT, Flugsrud K, Mykkelbost TC (1996) A balance of use of wood products in Norway. Report 96:04. Norwegian Pollution Control Authority SFT, 54p

Glover J, White DO, Langrish TAG (2002) Wood versus concrete and steel in house construction: a life cycle assessment. J For 100:34–41

Gorte RW, Ramseur JL (2008) Forest carbon markets: potential and drawbacks. CRS report for Congress 7-5700. Congressional Research Service. Available in www.crs.gov

Gower ST, McKeon-Ruedifer A, Reitter A (2006) Following the paper trial: the impact of magazine and dimensional lumber production on green house gas emissions: a case study. The H. John Heinz III Center for Science, Economics and the Environment, Washington, DC

Gret-Remaney A, Hendrick E, Hetsch S (2008) Challenges and opportunities of accounting for harvested wood products. Background paper to the workshop on harvested wood products in the context of climate change policies, Geneva, Switzerland, 9–10 September 2008. Available in http://www.unece.org/timber/workshops/2008/hwp/HWP_Background_Paper.pdf

Gustavsson L, Madlener R, Hoen HF (2006a) The role of wood material for green house gas mitigation. Mitig Adapt Strateg Glob Chang 11:1097–1127

Gustavsson L, Pingoud K, Sathre R (2006b) Carbon dioxide balance of wood substitution: comparing concrete and wood framed buildings. Mitig Adapt Strateg Glob Chang 11:667–691

Haahtela Y, Kiiras J (1995) Talonrakennusten Kustannustieto 1995 (cost information of building construction 1995, in Finnish). Rakennustietosaatio, Helsinki, 539p

Hall DO, Mynick HE, Wiliams RH (1991) Cooling the greenhouse with bioenergy. Nature 353:11–12

Harmon ME, Harmon JE, Ferrell WK (1996) Modeling carbon stores in Oregon and Washington forest products: 1900–1992. Climate Change 33:521–550

Heath L, Skog K (2004) Criterion 5, indicator 28: contribution of forest products to the global carbon budget. In: Darr D, coord. Data report- a supplement to the national report on sustainable forests- 2003. FS-766A. US Department of Agriculture, Washington, DC, 10p http://www.fs.fed.us/research/sustain/contents.htm

Heath L, Smith J (2004) Criterion 5, indicator 27: contribution of forest ecosystems to the total global carbon budget, including absorption and release of carbon. In: Darr, D., coord. Data report- a supplement to the national report on sustainable forests- 2003. FS-766A. US Department of Agriculture, Washington, DC, 7p. http://www.fs.fed.us/research/sustain/contents.htm

Heath LS, Birdsey RA, Row C (1996) Carbon pools and flux in US forest products. In: Apps MJ, Price DT (eds) Forest ecosystems, forest management and the global carbon cycle, NATO ASI series. Springer, Berlin, pp 271–278

Heath L, Smith J, Skog K (2011) Managed forest carbon estimates for the US greenhouse gas inventory, 1990–2008. J For 109:167–173

Hennigar CR, MacLean DA, Amos-Binks LJ (2008) A novel approach to optimize management strategies for carbon stored in both forests and wood products. For Ecol Manag 256:786–797

Hetsch S (2008) Workshop on harvested wood products in the context of climate change policies. Available in http://www.unece.org/timber/workshops/2008/hwp/Proceedings_28Oct08.pdf

High C, Skog K (1990) Current and projected wood energy consumption in the United States. In: Klass DL (ed) Energy from biomass and waste 23, Proceedings of IGT's conference, 13–17 February 1989, New Orleans LA, Institute of Gas Technology, Chicago, III, pp 229–260

Honey BG, Buchanan AH (1992) Environmental impacts of the New Zealand building industry, Civil Engineering Research Report 92-2. University of Canterbury, Christchurch, New Zealand

Huber GW, Chheda JN, Barrett CJ (2005) Production of liquid alkanes by aqueous-phase processing of biomass-derived carbohydrates. Science (Washington DC) 308:1446–1450

Ingerson A (2011) Carbon storage potential of harvested wood: summary and policy implications. Mitig Adapt Strateg Glob Chang 16:307–323

Johnson L, Lippke B, Marshall JD (2005) Life-cycle impacts of forest resource activities in the Pacific Northwest and Southeast United States. Wood Fiber Sci 37:30–46

Karjalainen T (1996) Dynamics of the carbon flow through forest ecosystem and the potential of carbon sequestration in forests and wood products in Finland.

Research Notes, Faculty of Forestry, University of Joensuu, Finland, 31p

Karjalainen T, Kellomaki S, Pussinen A (1994) Role of wood-based products in absorbing atmospheric carbon. Silva Fennica 28:67–80

Karjalainen T, Kellomaki S, Pussinen A (1995) Carbon balance in the forest sector in Finland during 1990–2039. Climate Change 30:451–478

Kellomaki S, Karjalainen T (1996) Sequestration of carbon in the Finnish boreal forest ecosystem managed for timber production. In: Apps MJ, Price DT (eds) Forest ecosystems, forest management and the global carbon cycle. Springer, Berlin, pp 173–182

Kline D (2005) Gate-to- gate lifecycle inventory of oriented strandboard production. Wood Fiber Sci 37:74–84

Koch P (1991) Wood vs. non-wood materials in US residential construction: some energy related international implications. CINTRAFOR working paper 36. College of Forest Resources, University of Washington, Seattle WA

Kohlmaier GH, Weber M, Houghton RA (1998) Carbon dioxide mitigation in forestry and wood industry. Springer, Berlin

Kurz WA, Apps MJ, Webb TM (1992) The carbon budget of the Canadian forest sector: Phase I. Inf. Rep. NOR-X-326. Forestry Canada, Edmonton, 93p

Lehtinen E (1990) The Koralli computational system for forecasting the quantity of building component renovations, vol 1. The building market. CIB conference, Building economics and Urban Development, Sydney, Australia, pp 159–170

Li Z, Kurz WA, Apps MJ (2003) Belowground biomass dynamics in the carbon budget model of the Canadian Forest Sector: recent improvements and implications for the estimation of NPP and NEP. Can J For Res 33:126–136

Lippke B, Edmonds L (2006) Environmental performance improvement in residential construction: the impact of products, biofuels and processes. For Prod J 56:58–63

Lippke B, Perez-Garcia J, Manriquez C (2003) Executive summary: the impact of forests and forest management on carbon storage, Rural Technological Initiative, College of Forest Resources. Box 352100. University of Washington, Seattle

Lippke B, Wilson J, Perez-Garcia J, Bowyer J, Meil J (2004) CORRIM: life-cycle environmental performance of renewable building materials. For Prod J 54:8–19

Lippke B, Wilson J, Meil J (2010) Characterizing the importance of carbon stored in wood products. Wood Fiber Sci 42:5–14

Liski J, Pussinen A, Pingoud K (2001) Which rotation length is favourable to carbon sequestration? Can J For Res 33:126–136

Mader S (2007) Climate project: carbon sequestration and storage by California forests and forest products. Technical memorandum. CH2M Hill Inc.

Malmsheimer RW, Heffernan P, Brink S (2008) Forest management solutions for mitigating climate change in the United States. J For 106(3):115–173, Special Issue

Manriquez C (2002) Carbon sequestration in the Pacific Northwest: a model. Master's thesis. University of Washington, Seattle, Washington, 158p

Marland G, Schlamadinger B (1995) Biomass fuels and forest management strategies: how do we calculate the greenhouse gas emission benefits? Energy 20:1131–1140

Marland G, Schlamadinger B (1997) Forests for carbon sequestration or fossil fuel substitution? A sensitivity analysis. Biomass Bioenergy 13:389–397

Matilainen J, Lehtinen E, Vainio T (1991) Korjausrakentaminen 1990, Osa 2: Korjausten syyt. (Repair and renovation work in the building sector. (Part 2: Reasons for repair and renovation, in Finnish). VTT Research notes 1300. Espoo, Kluwer Academic Publishers, 54p

Maugham CW, Clough PWJ (1986) The market for sawn timber and panel products (exotic softwoods) in New Zealand, Policy Paper no. 12. Centre for Agricultural Policy Studies, Massey University, New Zealand

McKeever DB (2002) Inventories of woody residues and solid wood waste in the United States, 2002. USDA Forest Service, Forest Products Laboratory, Madison, WI. Available in http://www.fpl.fs.fed.us/documents/pdf2004/fpl_2004_mckeever002.pdf

McKeever DB, Phelps RB (1994) Wood products used in new single family house construction. For Prod J 44:66–74

Meil J, Lippke B, Perez-Garcia J (2004) Phase I Final Report, Module J Environmental impacts of a single family building shell- from harvest to construction. Consortium for Research on Renewable Industrial Materials, University of Washington Seattle. Available in http://www.corrim.org/reports/2006/final_phase_1/index.htm

Micales JA, Skog KE (1997) The decomposition of forest products in landfills. Int Biodeterior Biodegrad 39:145–158

Milota MR, West CD, Hartley ID (2005) Gate-to-gate lifecycle inventory of softwood lumber production. Wood Fiber Sci 37:47–57

Miner R (2006) The 100-year method for forecasting carbon sequestration in forest products in use, Mitigation and adaptation strategies for global change. Springer, The Netherland. Published online at http://www.springerlink.com/content/2167272117366751/

Miner R (2007) Clearing the air about biomass carbon neutrality. Paper 360. http://findarticles.com/p/articles/mi_m1AHU/is_3_2/ai_n25003935?tag=artBody;col1

Miner R, Lucier A (2004) A value chain assessment of climate change and energy issues affecting the global forest based industry. Available from http://www.wbcsd.org/web/projects/forestry/ncasi.pdf

Mitchell C (2003) Estimation, reporting and accounting of harvested wood products. United Nations Framework Convention on Climate Change reports. FCCC/TP/2003/7. http://www.greenhouse.crc.org.au/crc/ecarbon/CoP9.pdf

Nabuurs GJ (1996) Significance of wood products in forest sector carbon balances. In: Apps MJ, Price DT (eds) Forest ecosystems, forest management and the global carbon cycle. Springer, Berlin, pp 245–256

Nabuurs GJ, Mohren GMJ (1993) Carbon fixation through forestation activities. IBN Res. Rep. 93/4. Institute for Forestry and Natural Resources, Wangeningen, The Netherlands, 205p

Nabuurs GJ, Sikkema R (2001) International trade in wood products: its role in the land use change and forestry carbon cycle. Climate Change 49:377–395

Nebel B, Zimmer B, Wegener G (2006) Life cycle assessment of wood floor coverings. Int J LCA 11:172–182

Nippala E, Jaakkonen L (1993) Asuinrakennusten perusparannustarve. ASPE-mallin menetelmakuvaus ja laskelma 1990-luvun Perusparannustar (need for renovation of residential buildings). Process description of the ASPE model and renovation need in the 1990s, in Finnish. National Housing Board, Housing Research and Policy Department, Housing Studies 21, Helsinki, 48p

Nippala E, Jaakkonen L (1996) Condition assessments and need for renovation. CIB 70. User oriented and cost effective management, maintenance and modernization of building facilities. Helsinki, Finland, pp 43–46

Nippala E, Skogberg M (1991) Asuinrakennusten korjaustarve 1990—luvulla (need for renovation of residential buildings in the 1990s, in Finnish). National Housing Board. Housing Research and Policy Department, Housing Studies 5, Helsinki, Finland, 50p

Nippala E, Heljo J, Jaakkonen L (1995) Rakennuskannan energiankulutus Suomessa (Energy consumption of the building stock in Finland, in Finnish). VTT Research Notes 1625, Espoo, Kluwer Academic Publishers, 61p

Office of Energy Efficiency Webpage (2002) Canada. Natural resources Canada. http://oee.rncan.gc.ca/industrial/opportunities/sectors/wood.cfm?attr=12

Pajakkala P, Niemi O (1989) Rakentamisen Teollistaminen. RATA 2000 Osaprojekti 1. (Industrialization in construction. RATA 2000, part 1, in Finnish). VTT Building production laboratory and Tampere University of Technology, Construction Economics, Rakentajain Kustannus Oy, ISBN 95 1-676-454-1. 116p

Penman J, Gytarsky M, Hiraishi T (eds) (2003) Good practice guidance for land use, land-use change and forestry. Institute for Global Environmental Strategies for the IPCC, Hayama

Perala AL, Nippala E (1998) Construction wastes and their utilization (in Finnish, abstract in English). VTT Research notes 1936, 67p. Available in http://www.inf.vtt.fi/pdf/tiedotteet/T1936.pdf

Perez-Garcia J, Lippke B, Comnick J (2004) Tracking carbon from substitution in the forest to wood products and substitution, In CORRIM Phase I Final Report Module N. Life cycle environmental performance of renewable building materials in the context of residential construction. University of Washington, Seattle, 27p. Available in http://www.corrim.org/reports/

Perez-Garcia J, Lippke B, Briggs D (2005a) The environmental performance of renewable building materials in the context of residential construction. Wood Fiber Sci 37:3–17

Perez-Garcia J, Lippke B, Comnick J (2005b) An assessment of carbon pools, storage and wood product market substitution using life-cycle analysis results. Wood Fiber Sci 37:140–148

Petersen AK, Solberg B (2003) Substitution between floor constructions in wood and natural stone: comparison of energy consumption, greenhouse gas emissions and costs over the life cycle. Can J For Res 33:1061–1075

Petersen AK, Solberg B (2004) Greenhouse gas emission and costs over the life cycle of wood and alternative flooring materials. Climate Change 64:143–167

Petersen AK, Solberg B (2005) Environmental and economic impacts of substitution between wood products and alternative materials: a review of micro-level analyses from Norway and Sweden. For Policy Econ 7:249–259

Petersen AK, Gobakken T, Hoen HF (2004) Avoided greenhouse gas emissions when forest products substitute competing materials- effect on carbon account and optimal forest management. A case study of Hedmark County in Norway. Scand For Econ 40:113

Pingoud K, Perala AL (2000) Studies on greenhouse impacts of wood construction. 1. Scenario analysis of potential wood utilization in Finnish New Construction in 1990 and 1994. 2. Inventory of carbon stock of wood products in the Finnish building stock in 1980, 1990 and 1995 (in Finnish, abstract in English). Espoo, Technical Research Center of Finland, VTT Julkaisuja Publikationer 840, 58p. Available in http://www.inf.vtt.fi/pdf/julkaisut/2000/J840.pdf

Pingoud K, Savolainen I, Seppala H (1996) Greenhouse impact of the Finnish forest sector including forest products and waste management. Ambio 25:318–326

Pingoud K, Perala AL, Pussinen A (2001) Carbon dynamics in wood products. Mitig Adapt Strateg Glob Chang 6:91–111

Pingoud K, Perala A, Soimakallio S (2003) Greenhouse gas impacts of harvested wood products: evaluation and development of methods. VTT Research Notes 2189. VTT Technical Research Center of Finland, Espoo. http://www.vtt.fi/inf/pdf/tiedotteet/2003/T2189.pdf

Pingoud K, Pohjola J, Valsta L (2010) Assessing the integrated climatic impacts of forestry and wood products. Silva Fennica 44:155–175

Puettmann ME, Wilson J (2005) Life-cycle analysis of wood products: cradle-to-gate LCI of residential building materials. Wood Fiber Sci 37(CORRIM Special Issue):18–29

Puettmann ME, Bergman R, Hubbard S (2010) Cradle-to-gate life-cycle inventory of US wood products production: CORRIM Phase I and Phase II products. Wood Fiber Sci 42(CORRIM Special Issue):15–28

Pussinen A, Karjalainen T, Kellomaki S (1997) Potential contribution of the forest sector to carbon sequestration in Finland. Biomass Bioenergy 13:377–387

Rivela B, Hospido A, Moreira MT (2006) Life cycle inventory of particleboard: a case study in the wood sector. Int J LCA 11:106–113

Rivela B, Moreira MT, Feijoo G (2007) Life cycle inventory of medium density fibreboard. Int J LCA 12:143–150

Row C, Phelps B (1996) Wood carbon flows and storage after timber harvest. In: Sampson RN, Hair D (eds) Forests and global change: volume 2. Forest management opportunities for mitigating carbon emissions. American Forests, Washington, DC, pp 27–58

Ruddell S, Sampson R, Smith M (2007) The role for sustainably managed forests in climate change mitigation. J For 105:314–319

Sathre R (2007) Life-cycle energy and carbon implications of wood-based products and construction. Mid Sweden University Doctoral Thesis 34, Ecotechnology and environmental science, Department of Engineering, Physics and Mathematics, Mid Sweden University. www.diva-portal.org/diva/getDocument?urn_nbn_se_miun_diva-50-2_fulltext[1].pdf

Sathre R, O'Connor J (2008) A synthesis of research on wood products and greenhouse gas impacts, Technical report TR-19. FP Innovations Forintek, Canterbury, 79p

Schlamadinger B, Marland G (1996) The role of forest and bioenergy strategies in the global carbon cycle. Biomass Bioenergy 10:275–300

Sedjo R, Amano M (2006) The role of forest sinks in a post-Kyoto world. Resources, Number 162. Resources for the future, Washington, DC, pp 19–22. http://www.rff.org/Documents/RFF-Resources-162_ForestSinksPostKyoto.pdf

Seidl R, Tammer W, Jager D (2007) Assessing trade-offs between carbon sequestration and timber production within a framework of multi-purpose forestry in Austria. For Ecol Manag 248:64–79

Sharai-Rad M, Welling J (2002) Environmental impacts and energy balances of wood products and major substitutes. Forestry Department, FAO, Rome

Skog K (2008) Sequestration of carbon in harvested wood products for the United States. Forest Prod J 58:56–72

Skog KE, Nicholson GA (2000) Carbon sequestration in wood and paper products. In: Joyce LA, Birdsey RA (eds) The impact of climate change on America's forests (chapter 5). General Technical Report RMRS-GTR-59. USDA-Forest Service, pp 79–88. http://www.fpl.fs.fed.us/documents/pdf2000/skog00b.pdf

Skog KE, Pingoud K, Smith JE (2004) A method countries can use to estimate change in carbon stored in harvested wood products and the uncertainty of such estimates. Environ Manag 33(Suppl 1):S65–S73

Skog K, Heath L, Smith J (2008) The green house gas and carbon profile of the US forest products sector. Special report no. 08-05. National Council on Air and Stream Improvement and USDA Forest Service

Smith JE, Heath LS, Skog KE (2006) Methods for calculating forest ecosystem and harvested carbon with standard estimates for forest types of the United States. General technical report no. 08-05. National Council on Air and Stream Improvement and USDA Forest Service

Stockmann KD, Anderson NM, Skog KE (2012) Estimates of carbon stored in harvested wood products from the United States forest service northern region, 1906–2010. Carbon Balance Manag 7:1

Suadicani K (2010) Carbon sequestration and emissions from harvested wood products- different approaches and consequences. Forest and landscape working papers no. 56-2010. Forest and Landscape Frederiksberg, Denmark. Available in www.sl.life.ku.dk

Subak S, Raskin P, Von Hippel D (1993) National greenhouse gas accounts: current anthropogenic sources and sinks. Climate Change 25:15–58

Taverna R, Hofer P, Werner F (2007) The CO_2 effects of the Swiss forestry and timber industry. Scenarios of future potential for climate-change mitigation. Environmental Studies 0739. Federal Office for the Environment, Bern, 102p

Tonosaki M (2009) Harvested wood products accounting in the post Kyoto commitment period. J Wood Sci 55:390–394

UNFCCC (2009) Copenhagen climate change conference – December 2009. http://unfccc.int/meetings/copenhagen_dec_2009/meeting/6295.php

Upton B, Miner R, Spinney M (2007) The greenhouse gas and energy impacts of using wood instead of alternatives in residential construction in the United States. Biomass Bioenergy 32:1–10

Vainio T, Lehtinen E, Pajakkala P (1998) Menestystuotteita rakennusalalle (toward successful business in construction sector, in Finnish). VTT Research notes 1931, Espoo, 88p. Available in http://www.inf.vtt.fi/pdf/tiedotteet/1998/T1931.pdf

Vainio T, Riihimaki M, Makela P (1999) Rakennuskustannusindeksi 2000 (building cost index 2000, in Finnish). VTT Research notes 2003, Espoo, 70p. Available in http://www.inf.vtt.fi/pdf/tiedotteet/1999/T2003.pdf

Valsta L, Lippke B, Perez-Garcia J (2008) Use of forests and wood products to mitigate climate change. In: Bravo F, LeMay V, Jandl R. and von Gadow K (eds) Managing forest ecosystems: the challenge of climate change, vol 17. Springer, pp 137–149

Werner F, Taverna R, Hofer P (2005) Carbon pool and substitution effects of an increased use of wood in buildings in Switzerland: first estimates. Ann For Sci 62:889–902

Wilson A, Boehland J (2005) Small is beautiful: US house size, resource size and the environment. J Ind Ecol 1(2):277–287

Wilson JB, Sakimoto ET (2005) Gate-to-gate life-cycle inventory of softwood plywood production. Wood Fiber Sci 37:58–73

Winistorfer P, Chen Z, Lippke B (2005) Energy consumption and green house gas emissions related to the use, maintenance and disposal of a residential structure. Wood Fiber Sci 37:128–139

Winjum J, Brown S, Schlamadinger B (1998) Forest harvests and wood products- sources and sinks of atmospheric carbon dioxide. For Sci 44:272–284

Woodbury PB, Smith JE, Heath LS (2007) Carbon sequestration in the US forest sector from 1990 to 2010. For Ecol Manag 241:14–27

Ximenes FA, Gardner WD, Cowie AL (2008) The decomposition of wood products in landfills in Sydney. Aust Waste Manag 28:2344–2354

Zerbe JI (2006) Thermal energy, electricity and transportation fuels from wood. For Prod J 56:6–14

Zhang D, Hui D, Luo Y (2008) Rates of litter decomposition in terrestrial ecosystems: global patterns and controlling factors. J Plant Ecol 1:85–93

Essential Oil: Its Economic Aspect, Extraction, Importance, Uses, Hazards and Quality

15

M. Preema Devi, S. Chakrabarty, S.K. Ghosh,
and N. Bhowmick

Abstract

Essential oils are not the same as fragrance oils or perfume where essential oils are derived from true plants. Out of the total world production, India produces 4 %, while in terms of value its share is much better with 21–22 %. But considering the total share in world export of essential oils and perfumery material, it is only 0.4 %. In India, the states having the highest production of rose and tuberose oil are UP and Andhra Pradesh (Hyderabad), respectively, while jasmine oil production is highest at TN, Karnataka and Kerala. Essential oils are produced by steam distillation (simple, saturated, hydro-diffusion and microwaves), solvent extraction, methods using oils and fats (enfleurage and pneumatic) and extraction by supercritical gasses. There is a huge opportunity existing in the aromatherapy sector. But one should bear in mind that not all essential oil is safe to be used in aromatherapy. Some of the oil can be hazardous as they can cause severe dermal irritation and even damage the mucous membranes and delicate stomach lining in undiluted form. The quality of an essential oil can be analysed by various methods, i.e. specific gravity, optical rotation, refractive index and gas chromatography. Since most essential oils exhibit deterioration through oxidation and polymerization upon prolonged exposure to air and light, so, in order to prevent browning, essential oils should be stored in cool dry cellars in a hermetically sealed amber glass container. This paper will review the economic aspect, extraction, importance, uses, hazards and quality of essential oil extracted from different plant parts.

M.P. Devi (✉) • S.K. Ghosh • N. Bhowmick
Department of Pomology and Post Harvest
Technology, Uttar Banga Krishi Viswavidyalaya,
Pundibari, Cooch Behar, West Bengal, India
e-mail: preema.horti@gmail.com

S. Chakrabarty
Department of Post Harvest Technology of
Horticultural Crops, Bidhan Chandra Krishi
Viswavidyalaya, Mohanpur, Nadia,
West Bengal 736165, India

A.B. Sharangi and S. Datta (eds.), *Value Addition of Horticultural Crops: Recent Trends and Future Directions*, DOI 10.1007/978-81-322-2262-0_15, © Springer India 2015

15.1 Introduction

Essential oils are liquid that are generally distilled from various parts of plant that have strong aromatic components such as from the leaves, stems, flowers and roots. For example, in roses, it is found in the flowers; in basil, it is in the leaves; in sandalwood, in the wood; and so on. Using the different technologies available, essential oils are sourced from over 3,000 plants of which approximately 300 are of commercial importance. These aromatic substances are made up of different chemical compounds that can be found naturally in the plant. For instance, alcohol, hydrocarbons, phenol, aldehydes, esters and ketones are some of the major components. Besides that, it may also contain hundreds of organic constituents, including hormones, vitamins and other natural elements. They are 75–100 times more concentrated than the oils in dried herbs. Furthermore, these aromatic characteristics of essential oils may provide various functions for the plant itself including attracting or repelling insects (odours of the flowers); while in plant metabolism, a few essential oils might be involved in this process. Leaf oils, wood oils and root oils may serve to protect against plant parasites or depredations by animals as well as antibacterial agent which is utilizing the hormone in the oil. Essential oils are not the same as fragrance oils or perfume where essential oils are derived from true plants. Perfume oils are unnaturally created fragrances since they contain artificial substances and they may also not offer the therapeutic benefit that essential oils do. Pure essential oils are very expensive, but fortunately, they are also highly effective since only a few drops at a time are necessary to achieve the desired effects. Moreover, it is widely used in three primary ways: as odorants (in cosmetics, perfumes, detergents, soap), as flavours (in bakery goods, candies, meat, soft drinks and also as food additives) and as pharmaceuticals (in dental products and group of medicines).

Today, we could also easily find synthetic essential oils in the market where the price would be cheaper than the pure ones. There are a few differences between synthetic essential oils and pure essential oils. Synthetic essential oils are produced by blending aromatic chemicals mostly derived from coal tar. These oils may duplicate the smell of the pure essential oils, but the complex chemical components of each essential oil created in nature determine its true aromatic benefits. While synthetic essential oils are not suitable for aromatherapy, they add an approximation of the natural scent to crafts, potpourri, soap and perfume at a fraction of the cost. The reason of these synthetic products is mainly to reduce the cost of production.

15.2 The Parts of Plants Yielding Essential Oils

The oil of the essential oils bears the name of the plant from which it was derived, for example, rose oil and bergamot oil. Such oils were called essential because they were thought to represent the very essence of odours and flavour. The odours and flavour of these oils are usually dependent upon their oxygenated compound. Many oils are terpenoids; few are benzene derivatives. Table 15.1 shows the important constituents of some common essential oils.

15.3 Economic Aspects of Essential Oil

According to estimates by Dr. Brian Lawrence, editor of the Journal of Essential Oil Research, global production of the 20 leading essential oils amounts to approximately 104,000 tonnes. This figure should be used solely as an indication, since the figure addresses not only the cosmetics industry but also other industries such as foods, pharmaceuticals and households. Moreover, next to these 20, there is a huge range of smaller oils, such as cumin oil, caraway oil, valerian oil, etc. The food industry includes the drink and beverages sector. In terms of fragrance and cosmetic industry, Western Europe represents a massive share of over 31 % of the global cosmetics and toiletries market followed by North America and the Asia Pacific. In pharmaceutical industry, it is

Table 15.1 Essential oils from some natural plants

Name	Part of the plant used	Botanical name	Important constituents	Uses
Lemongrass and citronella	Leaf	*Cymbopogon* spp.	Citral	Perfumery
			Citronella	Disinfectant
			Terpenes	
Eucalyptus	Leaf	*Eucalyptus globulus*	Cineale	Pharmaceutical (decongestant)
		Eucalyptus citriodora	Citronella	
		Eucalyptus dives	Terpenes	
Turpentine	Not mention	*Pinus* spp.	Terpenes	Paints
Lavender	Flower	*Lavandula intermedia*	Linalool	Perfumery
Sandalwood	Wood	*Santalum album*	Sanatols	Perfumery
Vetiver (khus grass)	Roots	*Vetiveria zizanioides*	Vetiverols	Cosmetics and aromatherapy
Ginger	Rhizome	*Zingiber officinale*	Zingiberene	Treatment of fractures, rheumatism, arthritis, bruising, carbuncles, nausea, hangovers, travel and sea sickness

Source: Rusli and Wahid (1990)

also dominated by huge multinational companies, and North America, EU and Japan are the three biggest markets. The demand for essential oils in this industry appears stable (www.eoai.in/).

The major producers of essential oils are the USA (24 %), China (20 %), Brazil (8 %), Mexico (5 %), Morocco (5 %), Indonesia (5 %), India (4 %) and Egypt (2 %). It is estimated that about 65 % of the world production emanates from developing countries. The major consumers are the USA (40 %), Western Europe (30 %) and Japan (7 %). On a global scale, the 18 most important essential oils represent nearly 75 % of the total production value. The concentration in terms of tonnes is even higher, as there is a trade in small volumes of products with high unit values (e.g. rose, jasmine, vetiver). Quite a few countries (the United State of America, France, the United Kingdom, Japan and Germany) dominate the world's import market in essential oils. The United State of America is the leading importer with a 14 % share or US$390.9 m worth of imports in 2005. The United State of America, France, India, the United Kingdom and Brazil are the leading five exporters of essential oils in the world. Together, they exported US$936.3 m worth of essential oil products representing about 40 % of world exports. Again, the US is the major exporter with US$351.7 m worth of exports or 17 % share of world total exports.

The US is followed by France, India, the UK and Brazil.

Out of the total world production, India produces 4 %, while in terms of value, its share is much better with 21–22 %. But considering the total share in world export of essential oils and perfumery material, it is only 0.4 %; thus, the future holds a great promise for India in the fast charging global economy as far as the production and trade of natural raw material is concerned. India is the largest producer of Japanese mint oil and Indian basil oil in the world. It is a major exporter of many essential oils; it also imports from several countries. In India, the states giving the highest production of rose and tuberose oil are UP and Andhra Pradesh (Hyderabad), respectively, while jasmine oil production is highest at TN, Karnataka and Kerala (Kapoor 1991). The prices of essential oil are enlisted as follows (Table 15.2).

15.4 Extraction of Essential Oils

Although some of the chemical components of essential oils are similar to "oils", essential oils are not greasy themselves, and they are light in weight. However, the high alcohol components in essential oils give higher volatility and faster evaporation rate. The selection of appropriate

Table 15.2 Prices of essential oil

Essential oils	Price (Rs./kg)
Rose	207,690–289,800
Jasmine	82,167
Tuberose	1.93,200–198,030
Basil	600–675
Ginger	3,800–4,200
Lemongrass	550–650
Pine	80–100
Turmeric	700–800
Sandalwood	60,000–65,000
Vetiver	27,000–32,000
Eucalyptus	400–600

Source: www.eoai.in

extraction method will determine the quality and quantity of essential oils. Other factors such as types of plant, chemical makeup of oils and location of oils within the plant (root, bark, wood, branch, leaf, flower, fruit and seed) are also needed to be considered prior to the extraction. Some plants like rose and jasmine contain very little essential oil. Their important aromatic properties are extracted using a chemical solvent. The end product, known as an absolute, contains essential oil along with other plant constituents. Though not a true essential oil, the absolute is commonly used for fragrance cosmetic products like fine perfumes. Some extraction methods are described herein.

15.4.1 Methods

15.4.1.1 Steam Distillation

In this process, the botanical material is placed in a still, and steam is forced over the material. The hot steam helps to release the aromatic molecules from the plant material. The molecules of these volatile oils are then escaped from the plant material and evaporate into the steam. The temperature of the steam therefore needs to be carefully controlled. The temperature should be just enough to force the plant material to release the essential oils, yet not too hot as it can burn the plant material or the essential oils. The steam which then contains the essential oil is passed through a cooling system to condense the steam,

which then forms a liquid from which the water and the essential oils are then separated. The steam is produced at a higher pressure (3 pounds psi) than the atmosphere, and therefore it boils at above 100 °C which facilitates the removal of the essential oil at a faster rate, and by doing so, it could prevent damage to the oil as well. Some oils like lavender are heat sensitive, and with this method, it could prevent it from damage, and ingredients like linalyl acetate will not decompose to linalool and acetic acid. Apart from simple steam distillation method, there are other methods, i.e. saturated, hydro-diffusion steam distillation method and steam distillation by microwaves under vacuum employed for distillating essential oils from plant parts. In saturated steam distillation method, the plant does not come into contact with water; steam is injected through the plant material placed on perforated trays. It is possible to operate under moderate pressure. The advantages are it limits the alteration of the constituents of the oil, shortens the duration of the treatment, conserves energy and can also be conducted on online in automated setups. In steam distillation method, pulses of steam are sent through the plant material at very low pressure from top to bottom. The advantages are it normally produces a product of high quality and saves time and energy.

In steam distillation by microwaves under vacuum, the plant is heated selectively by microwave radiation in a chamber inside which the pressure is reduced sequentially where fresh plants require no added water. This method is fast, consumes little energy and yields a product which is most often of a higher quality than the traditional steam distillation product.

15.4.1.2 Solvent Extraction

A hydrocarbon solvent is added to the plant material to help dissolve the essential oil. When the solutions are filtered and concentrated by distillation, a substance containing resin or a combination of wax and essential oil (concrete) remains. From the concentrate, pure alcohol is used to extract the oil, and when the alcohol evaporates, the oils are left behind. This is not considered the best method for extraction as the solvents can leave small amount of residue behind which could cause allergies and affect the immune system.

15.4.1.3 Enfleurage

When essential oil content of fresh plant parts such as flower petals is so small that oil removal is not commercially feasible by the aforementioned methods, enfleurage method is used (Robbers et al. 1996). Some flowers, such as jasmine or tuberose, have such low contents of essential oil or too delicate that heating them would destroy the blossoms before releasing the essential oils. In such cases, this method of extraction is sometimes used to extract the essential oils. Flower petals are placed on trays of odourless vegetable or animal fat, which will absorb the essential oils. Every day or every few hours, after the fat has absorbed the essential oils as much as possible, the depleted petals are removed and replaced with fresh ones. This procedure continues until the fat or oil becomes saturated with the essential oil. Adding alcohol to this mixture separates the essential oil from the fatty substance. Afterwards, the alcohol evaporates, and only the essential oils remain. And yet this process is a very labour-intensive way of extraction and needless to say a very costly way to obtain essential oil and nowadays, only sometimes used to extract essential oils from tuberose and jasmine.

Another similar method of extracting essential oil using oils and fat is pneumatic method which is similar in principle to the enfleurage process. It involves the passage of a current of hot air through the flowers. The air, laden with suspended (extracted) volatile oil, is then passed through a spray of melted fat in which the volatile oil is absorbed.

15.4.1.4 Supercritical Carbon Dioxide Extraction

Supercritical carbon dioxide extraction is an alternative process to avoid the degradation of heat-sensitive compounds and partial hydrolysis of water-sensitive compounds brought about by steam distillation or solvent extraction (Roy et al. 2007). Supercritical fluid extraction has received increasing attention in a variety of fields due to the following factors: (a) supercritical fluids provide satisfactorily solubility and improved mass transfer rates and (b) operation can be manipulated by changing the pressure or temperature (Roy et al. 2007). In practice, more than 90 % of all analytical supercritical fluid extraction (SFE) is performed with carbon dioxide for several practical reasons. Apart from having relatively low critical pressure of 73.8 bar and temperature of 31 °C, CO_2 is relatively non-toxic, non-flammable, available in high purity at relatively low cost and is easily removed from the extract. The main drawback of CO_2 is its lack of polarity for the extraction of polar analytes (Pourmortazavi and Hajimirsadeghi 2007).

15.5 Some Important Flowers Yielding High-Quality Essential Oil

A few flowers which yield high value essential oils are described hereunder:

15.5.1 Rose

15.5.1.1 Essential Oil of *Rosa bourboniana*

The fresh rose petals yield oil of 0.013–0.015 % and concrete 0.18 %. The oil is considered as inferior in content and odour. It is not used for the production of rose otto or attar, but largely used for rose water production. There are several varieties like Boule de Neige, Souvenir de la Malmaison, Zephirine, Edward, Louise Odier, etc. which are deliciously fragrant.

15.5.1.2 Essential Oil of *Rosa damascena*

Flowers are collected by hand-picking during early hours of the morning and should be plucked when they start opening. Collected flowers should be distilled as early as possible at the site of the plantation itself. If the distillation is delayed, then the flowers may be sprinkled with 5 % salt solution to check the fermentation evaporation of oil. Flowers lose their aroma after 24 h after their opening. The average recovery of rose oil through steam distillation ranges between 0.04 and 0.05 % in temperate climates, and in the subtropical climate, it is 0.02–0.03 % on fresh weight basis of flowers. Rose concretes are extracted by volatile solvents (petroleum ether, hexane and benzene) which yield 0.25 % flower concrete. The absolute is obtained by further extraction of concrete with alcohol. Rose oil is transparent, pale yellow and mobile liquid. The principal constituents of the oil are geraniol and citronellol. Other chemical compounds identified are ethanol, rose oxide, linalool, nerol, eugenol, carvone and phenyl ethyl alcohol (Leffingwell 2000).

15.5.1.2.1 Uses

Rose oil has a wide role in the perfumery industry and is a highly valued perfume material. Several other products are produced from damask rose. *Rose water* is produced by hydrodistillation of rose flower. More than 80 % of rose flowers are used for the production of rose water in India. It has been valued from ancient times for use in making syrups, medicinal preparation and also for sprinkling at social function. It is also used in eye lotions and eye drops for its soothing qualities.

Rose jam or Gulkand is prepared from rose petals by mixing one part of rose petals with two parts of sugar and kept in the sun for about a month for maturation. It is used in Unani medicine as a good tonic and mild laxative.

Rose attar is made by distillation of rose petals, and the distillation is collected over oil of sandalwood. It is used as flavouring tobacco and as a perfume in agarbattis. It is also used in perfumes.

Gul roghan is a type of hair oil made by modified enfleurage method, that is, by maceration of rose petals by sesame seed oil.

Pankhuri is shade-dried rose petals and is used for the preparation of cool summer drinks.

Bulgaria produces about 70 % of all rose oil in the world. Other significant producers are Morocco, Iran and Turkey. Recently, China has begun producing rose oil as well.

15.5.2 Tuberose (*Polianthes tuberosa*)

15.5.2.1 Essential Oil

The yield of fresh, loose flowers per hectare for the production of concrete in single tuberose varies from 10,000 to 14,000 Kg. In the first 2 years, flower yield is high and declines considerably in the third year. The 3 years total yield of 30,000 Kg of loose flowers gives 27.5 Kg of concrete per hectare. This concrete, in turn, yields 5.50 Kg of absolute. The extraction of essential oils from the flower is carried out by enfleurage or by use of volatile solvents, but the former process is almost abundant for tuberose in India. However, enfleurage yields about 15 times more volatile oils than extraction with petroleum ether. The concrete is extracted from tuberose flower with petroleum ether, which yields 0.08–0.14 % concrete. Generally, from 1,150 Kg of tuberose flowers, 1 Kg of concrete is obtained. Tuberose concrete is light to dark brown in colour, a waxy hard mass and partly soluble in high proof alcohol. The concrete yields 18–23 % of alcohol-soluble absolute. The absolute is very viscous or a semi-liquid brown mass. It possesses a delightful and lasting odour resembling that of living flowers.

Tuberose flower oil consists of geraniol, nerol, farnesol, benzyl alcohol, methyl benzoate, benzyl benzoate, methyl salicylate, methyl anthranilate, eugenol and butyric acid. Methyl vanillin and piperonal have also been identified.

Tuberose oil is used only for the highest grade of perfumes since it is more expensive. It is used in the heavier types of scent, floral as well as oriental. Specifically for Gardenia perfume, tuberose oil is an important base. The oil forms an important adjunct in the formation of high grade perfumes. The oil is used in items like non-alcoholic beverages (0.25 ppm), ice cream, ices, etc. (0.45 ppm), candy (1.5 ppm) and baked goods (1.7 ppm).

15.5.3 Gardenia (*Gardenia grandiflora*)

15.5.3.1   Essential Oil

The essential oil is extracted from the flowers by petroleum ether. Flower concrete of 1 Kg can be obtained from 2,500 kg of flowers, which yields 0.5 Kg of absolute. The oil is yellowish in colour and contains benzyl acetate, styrallyl acetate, linalool, linalyl acetate, terpineol and methyl anthranilate.

Gardenia oil is not commercially used in perfumery industry. The oil resembles the perfume of the living flowers, which possesses a very delicate fragrance. The Chinese use the flower for perfuming tea.

15.5.4 Lavender (*Lavandula latifolia*)

15.5.4.1   Essential Oil

Lavender oil is an essential oil obtained by distillation from the flower spikes of certain species of lavender. Two forms are distinguished: *lavender flower oil*, a colourless oil, insoluble in water, having a density of 0.885 g/mL, and *lavender spike oil*, a distillate from the herb *Lavandula latifolia*, having a density of 0.905 g/mL. Like all essential oils, it is not a pure compound; it is a complex mixture of naturally occurring phytochemicals, including linalool and linalyl acetate. Kashmir lavender oil is famous for being produced from lavender at the foothills of the Himalayas. As of 2011, the biggest lavender oil producer in the world is Bulgaria.

Lavender oil, which has long been used in the production of perfume, can also be used in aromatherapy. The scent has a calming effect which may aid in relaxation (Shiina et al. 2008) and the reduction of anxiety (Hongratanaworakit 2011) and stress (Hwang 2006).

15.5.5 Jasmine (*Jasminum auriculatum*)

15.5.5.1   Essential Oil

The high yielding varieties are Co-1 Mullai, Co-2 Mullai and Pari Mullai. Other strains are long point, long round, median point and short round. In *Jasminum auriculatum*, flower yield and concrete recovery percentage of different clones are Pari Mullai: 7,800 Kg/ha and 0.29 %, Co-1 Mullai 8,800 kg/ha and 0.34 %; and Co-2 Mullai: 11,200 Kg/ha and 0.29 %, respectively.

Other *Jasminum* species like *Jasminum officinale* and *Jasminum pubescens* yields concrete ranges from 0.4 to 0.5 % by solvent extraction method, while *Jasminum arborescens, Jasminum japonicum, Jasminum niloticum* and *Jasminum odorata* yield oil from 0.25 to 0.35 % depending on species.

15.6   Why Are Essential Oils So Essential

The importance of essential oils is provided in the list below:

1. They are antibacterial, antiviral, antifungal and antimicrobial.
2. Essential oils bypass the digestive system so they are beneficial for people with poor digestion/assimilation.
3. They require no refrigeration and require very little storage space.
4. They have the longest shelf life of any plant known to man.
5. Essential oils are highly oxygenating. It is said that oxygen and disease cannot exist in the same environment.
6. They are very cost-effective because they are one of the few substances that "the more you use them, the less you need them".
7. They are suitable for babies since they cannot swallow tablets and capsules thus essential oils provide a solution as they only come in liquid form.

15.7   Uses of Essential Oils

According to Johnston and Parsons (1999), in therapeutic aromatherapy, essential oils treat medical conditions. For example, they can fight infections, promote wound healing (Singh et al. 2002), heat the skin in a liniment, promote blood circulation and digestion and lessen sinus or lung congestion (Watt et al. 1962). Specifically, essential oils were used to prevent respiratory infections, promote mucus clearing, fight depression and promote sleep. In psycho-aromatherapy, they are used as a relaxant or stimulant to the brain. Some oils can have calming and tranquilizing effects; others are energizing and can help relieve depression (Bakkali et al. 2008). Essential oils can be administered as inhalations like eucalyptus oil, orally like peppermint oil, gargles and mouthwashes like thymol oil and like lavender oil and rosemary oil which are used in the practice of aromatherapy (Dutta et al. 2007). Oils with high phenol content like clover and thyme have antiseptic properties (Pratt and Youngken 1951; Sapeika 1963). Some volatile oils also show antispasmodic activity like *Melissa officinalis*, *Rosmarinus officinalis and Carum carvi* (Evans 2002). Volatile oils possess a carminative action, but others like eucalyptus and wintergreen have additional therapeutic properties (Seenivasan et al. 2006). But volatile oils are commonly used as flavourant (Robbers et al. 1996). Each essential oil has its own properties and uses which can be classified and identified accordingly to the type of plant it was derived.

15.8 Potential Hazards of Essential Oils

15.8.1 Toxicity

One should bear in mind that not all essential oil are safe to be used in aromatherapy even with or without the express administration by a qualified aromatherapy practitioner. This is due to the high toxicity levels that the essential oils might have. Some of the oil can be hazardous as they can cause severe dermal irritation and even damage the mucous membranes and delicate stomach lining in undiluted form. Hence, dermal application should be avoided as a general practice; it is advisable to use essential oils only for external remedies. Oils that fall under this category are cassia, pine (dwarf), sage (common), thyme (red) and wintergreen.

15.8.2 Phototoxicity

Some oils can cause skin pigmentation if the applied area is exposed to direct sunlight. So these oils should not be used either neat or on dilution on the skin, if the area will be exposed to direct sunlight.

15.8.3 Pregnancy

Essential oils should be used in half the usual stated amount during pregnancy, because of the sensitivity of the growing child (Nordin et al. 2004). Oils of basil, cedarwood (all types), citronella, clary sage, juniper, Spanish sage and thyme (white) should be totally avoided during pregnancy.

15.8.4 High Blood Pressure

Oils of rosemary, sage (Spanish and common) and thyme are to be avoided in case of high hypertension.

15.8.5 Dermal/Skin Irritation

Oils of basil (sweet), cedarwood (Virginian), eucalyptus, pine needle (scotch and longleaf), and thyme (white) especially if used in high concentration may cause irritation to the skin.

15.9 Quality of Essential Oils

To ensure that each of the oils is in highest quality as well as completely unadulterated, some scientific analyses are required on every essential oil. There are four major analyses that can be carried out.

15.9.1 Specific Gravity

The weight of essential oils is measured at 25 °C. Every oil is made up of unique constituents. At a given temperature, these constituents have a predictable weight. If the oil has been contaminated, the weight may be thrown off.

15.9.2 Optical Rotation

This technique measures the direction, left or right, and the degree to which light rays bend or rotate as they pass through the essential oil. If the oil has been contaminated, the speed and degree of refraction may be thrown off.

15.9.3 Refractive Index

This technique measures the speed at which light passing through the oil is refracted. If the oil has been contaminated, the speed and degree of refraction may be thrown off.

15.9.4 Gas Chromatography

This technique separates the individual constituents of an essential oil and measures the amount of each constituent present. It would positively confirm oil botanical identity by comparing the presence and amount of each constituent. This evaluator can screen for non-natural or missing constituents or constituents occurring in unnaturally high ratios, signifying adulteration.

15.10 Storage of Essential Oils

Most essential oils exhibit deterioration through oxidation and polymerization upon prolonged exposure to air and light; so, in order to prevent browning, the essential oils are stored in cool dry cellars in a hermetically sealed amber glass container. Without such precautions, an essence becomes less intense, grows darker and more viscose, develops a bleaching effect and eventually changes into brown, odourless resin. Storage

materials should be opaque and should be glass bottles, aluminium bottles and drums (used for expensive essential oils), lacquered and lined steel drums and plastic drums in high density polyethylenes, which are less expensive than lined steel drums. Before they are stored for shipment, oils should be dried by filtration or the use of anhydrous calcium sulphate. Headspace should be filled with nitrogen gas although carbon dioxide is cheaper and easier to source in developing countries. The danger with using carbon dioxide, though, is that it might react with residual moisture to form carbonic acid, which may in turn react with essential oil constituents. Before buyers of essential oils and essential oil products make a full order, it is a normal procedure that a sample is first shipped to them for assessment and quality testing. The sample shipment should show the producer name, date, sample contents, batch number and quantity represented. Plastic bottles are not suitable. Individual bottles of each sample should be sent in plastic bags to avoid or isolate leakages.

15.11 Organizations Offering Information Sources or Which Can Be Contacted to Obtain Information Are

1. Personal Care Products Council: http://www.personalcarecouncil.org
2. FOSFA – Federation of Oils, Seeds and Fats Associations: http://www.fosfa.org
3. IFEAT – International Federation of Essential Oils and Aroma Trades: http://www.ifeat.org
4. IFRA, International Fragrance Association: http://www.ifraorg.org/
5. Colipa – The European Cosmetic Toiletry and Perfumery Association http://www.colipa.com
6. Aromatherapy Trade Council: http://www.a-t-c.org.uk
7. Marketing of Organic Wild Products: http://www.intracen.org
8. FSC: http://www.fsc.org
9. USDA-FAS: http://www.fas.usda.gov/

References

Bakkali F, Averbeck D, Idaomar M (2008) Biological effects of essential oils – a review. Food Chem Toxicol 46(2):446–475

Dutta BK, Karmakar S, Naglot A, Aich JC, Begam M (2007) Anticandidial activity of some essential oils of a mega biodiversity hotspot in India. Mycoses 50(2):121–124

Evans W (2002) Trease and Evans pharmacognosy, 15th edn. Elsevier Limited, Toronto, China

Hongratanaworakit T (2011) Aroma-therapeutic effects of massage blended essential oils on humans. Nat Prod Commun 6(8):1199–1204

http://www.bellaonline.com/articles/art66737.asp

Hwang JH (2006) The effects of the inhalation method using essential oils on blood pressure and stress responses of clients with essential hypertension. Taehan Kanho Hakhoe Chi 36(7):1123–1134

Johnston L, Parsons P (1999) Healing with essential oils and aromatherapy. Aromatherapy. Retrieved on August 5, 2010

Kapoor JN (1991) Attars of India – a unique aroma. Perfumer and Flavorist. Jan/Feb, pp 21–24

Leffingwell JC (2000) Rose (Rosa damascena) – a part of our series on aroma materials produced by carotenoid degradation. Leffingwell Rep 1(3):1–3

Nordin S, Broman DA, Olofsson JK, Wulff M (2004) A longitudinal descriptive study of self-reported abnormal smell and taste perception in pregnant women. Chem Senses 29(5):391–402

Pourmortazavi S, Hajimirsadeghi S (2007) Supercritical fluid extraction in plant essential and volatile oil analysis. J Chromatogr 1163(1/2):2–24

Pratt R, Youngken H (1951) Pharmacognosy: the study of natural drug substances and certain allied products. J.B. Lippincott Company, Lippincott

Robbers J, Speedie M, Tyler V (1996) Pharmacognosy and pharmacobiotechnology. Williams and Wilkins, Baltimore

Roy B, Hoshino M, Ueno H, Sasaki M, Goto M (2007) Supercritical carbon dioxide extraction of the volatiles from the peel of Japanese citrus fruits. J Essent Oil Res 19:78–84

Rusli D, Wahid P (1990) Prospects of the essential oil development in Indonesia. Ind Crop Res J 2(2):24–29

Sapeika N (1963) Actions and uses of drugs. Pub A.A. Balkema, Capetown

Seenivasan P, Manickkam J, Savarimuthu I (2006) In vitro antibacterial activity of some plant essential oils. BMC Complement Altern Med 6:39

Shiina Y, Funabashi N, Lee K, Toyoda T, Sekine T, Honjo S, Hasegawa R, Kawata T, Wakatsuki Y, Hayashi S, Murakami S, Koike K, Daimon M, Komuro I (2008) Relaxation effects of lavender aromatherapy improve coronary flow velocity reserve in healthy men evaluated by transthoracic Doppler echocardiography. Int J Cardiol 129(2):193–197

Singh G, Kapoor IPS, Pandey SK, Singh UK, Singh RK (2002) Studies on essential oils: Part 10; Antibacterial activity of volatile oils of some spices. Phytother Res 16(7):680–682

Watt JM, Breyer B, Maria G (1962) The medicinal and poisonous plants of Southern and Eastern Africa. E & S Livingstone, Edinburgh

www.eoai.in

www.fao.org.in

www.tips.org.za/

K. Pradhan and S. Pradhan

Abstract

In the changing global scenario, the making of horticultural enterprise more economically vibrant through total quality assurance is a resilient concept and approach as the enterprise is emerging as an avenue for nutritional security, poverty alleviation and employment generation. Nowadays, horticulture crop production has moved from rural confines to commercial ventures and has attracted youth since it has proved to be intellectually satisfying and economically rewarding. In such a growth-oriented situation, the horticultural enterprise is starting the journey in the global market with its table products and postharvest products after complying with the need of the domestic market. But the main issue with the export of the horticultural produce lies in the total quality assurance of the marketed surplus. Quality, the degree of excellence or superiority, is a combination of attributes, properties or characteristics that give each commodity value in terms of its intended use. In the era of globalisation and to uplift the status of the horticultural crop growers and exporters, the need of the hour is to manage the total quality during the production and marketing process of the horticultural produce. That can be obtained through introducing the culture of quality management within all the actors present in the production and marketing chain or value chain.

K. Pradhan (✉)
Assistant Professor, Department of Agricultural
Extension, Uttar Banga Krishi Viswavidyalaya,
Pundibari, Cooch Behar, West Bengal, India
e-mail: kausik_pradhan@rediffmail.com

S. Pradhan, M.Sc(Ag)
Agricultural Extension, Bidhan Chandra Krishi
Viswavidyalaya, Mohanpur, Nadia,
West Bengal, India

16.1 Introduction

In the realm of globalisation and free trade liberalisation, the need of the hour is to develop the horticultural enterprise in conformity with the quality standards led down by different organisations and nations to assure the quality of the product developed by the horticultural enterprises for making the enterprise more economically

A.B. Sharangi and S. Datta (eds.), *Value Addition of Horticultural Crops: Recent Trends*
and Future Directions, DOI 10.1007/978-81-322-2262-0_16, © Springer India 2015

vibrant. India is bestowed with a varied agro-climate, which is highly favourable for growing a large number of horticultural crops such as fruits, vegetables, root tuber, aromatic and medicinal plants, spices and plantation crops like coconut, areca nut, cashew and cocoa. India has now emerged as the world's largest producer of coconut and tea and the second largest producer and exporter of tea, coffee, cashew and spices; exports of fresh and processed fruits, vegetables, cut flowers and dried flowers have also been picking up.

Over the years, horticulture has emerged as one of the potential agricultural enterprise in accelerating the growth of the economy. Its role in the country's nutritional security, poverty alleviation and employment generation programmes is becoming increasingly important. It not only offers a wide range of options to the farmers for crop diversification but also provides ample scope for sustaining a large number of agro-industries which generate huge employment opportunities. At present, horticulture is contributing 24.5 % of GDP from 8 % land area. As a result of a number of thoughtful research, technological and policy initiatives and inputs, horticulture in India, today, has become a sustainable and viable venture for the small and marginal farmers. Besides, this sector has also started attracting entrepreneurs for taking up horticulture as a commercial venture. Therefore, there is a great scope for the horticulture industry to grow and flourish. The changing scenario encourages private investment, to go for hi-tech horticulture with micropropagation, protected cultivation, drip irrigation, fertigation and integrated nutrient and pest management, besides making use of latest postharvest measures particularly in the case of perishable commodities. As a result, horticulture crop production has moved from rural confines to commercial ventures and has attracted youth since it has proved to be intellectually satisfying and economically rewarding.

Exports of horticultural products are largely dependent on factors such as domestic production and consumption, exportable surpluses, consumer preferences, varieties traded, quality, domestic and international prices and availability of infrastructure facilities for storage, postharvest handling, etc. Having regard to the social and economic importance of the horticultural sector, the export strategy of the government is based on the premise that foreign earnings from this sector should be enhanced, thereby leading to higher income to farmers, taking care to make horticultural products available at reasonable prices to the domestic consumers. On the basis of this foreign trading policy, the total quality assurance of the horticultural products is a grave challenge to the horticultural entrepreneurs of our country. In such a challenging situation, the proper conceptualisation of the quality terms, quality procedures, methodology for retaining quality and quality assurance modalities related to the development and export of horticultural value-added products is the prime concern to make the economically viable horticultural enterprise more knowledge and quality vibrant one.

16.2 Conceptual Framework

Quality, the degree of excellence or superiority, is a combination of attributes, properties or characteristics that give each commodity value in terms of its intended use. Today, there is no single universal definition of quality. Some people view quality as the performance to standards; other people view it as meeting the customer's need or satisfying the customer.

Commonly, quality (conformance to specification context) measures how well the product or service meets the targets or tolerances determined by its designers. Quality (fitness for use context) focuses on how well the product performs its intended function or use. Quality (value for price paid context) is the usefulness of the product or service for the price paid. Quality (support service context) is the support provided after the product or service is purchased. Quality (psychological context) is the judgmental evaluations of what constitutes product or service excellence.

In the context of horticulture, the relative importance of each quality component depends upon the product and how it is utilised and varies among producers, handlers and consumers. To the producers, the product must have a high yield and good appearance, must be easy to harvest and must withstand long-distance shipping to markets. Appearance, firmness and shelf life are important to the wholesalers and retailers. To the consumers, appearances, freshness at the time of purchase, nutritional quality and safety are the important quality perspectives. The viability, vigour and freedom from insects and diseases are the quality standards of planting materials. Each one is important in determining the overall quality and consumers' acceptance of a horticultural product. In simplified form quality may be the fitness of use of a commodity or product on the basis of the set standards.

Quality may be a combination of characteristics that have significance in determining the degree acceptability of the product to a consumer. The International Organisations of Standards (IOS) defined quality as the totality of features and characteristics of a product or service that bear on its ability to satisfy stated or implied needs.

16.2.1 Total Quality Management

Quality is not always maintained in the case of production or export chain in a separate way, but the maintenance is required throughout the value chain from production to export of the product. The totality of quality management may be viewed as total quality management (TQM). So, in the era of globalisation and to uplift the status of the horticultural crop growers and exporters, the need of the hour is to manage the total quality during the production and marketing process of the horticultural produce. That can be obtained through introducing the culture of quality management within all the actors present in the production and marketing chain. For realising the highest degree of impact in the case of quality management, the organisational skill building in terms of organisational structure and method of operation, stock procurement, cultivation including labour and chemical inputs, harvesting, produce handling chain, export chain, quality control procedure and strategic planning through constant

monitoring is required for a desired produce with assured quality in an assured market at a remunerative price.

An effective quality assurance (QA) system throughout the handling steps between harvest and retail display is essential to provide a consistently good-quality supply of fresh horticultural crops to the consumers and to protect the reputation of a given marketing label. QA starts with the selection of the genotype and its proper time to harvest for the best appearance, texture, flavour (taste and aroma) and nutritional (including phytonutrients) quality. Careful harvesting and handling are required to minimise physical injuries. Each postharvest handling step has the potential to either maintain or reduce quality and in a few cases (such as ripening of climacteric fruits) improve eating quality. The availability of low-cost microcomputers and solid-state imaging systems has resulted in increased use of computer-aided video inspection to sort many products into two or more quality grades before marketing. Objective and non-destructive methods of differentiating horticultural perishables on the basis of their flavour and nutritional quality are being tested and will become excellent QA tools as they become more reliable and efficient. Safety assurance can be part of QA, and its focus is on minimising chemical and microbial contamination during production, harvesting and postharvest handling of intact and fresh-cut fruits and vegetables.

16.2.2 Quality Attributes

The quality attributes can be judged with the help of external factors and internal factors. For marketing of a horticultural produce, the external quality standards play a pivotal role. The practices available in the market for increasing the external quality of horticultural produce are waxing, de-greening, use of coloured mess bags, etc. But always, the consumers prefer internal quality of the produce along with the external qualities. The important quality criteria for the consumers of the horticultural produce are appearance including size, colour and shape, absence of defects, texture, flavour, nutritional value and safety.

The quality of fresh produce includes appearance (size, shape, colour, gloss and freedom from defects and decay), texture (firmness, crispness, juiciness, mealiness and toughness, depending on the commodity), flavour (sweetness, sourness (acidity), astringency, aroma and off-flavours) and nutritive value (vitamins, minerals, dietary fibre, phytonutrients). The relative importance of each quality component depends on the commodity and the individual's interest (Kader 1992). Most postharvest researchers, producers and handlers are product oriented in that quality is described by specific attributes of the product itself, such as sugar content, colour or firmness. In contrast, consumers, marketers and economists are more likely to be consumer oriented in that quality is described by consumer wants and needs (Shewfelt 1999). Although consumers purchase fresh produce based on appearance and textural quality, their repeat purchases depend upon their satisfaction with the flavour (taste and aroma). They are also interested in the health-promoting attributes and nutritional quality of fresh fruits and vegetables (Kader 1988).

16.2.2.1 Appearance
According to the consumers of fresh fruit and vegetables, the visual first-time appearance of the product is the foremost quality parameter. During purchase, the fast visual analysis is always carried out by the buyers according to the shape, size, colour and freshness. A bright red apple, curved yellow banana, uniform-sized orange or cabbage, etc. are higher valued in the market.

16.2.2.2 Organoleptic Quality
The organoleptic quality includes sourness, sweetness, astringency, bitterness, aroma and flavours. Nutrition is also an important consideration for purchasing a fresh horticultural commodity. The fresh fruit avocados have a demand in the market as it contains no cholesterol.

16.2.2.3 Food Safety
Contamination of plant and human pathogens and microorganisms is the main drawback of food safety in the case of fresh fruit and vegetables.

16.2.3 Quality Standards

16.2.3.1 International Standards
The International Standards Organisation (ISO) produced an international quality standard in 1987 that was largely based on British Quality Standards (BS 5750). The categories of standards are ISO 9000 (Quality management and quality assurance standards – Guidelines for selection and use), ISO 9001 (Quality systems – Model for quality assurance in design, development, production, installation and servicing), ISO 9002 (Quality systems – Model for quality assurance in production, installation and servicing), ISO 9003 (Quality systems – Model for quality assurance in final inspection and test) and ISO 9004 (Quality management and quality system elements – Guidelines).

16.2.3.2 Indian Standards in Agricultural Products
Indian standards in the agricultural sector have been framed by various organisations of the government, most of which are under the Ministry of Agriculture. These organisations are responsible not only for production and product standards but also for their inspection and quality control. Some of the major organisations involved are given below in table.

16.2.4 Agreement on Technical Barriers to Trade

This agreement deals with the product standards related to physical parameters such as size, colour, appearance, maturity, packaging and labelling requirements. Presently, for most of the horticultural produce, the standards stating the above requirements are not mandatory for domestic produce, in lieu of which, the imported products are enjoying the liberty of following their own standards as applicable in their country. As a result several of the packaged imported products are having labels which are in foreign languages. However, as per our national legislation on packaged item, it should be made incumbent on all packaged products to have labels either in English or Hindi.

16.2.5 Organisation Involved in Standards for Agricultural Products

No.	Departments	Organisation	Products covered
I	Agriculture and Cooperation	Directorate of Marketing and Inspection (Agmark) under the Ministry of Agriculture	Fresh fruits and vegetables, walnuts, flowers and spices
		Directorate of Plant Protection, Quarantine and Storage under the Ministry of Agriculture	Issue of phytosanitary certificate for export of fresh fruits and vegetables, flowers and seeds
		State Seed Certification Agencies	Seeds of fruits, vegetables and flowers
		Food Products Order (FPO) under the Ministry of Agriculture, Dept. of Food Processing Industries (DFPI)	All processed food products specially processed fruits and vegetables including pickles and chutneys
II	Consumer Affairs	Bureau of Indian Standards	All agricultural products
III	Commerce	Agricultural and Processed Food Products Export Development Authority (APEDA)	Grading and packing standards for exports
IV	Health and Family Welfare	Directorate General of Health Services	Standards on food hygiene

Apart from the product standards, there is also a need to develop a procedure for implementation and monitoring of these standards from import point of view. As per the above agreement, all the standards should be clearly laid down so that they are published in a transparent manner and other member countries may utilise these in order to promote their exports. In order to meet these conditions, there is an urgent need to specify procedure in this regard.

these with ISO standards for different commodities. Standards for growing and package requirements of international markets are also not available in a large number of commodities. The standards developed by the Directorate of Marketing and Inspection are old and outdated. There is, therefore, an urgent need to fix standards to promote domestic as well as international trade in agricultural/horticultural commodities.

16.2.6 Harmonisation of Standards

There is also a multiplicity of standards in several horticultural products. There is, therefore, an urgent need to not only rationalise standards fixed by various organisations but also harmonise

16.2.7 Quality Issues

In the light of recent changes in the international trade, to exploit the potential for export of horticultural produce to its full capacity, sincere efforts are required to be made to develop a full-fledged quality assurance machinery. This proposal would affect business of all sizes engaged in the production and trade of products and services throughout the supply chain. In addition to compliance with national legislation, this machinery will ensure that the following issues are considered:

- To ensure that all stages of production and distribution for which they are responsible are carried out in such a manner as to comply with food safety and consumer protection requirement
- To inform the competent authority if it considers or suspects that a product placed on the market does not comply with the relevant food safety requirement and of the action taken to prevent risk to the final consumer
- To ensure that no product is exported which does not comply with the general requirement of the food laws
- To have a system in place to identify the supplier of the product and the buyers or traders to whom they have supplied such products
- To adequately label and identify produce to allow its traceability and withdraw the product if it considers or suspects that it is not complying with food safety requirements

The above machinery would also be useful for ensuring that the products imported into India comply with conditions outlined above. In the present situation, there are chances that the developed countries could dump their products at a throwaway price to capture the Indian market along with the pests and diseases. Similarly it will ensure that genetically modified organisms (GMO) and microorganisms presently not existing in Indian soil do not enter into the food chain of the country.

There is a need to have a national approach for ensuring the quality and safety of food items including domestic, imported and exported products. This necessitates the overhauling of the current monitoring and control mechanism to ensure that the products entering into the Indian market comply with the national food safety laws. In order to give a systematic approach to this important issue, a two-way action plan is suggested.

16.2.7.1 Quality Building Machinery

Includes activities targeted to upgrade the status of food processing industries including handling of fresh produce to achieve good-quality and safe food products, through creating awareness and training among food industries. This has to be ensured by the industry.

16.2.7.2 Quality Assurance Machinery

To ensure development and implementation of produce and process standards through monitoring and certification, thereby ensuring the quality and safety of the final product. This is primarily the role of the government in the current scenario.

16.2.8 Laying Down Standards

The task of laying down the standards and their notification and ensuring their adherence may be divided as follows:

16.2.8.1 Voluntary Standards

A National Committee for Product and Process Standards should be formed under the umbrella of the Ministry of Agriculture. Its role would be to form different subject-wise subcommittees under various departments of the Ministry of Agriculture. The other members of the subcommittee may include various promotional boards and state departments. These subcommittees may be constituted product-wise as follows:

16.2.8.2 Subcommittee on Fresh Fruits and Vegetables

This committee may be formed under the chairmanship of the Directorate of Marketing and Inspection for developing standards for fresh fruits and vegetables. These standards should include the following requirements:

- Name of the standard
- Scope
- Description
- Essential composition and quality factors
- Food additives and contaminants

- Hygiene
- Weights and measures
- Labelling
- Method of analysis and sampling

In order to achieve the goal of safety of the product, the final product testing is not a reliable criterion. In view of this, it is required that the practices, which are implemented at the production and harvesting level, should also be improved. In view of this, process standards should be developed which would identify the critical points, which are required to be monitored at the production and the harvesting level. It would also provide reference in terms of guidelines for production of fresh fruits and vegetables during primary production and packaging.

16.2.8.3   Subcommittee on Processed Products

Processed products would include processed fruits and vegetables, nuts, alcoholic and non-alcoholic beverages, coconut, spices, tea, coffee, honey, medicinal and herbal preparations, etc. Product-wise subcommittees should be formulated to develop product and process standard under the umbrella of the Ministry of Agriculture, i.e., Department of Food Processing Industries. These standards should also be harmonised in terms of the format prescribed under international standards such as Codex Standards. The different product orders such as the FPO specification which lays down requirements related to the location of the factory, personnel hygiene, equipment requirements, its maintenance, etc. are also required to be modified in order to incorporate the hygienic practices as prescribed under Codex, i.e., General Principles of Food Hygiene.

16.2.8.4   Mandatory Standards

The standards related to sanitary and phytosanitary (SPS) aspects listed below are referred to as mandatory standards:

- Minimum residue level (MRL) for pesticides
- Heavy metals
- Mycotoxins
- Microbiological requirements
- Any other requirements related to food safety of processed horticultural products

All the requirements may be notified under the Prevention of Food Adulteration Act (PFA) because the major objective of PFA formulation is to protect the health of the consumer. Thereby, the rule formulated under the PFA would be mandatory for all the commodities for domestic consumption import and export.

16.2.8.5   Monitoring and Inspection Mechanisms to Ensure Implementation of Product and Process Standards

Once the product and process standards for various commodities are laid down and notified under the various departments, the next aspect which requires attention is the monitoring and inspection of the above standards. Primarily, the monitoring and inspection is the role of the government. However, this task may be carried out through the accreditation of inspection agencies which would be accredited the task of carrying out the inspection on behalf of the government before granting approval for the processing of the respective food commodities. In order to achieve this objective of accrediting the inspection bodies which may include both private and government bodies, the following action is required to be taken:

- Laying down the guidelines for assessment and accreditation of inspection and certification bodies which would also cover the legislative framework and other infrastructure requirements including control programmes
- Laying down the guidelines to be followed by inspection bodies for inspection and monitoring of food processing industries for certification

The responsibility of ensuring the implementation of the above product and process standards and also laying down the guidelines for assessment, accreditation and inspection would lie with the departments which are chairing product-wise subcommittees. These departments may also accredit private agencies which would further identify the certification bodies to carry out the inspection of food industries for compliance to the product and process standards on behalf of each department.

The overall evaluation of the situation related to the implementation and assurance of product and process standards may be with the National Committee for Product and Process Standards. It would include the yearly review of the status of national standards with respect to the changes taking place in the international scenario related to the aspects of quality and safety in food trade.

16.2.9 Quality Control and Assurance

Quality control (QC) is the process of maintaining an acceptable quality level to the consumer. Quality assurance (QA) is the system whose purpose is to assure that the overall QC job is being done effectively (Hubbard 1999). QA and QC are often used interchangeably to cover the planning, development and implementation of inspection and testing techniques; they take time and a lot of training. A successful QA/QC system cannot be flexible, but it must be subject to constant review and improvement as conditions change (Hubbard 1999).

Several attempts have been made to develop portable instruments with sensors that detect volatile production by fruits as a way to detect maturity and quality. Other strategies include the removal of a very small amount of fruit tissue and measurement of total sugars or soluble solids content. Near-infrared detectors have a great potential for a non-destructive estimation of sugar content in fruits (Abbott 1999). Many attempts are currently being made to automate the separation of a given commodity into various grades and the elimination of defective units. The availability of low-cost microcomputers and solid-state imaging systems has made computer-aided video inspection on the packing line a practical reality. Solid-state video camera or light reflectance systems are used for the detection of external defects, and x-ray or light transmittance systems are used for detecting internal defects (Abbott et al. 1999; NRAES 1997). Further development of these and other systems to provide greater reliability and efficiency will be very helpful in quality control efforts.

An effective quality control and assurance system throughout the handling steps between harvest and retail display is required to provide a consistently good-quality supply of fresh horticultural crops to the consumers and to protect the reputation of a given marketing label. Quality control starts in the field with the selection of the proper time to harvest for maximum quality. Careful harvesting is essential to minimise physical injuries and maintain quality. Each subsequent step after harvest has the potential to either maintain or reduce quality; few postharvest procedures can improve the quality of individual units of the commodity (Cavalieri 1999; Kader 1988, 1992, 1999; Shewfelt and Prussia 1993) (Table 16.1).

Exposure of a commodity to temperature, relative humidity and/or concentration of oxygen, carbon dioxide and ethylene outside its optimum ranges will accelerate the loss of all quality attributes. The loss of flavour and nutritional quality of fresh intact or cut fruits and vegetables occurs at a faster rate than the loss of textural and appearance quality. Thus, QC/QA programmes should be based on all quality attributes and not only on appearance factors as often is the case. More research is needed to identify the reasons for the faster loss of flavour than appearance quality and to develop new strategies for extending postharvest life based on flavour to match that based on appearance.

The guidelines to minimise the microbial food safety hazards on fresh fruit and vegetable have been developed by United States Food and Drug Administration (FDA) during 1998 along with recommendations for Good Agricultural Practices (GAP) as part of Standard Operating Practices (SOP). This concern for food safety leads to changes in quality-assured handling practices.

16.2.10 Standardisation and Inspection

Grade standards identify the degrees of quality in a commodity that are the basis of its usability and value. Such standards, if enforced properly, are essential tools of quality assurance during marketing and provide a common language for trade

Table 16.1 Handling steps and quality assurance procedure of the horticultural produces

Handling steps	Quality assurance procedure
Harvesting	Training workers on proper maturity and quality selection, careful handling and protecting produce from sun exposure
Packing house operation	Checking product maturity, quality and temperature upon arrival
	Implementing an effective sanitation programme to reduce microbial load
	Checking packaging materials and shipping containers to ensure they meet specifications
	Training workers on proper grading by quality (defects, colour, size), packing and other packing house operations
	Inspecting a random sample of the packed product to ensure that it meets grade specification
	Monitoring product temperature to assure completion of the cooling process
	Maintaining effective communications with quality inspectors and receivers to correct any deficiencies as soon as they are identified
Transportation	Inspecting all transport vehicles before loading for functionality and cleanliness
	Training workers on proper loading and placement of temperature-recording devices in each load
	Keeping records of all shipments as part of the "trace-back" system
Handling at destination	Checking product quality upon receipt and moving it quickly to the appropriate storage area
	Shipping product from distribution centre to retail markets without delay and on a first in/first out basis unless its condition necessitates a different order

among growers, handlers, processors and receivers at terminal markets. Some production areas like California, USA, enforce minimum standards concerning produce quality, maturity, container, marking, size and packing requirements. This provides orderly marketing and equity in the marketplace and protects consumers from inedible and poor-quality produce. The California Department of Food and Agriculture is also responsible for enforcing provisions of laws governing the sale of foods labelled as organic.

The US standards for fresh fruit and vegetable grades are voluntary, except when required by industry marketing orders, by the buyer or for export marketing. The USDA Agricultural Marketing Service is responsible for developing, amending and implementing grade standards.

To ensure uniformity of inspection, (1) inspectors are trained to apply the standards, (2) visual aids (colour charts, models, diagrams, photographs and the like) are used whenever possible, (3) objective methods for determining quality and maturity are used whenever feasible and practical, and (4) good working environments with proper lighting are provided. Recently, the Fresh Products Branch of the USDA's Agricultural Marketing Service equipped inspectors with digital cameras and enhanced computer technology for taking and transmitting images of produce or containers. AMS is offering the images to applicants over the Internet as an additional resource in its fresh fruits and vegetables inspection service. Inspectors also use the imaging to confer with produce quality experts working in USDA's headquarters in Washington, D.C. The imaging provides a quick, visual confirmation of product appearance and defects, damage from shifted loads, brands and container markings and container condition. It can facilitate "ecommerce" (buying and selling produce via the Internet) and help the produce industry quickly resolve disputes over the quality or condition of shipments.

International standards for fruits and vegetables were introduced by the Organisation for Economic Cooperation and Development beginning in 1961, and now there are standards for about 40 commodities. Each includes three quality classes with appropriate tolerances: extra class = superior quality (equivalent to "US Fancy"), class I = good quality (equivalent to "US No. 1") and class II = marketable quality (equivalent to "US No. 2"). Class I covers the bulk of produce entering into international trade. These standards or their equivalents are mandatory in the European Union countries for imported and exported fruits and vegetables.

16.2.11 Hazard Analysis and Critical Control Point (HACCP)

HACCP (hazard analysis and critical control point) is an internationally accepted systematic approach for preventing microbiological, chemical and physical contamination along the food supply chain. The HACCP technique does this by identifying the risks, establishing critical control points, setting critical limits and ensuring control measures are validated, verified and monitored before implementation.

The effective implementation of HACCP will enhance the ability of companies to protect and enhance brands and private labels, promote consumer confidence and conform to regulatory and market requirements.

All businesses involved in the food supply chain from producers to retailers can use HACCP. Enterprises include, but are not restricted to, those linked with fruits and vegetables, dairy products, meat and meat products, fish and fishery products, spices and condiments, nuts and nut products, cereals, bakery and confectionary, restaurants, hotels, fast-food operations, etc. HACCP is a management system in which food safety is addressed through the analysis and control of biological, chemical and physical hazards from raw material production, procurement and handling to manufacturing, distribution and consumption of the finished product. It involves a system approach for identification of hazards, assessment of chances to occurrence of hazard during each phase, raw material procurement, manufacturing, distribution, usage of food products and defining the measures for food hazard control. It is not simply an inspection approach, but it helps in avoiding the drawbacks in microbial testing. The HACCP involves the following seven principles:

16.2.11.1 Conduct a Hazard Analysis (Principle 1)

The purpose of the hazard analysis is to develop a list of hazards which are of such significance that they are reasonably likely to cause injury or illness if not effectively controlled. Hazards that are not reasonably likely to occur would not require further consideration within an HACCP plan. It is important to consider in the hazard analysis the ingredients and raw materials, each step in the process, product storage and distribution and final preparation and use by the consumer. When conducting a hazard analysis, safety concerns must be differentiated from quality concerns. A hazard is defined as a biological, chemical or physical agent that is reasonably likely to cause illness or injury in the absence of its control. Thus, the word hazard as used in this document is limited to safety.

16.2.11.2 Determine Critical Control Points (CCPs) (Principle 2)

A critical control point is defined as a step at which control can be applied and is essential to prevent or eliminate a food safety hazard or reduce it to an acceptable level. The potential hazards that are reasonably likely to cause illness or injury in the absence of their control must be addressed in determining CCPs. Complete and accurate identification of CCPs is fundamental in controlling food safety hazards. Critical control points are located at any step where hazards can be either prevented, eliminated or reduced to acceptable levels. CCPs must be carefully developed and documented. In addition, they must be used only for purposes of product safety.

16.2.11.3 Establish Critical Limits (Principle 3)

A critical limit is a maximum and/or minimum value to which a biological, chemical or physical parameter must be controlled at a CCP to prevent, eliminate or reduce to an acceptable level the occurrence of a food safety hazard. A critical limit is used to distinguish between safe and unsafe operating conditions at a CCP. Critical limits should not be confused with operational limits which are established for reasons other than food safety.

Each CCP will have one or more control measures to assure that the identified hazards are prevented, eliminated or reduced to acceptable levels. Each control measure has one or more associated critical limits.

16.2.11.4   Establish Monitoring Procedures (Principle 4)

Monitoring is a planned sequence of observations or measurements to assess whether a CCP is under control and to produce an accurate record for future use in verification. Monitoring serves three main purposes. First, monitoring is essential to food safety management in that it facilitates tracking of the operation. If monitoring indicates that there is a trend towards loss of control, then action can be taken to bring the process back into control before a deviation from a critical limit occurs. Second, monitoring is used to determine when there is loss of control and a deviation occurs at a CCP, i.e., exceeding or not meeting a critical limit. When a deviation occurs, an appropriate corrective action must be taken. Third, it provides written documentation for use in verification.

An unsafe food may result if a process is not properly controlled and a deviation occurs. Because of the potentially serious consequences of a critical limit deviation, monitoring procedures must be effective. Ideally, monitoring should be continuous, which is possible with many types of physical and chemical methods. Assignment of the responsibility for monitoring is an important consideration for each CCP. Specific assignments will depend on the number of CCPs and control measures and the complexity of monitoring.

16.2.11.5   Establish Corrective Actions (Principle 5)

The HACCP system for food safety management is designed to identify health hazards and to establish strategies to prevent, eliminate or reduce their occurrence. However, ideal circumstances do not always prevail and deviations from

established processes may occur. An important purpose of corrective actions is to prevent foods which may be hazardous from reaching consumers.

Where there is a deviation from established critical limits, corrective actions are necessary. Therefore, corrective actions should include the following elements: (a) determine and correct the cause of non-compliance, (b) determine the disposition of non-compliant product, and (c) record the corrective actions that have been taken. Specific corrective actions should be developed in advance for each CCP.

16.2.11.6   Establish Verification Procedures (Principle 6)

Verification is defined as those activities, other than monitoring, that determine the validity of the HACCP plan and that the system is operating according to the plan. One aspect of verification is evaluating whether the facility's HACCP system is functioning according to the HACCP plan. An effective HACCP system requires little end-product testing, since sufficient validated safeguards are built early in the process. Therefore, rather than relying on end-product testing, firms should rely on frequent reviews of their HACCP plan, verification that the HACCP plan is being correctly followed, and review of CCP monitoring and corrective action records.

Another important aspect of verification is the initial validation of the HACCP plan to determine that the plan is scientifically and technically sound, that all hazards have been identified and that if the HACCP plan is properly implemented, these hazards will be effectively controlled. Information needed to validate the HACCP plan often include (1) expert advice and scientific studies and (2) in-plant observations, measurements and evaluations.

In addition, a periodic comprehensive verification of the HACCP system should be conducted by an unbiased, independent authority. Such authorities can be internal or external to the food operation. This should include a technical evaluation of the hazard analysis and each element of the HACCP plan as well as on-site review of all flow diagrams and appropriate records from operation of the plan. A comprehensive verification is independent of other verification procedures and must be performed to ensure that the HACCP plan is resulting in the control of the hazards.

16.2.11.7 Establish Record-Keeping and Documentation Procedures (Principle 7)

Generally, the records maintained for the HACCP system should include a summary of the hazard analysis, including the rationale for determining hazards and control measures, critical limits, monitoring, corrective actions, verification procedures and schedule. For the food industry in India, the introduction and implementation of HACCP and sanitary and phytosanitary (SPS) standard are becoming imperative to reach global standards and demonstrate compliance to regulations and customer requirements.

16.2.12 Improving Traceability and Food Safety Through ICT

Traceability is an increasingly common element of public and private systems for monitoring compliance with quality, environmental and other product and/or process attributes related to food. Small-scale farmers may lack the resources to comply with increasingly strict food safety standards, particularly traceability requirements. Given the role of traceability in protecting consumers, ensuring food safety and managing reputational risks and liability, it is vital to integrate and empower small-scale horticultural producers in the food supply chain through ICTs. Traceability can improve the management of hazards related to food safety and animal health, guarantee product authenticity and provide reliable information to customers, enhance supply-side management and improve product quality. The benefits of traceability for consumers, government authorities and business operators are widely recognised. Yet for small-scale farmers in developing countries, especially farmers producing horticultural and other fresh food products, traceability requirements can represent barriers to trade. The market for safe and traceable food can exclude small-scale agricultural producers who lack the resources to comply with increasingly strict standards, particularly requirements for tracking and monitoring environmental and supply chain variables through sophisticated technologies. A wider access to ICTs may lift some of these barriers. The proliferation of mobile devices, advances in communications and greater affordability of nanotechnology offer potential for small-scale producers to implement traceability systems and connect to global markets. Mobile phones, radio frequency identification (RFID) systems, wireless sensor networks and global positioning systems (GPS) make it possible to monitor environmental and location-based variables, communicate them to databases for analysis and comply with food safety and traceability standards. In the context of food safety and smallholders' participation in global markets, the module which includes detailed information on standards, technical solutions and innovative practices explores incentives for investing in traceability systems and the prospects for traceability to empower small-scale producers in the value chain.

16.3 Conclusion

In the realm of globalisation and free trade liberalisation, there is a need to introduce and implement total quality management strategies to ensure the quality perspective of horticultural produce for fetching foreign exchequer. Total quality management is not only a management strategy but also it is a holistic system perspective in this market-led horticulture era to empower the growers of horticultural produce through changing them from mere producer to high-value entrepreneur. Along with this, ICT can play a pivotal role to assure the market for small producers of horticultural produce by complying the food safety and traceability standards in total quality management.

References

Abbott JA (1999) Quality measurement of fruits and vegetables. Postharvest Biol Technol 15:207–225

Abbott JA, Lu R, Upchurch BL, Stroshine R (1999) Technologies for non destructive quality evaluation of fruits and vegetables. Hortic Rev 20:1–120

Cavalieri RP (ed) (1999) Fresh fruits and vegetable: quality and food safety. Postharvest Biol Technol 15:195–340

Hubbard MR (1999) Choosing a quality control system. Technomic Publ. Co., Lancaster, p 207

Kader AA (1988) Influence of pre harvest and postharvest environment on nutritional composition of fruits and vegetables. In: Quebedeaux B, Bliss FA (eds) Horticulture and human health, contributions of fruits and vegetables. Prentice-Hall, Englewood Cliffs, pp 18–32

Kader AA (ed) (1992) Postharvest technology of horticultural crops, 2nd edn. University of California, Division of Agriculture and Natural Resources, California, USA, Publ. 3311, p 296

Kader AA (1999) Fruit maturity, ripening, and quality relationships. Acta Hortic 485:203–208

Northeast Regional Agricultural Engineering Service (1997) Sensors for nondestructive testing – measuring the quality of fresh fruits and vegetables. In: International conference proceedings, NRAES – 97, NRAES, Ithaca, NY, p 440

Shewfelt RL (1999) What is quality? Postharvest Biol Technol 15:197–200

Shewfelt RL, Prussia SE (eds) (1993) Postharvest handling: a systems approach. Academic, San Diego, p 358

Supply Chain Management of Horticultural Crops

17

T.N. Roy

Abstract

The supply chain management (SCM) is the active management of supply chain activities to maximise customer value and achieve a sustainable competitive advantage. Horticulture sector is dominated mainly by small holdings. Horticultural markets are nearly imperfect in nature. The estimates of the marketed surplus for the year 2011–2012 indicate that the quantum of marketed surplus will be more for fruits and vegetables than for food grains and oilseeds (GoI 2007). Only 2 % of horticultural produce is processed, 0.4 % is exported and 22 % is lost or wasted in the market chain (Singh 2009). Studies on the present status of the supply chain and its management for horticultural produce suggest for more investment and extension activities. Market infrastructures in terms of storage (godown) facilities, cold storages, loading, weighing facilities, improvement in the road network and cold chain facilities constitute the basics of an efficient supply chain management. Integrating small farmers with high-value urban and export markets and provision of up-to-date market information through ICT would lead to the development and growth of the sector and can act as a model for SCM. Due to constraints in government investment, a private investment or public-private partnership (PPP) model may be encouraged. All these efforts may help to bring the horticultural sector into a systematic supply chain and logistics management for its growth and development.

17.1 Introduction

17.1.1 General

The supply chain management is the management of a network of interconnected businesses involved in the ultimate provision of product and service required by the end customer (Mentzer

T.N. Roy (✉)
Department of Agricultural Economics,
Uttar Banga Krishi Viswavidyalaya,
P.O. Pundibari, Cooch Behar 736165,
West Bengal, India
e-mail: tuhinnroy@rediffmail.com

A.B. Sharangi and S. Datta (eds.), *Value Addition of Horticultural Crops: Recent Trends and Future Directions*, DOI 10.1007/978-81-322-2262-0_17, © Springer India 2015

2007). The horticultural supply chain is the entire vertical chain of activities from the supply of input (seed, fertiliser, chemicals) through production, postharvest operations, distribution and retail.

The estimates of the marketed surplus from agri-commodities for the year 2011–2012 indicate that the quantum of marketed surplus will be more for fruits and vegetables than for food grains and oilseeds (FAO 2009). Only 2 % of horticultural produce is processed, 0.4 % is exported and 22 % is lost or wasted in the market chain (Singh 2009). The efforts to increase horticultural production without giving much emphasis on developing adequate market infrastructure may have unfavourable implications for farmers.

Presently, not many details are available in the public domain related to the supply chain management of agricultural commodities in India. There is a need for a new revolution to bring down the prices of agricultural produce for consumers through an efficient supply chain management and incentivise farmers to increase their production. The supply chain for horticultural products in India is highly fragmented and skewed away from producers for its inherent features like small landholding, illiteracy, poor access to organised finance, markets and information (Singh 2009). A number of supply chains are operating in India for the movement of commodities from the farm gate to the ultimate consumer.

The supply chain management (SCM) is the most recent addition in the jargon of management due to its capability to offer the best possible customer service in a cost-efficient manner and gain a comprehensive edge by a flexible approach towards movement (Agrawal 2010). The concept of the supply chain management is based on two core ideas. The *first* is that practically every product that reaches an end user represents the cumulative effort of multiple organisations. These organisations are referred to collectively as the supply chain. The *second* idea is that while supply chains have existed for a long time, most organisations have only paid attention to what was happening within their 'four walls'. Few businesses understood, much less managed, the entire chain of activities that ultimately delivered products to the final customer. The result was disjointed and often ineffective supply chains.

The supply chain management, then, is the active management of supply chain activities to maximise customer value and achieve a sustainable competitive advantage. It represents a conscious effort by the supply chain firms to develop and run supply chains in the most effective and efficient ways possible. Supply chain activities cover everything from product development, sourcing, production and logistics, as well as the information systems needed to coordinate these activities.

The organisations that make up the supply chain are 'linked' together through physical flows and information flows. Physical flows involve the transformation, movement and storage of goods and materials. They are the most visible piece of the supply chain. But just as important are information flows. Information flows allow the various supply chain partners to coordinate their long-term plans and to control the day-to-day flow of goods and materials up and down the supply chain.

Literally, the *supply chain management* (*SCM*) is the systemic, strategic coordination of the traditional business function within a particular company and across businesses within the supply chain, for the purposes of improving the long-term performance of the individual companies and the supply chain as a whole. In other words, SCM is an integrated management of various functions in the areas of materials, operations, distributions, marketing and services after sales with a customer focus in perspective so as to synergise various processes in the organisation with a view of optimising the total cost, i.e. it refers to a management process of a joint approach of all supply chain participants to design, develop and operate a system which responds to customer expectations by making available the right quantity and right quality products at the right time and place in the right physical form at a right cost. Hence, SCM facilitates to offer the best customer service in a cost-efficient manner (Agrawal 2008). OR, a *supply chain*, is a set of three or more companies directly linked by one or more of the upstream (supply of inputs or materials) and downstream (distribution) flows of products, services, finances and information from a source to a customer (Mentzer 2007). The supply chain is viewed as a single process. Responsibility for the various segments in the chain is not fragmented and relegated to functional areas such as manufacturing, purchasing, distribution and sales.

17.1.2 Supply Chain Management and Corporate Perspectives

Corporate Perspectives

- Corporations have turned increasingly to global sources for their supplies. This globalisation of supply management has forced companies to look for more effective ways to coordinate the flow of materials into and out of the company.
- Companies and distribution channels compete more today on the basis of time and quality. Providing a defect-free product to the customer faster and more reliably than the competition is no longer seen as a competitive advantage but simply a requirement to be in the market. Customers demand products consistently delivered faster, exactly on time and with no damage. Each of these necessitates closer coordination with suppliers and distributors.
- The global orientation and increased performance-based competition combined with rapidly changing technology and economic conditions all contribute to marketplace uncertainty. This uncertainty requires greater flexibility on the part of individual companies and distribution channels, which in turn

demands more flexibility in channel relationship.

All these factors have made the concept of the supply chain management (SCM) more important to business sectors. Despite the popularity of the term SCM, some authors define SCM in operational terms involving the flow of materials and products (Tyndall 1988); others view it as a management philosophy (Ellram and Cooper 1990), and still others view it in terms of management process (Koontz & Weihrich 2012).

Thus, the objective of managing the supply chain is to synchronise the requirements of the customer with the flow of materials from suppliers in order to effect a balance between what are often seen as conflicting goals of high customer service, low inventory and low unit cost (Mentzer 2007).

The term *logistics management* has become a popular key word in recent years, and often there is confusion between its meaning and that of SCM; in many cases, they are used interchangeably. However, logistics contrasts with SCM, in that it is '*the work required to move and position inventory throughout the supply chain*' (Bowersox et al. 2002). The recent (1998) definition from the Council of Logistics Management is '*logistics is that part of the supply chain process that plans, implements, and controls the efficient, effective flow and storage of goods, services and related information from the point of origin to the point of consumption in order to meet customer's requirements*'. Logistics has become a strategic issue for almost all the industrial sectors due to its significant mission of ready availability of goods, best possible customer service and minimum logistic costs. Logistics management has strategic, primary and supportive components which have a wide range of functions in production, procurement and physical distribution functions for continuous improvement. That is why logistics management nowadays contributes significantly at both the macro- and micro-level. It has the capability to create an unbeatable long-run competitive advantage in the marketplace marketplace (Reddy 2010).

17.1.3 Roles, Consequences and Driving Forces of the Supply Chain Management

Importance of the Supply Chain Management
- To compete in the global market and network economy.
- Inter-organisational supply network can be acknowledged as a new form of organisation.
- A network structure fits neither 'market' nor 'hierarchy' categories.

Roles of the Supply Chain Management
- A communicator of customer demand from point of sale to supplier
- A physical flow process that engineers the movement of goods

Benefits of the Supply Chain Management
- Reduction of product losses in transportation and storage.
- Increasing sales.
- Dissemination of technologies, advanced techniques, capital and knowledge among chain partners.
- Better information about the products, markets and technologies.
- Transparency in the supply chain.
- Tracking and tracing to the source.
- Better control of product safety and quality.
- Large investments and risks are shared among partners in the chain.

The Major Consequences of SCM
- Improved customer value and satisfaction
- Profitability (lower costs and/or increased revenue)
- Differential advantages

Since the beginning of the 1990s, Indian economy has experienced a paradigm change in the business scenario, mainly due to the liberalisation policy of various economics all over the world and revolutionary innovations in the field of science and technology. Particularly, an information technology and communication infrastructure has resulted into a continuous acceleration in the magnitude of competition. To sustain themselves in such an erratic environment, firms need to have core competency and productivity. This is why firms are perceived to have more systematised activities related to movement and storage of goods so as to make them available at a short notice with the lower inventory level.

17.1.4 Driving Forces of the Supply Chain

There are four main driving forces of a successful supply chain management, some of which require significant paradigm shifts in traditional organisational thinking. These are:
- Development of *strategic alliances* of firms with specialised skills
- Creation of *organisational structures* that facilitate communication, information sharing and transparency between partners
- *Human resource partnerships* with all levels of staff in all firms in the chain having a common vision and commitment to excellence
- Utilisation of advanced *information technology* systems including electronic media, barcoding, GPS systems and appropriate software that allows instantaneous and timely feedback

17.1.5 Supply Chain and Value Chain

17.1.5.1   Value Chain in SCM

Nowadays, corporate enterprises are to a large extent developing a supply chain system in their organisations in order to gain a competitive edge in the marketplace due to its value addition capability in cost-effective ways. In other words, SCM ensures superior customer value for core competency by a blend of better-quality logistical services at minimum costs. That is why the value chain of SCM focuses on speeder flow of goods, cash, value and related information within the complete supply chain process.

| Inbound Logistics | ▶ | Operations | ▶ | Outbound Logistics | ▶ | Marketing & Sales | ▶ | Services |

Michael Porter's Primary Value Chain Activities

17.1.5.1.1 Primary Value Chain Activities

1. *Inbound logistics* – receiving, warehousing and inventory control of raw materials (inputs) used in the manufacturing process
2. *Operations* – a *process* of transforming inputs into finished goods or services
3. *Outbound logistics* – the warehousing and transportation activities required to get the finished goods to the customer
4. *Marketing and sales* – identifying customer needs and generating sales of the product or service
5. *Service* – activities which enhance or maintain the product or service's value after the sale of the product (customer support, repair services, etc.)

The value chain is intrinsically tied to supply chain activities regarding inbound and outbound logistics. Value chain activity analysis requires the firm to look outside its four walls for solutions and consider the benefits of having a synchronised supply chain to increase service levels and/or lower the delivered cost of the product or service.

Basic differences between supply chains and value chains

Supply chains	Value chains
Talk about the downstream flow of goods and supplies from the source to the customer	Value flows the other way
Customer is the source of value	Value flows from the customer in the form of demand
Focus on integrating supplier and producer processes	Focus on creating value in the eyes of the customer

Distinction is lost in the business and research literature. Therefore, some suggest that the supply chain management focuses more on the internal process with an emphasis on the efficient flow of resources, such as materials, while the value chain management has similar aims with an additional concern for the external environment, such as the customer.

In this context, the concept of *marketing channel* also assumes much relevancy which indicates the route/path of movement of any produce from the point of production to the ultimate consumer (Kohl 1985). It only serves the statement of information about the direction of marketed produce.

Examples of marketing channel, supply chain and value chain in Indian situations

Rice paddy/pulses	Marketing channel
Sugarcane/milk/poultry	Supply chain
Processed fruit-food/organic vegetables	Value chain

17.2   Agricultural Supply Chains in India

17.2.1   General

The supply chain management (SCM) in agriculture poses enormous challenges to the processors, market intermediaries and policymakers (Woods 2004). A large part of the agri-supply chain, including its allied sectors like livestock, poultry, fisheries, dairy, etc., is still in an unorganised or semi-organised state entailing a highly inefficient supply chain system affecting the whole gamut of agribusiness in India. They are exposed to risks on various fronts such as input supply, production, marketing and policies.

Management of agricultural supply chain, through effectively integrating farmers with the input and output market, can greatly influence the well-being of people dependent on agriculture for livelihood. It also influences the actors who provide services to the agricultural sector such as input suppliers, equipment makers, credit providers and R&D institutions. While rising population, reduced cultivable land, increased cost of production and rising input costs and marketing costs possess enormous pressure on market players, the emergence of organised retail chains, the steady rise in the per capita expenditure, education and expanding middle-class and young population have greatly increased the demand of a variety of quality agriproducts, which calls for an efficient organised supply chain. Though agriculture is the mainstay of the Indian economy, it remained unaffected by global market forces before liberalisation. After changing the concept from agriculture to 'agribusiness', this sector assumed much economic interest. The entire value chain from the farm gate to food plate is dynamically changing. Corporates have started investing in the agri-chain and are getting closer to the farmer, and the farmer is now getting access to the value chain. With easier supply chains and use of modern technologies, the Indian farmer is competing on a global level. Indian agricultural sector is very big with huge cultivable land (52 % of the total land is under cultivation to global average of 11 %), favourable climatic conditions and cost-effective skilled labour.

17.2.2 The Supply Chain Matrix

The agribusiness sector consists of organisations which:
- Supply raw materials to agriculture
- Supply capital goods to agriculture
- Procure agricultural produce for selling
- Procure agricultural produce for processing
- Provide services to agriculture

Alongside this would become an impressive network of infrastructures from precoolers and pack houses to cold stores and refrigerated trucks. A modern supply chain comes at high cost, and the burden is borne by the consumer.

CRISIL Research has estimated the total avoidable supply chain costs in the F&G vertical in India at about Rs. 1 trillion. About 57 % of this

is due to avoidable wastage, and about 43 % is due to avoidable costs of poor storage and commissions. Reduced supply chain costs arising out of lower wastage and storage costs can be shared between producers and consumers. Realisations earned by the farmers at current levels are estimated at around Rs. 1.8 trillion. Assuming this segment shifts entirely to organised retailing and two thirds of the savings from reduced supply chain inefficiencies are passed on to the farmer, farm incomes could grow by more than 37 %. With 60 % of India's population engaged in agriculture for livelihood, this is economically very important.

17.2.3 Main Issues in the Supply Chain of Agricultural Produce

- Average of six intermediaries in the agri-supply chain.
- Lack of proper transport and cold storage infrastructure.
- Too many procedural bottlenecks.
- Agriculture is disjointed as there is no linkage between production of agricultural produce and its demand in the market.

17.2.4 Present Status of Agricultural Supply Chain

Agricultural supply chains assume much importance in Indian economy for its role of facilitating and supporting in the livelihood of millions of farmers as well as stakeholders and other people. These involve both backward and forward linkages among all the stakeholders, i.e. input companies, government institutions, market intermediaries, consumers and farmers. At the back end of the supply chain are private sector and public sector companies that manufacture, trade and export inputs like seeds, pesticides, fertilisers, farm machinery, etc. for the farmers' use. Information is also counted as an important input these days. Farmers need to have information on farming practices, weather, sowing and harvesting time, pest management, fertiliser use, etc. Farmers need information on new products and brands launched by these companies. Information is the most important input which can ensure the availability of all other required inputs at the right place and right time. Presently, due to the lack of proper information on inputs, advisory services and weather and climate information services, farmers are unable to align the quality and price of their products to the market standards (Narula 2009).

Marketing of agricultural products in India takes place through agricultural mandis, which are regulated by the Agricultural Produce Market Committee Act (APMC Act). The produce is brought to these mandis by farmers, where a long chain of intermediaries is involved. The price of the commodity is decided through behind-the-scene auctions by the Aartiyas[1] who function as intermediaries. During the peak season, the farmer sometimes has to wait for many days to get his produce unloaded. These mandis lack basic marketing infrastructures such as grading, standardisation and storage facilities. In the process, the quality of the produce deteriorates. It has been found that postharvest losses in mandis occur primarily due to the lack of marketing infrastructures and storage facilities. After the produce gets unloaded, the farmer is at the mercy of the Aartiyas, as they are the ones who decide the price of their produce. The farmer has to sell at whatever price is decided by these intermediaries. There is a lack of transparent weighing facilities in the mandis. All these constraints lead to inefficiencies in the value chain and result in a very small monetary share for the farmer.

---

[1] Aartiya is the common name for traditional middlemen in the agricultural marketing system in India, who procure agricultural produce from farmers and charge commission. They make the maximum margin in the value chain, but add very little value to it.

17.3 Present Status of Horticultural Supply Chain in India

17.3.1 General

There has been an emergence of more coordinated supply chains for fruits and vegetables in India catering to the export market and to the high-end domestic market. The coordinated supply chain involves structured relationships among producers, traders, processors and buyers whereby detailed specifications are provided as to what and how much to produce, the time of delivery, quality and safety conditions and the price. The coordinated supply chains fit well with the logistic requirement of modern food markets, especially for fresh and processed perishable foods. These chains can be used for the process control of safety and quality and are more effective and efficient than control only at the end of the supply chain (Raju 1998).

Recently, a terminal market for fruits and vegetables has been set up in Bangalore known as SAFAL (www.Safalindia.com). Physically, it can handle up to 1,600 tonnes of produce per day. It is linked to some 250 farmer associations and 40 collection centres that have been established in selected producing areas. The market receives sorted, graded and packaged produce from these associations and centres, and then this is auctioned at the market. SAFAL also has forward linkages to a number of retail outlets. The market has a modern infrastructure including temperature-controlled storage facilities and ripening chambers.

Normally, small and marginal fruit and vegetable growers are unable to get remunerative price for their produce due to the small marketable surplus and highly perishable nature of produce. To support such farmers is the need of the hour. For this, they should form fruit and vegetable cooperative societies/self-help groups (SHGs)/FIGs. The existing models for fruit and vegetable cooperative society is presented in

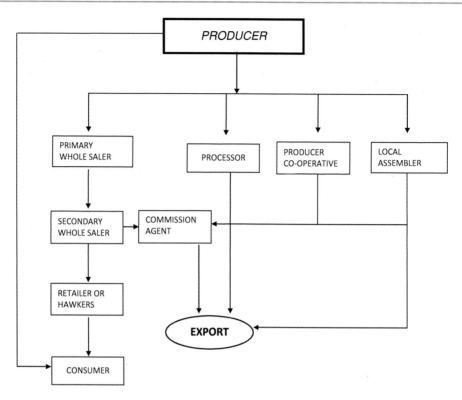

Fig. 17.1 Existing marketing channel (supply chain) of horticultural products in India

Fig. 17.2 Cross-border supply chain

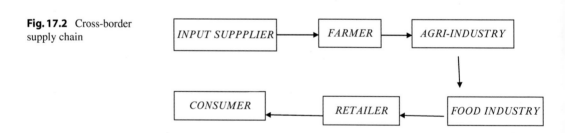

(Figs. 17.1 and 17.2). This will not only help in the marketing of small surplus of fruits and vegetables but also provide remunerative price of produce to the small and marginal farmers (Singh 2009).

The horticultural crops differ from the food crops like cereals with respect to certain natural characteristics like moisture content, texture, unit size, etc. that make them highly perishable, resulting in postharvest losses. These crops need special attention in ensuring time, space, form and possession utilities. This aspect should receive higher priority in order to improve the marketing system of horticultural crops. Since production centres are usually located at far remote areas, it is necessary to transport them to established market or urban market with efficient supply chain mechanism. Horticultural crops are also capital and labour intensive and need care in handling and transportation. Their bulkiness makes the handling and transportation a difficult task. The inefficiencies in the system lead to huge postharvest loss which is estimated at around Rs. 50,000–52,000 crore (25–30 % of the produce), imposing an additional burden on both the grower and the consumer (Mittal 2007) annually. Their seasonal production pattern results in frequent market gluts and associated price risk, thereby forcing the farmers into distress sale to preharvest contractors and commission agents.

In some areas, fruits are sold through the preharvest contractor (PHC) at the field much before they come to harvest. Very often, the PHC takes most of the production risks due to pests and diseases and also the cost of maintenance, while he makes his margin through bulking. Vegetables, barring cabbage and cauliflower are mainly sold through the commission agents at the market, who, in turn, transport the produce to the distant markets and make his/her margin; traditional flowers are self-marketed at the wholesale auction centres. Horticulture development is currently constrained by poor marketing arrangements. The gap between prices received by the farmers and those paid by urban consumers is large, reflecting inefficient marketing arrangements established (Hewett 2010).

The institutional arrangement through the regulated markets established by the government under the Agricultural Produce Market Committee Acts was to regulate trade practices, increase marketing efficiency by reducing marketing charges, eliminate intermediaries and protect the interests of the producer-seller. Horticultural produce is typically collected from farmers by market agents, who sell it in this organised market. Though regulated markets helped to reduce multiple charges to the producer-seller, the system failed to check trade malpractices, making such markets highly restrictive, inefficient, highly non-transparent and dominated by traders. The net result is much lower realisation of income by the farmer.

To overcome the defects of regulated markets, direct marketing by farmers was experimented by Apni Mandis in Punjab and Haryana, Rythu Bazars in AP and Uzahvar Santhaigal in TN. In the meantime, private players such as Cargill India, Mahindra, ITC e-Choupal, Bharti, etc. have emerged with sophisticated supply chain management systems and vertical co-ordination in India.

17.3.2 Marketing Channels of Horticultural Produces

Marketing channel is the chain (now called supply chain) of intermediaries through whom the various agricultural/horticultural food products pass from producers to consumers. The length of the channel varies from commodity to commodity, depending on the quantity to be moved, the form of consumer demand and degree of regional specialisation in production. The marketing channel of horticultural produce mainly fruits and vegetables varies from commodity to commodity and from producer to producer. In rural areas and small towns, many producers perform the function of retail sellers. Large producers directly sell their produce to the wholesalers or processing firms. The price spread along the marketing channel is directly proportional to the number of market intermediaries involved along the channel (Gupta and Rathode 1998). Some of the common marketing channels for vegetables and fruits are:

1. *Producer to consumer*
2. *Producer to primary wholesalers to retailers or hawkers to consumer*
3. *Producer to processors (for conversion into juice, preserves, etc.)*
4. *Producer to primary wholesaler to processor*
5. *Producer to primary wholesaler to secondary wholesaler to retailer or hawker to consumer*
6. *Producer to local* assembler to primary wholesaler to retailer or hawker to consumer

An important feature of marketing channels for fruits and vegetables is that these commodities just move to some selected large cities/centres and subsequently are distributed to urban population and other medium-sized urban market centres. The wholesale markets of these urban centres work as transit points and thus play an important role in the entire marketing channel for fruits and vegetables. *Large wholesale markets* for fruits and vegetables are concentrated in ten major cities, viz. Delhi, Kolkata, Bengaluru, Chennai, Mumbai, Jaipur, Nagpur, Vijayawada, Lucknow and Varanasi. These cities account for

75 % of vegetables marketed in major urban areas in India. Further, the transit trade takes place through the cities with more than 20 lakh population which account for 68 % of the fruits and vegetables grown in the respective regions. There are 65 urban wholesale markets for fruits and 81 for vegetables. Each market, on an average, serves a population of about more than 7 lakhs.

The efficiency of a supply chain depends upon the extent to which both our backward and forward linkages are integrated with all the functions in the supply chain so that all the stakeholders involved are benefitted. To reap the benefits of the existing opportunities, there is a need to circumvent the agricultural supply chains by removing the inefficiencies in the marketing system. The government is taking steps to improve the system in the country by permitting contract farming and direct procurement from farmers through amendments of the APMC Act. Several schemes have been launched such as the marketing infrastructure scheme, rural godown scheme and AGMARKNET (agricultural marketing network of Govt. of India) to empower the farmers. Besides government efforts, corporate giants, which have entered into food retailing, are also investing in the supply chain and are trying to build up an information network to facilitate the agricultural marketing trade in the country.

17.3.3 Major Players in Food and Vegetable Retailing in India

In India, the supply chain industry in the sector of food and vegetable, milk and fish is developing through the following emerging models:
- Cooperative (Hopcoms, Karnataka; Mother Dairy, Delhi)
- Exports with EurepGAP certification (Namdhari's Fresh, Bharti Airtel)
- Farm to Fork: complete chain (Godrej, Reliance, ITC)
- Wholesaling (Adani Agrifresh Ltd., METRO)
- Front end: convenience stores (Food Bazaar, 3Cs)
- Economy stores (Subhiksha)

17.3.4 Vegetable Retail Models

Distinct and primary routes adopted in the retail vegetable marketing have been revealed by this exploratory study. The study found three business models of vegetable retailing. Traditional retailers follow the *traditional retail model* (TRM), and organised retailers implement two different business models – 'hub-and-spoke model' (HSM) and *value chain model* (VCM). *Reliance Fresh* (Reliance Retail Ltd.) strategically deployed the value chain model, and the rest of the organised players in the industry go with the hub-and-poke model with minor modifications to fit in to their marketing and logistical strategies.

17.4 Institutions Involved in Marketing and Supply of Horticultural Produce

17.4.1 General

Marketing institutions are business organisations which have come up to operate the marketing machinery. In addition to individuals, corporate, cooperative and government institutions are operating in the field of agricultural marketing. They perform one or more of the marketing functions. They assume the role of one or more marketing agencies. Some important institutions in the field of agricultural marketing are:
(a) **Public Sector Institutions**
 1. *Directorate of Marketing and Inspection (DMI)*
 2. *Commission for Agricultural Costs and Prices (CACP)*
 3. *Specialised Commodity Boards*
  - *Rubber Board*
  - *Tea Board*
  - *Spices Board*
  - *Coconut Development Board*
  - *National Oilseeds and Vegetable Oils Board*
  - *Cardamom Board*
  - *Arecanut Board*
  - *Coir Board*
  - *National Horticulture Board (NBH)*

4. **Others**
   - *Central Warehousing Corporation*
   - *State Warehousing Corporation*
   - *Agricultural and Processed Food Products Export Development Authority (APEDA)*
   - *The Cashew Export Promotion Council of India (CEPCI)*
   - *Agricultural Produce Market Committee (APMC)*
   - *State Directorate of Agricultural Marketing*

(b) **Cooperative Sector Institutions**
   1. *National Cooperative Development Corporation (NCDC)*
   2. *National Agricultural Cooperative Marketing Federation (NAFED)*
   3. *Tribal Cooperative Marketing Development Federation (TRIFED)*
   4. *State Cooperative Marketing Federations*
   5. *Primary Agricultural Co-operative Marketing Societies*

17.4.2 Need for Supply Chain Management for Horticultural Crops/Products

India's horticultural sector is gaining much more importance than before for its contribution towards investment, production and export. This paradigm shift has occurred due to the operation of the following macroeconomic parameters:
- Natural resources for agro-based industry
- Population and demographic changes
- Integration of primary sector
- Secondary sector and tertiary sector
- Emergence of organised retail
- Emergence of technologies
- Globalisation
- Role of the government
- Agri-export zones (AEZs)
- Private sector initiatives

It is believed that the horticultural sector can be promoted as a means of agro-diversification for the second green revolution in India, providing the much-needed impetus to the growth of the agricultural sector, through the increase in trade, income and employment generation. During the past decade, the shift in the cropping pattern has been more towards the horticultural sector and commercial crops like cotton. The overall decline in area under food grains during the past one-and-a-half decade is very close to the magnitude of increase in area under horticulture. Within the horticultural sector, the shift in the total area under fruits and vegetables is 44.75 %. On the production front, fruits and vegetables together are the major contributors to the total horticultural production. India can be a market leader in the agricultural sector for horticultural produce if its markets function properly. Since the organised retail sector has started showing interest in fresh fruit and vegetable marketing and already some of them have entered into food retailing with huge investments, the supply chain management, i.e. from the farm to fork, is still in a very pathetic state.

Favourable and diverse climatic situations ensure competitive horticultural crops in India, growing all varieties of fresh fruits and vegetables. It ranks second in fruit and vegetable production in the world, after China. As per National Horticulture Database 2011 published by the National Horticulture Board, during 2010–2011, India produced 74.878 million metric tonnes of fruits and 146.554 million metric tonnes of vegetables (impressive 30 % rise during the last 5 years). The area under cultivation of fruits stood at 6.383 million hectares, while vegetables were cultivated at 8.495 million hectares. This has become a vital sector contributing 24 % of India's agriculture GDP. India produces almost 11 % of global vegetable output and 15 % of global fruit production. It also ranks among the lowest-cost producers of fruits and vegetables. However, the country's share in international trade remains low – almost 1.7 % in vegetables and 0.5 % in fruits.

17.4.3 Major Constraints Faced by the Horticulture Sector in India

- *Low productivity*
- *Lack of availability of latest varieties*
- *Outdated agro-practices/lack of modern agro-technologies (including mechanisation)*
- *Weak postharvest management infrastructure*
- *Poor processing capabilities*: The level of processing perishable products is 6 % (China has got 30 %). The $70 billion Indian food processing industry is dominated by small and medium enterprises, which do not have the capacity to undertake large-scale processing of fruits and vegetables. The World Bank believes that huge investments by the retail biggies in the supply chain infrastructure could usher in a service revolution that would shorten the distance that fresh produce travels to reach the consumer. In a supply chain analysis of 13 high-value commodities (HVC) that covered 1,400 farmers, 200 commission agents and 65 exporters across the country, the Bank found that high transport costs and multiple players in the linear supply chain were crippling horticulture. India is a large low-cost producer of fruits and vegetables but is unable to compete in the global market on account of what it terms the logistics tax on fresh farm produce. The inefficiencies in the system also mean that 25–30 % of the produce (valued at Rs. 50,000–52,000 crore) is wasted, imposing an additional burden on both the grower and the consumer (TCI 2012).

The biggest challenge the Indian agricultural sector faces is the inefficiency in the supply chain. These new initiatives are trying to strengthen the supply chain. At present, the horticultural supply chain has multilayered marketing channels and lacks market infrastructure. To improve efficiency, proper integration of postharvest technology into the marketing supply chain is crucial. Cultivable waste of about 25–30 % is reported because of inefficient management. Thirty percent of India's fruit and vegetable produce is wasted because of lack of cold chain. On the policy front, there is a need to integrate agricultural markets and supply chains. In the existing traditional system of wholesale marketing, the commission agents and traders dominate the supply chain and are the major price setters, and most of the time, farmers have to depend on them for credit. Small farmers lack marketing power and have low share in the final consumer price. Since the produce is marketed through the commission agents, there is no incentive for its quality. The wholesale markets are poorly designed with non-existence of infrastructure for packing, grading, sorting and cold storage.

There have been major changes in the supply chain in agro and food products over the last decade. The more liberalised international trade system, globalisation and transition towards the consumer-driven market have empowered this development. The international trade in fruits and vegetables has expanded at a higher rate over the last 25 years than the trade in agricultural commodities (Huang 2004). This global food system means opportunities for further growth of this sector. This will not be an easy challenge: worldwide competition for supply is very high. Understanding the market and consumer demand and building up business relation are the only requirement. There can be a profitable business for those who met the high demand for quality and safety of the products. Furthermore, in a specific consumer segment, other added value concepts are needed like healthy convenient products produced in a sustainable manner. Since chain performance is needed to be organised in an effective and efficient way, the use of state-of-the-art postharvest technology and supply chain management concepts contributes to 'making the difference'.

17.4.4 Horticulture: A Potential Agribusiness

The horticultural sector in India is being affected by various factors. In spite of that, the country offers immense potential for various agribusiness companies in this sector due to the following.

Table 17.1 Per capita monthly expenditure on fruits and vegetables in India (unit, Rs.)

	Rural			Urban		
Crops	1993–1994	2009–2010	% change	1993–1994	2009–2010	% change
Vegetable	17.00 (9.56)	57.20 (11.51)	+236.50	25.00 (9.99)	76.66 (10.54)	+206.6
Fruits	4.90 (2.76)	14.88 (2.99)	+203.70	12.20 (4.87)	37.37 (5.14)	+206.3

Source: Sharma (2011)

Figures in the parenthesis indicate percentage of total food consumption

- *Robust Demand*

  The increasing share of high-value commodities in the consumption basket of the households, higher incomes and urbanisation, changing lifestyles, market integration and trade liberalisation at a global level has led to an increase in the demand for horticultural products in India. Continuous increasing trend of domestic as well as external demand makes this sector more viable. Rising urban and rural income has aided demand growth. According to the Mckinsey Global Institute, India's aggregate consumer spending will quadruple to $1.4 trillion by 2025, and the country is set to become the world's fifth largest economy in terms of consumption, up from twelfth place in 2010. Domestic demand for horticultural crops mainly fruits and vegetables has shown an increasing trend which has been evident from the following table (Table 17.1).

  The improvement in economic access to food due to increased income did not result in a higher consumption of cereals but has increased the consumption of vegetables, fruits and nuts and livestock products, especially milk and eggs. The share of expenditure on vegetables, fruits and nuts witnessed an increasing trend in both rural and urban India. The total monthly per capita food expenditure was higher in urban areas due to higher income, urbanisation and varied preferences. Price elasticities of demand of vegetables and fruits are found to be much higher compared to the overall food in both rural ($E_p = 1.05$ and 0.91) and urban ($E_p = 0.73$ and 0.72) areas. These results indicate a higher demand for horticultural products in the future. Increase in income, education and easy availability of ready-to-eat foods may bring about enormous changes in the food consumption pattern in the near future. Therefore, production, processing and distribution of processed foods should be a priority in the policies of the government. Cereals being an essential component of food, their price rise needs to be kept under control.

- *Attractive Opportunities*

  The introduction of Horticultural Mission in India offers tangible increase in horticultural production. This leads to an increase in demand for agricultural inputs such as hybrid seeds/quality planting material, fertilisers, plant protection measures, etc. Promising opportunities in storage capacities as adequate cold storage facilities are available for just about 10 % of India's horticultural production. There is a potential storage capacity expansion of 35 million tonnes under a current five year plan. Besides, in view of maintaining standardisation, investment in quality control measures may appear.

- *Foreign Direct Investment (FDI)*

  FDI up to 100 % equity is permitted under the automatic route in food and infrastructures like food parks and cold chains. There are many areas for investment in this sector which include mega food parks, agri-infrastructure, supply chain aggregation, logistics and cold chain infrastructure, fruit and vegetable products, animal products, meat and dairy, fisheries and seafood and cereals, consumer foods/ ready-to-eat foods, machinery/packaging, etc. FDI policy for agriculture was amended to allow 100 % FDI in the agricultural sector including seeds, plantation, horticulture and cultivation of vegetables.

Fig. 17.3 Drivers of
supply chain performance

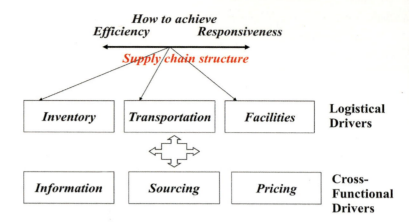

• *Organised Retail Space*

  The Indian retail industry is one of the largest in the world and has experienced high growth over the last decade with a noticeable shift towards organised retailing formats. On an average, 10–15 % of the total organised retail space is captured by imported food produce in India. As per 2010–2011 Indian import data, dairy (163 %), wine (58 %), packaged food (45 %) and fruits and nuts (11 %) have shown good growth. These information offer an idea how, in the best way, horticultural sector could take up this opportunity with appropriate initiatives.

  Keeping the importance of agricultural production and growth, the current Five year plan emphasises for the diversification into horticulture and floriculture which in turn imply structural changes in the relation between agriculture and nonagriculture. Diversification requires effective marketing linkages, supported by modern marketing practices including introduction of grading, postharvest management, cold chains, etc.

17.4.5 Drivers of Supply Chain Performance

The mission of the supply chain management for any commodity or product is to offer the best customer service at least costs by making goods/services available as per requirements and specification in time and intact. There are some elements or drivers of the supply chain management, and to accomplish the mission, efficient management of these elements is highly needed. The drivers of the supply chain ((1) inventory, (2) transportation, (3) facilities, (4) information, (5) sourcing and (6) pricing) and the way they collectively determine the performance are presented in Fig. 17.3.

17.5 Strategies for Efficient Horticultural Supply Chain Management

17.5.1 General

The challenges relating to the supply chain in Indian agriculture are non-availability of good quality seeds, lack of soil testing facilities and extension staff, poor access to credit, lack of information, huge postharvest losses, lack of infrastructures like roads, cold storage, etc., poor market intelligence, high transportation cost, etc. (Mittal 2007).

  Horticultural marketing encompasses all the activities involved in the flow of the product from producer to consumer. It is system of mutifunctionaries with the presence of different channels, middlemen, functionaries, etc. which make it more important with the advancement of specialisation, urbanisation and other socioeconomic trends.

India's horticultural sector is characterised by many small producers, often in dispersed regions throughout the country with many different market outlets, ranging from roadside sales to domestic market stores and to international chains of supermarkets in diverse countries. The supply chain is complex and challenging with numerous parties involved that are often not having knowledge of best practices to optimise quality. Growers tend to be of an independent mindset, selling their products to agents offering the best price. Horticulture is an industry that experiences fluctuations in volume and quality between seasons, much of which are driven by variations or extremes in weather patterns (such as frosts, floods or drought).

An integrated approach is required for an effective SCM. But it is observed that relationships within the industry are often poor, segmented and suspicious. Every grower knows that they produce perfect produce and consequently expect to receive optimum prices. When they don't, they tend to blame the factors beyond the orchard gate such as poor retailing, inadequate cooling or rough transportation; somewhere or someone down the chain is responsible for 'cheating' them out of their due rewards, by poor handling, inadequate promotion or marketing or untimely sales, etc. Those responsible for purchasing the product (for supermarket chains) have to buy the product at a price which is seemed to be the lower prices as perceived by the producer-seller. The process could be satisfactory only when the two parties get together and talk do they begin to understand the realities of their respective businesses. The buyer appreciates the true cost of production and the breakeven point for the grower (often above current market prices) and the devastating effects of extreme weather patterns. The grower also develops awareness of the demands of the customer, the importance of product quality and the maintenance of optimal postharvest shelf life conditions to ensure maintenance of quality for the marketing period.

In fact, it is unusual for our horticultural industries that greater mutual dialogue does not occur on a regular basis among all those involved in the supply chain from the production centre to the consumer. If more meetings of this kind occurred in different areas, then it is possible that there could be an emergence of the true supply chain management incorporating all players within and/or between states and countries.

17.5.2 Principal Changes in India's Agricultural Supply Chain

India's horticultural sector aims at having efficiencies and being competitive in respect of quality production to meet up the demand of domestic as well as international consumers. Accordingly, there have been major changes in the supply chain in agro and food product over the last decade.

There have been four principal changes in India's agri-supply chain:

1. *First*, the food supply chain's volume has tripled in the past three decades: urban food expenditures have tripled (in real terms) over the past 35 years.
2. *Second*, the food supply chain's composition has diversified over the past three decades. Nongrain foods (dairy, pulses, fruits, vegetables, meat and fish) are 71 % of India's food consumption and are important sources of calories, proteins and vitamins. These foods share the centre stage with grains for food security.
3. *Third*, the government's direct role in the marketing of grain output doubled over the past three decades.
4. *Fourth*, the private part of India's food economy, which constitutes 93 % of the food economy, has been structurally transforming rapidly especially over the past decade. It is on this transformation that the rest of the paper focuses; therefore, we provide scant detail here in the introduction. In comparison with traditional market channels (fragmented, small traditional processors, shops, wet markets, hawkers and village brokers), all *modern* market channels – both private sector led (modern retail, food processing, food service industry) and public sector led (parastatal wholesaling, processing and retail) – show

higher annual growth rates than do overall urban food expenditures.

The attainment of efficiency, hence propelled growth in the industry, is only going to be possible through effective integration of supply chain management function. Integration is the central theme in SCM which leads to building synergies by integrating business functions, departments and companies. Components of SCM integration include the following aspects:

- Planning and control
- Work structure
- Organisation structure
- Product flow facility structure
- Information flow facility structure
- Management methods
- Power and leadership structure
- Risk and reward structure
- Culture and attitude

In achieving that objective, the research will establish the current structure of the supply chain in the industry which will be key in developing an appropriate model. The need of the research is to establish proper linkage from the sourcing of the inputs.

It is necessary from a strategic point of view to list out the preliminary activities of the supply chain management which include:

- Determine channel strategy and level of distribution intensity.
- Manage relationships in the supply chain.
- Manage the logistical components of the supply chain.
- Balance the costs of the supply chain with the service level demanded by the customer.

Before taking an appropriate strategy for effective SCM of horticultural crops, an entrepreneur should adhere to the following suggestions for the best fitting of the supply chain (SC) of their products with the customer or vice versa:

- Understand the customer wishes.
- Understand the capabilities of your SC.
- Match the wishes with the capabilities.
- Challenge: how to meet extensive wishes with limited capabilities?

17.5.3 Flows Governing Optimal Chain Functioning

Three major flows within horticultural supply chains govern their optimal functioning. Figure 17.4 shows the flow of the supply chain of a product in general:

- *Flow of produce* – fresh produce flows in one direction through the chain, from input supply (seeds, fertilisers, etc.) to the retailer, who makes the produce available to consumers.
- *Financial flow* – financial flow takes place in the opposite direction of the produce flow, whereby payments are made to suppliers as produce moves from the producer through the various customers within the chain to the consumer. Financial flow is generated through the willingness of the consumer to pay for produce that meet their requirements.
- *Information flow* – information flows in both directions throughout the supply chain. Information flow is very important in coordinating activities and practices at the different steps of the chain in order to assure that these activities and practices satisfy market requirements. It facilitates planning and coordination of supply. Information related to the identity of produce (e.g. origin, variety, orchard block from which produce is harvested), treatment at the packing house and handling (e.g. the temperature and relative humidity during distribution) through the chain can be recorded and stored at the different steps of the chain.

Fig. 17.4 Flow of supply chain

The flows resemble a chain reaction.

Stored data is used in providing traceability (tracing and tracking of produce) in the chain. Stored data may also serve during the later verification of compliance with protocols such as those related to the application of good agricultural practice (GAP) and good manufacturing practice (GMP).

17.5.3.1 Enabling Environment

Optimal functioning of horticultural supply chains hinges upon a number of factors that are external to the chain and which constitute an enabling environment. These include:

- Enabling policies and regulations
- The infrastructural support base to facilitate chain operations
- Business development support services, which include banks that provide loans

17.5.4 Marketing Strategy

Production and marketing strategies are the most crucial in strategy development. The development strategy should be based on innovation. Production innovations initially focused on efficiency and effectiveness in order to increase yields and lower costs. The customer in the supply chain is the grower, the trader, the processor and the retailer. The whole objective of supply chain management is to provide or supply the raw materials or final products to the customer profitably. So an appropriate marketing strategy occupies the most important place in this perspective. The aim of marketing strategy of any product is to create and retain satisfied customers in the market. This indicates that traditional business management aims at customer satisfaction with the products and services. In the case of supply chain management, it is a relationship management for all the participants involved in the supply of raw materials, goods and services and to develop integration among them for efficient distribution to make customers satisfied. Figure 17.5 will give an idea about the relationship mechanism of the supply chain and business strategy.

Right now, the Indian government is considering liberalising the country's $400 billion retail market. A large portion of this market stems from agriculture. The liberalisation would pave the way for multinational corporations like Wal-Mart to own up to '51 percent of a multi-brand retailer if they invested at least $100 million, with half

Fig. 17.5 Linking supply chain and business strategy

spent on infrastructure development in India'. Here, infrastructure development would include supply chain investments as Wal-Mart would want to be able to transport its perishable goods throughout the country. Rural farmers and small business owners have become understandably concerned with regard to how a major retailer could displace jobs and exert unfair buying pressures on single farmers and storeowners.

On the other end of the spectrum, social enterprises like Jagriti Agro Tech work with farmers directly to consolidate the 'best practices developed by grassroot organisations and organic farming experts, along with IT tools to make farm the monitoring, production and supply chain management efficient'. In doing so, the organisation has enabled its farmers to enjoy a 25 % increase in prices.

These anecdotes suggest that both approaches may be needed to fully address the scope and scale of India's supply chain issues. While working with farmers directly provides them with the tools and information available to receive better prices and avoid exploitative middlemen, it is unclear that without multinational help and expertise, the supply chains can nationalise and the supply and demand of perishable goods could rationalise.

17.6   Infrastructure for Horticultural Supply Chain Management

17.6.1 General

First, the current structure of the supply chain allows for too many middlemen who distort prices, exploit farmers and prevent produce and other goods from reaching places where demand exists. An Indian School of Business working paper states: In India, the infrastructure connecting these partners is very weak. Each stakeholder, farmers, wholesalers, food manufacturers, retailers all work in silos. Also, demand forecasting is totally absent and the farmers try to push what they produce in to the market. Data integration, financial flow management, supply-demand

matching, collaborative forecasting, information sharing, goods movement synchronization through efficient transport scheduling, are very well practiced in high technology industries with immense benefits. Each industry within the agricultural sector, e.g. vegetables, fisheries, poultry, etc., will have unique structures. However, the underlying causes for the inefficiencies are likely to be shared across all of them.

Second, there has been an insufficient investment in necessary technologies that would extend the shelf life of these perishable goods. Produce spoils rapidly and requires special cold storage to increase the chance it may be sold before spoilage. More than 30 % of fruits and vegetables rot before being sold today. ISB identifies the following, collectively called cold storage, as vital to a successful supply chain:

Cold Chain Facilities
- Industrial cold stores in the production zones
- Refrigerated vehicles for long distance transport
- Refrigerated room in wholesale markets
- Field heat facilities
- Product temperature upon harvest
- Major fraction of the heat load to be removed
- Influence the rate of reactions during storage

Precooling Facilities
- Necessary for removal of field heat
- Slow down the temperature-dependent metabolic processes
- Ensure longer storage and better product quality

Refrigerated Cooling
Hydro-cooling
Vacuum Cooling

17.6.2 Major Obstacles for Operation of Efficient Supply Chain

- *Supply chain is big*
  - Variety of products/services
  - Spoiled customer

- Multiple owners (procurement, production, inventory, marketing)/multiple objectives
- Globalisation
- *Dealing with multiple owner/local optimisation.*
  - Lack of coordination and misleading reliance on metrics
- *Instability and randomness.*
  - Increasing product variety
  - Shrinking product life cycle
  - Customer fragmentation
  - Fragmentation of the supply chain ownership – globalisation
- *Poor inventory status information.*
- *Poor delivery status information.*
- *Poor IT design.*
- *Poor integration.*
- *Ignoring uncertainties.*
- *Internal customer discrimination.*
- *Supply chain-insensitive product design.*

17.6.3 Integration of Small Farmers into Horticultural Chains

One major challenge to fixing the supply chain is to do so in a way that does not disrupt and/or destroy the livelihoods of individual farmers. The farmer's plight has taken the centre stage in an ongoing discourse over the best way to address these problems. With the commercialisation of horticulture, new and diversified market opportunities are becoming available, information and communication systems have expanded with mobile phone communications and Internet services are now accessible and affordable even in rural areas. High-quality seeds and improved planting materials, tools and equipment are increasingly becoming available. At the same time, many small farmers in the region are becoming increasingly vulnerable to marginalisation because of low productivity, poor-quality outputs and limited market access. Averting this risk necessitates development of their skills and capacities in order to empower them to adopt new attitudes and approaches to farming and market-ing. To meet this end, small farmers must also increasingly work collaboratively with each other either through networks or cooperative groupings in order to leverage on each other for more efficient production capacity. They must seek to identify niche opportunities for their fresh produce outputs, create and develop those opportunities that become available to them and then promote markets for their outputs.

17.6.3.1 Horizontal Collaboration in Horticultural Chains

The prospects of competitiveness work differently for a small farm than for a large farm with available resources. In order to achieve the competitive advantage usually associated with large farms, small farms must collaborate with the other stakeholders in order to survive. Greater benefit is derived through horizontal collaboration among small farmers. *Horizontal collaboration* may involve small farms of similar sizes pooling together their resources and outputs and working collaboratively with service and technology providers. Government programmes designed to promote horizontal collaboration could either reposition existing farms or direct the location of new farms, meanwhile bringing associated service providers into a designated geographical area through a *cluster initiative* which includes:

- Development of economies of scale
- Increased access to information and improved information flows
- Improved adaptation of information and improved innovation
- Improved personal relationships among small farmers

17.6.4 Models of Supply Chain

The following is the proposed supply chain as suggested by Singh (2011) (Fig. 17.6).

The following is the emerging fruit and vegetable supply chain model as given by Halder and Pati (2011) (Fig. 17.7).

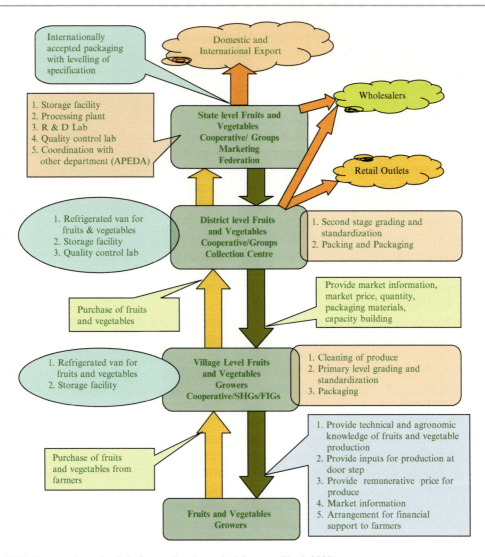

Fig. 17.6 Suggested supply chain for small and marginal farmers (Singh 2009)

17.7 Conclusion

Supply chain management is the active management of supply chain activities to maximise customer value and achieve a sustainable competitive advantage. Studies on the present status of the supply chain and its management for horticultural produce indicate that it deserves much more improvement, and accordingly, the government can play influential roles to create mutually beneficial relationship between farmers and organised sectors through investment in infrastructure and the development of extension activities.

There is no doubt that production and marketing strategies are the most crucial in strategic development. The development strategy should be based on innovation. Production innovations initially focused on efficiency and effectiveness in order to increase yields and lower costs.

Horticulture sector is dominated mainly by small holdings. So integrating small farmers with high-value urban and export markets would lead to the development and growth of the rural sector.

The agricultural development requires a minimum set of basic production factors and further

Fig. 17.7 Emerging F&V
supply chain model
(Halder and Pati 2011)

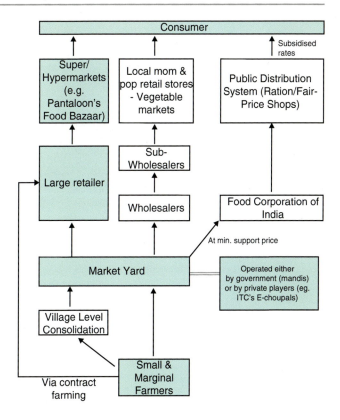

requires an optimal crop management and developing postharvest infrastructure, entrepreneurial management and expertise, logistical infrastructure and supporting financial infrastructure.

Resource-poor farmers need credit for meeting up the operational cost as the marketing cost of fruits and vegetables is almost 50 % of the total cost of production; thus, there is a need to set up institutional agencies that can advance credit to farmers and motivate them to market the produce themselves.

India has gained capability to generate surpluses for exports if potentially small farmers are fully tapped. Assistance from APEDA and exporters association as well as training to the farmers is necessary.

Horticultural markets are nearly imperfect in nature. Policy for market regulations and supervision should be given much more priority. Besides, market infrastructures in terms of storage (godown) facilities, cold storages, loading

and weighing facilities assume much importance which are the basics of efficient SCM. Improvement in the road network and cold chain facilities are also of substantial importance.

The market integration and efficiency can also be improved by making up-to-date market information available to all participants through various means, including good market information systems, good Internet service and good telecommunication facilities at the markets (Halder and Pati 2011).

Due to constraints in the government investment, it is pertinent that a private investment or public-private partnership (PPP) model may be encouraged for extending further facilities in this high-value livelihood sector.

Thus, all these efforts, if addressed properly, will help to bring horticultural sector into a systematic supply chain and logistics management for its growth and development.

References

Agrawal DK (2010) A textbook of logistics and supply chain management (Book). Macmillan Publishers India Limited, Kolkata

FAO (2009) Horticultural chain management for countries of Asia and the Pacific region: a training package. Available through www.fao.org

Government of India (2007) Ministry of Agriculture & Cooperation, New Delhi. Available through www.agricoop.nic.in

Gupta SP, Rathore NS (1998) Marketing of vegetables in Raipur district of Chhattisgarh state: an economic analysis. Indian J Agric Econ 53(3), July

Halder P, Pati S (2011) A need for paradigm shift to improve supply chain management of fruits & vegetables in India. Asian J Agric Rural Dev 1(1):1–20

Hewett EW (2010) Perceptions of supply chain management for perishable horticultural crops: an introduction. www.isb.edu/.../....Agriculture-Supply-Chain-..pdf

Kaul GL (1997) Horticulture in India – production, marketing and processing. Indian J Agric Econ 52(3):57

Kohls RL (1985) Marketing of agricultural products (Book). McMillan Publishing Company, New York

Koontz H, Weihrich H (2012) Essentials of management (Book). Tata McGraw Hill Education Private Limited, New Delhi

Kotler P (1984) Principles of marketing (Book). Prentice Hall, New Delhi

Mentzer J (2007) Supply chain management (Book), Response Book. Sage Publication India Private Limited, New Delhi

Mittal S (2007) Can Horticulture be a success story in India, Working paper no.197, Indian Council for Research on International Economic Relation. Available from http://www.icrier.org/

Narula SA (2009) ICTs and agricultural supply chains – opportunities and strategies for successful implementation. Available through www.iimahd.com

National Horticultural Board (2013) Govt. of India

Porter ME (1990) The competitive advantage of nations (Book). Free Press, New York

Reddy GP et al (2010) Value chains and retailing of fresh vegetables and fruits, Andhra Pradesh. Agric Econ Res Rev 23(Conference Number):455–460 (www.Safalindia.com).

Senam Raju MS (2002) Fruit marketing in India (Book). Daya Publishing House, New Delhi

Sharma VK (2011) International Referred Research Journal, Sept., 2011, ISSN-0975-3486, RNI: RAJBIL 2009/30097, Vol-2(24)

Singh HP (2009) Triggering agricultural development through horticultural crops. www.indiaenvironmentportal.org.in/Keynote_paper-H.P._Singh.doc

TCI (2012) India's agri supply chain: issues and opportunities

Woods E (2004) Supply chain management: understanding the concept and its implications in developing countries. Proceedings of Agriproduct supply-chain management in developing countries workshop, Bali, ACIAR Proceedings No. 119, pp 18–26

Development of Women Entrepreneur Through Value Addition of Horticultural Crops

P.K. Pal and R. Chatterjee

Abstract

Although women play a vital role in the farm and home system and contribute substantially in the physical aspect of farming, livestock management, postharvest and allied activities, but they are socially suppressed, are educationally at a disadvantageous position and have little say in the decision-making process due to lack of independent income source in hand. Entrepreneurship development among rural women cannot only enhance national productivity and generate employment but also help to develop economic independence and personal and social capabilities. The present paper reviews the concept of women entrepreneurship, their characteristics, problems and challenges and factors responsible for the development of women entrepreneurship in Indian social-political system. The paper emphasised on different fields of value addition to be undertaken by women to develop their own business and also discussed the steps to be followed to develop it with an elaborate list of schemes and institutions which supports entrepreneurship for women in different levels. Finally, the paper advocated a comprehensive extension strategy to promote entrepreneurship among women and concluded that women can do wonders by their effectual and competent involvement in entrepreneurial activities if they are supported and motivated by skill development, financial assistance and removing social stigma.

P.K. Pal (✉)
Department of Agricultural Extension,
Uttar Banga Krishi Viswavidyalaya,
PO. Pundibari, Cooch Behar
736101, West Bengal, India
e-mail: pkpalubkv@gmail.com

R. Chatterjee
Department of Vegetable and Spice Crops,
Uttar Banga Krishi Viswavidyalaya,
PO. Pundibari, Cooch Behar
736101, West Bengal, India

18.1 Introduction

Women contribute around half of the total world population and play a vital role in the farm and home system. They contribute substantially in the physical aspect of farming, livestock management, postharvest and allied activities. Their direct and indirect contribution at the farm and home level along with livestock management

A.B. Sharangi and S. Datta (eds.), *Value Addition of Horticultural Crops: Recent Trends and Future Directions*, DOI 10.1007/978-81-322-2262-0_18, © Springer India 2015

operations has not only helped to save but has also led to an increase in family income. They perform various farm, livestock, postharvest and allied activities and possess skills and indigenous knowledge in these areas. This is in spite of the fact that they are socially suppressed and educationally at a disadvantageous position and having little say in the decision-making process due to lack of independent income source in hand. Entrepreneurship development among rural women not only enhances national productivity and generates employment but also helps to develop economic independence and personal and social capabilities. According to Sidhu and Kaur (2006), the following are some of the personal and social capabilities which are developed as a result of taking up enterprise among rural women:

• Economic empowerment
• Improved standard of living
• Self-confidence
• Enhanced awareness
• Improvement in decision-making status
• Sense of achievement
• Increased social interaction
• Improvement in leadership quality

18.2 Concept of Women Entrepreneurs

In a conference on entrepreneurship held in the United States, the term "entrepreneurship" was defined as "the attempt to create value through recognition of business opportunity, the management of risk-taking appropriate to the opportunity, and through the communicative and management skills to mobilize human, financial and material resources necessary to bring a project to function" (In: Khanka 1990, 1999).

Cole (1959) opined that entrepreneurship is the purposeful activity of an individual or a group of associated individuals undertaken to initiate, maintain or aggrandise profit by production or distribution of economic goods and services.

Taking light from the above and as discussed and explained by Goyal and Parkash (2011),

women entrepreneurs may be defined as women or a group of women who initiate, organise and operate a business enterprise. The Government of India has defined women entrepreneurs as "an enterprise owned and controlled by women having a minimum financial interest of 51 % of the capital and giving at least 51 % of the employment generated in the enterprise to women". They also opined that women entrepreneurs engage in business due to push and pull factors which encourage women to have an independent occupation and stand on their own feet. A sense towards independent decision-making on their life and career is the motivational factor behind this urge. Saddled with household chores and domestic responsibilities, women want to get independence. Under the influence of these factors, women entrepreneurs choose a profession as a challenge and as an urge to do something new. Such a situation is described as a pull factor. On the other hand, in push factors women engage in business activities due to family compulsion, and the responsibility is thrust upon them.

18.3 Characteristics of Women Entrepreneurs

Women entrepreneurs are not different from men entrepreneurs. All entrepreneurs have the following common characteristics. Muhammad Yunus the maiden Peace Nobel Laureate of Bangladesh said, "All people are entrepreneurs, but many don't have the opportunity to find that out". There are some personal characters of some of the people which provoke them to search these opportunities to be an entrepreneur. Khanka (1999) enumerated the following characteristics of an entrepreneur in his book *Entrepreneurial Development*:

Willingness for Hard Work
Most of the successful entrepreneurs work hard endlessly, especially in the beginning, and the same becomes their whole life. Willingness to work hard distinguishes a successful entrepreneur from an unsuccessful one.

Desire for High Achievement

High achievement motive strengthens an entrepreneur to surmount the obstacles, suppress anxieties, repair misfortunes and devise expedients.

Highly Optimistic

Successful entrepreneurs are always optimistic and think that the situations will become favourable to business in the future. They are not motivated by the present problems.

Independence

One of the common characteristics of successful entrepreneurs has been that they do not like to be guided by others. They like to be independent in the matters of their business.

Foresight

The entrepreneurs have a good foresight to know about future business environment. In other words they well visualise the likely changes to take place in the market, consumers' attitude, technological developments, etc. and take timely actions accordingly.

Good Organiser

Different resources required for production are divorced from each other. It is the ability of the entrepreneurs that brings together all resources required for starting up an enterprise and then to produce goods.

Innovative

The entrepreneurs initiate research and innovative activities to produce goods to satisfy the customers' changing demands for the products.

18.4 Factors Responsible for Entrepreneurship Development for Women

Women's entry into entrepreneurship in India is a new phenomenon. Women enter into business by both pull and push factors. Pull factors emerge from inside the minds of women, *viz*. the internal motivation that encourages women to start something on their own. Push factors are those which compel women to enter into the business to overcome the economic factors. There are many pull and push factors, economic, social, political or psychological, which influence the emergence of a successful entrepreneurship. Khanka (1990) in his book *Entrepreneurship in Small Scale Industries* has analysed the opinions of different authors like Wilken (1979), Landes (1969), Schumpeter (1939), Katzin (1964), Hoselitz (1957, 1960), Hagen (1968), Cole (1959), Rostow (1956), etc. and forwarded the following factors as responsible for successful development of entrepreneurship:

Economic Factors

From an economic view point, it is suggested that with an increase in capital investment, capital-output ratio also tends to increase. This results in an increase in profit which ultimately goes to capital formation. This suggests that as capital supply increases, entrepreneurship also increases. The quality rather than quantity of labour is another factor which influences the emergence of entrepreneurship. Adequate supply of raw materials; other factors being unchanged; also influence the emergence of entrepreneurship. The size and composition of the market both influence entrepreneurship in their own ways. It appears that whether or not the market is expanding, the rate at which it is expanding is the most significant characteristic of the market for entrepreneurial emergence. Indian rural women mostly lack the accessibility of family economic capitals and are also constrained with the lack of courage to manage the economic factors for running a business enterprise. This is mainly due to the less or no access to economic capital in both household and social levels.

Noneconomic Factors

Social system, norms and values within a sociocultural setting determine the level of social sanction (degree of approval or disapproval granted for an entrepreneurial behaviour) which ultimately influences the emergence of any entrepreneurship. Social mobility (social mobility is the degree of movement, both social and geographical) is another crucial factor for entrepreneurial emergence. The degree of social mobility determines the flexibility or rigidity of a social system. Brozen (1954) emphasises that if a social system is too flexible, the individuals will gravitate towards other roles, if it is too rigid, entrepreneurship will be restricted along with other activities. Psychological factors like the high need for achievement, determination for unique accomplishment, self-confidence, creativity, vision, leadership, etc. promote entrepreneurship among individuals. On the other hand, psychological factors like security orientation, conformity and compliance, high need for affiliation, etc. inhibit promotion of entrepreneurship. In Indian society, women are traditionally believed to be confined in the four walls of the household and perform the responsibility of perpetuating the in-home activities. Such social stigma is the strongest factor to be overcome by the rural women to be a successful entrepreneur.

Political and Government Actions

Political system which promotes free market, individual freedom and private enterprise will promote entrepreneurship. The economic policies of the government and other financial institutions and the opportunities available in a society as a result of such policies play a crucial role in exerting direct influence on entrepreneurship. The political system and sanction are the ultimate factors in our country which can motivate rural women by enacting laws (many bills and laws have already been passed to empower women in the country) to come into entrepreneurship.

18.5   Steps for Development of Entrepreneurship for Women

Right business opportunity is the key to enterprise success. Opportunity sensing is essential for new as well as existing entrepreneurs; once existing entrepreneurs reach the comfort stage in their

project, new business opportunities need to be identified for enterprise sustenance, growth and development. An entrepreneur needs to scan the emerging environment before finalising the most optimal opportunity that maximises gains from limited investment and resources. Development of new enterprise or expansion of the existing one starts through project identification, selection and then formulation of the new project.

18.5.1   Project Identification, Selection and Formulation

A project is a well-evolved work plan designed to achieve specific objectives within a specified period of time. The process of identifying a product idea for developing into a project is called *project identification*. This is a systematic process. Project ideas can be generated from observation, market analysis, trade and professional magazines, bulletin of research organisations, government sources, experienced persons, etc. *Project selection* is a careful study of each project idea in detail by strength, weakness, opportunity and threat (SWOT) analysis and choosing one of them for further consideration and development. A project must be selected in the background of available technology, required equipment, amount of investment, location character, availability of market and its characteristics. *Project formulation* is the process of studying a selected project further or developing a new project with reference to investment decisions. It considers issues such as relevance and feasibility of the project. It involves a step-by-step procedure to investigate and develop a project report.

The Planning Commission of India has given some guidelines for preparing a project and advocated that the following points should be included in the project report:

- General information
- Preliminary analysis of alternatives
- Project description
- Marketing plan
- Capital requirement and cost
- Operating requirements and cost
- Financial analysis

- Economic analysis
- Miscellaneous information

18.6 Successful Indian Businesswomen: Some Worth Mentioning (Adopted from Shanta 2014)

Indra Nooyi
This brilliant corporate woman started her career in Boston Consulting Group. She joined Pepsi Company in 1994 and turned the company into a bold risk taker. The *Wall Street Journal* included her name in The 50 Women to Watch: 2005. Simultaneously *Fortune Magazine* also declared her 11th most powerful women in business.

Dr. Kiran Mazumdar-Shaw
She is the chairman and managing director of Biocon Ltd. who became India's richest woman in 2004. She founded Biocon India with a capital of 10,000 in her garage in 1978. Today her company is the biggest biopharmaceutical firm in the country.

Naina Lal Kidwai
She was the first Indian woman to graduate from Harvard Business School. *Fortune Magazine* listed Kidwai among the world's top 50 corporate women from 2002 to 2003. According to the *Economic Times*, she is the first woman to head the operations of foreign banks in India. Also she was awarded the Padmashree.

Vidya Manohar Chhabria
The wife of the late Manohar Rajaram Chhabria, she is now leading Jumbo Group, a Dubai-based 1.5 billion dollar business conglomerate. She was ranked 38th most powerful women by the *Fortune Magazine* in 2003.

Shahnaz Husain
She is another successful woman entrepreneur of India. She popularised herbal treatments for beauty and health problems. Her company Shahnaz Husain Herbals was the largest of its kind in the world and had a strong presence in over a hundred countries, from the United States to Asia.

Ekta Kapoor
She is popularly known as the "soap queen" and creative director of Balaji Telefilms and is credited for bringing about a revolution in the Indian small screen industry. She is a rare combination of beauty and brains and a great inspiration for budding entrepreneurs.

18.7 Problems Faced by Women Entrepreneurs

Shanta (2014) enlisted the following problems which were also supported by Bulsara (2007):

1. Lack of working capital: Being women doing something on their own becomes quite difficult for them because of the lack of access to funds, as women do not process any tangible security.
2. Motivational factor: Self motivation and the attitude to take up risk and behaviour towards the business society, family support, financial assistance from public and private institutions and also an environment suitable for women to establish business units.
3. The family structure is generally male dominated; hence, the male members think it's a big risk financing the ventures run by women. The greatest deterrent to women entrepreneurs is that they are women.
4. The financial institutions are sceptical about the entrepreneurial abilities of women. The bankers consider women loonies as higher risk than men loonies. The bankers put unrealistic and unreasonable securities to get loan to women entrepreneurs.
5. Women's family obligations also bar them from becoming successful entrepreneurs in both developed and developing nations. "Having primary responsibility for children, home and older dependent family members, few women can devote all their time and energies to their business".

6. The male–female competition is another factor which develops hurdles to women entrepreneurs in the business management process. Despite the fact that women entrepreneurs are good in keeping their service prompt and delivery in time, due to lack of organisational skills compared to male entrepreneurs, women have to face constraints from competition.

18.8 Value Addition and Development of Women Entrepreneurship

Value addition is the enhancement added to a product or service by a company before the product is offered to the customers. Value added refers to the additional value of a commodity over the cost of commodities used to produce it from the previous stage of production (*wikipedia.com*).

A broad definition of value addition is to economically add value to a product and form characteristics more preferred in the market place.

There are three ways in which value addition to farm produce is possible: the primary level involves cleaning, grading and packaging of fruits, vegetables and other horticultural crops; the secondary level includes basic processing, packaging and branding, e.g. packed *atta*, *suji*, etc.; and the tertiary level includes high-end processing which requires supply chain management, processing technology, packaging of processed foods, branding, marketing, etc., e.g. potato chips. The term "value chain" refers to the full range of activities needed to bring a product or service from conception through production and delivery to final consumers.

18.9 Areas for Microenterprise Development in the Field of Value Addition of Horticultural Crops

Ali (1997) in his paper on postharvest processing of agricultural produce stated that value addition includes processes like sorting, grading, cutting, seeding, shelling and quality packaging, etc. Growing organic vegetables, flowers, and oilseeds; and commercial seed production are some of the areas in the field of value addition in horticultural fields. Some more areas can be like dehydration of fruits and vegetables, canning or bottling of pickles, chutneys, jams, squashes, dairy and other products which are ready to eat.

The following areas may also be explored for entrepreneurship development in horticulture:
1. Seed production of horticultural crops and horticulture nursery
2. Tissue culture
3. Production, processing and marketing of vegetables, fruits, flowers, spices, medicinal aromatic and plantation crops, etc.
4. Organic farming for horticultural crops
5. Food processing
6. Horti-clinic
7. Orchid nursery and marketing
8. Landscaping

NAIP-Domestic and Export Market Intelligence Cell (http://www.tnagmark.tn.nic.in) reported the following value-added products with specification which can be explored by women for developing entrepreneurship:
1. Fruit juice: It is a natural juice obtained by pressing out the fruits. Fruit juices may be sweetened or unsweetened.
2. Fruit juice powder: The fruit juice is converted into highly hygroscopic powder. These are kept freeze dried and used for fruit juice drinks by reconstituting their composition.
3. Fermented fruit beverages: These are prepared by alcoholic fermentation by yeast of fruit juice. The product thus contains varying amounts of alcohols, e.g. grape wine, orange wine and berry wines from strawberry, blackberry, etc.
4. Jam: Jam is a concentrated fruit pulp processing a fairly heavy body form rich in natural fruit flavour. It is prepared by boiling the fruit pulp with sufficient quantity of sugar to a reasonably thick consistency to hold tissues of fruit in position.
5. Jelly: Jelly is a semi-solid product prepared by cooking clear fruit extract and sugar.

6. Marmalade: It is usually made from citrus fruits and consists of jelly containing shreds of peels suspended.
7. Tomato ketchup: It is prepared from tomato juice or pulp without seeds or pieces of skin. Ketchup should contain not less than 12 % tomato solids and 28 % total solids.
8. Pickles: Food preserved in common salt or in vinegar is called pickle. Spices and oil may be added to the pickle.

18.10 Assistance for Entrepreneurship Development for Women

The following is a list of institutions and projects which assist for entrepreneurship development. Women interested to develop entrepreneurship on value addition of horticultural crops may take help of such institutions and projects for finance, technical consultancy and project development:

1. Ministry of Small Scale Industries (MSSI)
2. Small Industries Development Organisation (SIDO)
3. The National Small Industries Corporation Limited (NSIC)
4. National Institute for Small Industry Extension Training (NISIET)
5. The Indian Institute of Entrepreneurship (IIE)
6. Small Industries Development Bank of India (SIDBI)
7. The Khadi and Village Industries Commission (KVIC)
8. Prime Minister's Rozgar Yojana (PMRY)
9. Entrepreneurship Development Programmes (EDPs)
10. Women's Development Corporations (WDCs)
11. Trade Related Entrepreneurship Assistance and Development (TREAD)
12. Working Women's Forum
13. Indira Mahila Yojana
14. Indira Mahila Kendra
15. Mahila Samiti Yojana
16. Mahila Vikas Nidhi
17. Micro Credit Scheme
18. Rashtriya Mahila Kosh
19. SIDBI Mahila Udyam Nidhi
20. SBI Stree Shakti Scheme
21. NGO's Credit Schemes
22. Micro & Small Enterprises Cluster Development Programmes (MSE-CDP)
23. National Banks for Agriculture and Rural Development Schemes
24. Rajiv Gandhi Mahila Vikas Pariyojana (RGMVP)
25. Priyadarshini Project – A Programme for Rural Women Empowerment and Livelihood in Mid Gangetic Plains

Different banks have also their credit schemes for women which can also be explored for entrepreneurship development for value addition.

18.11 Self-Help Groups (SHGs) and Women Entrepreneurship

A SHG is a small economically homogeneous affinity group of the rural poor voluntarily coming together to save a small amount regularly, which are deposited in a common fund to meet members emergency needs and to provide collateral-free loans decided by the group. The concept of SHG is originated from the concept of Grameen Bank of Bangladesh, which was the brainchild of Muhammad Yunus. SHGs were started and formed in 1975. In India NABARD had first initiated it in 1986–1987. But the real effort was taken after the recommendation of a group approach in rural development by Hashim Committee in the 1990s from the linkage of SHGs with the banks. The GOI made linking SHGs with banks a national priority from 1999, and NABARD continues to nurture the expansion of the outreach of the programme by providing umbrella support to various stakeholders. The credit-linked rural entrepreneurial development programme of NABARD is getting momentum in rural areas and has helped promote entrepreneurship, particularly among women. Organisations (both government and nongovernment) other than NABARD are also forming SHGs nowadays to empower rural women through entrepreneurship development. During 2003, out of 75,000 SHGs that received financial assistance, 3,628 have become micro-entrepreneurs (<5 %). Out of 3,628 μ-entrepreneurs, 2,476 were women (68 %). Almost all the people involved in various activities were earning over Rs.10,000 per month and had enhanced their business and marketing skills (Murali 2011). A study undertaken by different authors concluded that entrepreneurial skills of women are progressively enhanced with the participation of SHGs (Singh and Mathur 2005; Nagar et al. 2005; Dhanotiya et al. 2013). Both the group and the members had equally availed the given opportunities. So, SHG may be the crucial agent to establish women entrepreneurship for value addition of horticultural crops.

18.12 Extension Strategies to Promote Entrepreneurship on Value Addition

Communities or individuals cannot become entrepreneurs overnight. Small-scale rural producers need improved and equitable access to productive resources, in particular, land, water, credit, market, market information as well as social and productive services. They also need capacity building support on leadership and management (Prasad et al. 2009). Sidhu and Kaur (2006) advocated the following extension strategies to promote entrepreneurship:

Creating Awareness
The government, semi-government or nongovernment organisations should create awareness regarding entrepreneurship development. The printed media can be effectively put to use for the purpose.

Motivating Entrepreneurs
Psychological stimulation is the prerequisite for putting any idea virtually into action. For proper motivation of rural women, the economic, social and health benefits of various possible enterprises should be highlighted. The use of farm visits,

video film shows, dramas, puppet shows, group meeting, etc. will help in motivating the potential group.

Expertise Development
After awakening and motivating, the next step in the development of an enterprise is the acquisition of knowledge and skill upgradation, and polishing of existing knowledge and skills in production, processing, packaging and marketing techniques is the basic requirement. In addition to this, knowledge regarding accessibility to loans, various funding agencies, procedures regarding certification, etc. should be provided. Lectures, printed materials, discussions and institutional and non-institutional skills training for imparting first-hand technical knowledge in production, processing, procurement and management should be provided to rural people who are interested or already engaged in various enterprises. Education in direct and indirect marketing of the produce and finance management should be in-built component of future training programmes for would-be or new entrepreneurs.

Continuous Follow-Up
Constant follow-up should be ensured for the sustainability of a microenterprise. During this phase, various constraints such as personal, social, economic, marketing, etc. faced by entrepreneurs should be addressed. Possible help in the form of knowledge, technical skills and inputs should be provided to enable them to solve their problems.

18.13 Conclusion

Entrepreneurship raises the position of women in the society and also transforms the role of women in the society. In the transformed role structure, women entrepreneur have to manage both the household economic activities. Moreover, they have to face many other obstacles in marketing their product, holding property and entering contracts. However, if these obstacles are overcome and their participation in microenterprises is encouraged and aggravated, women can do wonders by their effectual and competent involvement in entrepreneurial activities. Rural women have to be provided with basic knowledge, skill, potential and resources to establish and manage enterprise. Knowledge regarding accessibility to loans, various funding agencies, procedure regarding certification, awareness on government welfare programmes, motivation, technical skill and support from family, government and other organisations; are the needs of the day to motivate women to be an efficient and competent entrepreneur to contribute in the family income and national prosperity.

References

Ali N (1997) Post harvest processing of agriculture produce for high quality and export. Theme paper published in Souvenir of XXXI annual convention of society of agricultural engineers, Agricultural Engineering. PAU, Ludhiana

Brozen Y (1954) Determinants of entrepreneurial ability. Soc Res 21(3):339–364

Bulsara HP (2007) Developing women entrepreneurship: a challenge. In: Jyothi P (ed) Entrepreneurship: issues and challenges. Allied Publications Pvt. Ltd, New Delhi. pp 387–406. ISBN: 81-8424-202-6

Cole AH (1959) Business enterprise in its social setting. Harvard University Press, Cambridge, MA

Dhanotiya B, Choudhary S, Swarnakar VK (2013) Study of women entrepreneurial behavior in self help group through Krishi Vigyan Kendra, Kasturba Gram, Indore District of M.P. Int J Humanit Soc Sci 2:44–46

Goyal M, Parkash J (2011) Women entrepreneurship in India-problems and prospects. Int J Multidiscip Res 1(5):5

Hagen EE (1968) The economics of development. Homewood, Dorsey

Hoselitz BF (1957) Economic growth and development: non-economic factors in economic development. Am Econ Rev 2(47):36

Hoselitz BF (1960) Sociological aspects of economic growth. Free Press, Glencoe

Katzin MP (1964) The role of the small entrepreneur. In: Herskkovitz MJ, Horwitx M (eds) Economic transition in Africa. Northwestern University, Evanston

Khanka SS (1990) Entrepreneurship in small scale industries. Himalaya Publishing House, New Delhi

Khanka SS (1999) Entrepreneurial development. S. Chand & Company Ltd, Ram Nagar

Landes DS (1969) The unbound Prometheus: technological change and industrial development in Western Europe from 1750 to the present. Cambridge University Press, Cambridge

Murali P (2011) Women entrepreneurship through self help groups. Indian Streams Res J 1(8):1–4

Nagar RK, Singh M, Verma AK (2005) Rural women empowerment through self help group. Unnat Krishi 44(1):15

Prasad KVS, Radha TM, Balamatti A, Kandi P (2009) Editorial LEISA INDIA 11(2)

Rostow WW (1956) The take-off into self-sustained growth. Econ J 1(66):25

Schumpeter JA (1939) Business cycles. McGraw-Hill Book Co., New York

Shanta AV (2014) Women entrepreneurship in India. J Res Arts Educ 1(3):1–12. www.abhinavjournal.com

Sidhu K, Kaur S (2006) Development of entrepreneurship among rural women. J Soc Sci 13(2):147–149

Singh S, Mathur P (2005) Self help groups – a successful approach for women empowerment. Agric Ext Rev 17(6):27

Wilken PH (1979) Entrepreneurship: a comparative historical study. Ablex Publishing Corporation, Norwood

Phytophthora: A Member of the Sixth Kingdom Revisited as a Threat to Food Security in the Twenty-First Century

19

S. Guha Roy

Abstract

This genus *Phytophthora* has a long history in modern science. The re-emergence of *Phytophthora* spp. causing damages in almost all ecological niches along with advancement in molecular technologies and discovery of new species has led to a renewed interest in *Phytophthora* spp. All of which has led to significant changes in the way the taxonomy of *Phytophthora* is now being studied. The genus *Phytophthora* infects an array of spices and plantation crops, and this scenario vis-à-vis *Phytophthora* spp. has been discussed here with special reference to black pepper, onion, garlic, leek, chilli, cocoa, coconut and rubber. Possible approaches for management of these diseases using databases derived from population characterisation through a correlation of their phenotypic and genotypic diversity have been discussed. Molecular tools that can be used for the production of such databases have also been discussed.

19.1 Introduction

The existence of man on Earth is mostly dependent on the ability of plants to harness light and produce oxygen and organic matter. Domestication of plants for agriculture resulted in many great civilizations of the past: Asian civilizations based on rice, Middle Eastern on wheat and barley and American on maize and potato. Like in the past centuries, the staple food of the world population today also depends on only a few major crops: wheat, rice, maize and potato (FAO 2002). However, mankind alone is not in the need to live off plants; a large number and different types of pathogens attack plants and, having 'fine-tuned' their ability to parasitize the living plants throughout evolutionary history, are at considerable advantage in competing to obtain nutrients from this primary food source and therefore are our competitors and enemies too (Strange and Scott 2005)! Worldwide crop loss due to pathogenic diseases, insects and weeds accounts for 31–42 % of the potential crop production capacity; without protective measures, this loss would be greater than 50 % (Agrios 2005).

S. Guha Roy (✉)
Department of Botany, West Bengal State University,
Barasat, Kolkata 700126, WB, India
e-mail: s_guharoy@yahoo.com

A.B. Sharangi and S. Datta (eds.), *Value Addition of Horticultural Crops: Recent Trends and Future Directions*, DOI 10.1007/978-81-322-2262-0_19, © Springer India 2015

Approximately 10,000 fungal species are considered to be plant pathogenic (Farr et al. 1989; Agrios 2005) out of the 72,000–100,000 fungal species identified till date, but, considering that this represents only a small fraction of the fungal kingdom, estimated to include ~1.5 million species (Hawksworth 1991, 2001; Hawksworth and Rossman 1997), the actual number of plant pathogenic fungi is likely to be much greater than 10,000!

Diseases caused by fungi are well established as major constraints of food, fibre and crop production. Equally devastating, if not more, among the fungi is a group of organisms which differ from the 'true fungi' in many characteristics (chitin-less cell wall (Bartnicky-Garcia and Wang 1983), sterol metabolism (Warner and Domas 1987; Wete 1989; Köler 1992; Griffith et al. 1992), other metabolic pathways (Hendrix 1970; Wang and Bartnicky-Garcia 1973; Elliot 1983), storage compounds (Wang and Bartnicky-Garcia 1974; Coulter and Aronson 1977; Bartnicky-Garcia and Wang 1983; Rast and Pfyffer 1989; Pfyffer et al. 1990; Griffith et al. 1992), tubular cristae in mitochondria (Alexopoulos et al. 1996), differential sensitivity to monomeric aluminium (Fichtner et al. 2006) and motile heterokont zoospores (Desjardins et al. 1969)) referred to as 'pseudofungi' by most mycologists and is placed into a new domain of life called Stramenopila (Cavalier-Smith 1987; Leipe et al. 1994; Beakes 1998).

The Stramenopila includes *Oomycetes* such as 'phytophthoras', 'downy mildews' and 'Pythia', which form a unique branch of eukaryotic plant pathogens with an independent evolutionary history (Kamoun et al. 1999) showing a distant evolutionary relationship with true fungi (Gunderson et al. 1987; Förster et al. 1990; Baldauf et al. 2000). The fact that oomycetes are not related to fungi is particularly relevant for heterologous expression of genes and comparative genomics and genetics in general. Still, despite their different evolutionary origins, the morphology of the hyphae, their mycelium-like growth and the airborne spores show remarkable resemblance to fungi. Oömycetes and fungi are proba-

bly one of the best examples of convergent evolution.

The *Oomycetes* contain some of the most destructive of plant pathogens; among them, some species of the genera *Phytophthora*, *Pythium* (rots, blights and damping off) and *Peronospora* and *Plasmopara* (downy mildews) stand out; in fact, the name *Phytophthora* means 'plant destroyer'. The genus *Phytophthora* with more than 108 members is presently placed in the kingdom Stramenopila (Belbahri et al. 2006), under the phylum Heterokonta, subphylum Pernosporomycotina, class Pernosporomycetes (Oömycetes), subclass Pernosporomycetidae, order Pythiales and family Pythiaceae. The taxonomy of *Phytophthora* has undergone an evolution in the way it has been studied, from the era of six morphospecies groups (Waterhouse 1963; Newhook et al. 1978; Stamps et al. 1990) to the era of phylogenetic clades (Cooke et al. 2000) and ITS fingerprinting-based keys (Gallegly and Hong 2008) and now to a new era of an integrated morphological and phylogenetic key (Ristaino 2011), and has been reviewed by Guha Roy and Grünwald 2014.

This genus *Phytophthora* has a long history in modern science. The scientific discipline of plant pathology was born in the early 1860s when Anton de Bary recognised *Phytophthora infestans* as the pathogen causing potato late blight responsible for the Irish potato famine in the 1840s (Abad and Abad 2003; Aragaki and Uchida 2001). It also brought about the first formulated use of a fungicide. In addition to this substantial social and historical impact, even considering merely a handful of *Phytophthora* spp. (e.g. *P. sojae*, *P. infestans*) documented to cause significant economic impacts, costs amount to anywhere between two and seven billions of dollars per crop per year worldwide in combined crop losses and management costs not even considering the less quantifiable but equally large impacts to natural ecosystems severely affected by some species (e.g. *P. cinnamomi*, *P. ramorum*).

Phytophthora pathogens also have a large impact on native ecosystems, forests and agricultural crops. The past decade has seen the discovery of a large number of phytophthoras especially from aquatic and forest ecosystems, and a wealth of information is now available on some of the phytophthoras attacking crops and plantations. Only the advances and their possibilities in horticultural and spice crops will be discussed here as exhaustive reviews are already available on the other ecosystem niches (Lamour and Kamoun 2009; Guha Roy 2008; Guha Roy and Grünwald 2014).

Phytophthora is now considered as one of the most important and destructive genera of plant pathogens in temperate and tropical regions, causing annual damages of billions of dollars (Drenth and Guest 2004) due to its high virulence and epidemiological ability to spread rapidly throughout the world.

19.2 The Spice Scenario vis-à-vis Phytophthora

The genus *Phytophthora* infects an array of spices and plantation crops. Some of the important diseases caused by *Phytophthora* are azhukal disease of cardamom, foot rot of black pepper, white tip of leek, leaf blight of onion and garlic, root rot of chilli, bud rot of coconut, abnormal leaf fall of rubber, wilt of *Piper betle*, koleroga of arecanut and some diseases of cocoa. Some of them have been discussed in the following paragraphs:

19.2.1 Phytophthora in Black Pepper

Phytophthora foot rot and leaf rot of black pepper are a serious problem in all black pepper-growing countries like India, Indonesia, Vietnam, Malaysia, Brazil, Thailand, Madagascar, etc. India and Indonesia are the main producers of pepper and account for more than 50 % of the world production. On a global scale, losses due to this disease have been estimated to be US\$ 4.5–7.5 million per annum (de Waard 1979). The disease was first reported in 1885 in Lampung, Indonesia and has been called foot rot disease since 1928 (Muller 1936). The disease starts as

dark brown spots on tender leaves at the lower region of the bush which enlarge rapidly covering the major area of leaf. These leaf spots have characteristic 'fimbriate margins' and infected leaves drop off prematurely. The fungus also infects green stems and branches causing rotting. In the case of root rot, infection that starts on the fibrous root system reaches the main root and ultimately the collar region or foot region of the bush (http://iisr.agropedia.in/content/phytophthora-disease-black-pepper).

In India, Kerala accounts for over 97 % of the area under pepper cultivation where it is a serious and dreaded disease even affecting coconut, arecanut and rubber plantations in the vicinity (Pruthi 1993; Chowdappa et al. 2003a). Different *Phytophthora* spp. are found to be associated with *Phytophthora* foot rot and leaf rot of black pepper in different geographical regions. In India, the disease was first reported way back in 1902 (Barber 1902; Butler 1906), and currently, *P. capsici* and *P. tropicalis* and isolates having similarity to both have been molecularly identified as being associated with the disease (Chowdappa et al. 2003b; Annual Report IISR 2012–2013); in Indonesia, where it has a long history of cultivation as Hindu migrants probably took pepper to Java between 100 BC and AD 600 (Purseglove et al. 1981), it is affected by *P. capsici* where it causes 52 % yield reduction (Purwantara et al. 2004). In Vietnam, which is now the world's second largest pepper exporter, though the disease was first reported in 1952, the identity of the causal agent was only recently conclusively determined as *P. capsici* (Truong et al. 2008) with low genetic diversity (Truong et al. 2010) but more adapted to black pepper hosts (Truong et al. 2012). Disease losses in Vietnam account to 15–20 % (Drenth and Sendall 2004). In Malaysia, pepper cultivation started with the British organised plantings of pepper in the early nineteenth century (Purseglove et al. 1981). Malaysia is now the fourth largest producer of black pepper in the world (PMB 2001), and currently, 95 % of the pepper produced in Malaysia is grown in Sarawak (PMB 2001) where *P. capsici* is the causal agent and the rest in Johor from where *P. nicotianae* has also been isolated (Lee and Lum 2004). On the other hand, the causal agents reported from Thailand are *P. palmivora*, *P. nicotianae* and *P. capsici* (Sangchote et al. 2004), while only *P. capsici* has been reported from Turkey on black pepper where it is very destructive and has the greatest economic importance. *P. capsici* is very dangerous for pepper spice and pepper paste production because it causes up to 100 % drying and killing of pepper plants under conditions of poor drainage and incorrect irrigation practices (Biçici and Çinar 1990).

- For the management of *Phytophthora* foot rot in black pepper, crop should be sprayed with 0.25 % Ridomil Gold 68 (WP) or 0.3 % potassium phosphonate during June and August and also drenching the soil with 3 l per vine and 1 kg of neem cake with 50 g of *Trichoderma harzianum* to the root zone of vines twice in a year.
- Biocontrol agent 1 % *Pseudomonas fluorescens* application to the vine as spraying (@ 2 l/vine) and drenching (@ 3 l/vine) during June and the second week of August for management of *Phytophthora* root rot was helpful [http://uhsbagalkot.edu.in/AICRP_sirsi.aspx].

19.2.2 Phytophthora in Onion, Garlic and Leek

(a) *Phytophthora porri* affects different alliums like leek, onion and garlic. White tip is one of the important foliar diseases of leek in Western Europe. The disease has been mainly reported from Europe, Canada and Japan. In Japan, a loss of 70 % or more is found in onion. The disease spreads rapidly on cool, wet weather. On onion and garlic, this pathogen causes water-soaked leaf blight and root rot symptoms. Early symptoms of leek leaves consist of irregularly shaped water-soaked lesions. Older lesions develop a bleached white centre with water-soaked margin which disappears in dry condition.

(b) *Phytophthora nicotianae* causes damping off of green onion seedlings and leaf blight and rot of green onion. On onion leaves, spots begin as small, irregularly shaped, water-soaked lesions on the young and mature leaves. In a few days, these spots expand, girdling the leaf and causing the tissue above the infection point to wither. As the infection progresses, healthy tissue is invaded, eventually killing the entire leaf. Infected plants have a mix of healthy and withered leaves with some leaves showing a characteristic of half-infected, half-healthy symptom. Splashing water from raindrops or irrigation helps to move spores from infected plants to nearby healthy plants. Effective disease control begins with prevention and proper water management to minimise excess moisture on the plants. Ridomil 81 W can be applied to green onion up to 14 days prior to harvest. Also, label directions are to be read carefully and chemicals applied responsibly [http://www.extento.hawaii.edu/kbase/crop/type/p_nicoti.htm].

19.2.3 Phytophthora in Chilli

Phytophthora blight, caused by *Phytophthora capsici*, is a devastating disease on both bell and non-bell peppers. The major symptom is root rot and wilt. However, more precisely, *Phytophthora capsici* infects roots, crowns, stems, leaves and fruit, causing seedling damping off, stem lesion, stem blight, leaf spot and fruit rot. It is a soil-borne pathogen which can produce several types of spores, enable it to spread throughout the field and to persist in the field between crops. *P. nicotianae* has also been found to be pathogenic on chilli pepper in Tunisia, NW and Western Spain, but the symptoms described are that of collar and root rot in contrast to fruit rot reported from the Indian subcontinent and those of seedling blight as the isolates were suggested to be different and this *P. nicotianae* is more adapted to their hosts (Andrés et al. 2003; Darinea et al. 2007; Saadoun and Allagui 2008; Rodríguez-Molina et al. 2010).

No single strategy should be used to control Phytophthora blight of pepper. A combination of methods is needed to effectively control this disease. The following practices can help to manage Phytophthora blight in pepper fields:

1. Fields to be selected with no history of Phytophthora blight, if possible.
2. Select fields that did not have peppers, cucurbits, eggplants or tomatoes for at least 3 years.
3. Selected fields are to be well isolated from infested fields with *P. capsici*.
4. Well-drained fields are to be chosen. Low areas or the areas which do not drain well are to be avoided.
5. Excessive irrigation is to be avoided.
6. Seeds should not be saved from a field where Phytophthora blight occurred.
7. Resistant varieties are to be planted, whenever it is possible.
8. Fungicides may be used to reduce Phytophthora infection in pepper fields.

19.2.4 Phytophthora in Cocoa

Phytophthora pathogens are responsible for some of the most serious diseases of cocoa including *Phytophthora* pod rot (PPR) or black pod, stem canker, leaf and seedling blight, chupon wilt and flower cushion infections. PPR causes 10–30 % annual losses in the production of cocoa beans globally and much higher losses locally in particularly wet and humid conditions. The estimated losses in the production in Asia, Africa and Brazil are 450,000 t annually, worth an estimated value of US$ 423 million (Drenth and Sendall 2004). Stem canker causes further losses and also tree deaths. Eight species of *Phytophthora* have been isolated from diseased cocoa, but most losses in the production are caused by *Phytophthora palmivora*, *P. megakarya*, *P. capsici* and *P. citrophthora*, and these vary in both their aggressiveness and the level of crop loss caused (Appiah et al. 2004). *Phytophthora megakarya* is the most aggressive and can cause between 60 and 100 % crop loss (Djiekpor et al. 1981; Dakwa 1988). In contrast, *P. palmivora* is less aggressive and can cause crop losses of 4·9–19 % (Blencowe and Wharton 1961; Dakwa 1984); this species is more aggressive than *P. capsici* (Lawrence et al. 1982). *Phytophthora*

citrophthora is more aggressive than *P. palmivora* or *P. capsici* and requires less time for zoospore germination and penetration on unwounded, detached pods (Campêlo et al. 1982). Sequence analysis showed that the four main cocoa-associated species formed two distinct groups, one comprising *P. capsici* and *P. citrophthora* and the other *P. palmivora* and *P. megakarya* (Appiah et al. 2004). Single reports of other *Phytophthora* species causing black pod include *P. botryose* (Chee and Wastie 1970), *P. heveae* (Lozano and Romero 1984) and *P. katsurae* (Liyanage and Wheeler 1989a) and *P. megasperma* (Zentmyer 1988), although these are not considered major problems for cocoa production.

The relative impact of each of these species of *Phytophthora* varies from region to region. In India, both *P. palmivora* and *P. capsici* cause black pod disease, but the *P. palmivora* isolates are of a single clonal lineage also infecting the coconut plantations below which it is cultivated as an understory crop, while on the other hand, the *P. capsici* isolates from cocoa belong to two genetic subgroups (Chowdappa et al. 2003b). A detailed sequence analysis of worldwide collection of *P. capsici* isolates from cocoa as well other hosts and comparison with published literature suggested that *P. capsici* isolates from cocoa may be closely related to *P. tropicalis* (Appiah et al. 2004). In Southeast Asia, *P. palmivora* seems to be the principal pathogen, while *P. megakarya* has only been found in West Africa (Brasier et al. 1981). In Africa, *P. megakarya* tends to be the principal pathogen, while in the Americas, *P. capsici* and *P. citrophthora* are the main causal agents of pod rot (Erwin and Riberio 1996) worldwide. *P. palmivora* is one of the most serious pathogens on cocoa, and in Southeast Asia, this species accounts for almost all of the Phytophthora diseases of cocoa. The most effective control measures are the introduction of resistant cocoa genotypes and farm management practices such as the removal of infected pod husks, proper pruning of the canopy and judicious selection of shade species and associated crops (McMahon and Purwantara 2004).

19.2.5 Phytophthora Diseases of Coconut

Coconut (*Cocos nucifera*) is one of the most valuable plant species in the tropics, providing oil, coconut milk, fibre from the husk, palm wine and timber for furniture and construction with its primary centre of origin in Asia and some secondary centres of origin in Central and South America. Rots caused by *Phytophthora* spp. lead to palm death (by bud rot) and/or yield reduction (by premature nut fall) (Waller and Holderness 1997) and are prevalent in all coconut-growing regions of the world. The disease though sporadic in nature causes severe losses. The earliest visible symptoms are the paling of leaves in the inner whorl followed by collapse of the spear leaf. Bud rot causes a total loss of the palm, since the apical portions are destroyed and will not regenerate further. The principal causal agent in India, Philippines, Indonesia and Malaysia is *P. palmivora*; In Indonesia, *P. arecae* and *P. nicotianae* have also been found in association with these diseases (Thevenin 1994). *Phytophthora nicotianae* is rarely encountered, and it is usually associated with cocoa and infested soil (Waroka and Thevenin 1992).

The economic impact of the disease on a country's economy can be exemplified by what has happened in the Philippines in the recent past. The Philippines was the number one coconut producer in the world during 1976–1986, but the average productivity has declined in the past decade (1991–2000) with an average production of 669 kg/ha. Although *Phytophthora palmivora* was known to cause bud rot and fruit and immature nut fall in the Philippines, the disease losses were relatively low. This changed dramatically after the introduction of highly susceptible MAWA hybrids, which are a cross between Malayan yellow dwarf and West African tall, both of which are known to be susceptible to Phytophthora. The Philippines now lags behind India, which produces, on average, 732 kg/ha, and Indonesia with an average production of 1,041 kg/ha. This lower productivity can be attributed to a number of factors, but all of which are related to disease outbreaks. Bud rot and fruit rot were major causes of the large loss of coconut trees and the significant decrease in the production (Concibido-Manohar 2004).

19.2.6 Phytophthora Diseases of Rubber

The major rubber-grower countries are Indonesia, Thailand, Malaysia, China and India, each with more than a million hectares. There are several different types of symptoms caused by *Phytophthora* spp. on rubber trees: (1) abnormal leaf fall, (2) black stripe of the tapping panel, (3) stem or patch canker and (4) pod rot (Sdoodee 2004). Abnormal leaf fall was recorded from Kerala in India as early as 1910 (McRae 1918) and can reduce 30–50 % of the production (Pillai 1982). Black stripe disease was first noted in Sri Lanka and is widespread in Southeast Asia as well as Africa and America. Other Phytophthora diseases are also common throughout most rubber-growing areas. Black stripe and leaf fall cause serious damage, but economically important outbreaks are confined to areas with long periods of high rainfall. Although patch or stem canker is widespread, recent records of high economic impact are few. At least six species of *Phytophthora* have been reported to be associated with diseases of rubber including *P. botryosa*, *P. heveae*, *P. meadii*, *P. palmivora*, *P. capsici* and *P. nicotianae*. However, *P. palmivora* and *P. meadii* are isolated most frequently as the causal agents (Sdoodee 2004). The identity of the species varies with geographical regions; in India, Myanmar and Sri Lanka, it is predominantly *P. meadii* (Liyanage 1982; Kochuthresiamma et al. 1988; Johnston 1989; Chowdappa et al. 2003a), whereas in Malaysia, Thailand and Vietnam, *P. palmivora* and *P. botryosa* are implicated (Chee 1969; Tsao et al. 1975; Duong et al. 1998). In China, although the main species involved appears to be *P. citrophthora*, other species including *P. palmivora, P. meadii, P. nicotianae* and *P. capsici* were also found to infect rubber (Zeng and Ward 1998). There are also reports of *P. citrophthora* infecting rubber in Indonesia (Liyanage and Wheeler 1989a, b). In Brazil, *P. capsici* was reported to be the main species associated with black stripe and stem canker, but *P. palmivora* and *P. citrophthora* were also isolated from diseased rubber (Dos Santos et al. 1995).

19.3　What Needs to Be Done?

Quality enhancement in spice crops/horticultural crops must necessarily include incorporation of improved control strategies along with those of quality control and agronomical practices. There has been an evolution in the way control strategies were thought of – from the days of chemical to biological to IPM strategies to that of decision support systems (DSS) and databases for specific crops and/or pathogens. Recently, such databases are being increasingly advocated and recommended by extension workers worldwide for timely diagnostic and mitigatory advice to achieve higher productivity (http://iapps2010. w o r d p r e s s . c o m / 2 0 1 3 / 1 1 / 0 7 / plant-protection-clinics-in-asia-3/).

For the latter to work, the first prerequisite is an accurate identification and sensitive detection of the pathogen. This is critical for regulatory action and disease management; especially faced with agricultural security concerns, the importance of diagnostic capacity cannot be overemphasised. Traditional culture-based detection and diagnostic methods for *Phytophthora* are inadequate as classical taxonomy of the genus is still based on often inconsistent morphological markers (Duncan and Cooke 2002). Combined with this is the fact that there is display of considerable morphological plasticity within some taxa limits (Brasier and Griffin 1979; Erwin and Riberio 1996; Appiah et al. 2003), and also the need for specialised expertise and time makes species identification based on morphological criteria difficult (Brasier et al. 1981; Erwin and Riberio 1996). This often leads to misidentification (Hall 1998), which, in turn, is detrimental to both practical control and clear scientific communication. Moreover, various reports on molecular identification of the *Phytophthora* pathogen have proved that in the past, new species have been wrongly assigned to current taxa and conversely, morphological variants of the existing taxa incorrectly assigned as new disease threats

when the identifications were solely based on morphological criteria (Chowdappa et al. 2003a, b; Mirabolfathy et al. 2001; Guha Roy et al. 2006). Also, several species 'complexes' can be observed in phylogenetic trees showing the presence of potential cryptic species. Presently, powerful molecular techniques combined with morphological characterisation and a renewed interest in probing of the environments have led to discovery of new species, novel variants within species and hybrids and provided a better resolution of species 'complexes' with differentiation of the species therein (Grünwald and Goss 2011).

In the last decade, traditional detection methods have been complemented by various molecular methods for *Phytophthora* (Martin and Tooley 2003, 2004; Schaad et al. 2003; Kong et al. 2003a, c), especially involving PCR amplification of pathogen-specific nucleic acid targets and serological detection of specific pathogen proteins (Benson 1991; Hardham and Cahill 1993). These available diagnostic techniques are effective but can detect only single target pathogen per assay. For parallel detection of multiple phylogenetically diverse organisms simultaneously, as is present in nature, microarray-based diagnostics have been developed (Fessehaie et al. 2003; Bodrossy and Sessitsch 2004; Lievens et al. 2005a, b, 2006). A specific microassay-based diagnostic method using padlock probes (PLP) (Szemes et al. 2005) detects the presence of *Phytophthora* from leaves, roots, soil and stream water and even from air in the multi-institutional Global Phytophthora Network (GPN) project.

While assessment of the diversity, distribution and dynamics of *Phytophthora* in nature requires the deployment of accurate diagnostic methods, implementation of effective control strategies also requires more knowledge about the genetic structure of population of plant pathogens (Wolfe and Caten 1987), as control strategies must target a population instead of an individual if they are to be effective. Defining the genetic structure of a population is a logical first step in studies of fungal population genetics because the genetic structure of a population reflects its evolutionary history and its potential to evolve: aspects important for formulating disease management strategies. 'Genetic structure' refers to the amount and distribution of genetic variation within and among populations (McDonald 1997).

In fungi that undergo both asexual and sexual reproduction, it is necessary to differentiate between diversity at individual locus, 'gene diversity', and diversity based on the number of genetically distinct individuals in a population, 'genotype diversity'. Taken together, gene and genotype diversity constitute genetic diversity (McDonald 1997). It is, however, important to distinguish between studies of population diversity and population genetics; the former yield the raw data, to which the latter can be applied to answer questions on the fundamental mechanisms and process of genetic change in populations (Cooke and Lees 2004).

Detection of diversity is usually done through the use of phenotypic and genotypic markers that are selectively neutral, highly informative, reproducible and relatively easy and inexpensive to assay. It is clear, however, that no single marker system would be adequate for all aspects of research on the diversity of *Phytophthora* species (Milbourne et al. 1997). The choice of genetic marker can have a substantial impact on the analysis and interpretation of data. As *Phytophthora* reproduce mainly asexually, producing a population structure that is largely composed of clonal lineages, a neutral marker such as a DNA fingerprint, may be used to address both questions relating to roles played by population size, mating systems and gene flow and also for questions relating to effects of selections, for which usually selective markers are used, assuming there is complete correspondence between genotype (DNA fingerprint) and phenotype (e.g. pathotype) (McDonald 1997). However, such assumption may not be valid as variable pathotypes can arise within the same clonal lineages (Drenth et al. 1996; Goodwin et al. 1995; Abu-El Samen et al. 2003).

Though it is best to use a widest practical array of genetic markers, combining a mixture of selected and neutral unlinked markers encompassing the nuclear (and mitochondrial) genome(s) distributed across many chromosomes, the number of marker loci assayed varies with the objective and resources available to the investigator. However, choicest markers are multilocus and have deeper resolving power like those of simple sequence repeats (SSRs) or single nucleotide polymorphisms (SNP). High throughput and massive parallel computing power have allowed multiplexing of these primers in recent times allowing assessment of field level diversity possible in shorter periods of time. Very recently, next-generation sequencing (NGS) has opened up the possibilities of looking at transcriptional polymorphisms of the pathogen population in the field allowing us to detect whether the isolates are under selection pressure and rapidly evolving and also the variability of the pathogen transcribed effectors that are the key in inducing the disease. These information would allow us to make informed choices by predicting about whether a particular fungicide or new host lines/varieties would be effective against the pathogen population.

In the coming decade, these next-generation technologies of sequencing, proteomics and metabolomics will have increased throughput and decreased costs. The upcoming fourth-generation technologies like single molecule real-time sequencing technology (SMRT) which is already available commercially are supposed to bring down the costs to about approx. Rs. 1.50 (3 cents) per Mb, and sequencing of a whole field population of *Phytophthora* spp. will come down to about Rs. 30, 000/- ($500)! Community and population cellular components can then be measured dynamically over space and time.

However, in the Indian scenario while that will be sometime in coming, we can as of now create databases from pathogen population data collected from neutral markers (like SSRs) and associate them with field data like germination temperature, mating type, fungicide sensitivity, morphological phenotypes and others like effector diversity and geographical location. Each cluster formed as a result of use of genetic markers can then be completely characterised. Once such a database is formed, it can be useful as a diagnostic aid to predict the characteristic of the pathogen population and prescribe control measures against that pathogen population. Since Phytophthora populations vary geographically and devastation times are typically 48–72 h, there is very little time to run series of traditional tests and then prescribe measures. In a typical scenario, once an infected sample is brought to the diagnostic centre/plant health clinic, a single quick molecular marker assay (in a few hours time) would assign the pathogen to a previously characterised pathogen population cluster. Once the match is done, it would become very easy to accurately prescribe control measures. The same information will be very helpful to create DSS specific for crops and their pathosystems.

19.4 Conclusion

The twenty-first century has already seen a major paradigm shift in our understanding of the biology, evolution and genetics of the genus *Phytophthora* as well as the tools and approaches used to develop novel approaches for disease management. The combination of novel tools and approaches provided by the convergence of genome sequencing, computing power and novel genomic/biotechnological tools paints a promising picture of the future of Phytophthora disease management. At the same time, *Phytophthora* pathogens continue to emerge at an accelerated rate due to increased global travel and trade (Guha Roy and Grünwald 2014). Also, selection pressures of random fungicide usage, changing climatic conditions and misdirected (due to little or no knowledge about pathogen populations) control measures have increased coevolutionary rates as evidenced from transcriptional profiling of effectors. The only way we can tackle this menace is if we also 'coevolve' in our approach and adopt novel tools and approaches facilitated by newer technologies to combat it.

References

Abad JG, Abad JA (2003) Advances in the integration of morphological and molecular characterization in the genus *Phytophthora*: the case of *P. niederhauseria* sp. nov. Phytopathology 93:S1

Abu-El Samen FM, Secor GA, Gudmestad NC (2003) Variability in virulence among asexual progenies of *Phytophthora infestans*. Phytopathology 93:293–304

Agrios GN (2005) Plant pathology, 5th edn. Academic, San Diego

Alexopoulos CJ, Mims CW, Blackwell M (1996) Introductory mycology, 4th edn. Wiley, New York, USA

Andrés JL, Rivera A, Fernández J (2003) *Phytophthora nicotianae* pathogenic to pepper in northwest Spain. J Plant Pathol 85(2):91–98

Appiah AA, Flood J, Bridge PD, Archer SA (2003) Inter and intraspecific morphometric variation and characterization of *Phytophthora* isolates from cocoa. Plant Pathol 52:168–180

Appiah AA, Flood J, Archer SA, Bridge PD (2004) Molecular analysis of the major *Phytophthora* species on cocoa. Plant Pathol 53:209–219

Aragaki M, Uchida JY (2001) Morphological distinctions between *P. capsici* and *P. tropicalis* sp. nov. Mycologia 93:137–145

Baldauf SL, Roger AJ, Wenk-Siefert L, Doolittle WF (2000) A kingdom-level phylogeny of eukaryotes based on combined protein data. Science 290:972–977

Barber CA (1902) Annual report for 1901–1902. Department of Agriculture, Madras

Bartnicky-Garcia S, Wang MC (1983) Biochemical aspects of morphogenesis in *Phytophthora*. In: Erwin DC, Bartnicki-Garcia S, Tsao P (eds) *Phytophthora*: its biology, taxonomy, ecology and pathology. American Phytopathological Society Press, St. Paul, pp 121–138

Beakes GW (1998) Evolutionary relationship among protozoa. In: Coombs GH, Vickerman K, Sleigh MA, Warren A (eds) The systematics association special volume series 56. Kluwer Academic Publishers, Dordrecht

Belbahri L, Moralejo E, Calmin G, Oszako T, Garcı'a JA, Descals E, Lefort F (2006) *Phytophthora polonica*, a new species isolated from declining *Alnus glutinosa* stands in Poland. FEMS Microbiol Lett 261:165–174

Benson DM (1991) Detection of *Phytophthora cinnamomi* in azalea with commercial serological assay kits. Plant Dis 75:478

Biçici M, Çinar A (1990) A review of *Phytophthora* diseases of different Mediterranean crops in Turkey. EPPO Bull 20(1):101–105. doi:10.1111/j.1365-2338.1990.tb01185.x. Article first published online: 28 JUNE 2008

Blencowe JW, Wharton AL (1961) Black pod disease in Ghana, incidence of disease in relation to levels of productivity. In: Report of the 6th commonwealth mycology conference. Cocoa, Chocolate and Confectionery Alliance, London. pp 139–147

Bodrossy L, Sessitsch A (2004) Oligonucleotide microarrays in microbial diagnostics. Curr Opin Microbiol 7(3):246–255

Brasier CM, Griffin MJ (1979) Taxonomy of *Phytophthora palmivora* on cocoa. Trans Br Mycol Soc 72:111–143

Brasier CM, Griffin MJ, Maddison AC (1981) Cocoa black pod Phytophthoras. In: Gregory PH, Maddison

AC (eds) Epidemiology of Phytophthora on cocoa in Nigeria, Phytopathological paper no. 25. Commonwealth Mycological Institute, Kew, pp 18–30

Butler EJ (1906) The wilt of pigeon pea and pepper. Agric J India 1:25

Campêlo AMFL, Luz EDMN, de Resnick FCZ (1982) Podridão-parda do cacaueiro, nos Estados da Bahia, Brasil.1 – virulencia das especies de Phytophthora. Revisita Theobroma, CEPEC, Itabuna-Brasil, Ano 12, pp 1–6

Cavalier-Smith T (1987) The origin of fungi and pseudo-fungi. In: Rayner ADM, Brasier CM, Moore D (eds) Evolutionary biology of the fungi. Cambridge University Press, Cambridge, pp 339–353

Chee KH (1969) Hosts of Phytophthora palmivora. Rev Appl Mycol 48:337–344

Chee KH, Wastie RL (1970) Black pod disease of cacao. Planter, Kuala Lumpur 46:294–297

Chowdappa P, Brayford D, Smith J, Flood J (2003a) Identity of Phytophthora associated with areacanut and its relationship with rubber and cardamom isolates based on RFLP of PCR-amplified ITS region of rDNA and AFLP fingerprints. Curr Sci 85:585–587

Chowdappa P, Brayford D, Smith J, Flood J (2003b) Molecular discrimination of Phytophthora isolates on cocoa and their relationship with coconut, black pepper and bell pepper isolates based on rDNA repeat and AFLP fingerprints. Curr Sci 84:1235–1238

Concibido-Manohar E (2004) Phytophthora diseases of coconut in the Philippines. In: Drenth A, Guest DI (eds) Diversity and management of Phytophthora in Southeast Asia, Monograph no. 114. ACIAR, Canberra, pp 7–9, p 238

Cooke DEL, Lees AK (2004) Markers, old and new, for examining Phytophthora infestans diversity. Plant Pathol 53:699–704

Cooke DEL, Drenth A, Duncan JM, Wagels G, Brasier CM (2000) A molecular phylogeny of Phytophthora and related oomycetes. Fungal Genet Biol 30:17–32. doi:10.1006/fgbi.2000.1202

Coulter DB, Aronson JM (1977) Glycogen and other soluble glucans from chytridiomycete and oomycete species. Arch Microbiol 15(3):317–322

Dakwa JT (1984) Nation-wide black pod survey. Joint CRIG/Cocoa production division project. In: Annual report of the Cocoa Research Institute, Ghana, 1976/77–1978/79. Tafo (Akim Abuakwa). Cocoa Research Institute, Ghana

Dakwa JT (1988) Changes in the periods for attaining the cocoa black pod disease infection peaks in Ghana. In: Proceedings of the 10th international Cocoa Research Conference, Santo Domingo, Dominican Republic. Cocoa Producers' Alliance, Lagos, pp 427–436

Darinea T, Allaguia MB, Rouaissia M, Boudabbous A (2007) Pathogenicity and RAPD analysis of Phytophthora nicotianae pathogenic to pepper in Tunisia. Physiol Mol Plant Pathol 70:142–148

de Liyanage AS (1982) Annual review of the Plant Pathology Department 1980. Rubber Research Institute, Sri Lanka

de Waard PWF (1979) Evaluation of the results of research on eradication of Phytophthora foot rot of black pepper (Piper nigrum. L) circulated during the first meeting of the pepper community permanent panel on techno-economic studies, 31st Jan–4th Feb 1979, Cochin, pp 1–47

Desjardins PR, Zentmeyer GA, Reynolds DA (1969) Electron microscopic observations of flagellar hairs of Phytophthora palmivora zoospores. Can J Bot 47:1077–1079

Djiekpor EK, Goka K, Lucas P, Partiot M (1981) Brown rot of cocoa pod due to Phytophthora species in Togo, evaluation and control strategy. Café Cacao Thé 25:263–268

Dos Santos AF, Matsuoka K, Alfenas AC, Maffia LA (1995) Identification of Phytophthora species that infect Hevea sp. Fitopatol Bras 20:151–159

Drenth A, Guest DI (2004) Introduction. In: Drenth A, Guest DI (eds) Diversity and management of Phytophthora in Southeast Asia. ACIAR, Canberra, pp 7–9. Monograph No. 114, P 238

Drenth A, Sendall B (2004) Economic impact of Phytophthora diseases in Southeast Asia. In: Drenth A, Guest DI (eds) Diversity and management of Phytophthora in Southeast Asia. Australian Centre for International Agricultural Research, Canberra, pp 227–231

Drenth A, Whission SC, Maclean DJ, Irwin JAG, Obstr NR, Ryley MJ (1996) The evolution of races of Phytophthora sojae in Australia. Phytopathology 86:163–169

Duncan JM, Cooke DEL (2002) Identifying, diagnosing and detecting Phytophthora by molecular methods. Mycologist 16:59–66

Duong N, Thanh HV, Doan T, Yen N, Tam TTM, Dung-Phan T, Phuong LTT, Duong NH, Thanh HN, Yen NT, Dung PT (1998) Diseases and pests of Hevea brasiliensis in Vietnam. In: Symposium on natural rubber (Hevea brasilliensis), vol 2, Ho Chi Minh City, Vietnam, pp 80–91

Elliot CG (1983) Physiology of sexual reproduction in Phytophthora. In: Erwin DC, Bartnicki-Garcia S, Tsao P (eds) Phytophthora: its biology, taxonomy, ecology and pathology. American Phytopathological Society Press, St. Paul

Erwin DC, Riberio OK (1996) Phytophthora diseases world wide. The American Phytopathological Society, St. Paul

FAO (2002) Statistical databases. http://apps.fao.org

Farr DF, Bills GF, Chamuris GP, Rossman AY (1989) Fungi on plants and plant products in the United States. APS Press, St. Paul

Fessehaie A, De Boer SH, Lévesque CA (2003) An oligonucleotide array for the identification and differentiation of bacteria pathogenic on potato. Phytopathology 93:262–269

Fichtner EJ, Hesterberg DL, Smyth TJ, Shew HD (2006) Differential sensitivity of Phytophthora parasitica var. nicotianae and Thielaviopsis basicola to monomeric aluminum species. Phytopathology 96(6):212–220

Förster H, Coffey M, Elwood H, Sogin ML (1990) Sequence analysis of the small subunit ribosomal RNAs of three zoosporic fungi and implications for fungal evolution. Mycologia 82:306–312

Gallegly ME, Hong CX (2008) *Phytophthora*: identifying species by morphology and DNA fingerprints. APS Press, St Paul, 158p

Goodwin SB, Sujkowski LS, Fry WE (1995) Rapid evolution of pathogenicity within clonal lineages of the potato late blight disease fungus. Phytopathology 85:669–676

Griffith JM, Davis AJ, Grant BR (1992) Target sites of fungicides to control oomycetes. In: Köler W (ed) Target sites of fungicide action. CRC Press, Boca Raton, pp 69–100

Grünwald NJ, Goss EM (2011) Evolutionary and population genetics of exotic and re-emerging pathogens: traditional and novel tools and approaches. Annu Rev Plant Physiol Plant Mol Biol 49:249–267

Guha Roy S, Bhattacharyya S, Mukherjee SK, Mondal N, Khatua DC (2006) *Phytophthora melonis* associated with fruit and vine rot disease of pointed gourd in India as revealed by RFLP and sequencing of ITS region. J Phytopathol 154(10):612–615

Guha Roy S (2008) Progress in *Phytophthora* research: identification, species diversity and population diversity. J Mycopathol Res 46:163–184

Guha Roy S, Grünwald NJ (2014) The plant destroyer genus Phytophthora in the 21st century. Rev Plant Pathol 6:387–412.

Gunderson JH, Elwood HJ, Ingold A, Kindle K, Sogin ML (1987) Phylogenetic relationships between chlorophytes, chrysophytes and oomycetes. Proc Natl Acad Sci U S A 84:5823–5827

Hall GS (1998) Examination of some morphologically unusual cultures of *Phytophthora* species using a mitochondrial DNA miniprep technique and a standardized sporangium caducity assessment. Mycopathologia 140:141–147

Hardham AR, Cahill DM (1993) Detection of motile organisms in a sample. Australian patent no. 48117/93(1 May 1997). US patent no. 5817472 (6 October 1998) European patent under examination. RSBS, Australia's National University

Hawksworth DL (1991) The fungal dimension of biodiversity: magnitude, significance, and conservation. Mycol Res 95:641–655

Hawksworth DL (2001) The magnitude of fungal diversity: the 1.5 million species estimate revisited. Mycol Res 105:1422–1432

Hawksworth DL, Rossman AY (1997) Where are all the undescribed fungi? Phytopathology 87:888–891

Hendrix JW (1970) Sterols in growth and reproduction of fungi. Annu Rev Plant Physiol Plant Mol Biol 8:111–113

Johnston A (1989) Diseases and pests. In: Webster CC, Baulkwil WJI (eds) Rubber. Longman Scientific and Technical, New York, pp 415–458

Kamoun S, Huitema E, Vleeshouwers V (1999) Resistance to oomycetes: a general role for the hypersensitive response? Trends Plant Sci 4:196–200

Kochuthresiamma J, Kothandaraman R, Jacob M (1988) Actinomycetes population of rubber growing soil and its antagonistic activity against *Phytophthora meadii* (McRal). Indian J Nat Rubber Res 1:27–30

Köler W (1992) Antifungal agents with target sites in sterol functions and biosynthesis. In: Köler W (ed) Target sites of fungicide action. CRC Press, Boca Raton, pp 119–206

Kong P, Hong CX, Richardson PA (2003a) Rapid detection of *Phytophthora cinnamomi* using PCR and primers derived from the Lpv storage protein genes. Plant Pathol 52:681–693

Kong P, Hong CX, Richardson PA, Gallegly ME (2003b) Single-strand-conformation polymorphism of ribosomal DNA for rapid species differentiation in genus *Phytophthora*. Fungal Genet Biol 39:238–249

Lamour K, Kamoun S (2009) Oomycete genetics and genomics: diversity, interactions and research tools. Wiley-Blackwell, Hoboken, p 582

Lawrence JS, Luz EDMN, Resnik FCZ (1982) The relative virulence of *Phytophthora palmivora* and *P. capsici* on cacao in Bahia, Brazil. In: Proceedings of the 8th international Cocoa Research Conference, 1981, Cartegena, Colombia. Cocoa Producers' Alliance, Lagos, pp 395–400

Lee BS, Lum KY (2004) *Phytophthora* diseases in Malaysia. In: Drenth A, Guest DI (eds) Diversity and management of Phytophthora in Southeast Asia. Australian Centre for International Agricultural Research, Canberra, pp 227–231

Leipe DD, Wainright PO, Gunderson JH, Porter D, Patterson DJ, Valois F, Himmerich S, Sogin ML (1994) The stramenopiles from a molecular perspective: 16S-like rRNA sequences from *Labyrinthuloides minuta* and *Cafeteria roenbergensis*. Phycologia 33:369–377

Lievens B, Brouwer M, Vanachter A, Levesque CA, Cammue BPA, Thomma B (2005a) Quantitative assessment of phytopathogenic fungi in various substrates using a DNA macroarray. Environ Microbiol 7:1698–1710

Lievens B, Grauwet TJMA, Cammue BPA, Thomma BPHJ (2005b) Recent developments in diagnostics of plant pathogens. In: Recent research developments in microbiology, S.G.S.G. Pandalai, Kerala, India, pp 57–79

Lievens B, Claes L, Vanachter ACRC, Cammue BPA, Thomma BPHJ (2006) Detecting single nucleotide polymorphisms using DNA arrays for plant pathogen diagnosis. FEMS Microbiol Lett 255(1): 129–139

Liyanage A de S (1982) Annual review of the plant pathology department 1980. Rubber Research Institute, Sri Lanka

Liyanage NIS, Wheeler BEJ (1989a) *Phytophthora katsurae* from cocoa. Plant Pathol 38:627–629

Liyanage NIS, Wheeler BEJ (1989b) Comparative morphology of *Phytophthora* species on rubber. Plant Pathol 38:592–597

Lozano TZE, Romero CS (1984) Estudio taxanomico de aislamientos de *Phytophthora* patogenos de cacao. Agrociencia 56:176–182

Martin FN, Tooley PW (2003) Phylogenetic relationships among *Phytophthora* species inferred from sequence analysis of mitochondrially encoded cytochrome oxidase I and II genes. Mycologia 95:269–284. doi:10.2307/3762038

Martin FN, Tooley FW (2004) Identification of *Phytophthora* isolates to a species level using restriction fragment length polymorphism analysis of a polymerase chain reaction-amplified region of mitochondrial DNA. Phytopathology 94:983–991

McDonald BA (1997) The population genetics of fungi: tools and techniques. Phytopathology 87:448–453

McRae H (1918) *Phytophthora meadii* n. sp. on *Hevea brasiliensis*. Mem Dep Agric India Bot Ser 9:219–273

Milbourne D, Meyer R, Bradshaw JE, Baird E, Bonar N, Provan J, Powell W, Waugh R (1997) Comparison of PCR based marker systems for the analysis of genetic relationships in cultivated potato. Mol Breed 3:127–136

Mirabolfathy M, Cooke DEL, Duncan JM, Williams NA, Ershad D, Alizadeh A (2001) *Phytophthora pistacae* sp.nov and *melonis*. The principal cause of pistachio gummosis in Iran. Mycol Res 105:1166–1175

Muller HRA (1936) The Phytophthora foot rot of black pepper (*Piper nigrum* L.) in the Netherlandish Indies. Cited in: Review of Applied Mycology 1937, 16:559

McMahon P, Purwantara A (2004) Phytophthora on cocoa. In: Drenth A, Guest DI (eds) Diversity and management of Phytophthora in Southeast Asia, ACIAR monograph 114 (printed version published in 2004), Canberra, Australia, pp 104–114

Newhook FJ, Waterhouse FM, Stamps DJ (1978) Tabular key to the species of *Phytophthora* de Bary, Mycological paper no. 3. Commonwealth Mycological Institute, Kew, England, p 20

Pfyffer GE, Boraschi-Gaia C, Weber B, Hoesch L, Orpin CG, Rast DM (1990) A further report on the occurrence of acyclic sugar alcohols in fungi. Mycol Res 94:219–222

Pillai PNR (1982) Abnormal leaf fall of rubber caused by Phytophthora spp. In: Nambiar KKN (ed) Proceedings of the workshop on *Phytophthora* diseases of tropical cultivated plants. Central Plantation Crop Research Institute, Kasargod, p 284

PMB (Pepper Marketing Board) (2001) Sarawak black pepper. Official website of the Department of Agriculture, Sarawak. On the Internet: http://www.doa.sarawak.gov.my/modules/web/page.php?id=138. Accessed 24 Mar 2014

Pruthi JS (1993) Major spices of India crop management post-harvest technology. ICAR, New Delhi, p 514

Purseglove JW, Brown EG, Green CL, Robbins SRJ (1981) Pepper. In: Spices, vol 1. Longman Scientific and Technical, London

Purwantara A, Manohara D, Warroka JS (2004) *Phytophthora* diseases in Indonesia. In: Drenth A, Guest DI (eds) Diversity and management of Phytophthora in Southeast Asia. Australian Centre for International Agricultural Research, Canberra, pp 227–231

Rast DM, Pfyffer GE (1989) Acyclic polyols and higher taxa of fungi. Bot J Linn Soc 99:39–57

Ristaino JB (2011) Key for identification of common *Phytophthora* species. APS Press, St. Paul

Rodríguez-Molina MC, Morales-Rodríguez MC, Palo Osorio C, Palo Núñez E, Verdejo Alonso E, Duarte Maya MS, Picón-Toro J (2010) *Phytophthora nicotianae*, the causal agent of root and crown rot (*Tristeza* disease) of red pepper in La Vera region (Cáceres, Spain). Span J Agric Res 8(3):770–744

Saadoun M, Allagui MB (2008) Pathogenic variability of *Phytophthora nicotianae* on pepper in Tunisia. J Plant Pathol 90:351–355

Sangchote S, Poonpolgul S, Sdoodee R, Kanjanamaneesathain M, Baothong T, Lumyong P (2004) Phytophthora diseases in Thailand. In: Drenth A, Guest DI (eds) Diversity and management of Phytophthora in Southeast Asia. Australian Centre for International Agricultural Research, Canberra, pp 227–231

Schaad NW, Frederick RD, Shaw J, Schneider WL, Hickson R, Petrillo MD, Luster DG (2003) Advances in molecular-based diagnostics in meeting crop biosecurity and phytosanitary issues. Annu Rev Plant Physiol Plant Mol Biol 41:305–324

Sdoodee R (2004) Phytophthora diseases of rubber. In: Drenth A, Guest DI (eds) Diversity and management of Phytophthora in Southeast Asia, ACIAR monograph 114 (printed version published in 2004), Canberra, Australia, pp 136–142

Stamps DJ, Waterhouse GM, Newhook FJ, Hall GS (1990) Revised tabular key to the species of Phytophthora, Mycological papers 162. CAB International Mycological Institute, Kew

Strange RN, Scott PR (2005) Plant disease: a threat to global food security. Annu Rev Plant Physiol Plant Mol Biol 43:83–116

Szemes M, Bonants P, de Weerdt M, Baner J, Landegren U, Schoen CD (2005) Diagnostic application of padlock probes—multiplex detection of plant pathogens using universal microarrays. Nucleic Acids Res 33(8):e70

Thevenin JM (1994) Coconut diseases in Indonesia – etiological aspects. Paper presented at the Coconut Phytophthora Workshop, Manado, Indonesia. Cited in: Waller and Holderness (1997)

Truong NV, Burgess LW, Liew ECY (2008) Prevalence and aetiology of *Phytophthora* foot rot of black pepper in Vietnam. Australas Plant Pathol 37:431–442

Truong NV, Liew ECY, Burgess LW (2010) Characterisation of *Phytophthora capsici* isolates from black pepper in Vietnam. Fungal Biol 114:160–170

Truong NV, Burgess LW, Liew ECY (2012) Cross-infectivity and genetic variation of *Phytophthora capsici* isolates from chilli and black pepper in Vietnam. Australasian Plant Pathol 41(4):439–447

Tsao PH, Chew-Chin N, Syamananda R (1975) Occurrence of *Phytophthora palmivora* on *Hevea* rubber in Thailand. Plant Dis Rep 59(12):955–958

Waller JM, Holderness M (1997) Beverage crops and palms. In: Hillocks RJ, Waller JM (eds) Soilborne diseases of tropical crops. CAB International, Wallingford

Wang MC, Bartnicky-Garcia S (1973) Novel phosphoglucans from the cytoplasm of *Phytophthora palmivora* and their selective occurrence in certain life cycle stages. J Biol Chem 248:4112–4118

Wang MC, Bartnicky-Garcia S (1974) Mycolaminarins: storage (1-3)-β-D-glucans from the cytoplasm of the fungus *Phytophthora palmivora*. Carbohydr Res 37:331–338

Warner SA, Domas AJ (1987) Biochemical characterization of zoosporic fungi: the utility of sterol metabolism as an indicator of taxonomic affinity. In: Fuller MS, Jaworski A (eds) Zoosporic fungi in teaching and research. Southeastern Publishing Corporation, Athens, pp 202–208

Waroka JS, Thevenin JM (1992) *Phytophthora* in Indonesian coconut plantations: populations involved. Paper presented at the Coconut *Phytophthora* Workshop, Manado, Indonesia

Waterhouse GM (1963) Key to the species of *Phytophthora* de Bary, Mycological paper no. 92. CMI, Kew, pp 1–22

Wete JD (1989) Structure and function of sterols in Fungi. Adv Lipid Res 23:115–167

Wolfe MS, Caten CE (1987) Populations of plant pathogens: their dynamics and genetics. Blackwell Scientific Publications, Oxford

Zeng FC, Ward E (1998) Variation within and between *Phytophthora* species from rubber and citrus trees in China, determined by polymerase chain reaction using RAPDs. J Phytopathol 146(2–3):103–109

Zentmyer GA (1988) Taxonomic relationships and distribution of *Phytophthora* causing black pod of cocoa. In: Proceedings of the 10th international Cocoa Research Conference, 1987, Santo Domingo, Dominican Republic. Cocoa Producers' Alliance, Lagos, pp 391–405

Future Directions

20

A.B. Sharangi and S.K. Acharya

Abstract

The future directions of the value addition process must have a spillover effect from its present and past performances, structural and functional. The transformation of peasantry into farming and farming into farm enterprise has got unique sequels and can be configured as a genology of agrarian transformation in the world. The feudal agrarian production system promotes peasantry where the peasants were offered with hungers and the productions were robed off. A farm is a consolidated mass of lands which caters a unit of farm production and capable of generating marketable and marketed surplus. The growth of farming has got a congenital pattern along with the growth in industry and entrepreneurship. Farming to entrepreneurship is basically the process of transformation, very simply production to product, and of course through value addition. So value addition is the catalytic component in bringing products from farms to the doorsteps of households, in the form and with the taste the potential consumers demand the best. This chapter describes the new-age value addition and addresses the future value addition and quality issues in the light of new economic policy, future trading and climate change issues.

The future directions of value addition process must have a spillover effect from its present and past performances, structural and functional. The constant changes of consumer's behavior and the subsequent transformations in their consumption behavior include the table text of food and demand for nutraceuticals. The mammoth pressure of population, going to be nine billion by 2050, certainly has a telling impact on the quantity of food available along with its nutritional aspects as well. We have to be cautious for the dichotomy wherein quantity may compromise with quality or quality issue may spare the issue of quantity requirement per capita. The elasticity of this relationship will certainly lay the founda-

A.B. Sharangi (✉)
Department of Spices & Plantation Crops, Bihan Chandra Krishi Viswavidyalaya,
Mohanpur, Nadia 741252, West Bengal, India
e-mail: dr_absharangi@yahoo.co.in

S.K. Acharya
Department of Agricultural Extension, Bihan Chandra Krishi Viswavidyalaya,
Mohanpur, Nadia 741252, West Bengal, India

tion of dynamics and dictum of value addition process.

20.1 The Transformation of Peasantry into Farming and Farming into Farm Enterprise

The transformation of peasantry into farming and farming into farm enterprise has got unique sequels and can be configured as a genealogy of agrarian transformation in the world. Peasantry is basically a kind of agriculture which meets up the subsistence, just the amount of production to support the existence of a family. The feudal agrarian production system promotes peasantry because land relation and means of production are having a catastrophic relationship. During peasantry, the farmers are to suffer from the statute through an extorting *zamindari* system. The peasants were offered with hungers and the productions were robed off.

The concept of farming is basically contributed by the US school of farm management wherein the agriculture has been organized both with sustenance and commerce and operationally supported by modern agricultural technologies and management. A farm is a consolidated mass of lands which caters a unit of farm production and capable of generating marketable and marketed surplus. The growth of farming has got a congenital pattern along with the growth in industry and entrepreneurship. The spurt of industry needs an equivalent growth in agriculture not only to support industrial growth but also to create an isochronous market to absorb what we may call industrial surplus as well as consumer products.

The farming to entrepreneurship is basically the process of transformation, very simply production to product, and of course through value addition. *So value addition is the catalytic component in bringing products from farms to the doorsteps of households, in the form and with the taste the potential consumers demand the best.*

20.2 The Future Value Addition and Quality Issues

The concept of value addition process, as it is now, has to undergo a change as well as metamorphosis, of course with positive notes and dicta. The conventional value addition is concerning about addition and upgradation of quality of the product just to fetch higher return and better market response. The dent has so far been a multiplication of initial investment into a generation of surplus values well commensurating with projected return. While it is just rupee, a value at the micro-level can be a growth of GDP in the national level.

The new age value addition, in addition to its monetary benefit, has certainly to consider the following imperatives and implications:
1. Value addition and quality issues related to good agricultural practices (GAP)
2. Value addition for sustainable livelihood generation (SLG)
3. Value addition for energy renewals and conservation
4. Creation of value-added technology and management for soil, water, and biodiversity conservation
5. Value addition for climate resilient production system
6. Value addition aiming at ensuring securities on food, gender, and social voices

20.3 A Future Value Addition Process in Line with and in Compliance with Millennium Development Goals (MDGs)

In 2000, 189 nations made a promise to free people from extreme poverty and multiple deprivations. This pledge turned into the eight Millennium Development Goals.

20.3.1 Eradicate Extreme Poverty and Hunger

This is the most crucial part of millennium declaration wherein value addition and entrepreneurships will serve the most for the farmers, who are economically disabled, ecologically vulnerable, and socially cryptic.

20.3.2 Achieve Universal Primary Education

For sustainable economic growth in agricultural and rural sector, primary education along with vocational skills needs to reach every finger and every mind of millions of farmers across the globe. Entrepreneurships cannot be possible without innovation, and innovation is a meaningless proposition without education.

20.3.3 Promote Gender Equality and Empower Women

Only an inclusive growth and cross-community and cross-gender farm skills can effectively contribute to both gender equality and empowerment.

20.3.4 Reduce Child Mortality

Here lies the importance of nutraceuticals for the child, both at accessible amount and affordable price. Low-cost baby food enriched with huge natural nutritional sources and forms can save the child from an untimely demise.

20.3.5 Improve Maternal Health

Both mother and child care has to go simultaneously by imbibing entrepreneurship with social responsibility and the both, with the livelihood generation process.

20.3.6 Combat HIV/AIDS, Malaria, and Other Diseases

A huge campaign and a humongous commitment, both from the government and people, can generate the desired result. A value-based education with perceptual skills can go a long way in combating this menacing threat.

20.3.7 Ensure Environmental Sustainability

Through value-added agricultural education and proper sensitization for creating stewardship, this will help make our good earth happily enduring over the optimal depletion.

20.3.8 Develop a Global Partnership for Development

A global vision only can make a local mission behave and perform operationally elegant and behaviorally sustainable, beyond limitations of resource cost and time. The application of ICT and mobile telephony can declare the death of the distance both of mind and geography.

20.4 Climate Change Agenda

The reproductive cycles of different horticultural crops are going to be adversely affected by the vagaries of climate change. So as a mitigation strategy, postharvest operations and value addition plans need to be accentuated with the changed scenario. The delay in sowing time or the retardation of harvesting time would count a lot on the quality or value addition process. The changing pattern of rainfall, uprising mean temperature, or humidity behavior of the atmospheric echelons would impact both the quantity and quality of the agricultural and horticultural products. The withdrawal of pollinators and with the appearance of menacing pest and diseases shall certainly impact on the product behaviors and its accessibility as well.

20.5 Role of Conventional and Nondescript Horticulture

The emerging domain for value addition process in agriculture or horticulture shall certainly be its nondescript components. The traditional rustic fruits, nuts, creepers, weeds, shrubs, etc. can earn a new focus, especially in herbal medicines, nutraceuticals, etc. The huge biodiversity, especially its hitherto overlooked diasporas, can be well taken for a renewed research intervention to explore the hidden properties for the benefit of humanity.

20.6 Future Trading, Contract Farming, and Horticulture

Forward trading and contract farming have become two important areas of global marketing and are presenting both promises and threats as well; nevertheless, the trends are well cognizable. The value addition process, in the case of contract farming, is being determined by the external entrepreneurs and gets characterized my market behavior across the borders.

20.7 New Economic Policy and Horticulture

While a nation has to go with value-added economy, the contribution of horticulture comes in a swashbuckling way. The most effective recipients of value addition processes are fruits, vegetables, flowers, roots, and corms. India's contribution to value-added horticulture has got a bleak stature, less than 2 % in the global market, to present a serious paradox between green production and processed volume. A mammoth volume of horticultural produce, approximately equalizing a loss of 40,000 crores of rupees per year, is being perished due to poor or nonexistent supply chain and logistic infrastructure.

With community planning for stewardship and leveraging while natural resources are the basic of human civilization, the living populace, the primary consumers of natural resources, must have to play the role of stewardship and mentoring. The organic farming has to play a greater role to protect the resource base in one hand and to usher the nutritional security on the other hand. The amount of crop residues, bioorganics from livestock and poultry birds, is being steered back to the shrink of production by the conscious farmers that can estimate their role for catering stewardship and mentoring.

Printed by Printforce, the Netherlands